Lecture Notes in Computer Science   12081

More information about this series at http://www.springer.com/series/7409

Selma Boumerdassi · Éric Renault ·
Paul Mühlethaler (Eds.)

Machine Learning for Networking

Second IFIP TC 6 International Conference, MLN 2019
Paris, France, December 3–5, 2019
Revised Selected Papers

 Springer

Editors
Selma Boumerdassi
Conservatoire National des Arts Métiers
Paris Cedex 03, France

Éric Renault
ESIEE Paris
Noisy-le-Grand, France

Paul Mühlethaler
Inria
Paris, France

ISSN 0302-9743   ISSN 1611-3349   (electronic)
Lecture Notes in Computer Science
ISBN 978-3-030-45777-8   ISBN 978-3-030-45778-5   (eBook)
https://doi.org/10.1007/978-3-030-45778-5

LNCS Sublibrary: SL3 – Information Systems and Applications, incl. Internet/Web, and HCI

This Springer imprint is published by the registered company Springer Nature Switzerland AG
The registered company address is: Gewerbestrasse 11, 6330 Cham, Switzerland

Preface

The rapid development of new network infrastructures and services has led to the generation of huge amounts of data, and machine learning now appears to be the best solution to process these data and make the right decisions for network management. The International Conference on Machine Learning for Networking (MLN) aims to provide a top forum for researchers and practitioners to present and discuss new trends in deep and reinforcement learning; pattern recognition and classification for networks; machine learning for network slicing optimization; 5G systems; user behavior prediction; multimedia; IoT; security and protection; optimization and new innovative machine learning methods; performance analysis of machine learning algorithms; experimental evaluations of machine learning; data mining in heterogeneous networks; distributed and decentralized machine learning algorithms; intelligent cloud-support communications; resource allocation; energy-aware communications; software defined networks; cooperative networks; positioning and navigation systems; as well as wireless communications; wireless sensor networks, and underwater sensor networks. In 2019, MLN was hosted by Inria Paris, a top-level research center in France.

The call for papers resulted in a total of 75 submissions from all around the world: Algeria, Canada, Chile, China, Colombia, Ecuador, France, Honduras, India, Ireland, Japan, Lebanon, Mauritius, Morocco, Norway, Peru, Portugal, Saudi Arabia, Senegal, Serbia, Singapore, South Africa, South Korea, Spain, Sri Lanka, Tunisia, the UK, and the USA. All submissions were assigned to at least three members of the Program Committee for review. The Program Committee decided to accept 26 papers. There were two intriguing keynotes by Jean-Claude Belfiore, Huawei, France, and Alberto Conte, NOKIA Bell Labs, France; two tutorials: the first one by Franck Gaillard, Microsoft, France, and the second one by Marie Line Alberi Morel, NOKIA Bell Lab, France; and a talk by? Kamal Singh, Télécom Saint-Etienne/Jean Monnet University, France, which completed the technical program.

We would like to thank all who contributed to the success of this conference, in particular the members of the Program Committee and the reviewers for carefully reviewing the contributions and selecting a high-quality program. Our special thanks go to the members of the Organizing Committee for their great help.

Thursday morning was dedicated to the Second International Workshop on Networking for Smart Living (NSL 2019). The technical program of NSL included five presentations and a keynote by Kevin Curran, Ulster University, UK.

We hope that all participants enjoyed this successful conference, made many new contacts, engaged in fruitful discussions, and had a pleasant stay in Paris, France.

December 2019

Paul Mühlethaler
Éric Renault

Organization

MLN 2019 was jointly organized by the EVA Project of Inria Paris, the Wireless Networks and Multimedia Services (RS2M) Department of Télécom SudParis (TSP), a member of Institut Mines-Télécom (IMT) and University Paris-Saclay, ESIEE Paris, and CNAM Paris.

General Chairs

Paul Mühlethaler   Inria, France
Éric Renault   ESIEE Paris, France

Steering Committee

Selma Boumerdassi   CNAM, France
Éric Renault   ESIEE Paris, France

Tutorial and Workshop Chair

Nardjes Bouchemal   University Center of Mila, Algeria

Publicity Chair

Rahim Haiahem   ENSI, Tunisia

Organization Committee

Lamia Essalhi   ADDA, France

Technical Program Committee

Claudio A. Ardagna   Università degli Studi di Milano, Italy
Maxim Bakaev   NSTU, Russia
Mohamed Belaoued   University of Constantine, Algeria
Paolo Bellavista   University of Bologna, Italy
Aissa Belmeguenai   University of Skikda, Algeria
Indayara Bertoldi Martins PUC Campinas, Brazil
Luiz Bittencourt   University of Campinas, Brazil
Naïla Bouchemal   ECE, France
Nardjes Bouchemal   University Center of Mila, Algeria
Selma Boumerdassi   CNAM, France
Marco Casazza   Università degli Studi di Milano, Italy

Alberto Ceselli	Università degli Studi di Milano, Italy
Hervé Chabanne	Télécom ParisTech, France
De-Jiu Chen	KTH, Sweden
Longbiao Chen	Xiamen University, China
Antonio Cianfrani	University of Rome Sapienza, Italy
Domenico Ciuonzo	Network Measurement and Monitoring (NM2), Italy
Alberto Conte	NOKIA Bell Labs, France
Pradipta De	Georgia Southern University, USA
Tamer ElBatt	Cairo University, Egypt
Vincenzo Eramo	University of Rome Sapienza, Italy
Paul Farrow	BT, UK
Arsham Farshad	Lancaster University, UK
Hacène Fouchal	Université de Reims Champagne-Ardenne, France
Kaori Fujinami	Tokyo University of Agriculture and Technology, Japan
Chenfei Gao	AT&T Labs, USA
Aravinthan Gopalasingham	NOKIA Bell Labs, France
Jean-Charles Grégoire	INRS, Canada
Bjarne Helvik	Norwegian University of Science and Technology (NTNU), Norway
Andreas J. Kassler	Karlstad University, Sweden
Donghyun Kim	Kennesaw State University, USA
Lito Kriara	Roche Pharmaceutical Research and Early Development, Switzerland
Cherkaoui Leghris	Hassan II University, Morocco
Feng Liu	Huawei, Germany
Wei Liu	National Institute of Information and Communications Technology, Japan
Diego Lopez	Telefonica, Spain
Victor Lopez	Telefonica, Spain
Olaf Maennel	Tallinn University of Technology, Estonia
Ilaria Malanchini	Nokia Bell Labs, Germany
Lefteris Manassakis	FORTH, Greece
Zoltán Mann	University of Duisburg-Essen, Germany
Fabio Martignon	University of Orsay, France
Saurabh Mehta	Vidyalankar Institute of Technology, India
Ruben Milocco	Universidad Nacional des Comahue, Argentina
Paul Mühlethaler	Inria, France
Francesco Musumeci	Politecnico di Milano, Italy
Kenichi Ogaki	KDDI Corporation, Japan
Satoshi Ohzahata	The University of Electro-Communications, Japan
Frank Phillipson	TNO, The Netherlands
Paulo Pinto	Universidade Nova de Lisboa, Portugal
Mahshid Rahnamay-Naeini	University of South Florida, USA
Sabine Randriamasy	NOKIA Bell Labs, France
Éric Renault	ESIEE Paris, France

Jihene Rezgui   LRIMA Lab, Canada
Sandra Scott-Hayward   Queen's University Belfast, UK
Lei Shu   University of Lincoln, UK
Wouter Tavernier   Ghent University, Belgium
Van Long Tran   Hue Industrial College, Vietnam
Daphne Tuncer   Imperial College London, UK
Vinod Kumar Verma   SLIET, India
Corrado Aaron Visaggio   Università degli Studi del Sannio, Italy
Krzysztof Walkowiak   Wroclaw University of Science and Technology,
                                 Poland
Bin Wang   Wright State University, USA
Haibo Wu   Computer Network Information Center, CAS, China
Kui Wu   University of Victoria, Canada
Miki Yamamoto   Kansai University, Japan
Sherali Zeadally   University of Kentucky, USA
Jin Zhao   Fudan University, China
Wei Zhao   Osaka University, Japan
Thomas Zinner   TU Berlin, Germany

Sponsoring Institutions

CNAM, Paris, France
ESIEE, Paris, France
Inria, Paris, France
IMT-TSP, Évry, France

Contents

Network Anomaly Detection Using Federated Deep Autoencoding Gaussian Mixture Model

Yang Chen⬛, Junzhe Zhang⬛, and Chai Kiat Yeo$^{(\boxtimes)}$⬛

School of Computer Science and Engineering, Nanyang Technological University,
50 Nanyang Avenue, Singapore 639798, Singapore
asckyeo@ntu.edu.sg

Abstract. Deep autoencoding Gaussian mixture model (DAGMM) employs dimensionality reduction and density estimation and jointly optimizes them for unsupervised anomaly detection tasks. However, the absence of large amount of training data greatly compromises DAGMM's performance. Due to rising concerns for privacy, a worse situation can be expected. By aggregating only parameters from local training on clients for obtaining knowledge from more private data, federated learning is proposed to enhance model performance. Meanwhile, privacy is properly protected. Inspired by the aforementioned, this paper presents a federated deep autoencoding Gaussian mixture model (FDAGMM) to improve the disappointing performance of DAGMM caused by limited data amount. The superiority of our proposed FDAGMM is empirically demonstrated with extensive experiments.

Keywords: Anomaly detection · Small dataset · Privacy-preserving · Federated learning · Deep autoencoding Gaussian mixture model · Network security

1 Introduction

Deep learning has been providing a lot of solutions which have previously posed big challenges to the artificial intelligence. It has been deployed in numerous applications such as computer vision, natural language processing and many other domains [1] thanks to the great advancement in computing power and the availability of vast amount of data. Much more complex and advanced algorithms can now be trained [2].

In the cybersecurity domain, anomaly detection is a critical mechanism used for threat detection and network behavior anomaly detection is a complementary technology to systems that detect security threats based on packet signatures. Network anomaly detection is the continuous monitoring of a network for unusual events or trends which is an ideal platform to apply deep learning. Deep anomaly

This project is supported by Grant No. NTU M4082227.

S. Boumerdassi et al. (Eds.): MLN 2019, LNCS 12081, pp. 1–14, 2020.
https://doi.org/10.1007/978-3-030-45778-5_1

detection [3] has thus seen rapid development such as self-taught learning based deep learning [4] and deep autoencoding Gaussian mixture model (DAGMM [5]. Deep learning significantly improves the model complexity resulting in substantial improvement to the detection accuracy. However, deep learning requires the availability of tremendous amount of data to be well trained and supervised learning not only requires large amount of data but they must also be labelled.

The DAGMM proposed in [5] significantly improves the F1 score compared to other state-of-the-art methods including deep learning methods. It employs dimensionality reduction and feature embedding via the AutoEncoder, a compression network, and then performs density estimation via an estimation network. The entire process is unsupervised and hence no labelled data is needed for training the networks. However, DAGMM still requires a huge amount of data to train its models.

Data availability poses a huge challenge for deep learning system. Compounding the problem is that not every one has amassed huge amount of data and even if they have, the data may not be labelled or they are not to be shared collectively due to privacy and security issues. Herein lies the interest in federated learning, where model parameters instead of training data are exchanged through a centralized master model in a secured manner [6] thereby preserving the privacy of individual datasets as well as alleviating the challenge of limited datasets.

This paper proposes the federated learning assisted deep autoencoding Gaussian mixture model (FDAGMM) for network anomaly detection under the scenario where there is insufficient data to train the deep learning models. FDAGMM can thus improve the poor performance of DAGMM caused by limited dataset and its superiority is empirically demonstrated with extensive experiments. In industry scenarios, the presented solution is expected to solve the problem of lacking training data that each organizations are not willing to share in a centralized mode.

2   Related Work

2.1   Anomaly Detection

Network Anomaly Detection, also called Network Intrusion Detection, has been studied for over 20 years [7]. Network intrusion detection systems are either signature (rule)-based or anomaly-based. The former uses patterns of well-known attacks or weak spots of the system to identify intrusions whereas the latter uses machine learning methods to determine whether the deviation from the established normal usage patterns can be flagged as intrusions [8].

Machine learning methods being applied for network anomaly detection, include genetic algorithms, support vector LLmachines (SVM), Self-organizing map (SOM), random forests, XGBoost, KNN, Naive Bayes networks, etc. However, many suffer low accuracy and high False Positive Rate (FPR). More recently, there are research in deep anomaly detection such as [3] and [4]. The use of deep learning significantly improves the model complexity and results in substantial performance improvement in terms of the various metrics, such as

F1 score, accuracy, precision, score and False Positive Rate (FPR). However, the more complex the model, the more labelled training data it needs to be well trained.

Deep Autoencoding Gaussian Mixture Model (DAGMM) is recently proposed in [5] and it produces good results with no need to label the training data. The model consists of a compression network and an estimation network which are trained end-to-end instead of using decoupled two-stage training and the standard Expectation-Maximization (EM) algorithm. The compression network is an autoencoder that embeds the feature into a low-dimension representation and meanwhile yields the reconstruction error for each input data point, which is further fed into the estimation network which acts as a Gaussian Mixture Model (GMM). It outperforms many state-of-the-art methods in terms of F1 score. However, even though it does not require labelled training data, it demands a large amount of unlabeled data in which normal users do not have or are unwilling or unable to share.

KDDCUP 99 [9] has been the most widely used dataset for the evaluation of anomaly detection methods since it was prepared by [10] in 1999. There are 5 million simulated tcpdump connection records, each of which contains 41 features and is labelled as either normal or an attack, with exactly one specific attack type. It covers attacks falling into the following 4 main categories:

- **DoS attack**: denial-of-service, e.g. smurf;
- **R2L**: unauthorized access from a remote machine, e.g. guessing password;
- **U2R**: unauthorized access to local superuser (root) privileges, e.g. various "buffer overflow" attacks;
- **Probing**: surveillance and other probing, e.g. port scanning.

2.2   Federated Learning

Federated learning is proposed by Konečný et al. [11,12] to use the availability of privacy sensitive data stored in mobile devices to boost the power of various deep/machine learning models. In typical federated learning (FL), clients, e.g. smart phones, suffer from unstable communication. In addition, their data are unbalanced, Non-IID and massively distributed. These features distinguish FL from conventional distributed machine learning [13,14].

As shown in Fig. 1, **Federated Learning (FL)** involves two components, i.e. central and local training. K clients indexed by k. The whole process is divided into communication rounds, in which clients are synchronously trained with local stochastic gradient descent (SGD) on their datasets $\mathcal{P}_k$. In the central server, parameters ω^k come from the local clients, where $k \in S$, and S refers to the participating subset of m clients in each communication round. These updated parameters are then aggregated. [11,12,15]

The setting of federated learning follows the principle of focused collection or data minimization proposed in the White House report [16]. In this setting, local models are trained upon data that are stored in the clients. The server does not need training data which contains private information. Only client parameters

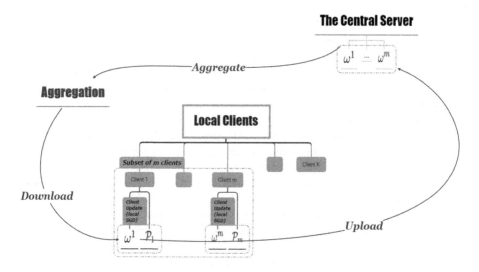

Fig. 1. Overview of federated learning.

are sent to the server, and they are aggregated to obtain the central model. After communication, clients receive the aggregated updates as initial parameters of the subsequent training. Then, their models are retrained on privacy data with local stochastic gradient descent (local SGD) [11,12,17,18].

In the typical FL that is executed on mobile devices, clients suffer from unstable communication. Hence it is reasonable to assume that only a subset of all clients, i.e. the aforementioned participating subset, is ready to get involved in each communication round. However, in our proposed federated deep autoencoding Gaussian mixture model (FDAGMM) for anomaly detection, the assumption does not hold anymore since the clients here commonly refer to companies or organizations with cutting-edge equipment and facilities.

2.3 Deep Autoencoding Gaussian Mixture Model

Two-step approaches that sequentially conduct dimensionality reduction and density estimation are widely used since they well address the curse of dimensionality to some extent [19]. Although fruitful progress has been achieved, these approaches suffer from a decoupled training process, together with inconsistent optimization goals, and the loss of valuable information caused by step one, i.e. the dimensionality reduction. Motivated by these, Zong et al. proposed a **Deep Autoencoding Gaussian Mixture Model (DAGMM)** [5].

As shown in Fig. 2, a *Compression Network* and an *Estimation Network* constitute DAGMM. It works as follows:

– **Dimensionality Reduction**: Given the raw features of a sample **x**, the compression network which is also a deep autoencoder conducts dimensionality

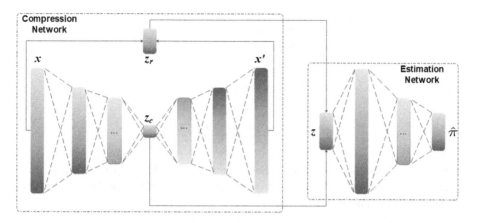

Fig. 2. Overview of deep autoencoding Gaussian mixture model.

reduction to output its low-dimensional representation $\mathbf{z}$ as follows:

$$
\begin{aligned}
\mathbf{z}_c &= h\left(\mathbf{x}; \theta_e\right) \\
\mathbf{x}' &= g\left(\mathbf{z}_c; \theta_d\right) \\
\mathbf{z}_r &= f\left(\mathbf{x}, \mathbf{x}'\right) \\
\mathbf{z} &= \left[\mathbf{z}_c, \mathbf{z}_r\right]
\end{aligned}
\tag{1}
$$

where θ_e and θ_d are the parameters of the decoder and encoder respectively, $\mathbf{x}'$ is the reconstruction of $\mathbf{x}$ generated by the autoencoder, $\mathbf{z}_c$ is the reduced/learned low-dimensional representation, $\mathbf{z}_r$. $h(\cdot)$, $g(\cdot)$, and $f(\cdot)$ denote the encoding, decoding and reconstruction-error calculation function respectively.

– **Density Estimation**: The subsequent estimation network takes $\mathbf{z}$ from the compression network as its input. It performs density estimation with a Gaussian Mixture Model (GMM). A multi-layer neural network, denoted as $MLN(\cdot)$, is adopted to predict the mixture membership for each sample as follows:

$$
\begin{aligned}
\mathbf{p} &= MLN\left(\mathbf{z}; \theta_m\right) \\
\hat{\gamma} &= \mathrm{softmax}(\mathbf{p})
\end{aligned}
\tag{2}
$$

where θ_m corresponds to parameters of MLN, K indicates the number of mixture components, and $\hat{\gamma}$ is a K-dimensional vector for the soft mixture-component membership prediction. Given the batch size N, $\forall 1 \leq k \leq K$, parameter estimation of GMM is further conducted as follows:

$$\hat{\phi}_k = \sum_{i=1}^{N} \frac{\hat{\gamma}_{ik}}{N}$$

$$\hat{\mu}_k = \frac{\sum_{i=1}^{N} \hat{\gamma}_{ik} \mathbf{z}_i}{\sum_{i=1}^{N} \hat{\gamma}_{ik}} \tag{3}$$

$$\hat{\mathbf{\Sigma}}_k = \frac{\sum_{i=1}^{N} \hat{\gamma}_{ik} (\mathbf{z}_i - \hat{\mu}_k)(\mathbf{z}_i - \hat{\mu}_k)^T}{\sum_{i=1}^{N} \hat{\gamma}_{ik}}$$

where $\hat{\gamma}_i$ is the membership prediction, and $\hat{\phi}_k$, $\hat{\mu}_k$, $\hat{\mathbf{\Sigma}}_k$ are the mixture probability, mean, and covariance for component k in GMM respectively. Furthermore, sample energy can be inferred as:

$$E(\mathbf{z}) = -\log \left(\sum_{k=1}^{K} \hat{\phi}_k \frac{\exp\left(-\frac{1}{2}(\mathbf{z} - \hat{\mu}_k)^T \hat{\mathbf{\Sigma}}_k^{-1}(\mathbf{z} - \hat{\mu}_k)\right)}{\sqrt{\left|2\pi\hat{\mathbf{\Sigma}}_k\right|}} \right) \tag{4}$$

where $|\cdot|$ denotes the determinant of a matrix.

Based on the three components, i.e. reconstruction error of autoencoder $L(\mathbf{x}_i, \mathbf{x}_i')$, sample energy $E(\mathbf{z}_i)$, and a penalty term $P(\hat{\mathbf{\Sigma}})$, the objective function of DAGMM is then constructed as:

$$J(\theta_e, \theta_d, \theta_m) = \frac{1}{N} \sum_{i=1}^{N} L(\mathbf{x}_i, \mathbf{x}_i') + \frac{\lambda_1}{N} \sum_{i=1}^{N} E(\mathbf{z}_i) + \lambda_2 P(\hat{\mathbf{\Sigma}}) \tag{5}$$

3 Federated Deep Autoencoding Gaussian Mixture Model

As discussed in the previous sections, limited data samples lead to the performance deterioration of DAGMM [5]. Therefore, the motivation of the proposed FDAGMM is to address this issue by extending the data sources while preserving the data privacy of the individual clients. Under the framework of federated learning (FL), not only can FDAGMM improve its performance with more data, but privacy is appropriately protected.

The rest of Sect. 3 is divided into two subsections, namely, *server execution* and *client update*. The two main components of FDAGMM are introduced in the form of pseudo-codes. FDAGMM shares most of the notation as FL except that ω is replaced with θ to be consistent with DAGMM as shown:

$$\theta = \{\theta_{\{e,d\}}, \theta_m\} \tag{6}$$

where the parameters of the autoencoder include those of the encoder and the decoder, i.e. $\theta_{\{e,d\}} = \{\theta_e, \theta_d\}$, and those of the estimation network is denoted as θ_m. Moreover, superscripts and subscripts are adopted to specify client k and communication round t that the parameters belong to, i.e. θ_t^k.

3.1 Server Execution

Algorithm 1 shows the *Server Execution* that is carried out on the central server. It consists of an initialization operation followed by T communication rounds. In initialization (Line 2), ω_0 is initialized.

Under the FL framework, the training process consists of the communication rounds that are indexed with t (Line 3). Lines 4–6 call sub-function *Client Update* for K clients in parallel. Then in line 7, the aggregation is performed to update θ, which is the parameter of the centre model. n and n_k indicate the number of instances belonging to client k and the total number of all involved samples, respectively.

Algorithm 1. Server Component of FDAGMM

1: **function** SERVEREXECUTION(if_two_phase)   ▷ Run on the server
2:     initialize θ_0
3:     **for** each round $t = 1, 2, ..., T$ **do**
4:         **for** each client $k \in \{1, ..., K\}$ **in parallel do**
5:             $\theta^k \leftarrow$ ClientUpdate($k, \theta_t, flag$)
6:         **end for**
7:         $\theta_{t+1} \leftarrow \sum_{k=1}^{K} \frac{n_k}{n} * \theta^k$
8:     **end for**
9: **end function**

3.2 Client Update

Client Update (Algorithm 2) takes k and θ as its input. k indexes a specific client and θ denotes the parameters of the central model in the current round. E denotes the local epoch. Line 2 splits data into batches, whereas Lines 3–7 train the local DAGMMs by batch with private data stored on each client. η denotes the learning rate; $J(\cdot)$ is the loss function. Its definitions is detailed in Equation (5). Line 8 returns the updated local parameters.

Algorithm 2. Client Component of FDAGMM

1: **function** CLIENTUPDATE($k, \theta, flag$)   ▷ Run on client k
2:     $\mathcal{B} \leftarrow$ (split $\mathcal{P}_k$ into batches of size N)
3:     **for** each local epoch i from 1 to E **do**
4:         **for** batch $b \in \mathcal{B}$ **do**
5:             $\theta \leftarrow \theta - \eta * \bigtriangledown J(\theta; b)$
6:         **end for**
7:     **end for**
8:     return θ to server
9: **end function**

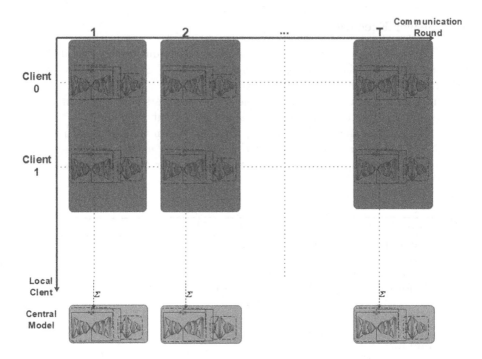

Fig. 3. Aggregation of federated deep autoencoding Gaussian mixture model.

Figure 3 shows an example illustrating the aggregation of FDAGMM. The abscissa and ordinate denote the communication round and the local client respectively. Two local devices, i.e. *Client 0, Client 1*, and a *server* are involved. The cross located at (*Clint 0; T*) indicates that *Clint 0* is participating in updating the central model in round T.

4   Experimental Results and Analysis

4.1   Experimental Design

Due to the lack of public datasets for anomaly detection, especially the Intrusion Detection Systems (IDSs), very few of them can be directly adopted in the evaluation of the proposed FDAGMM.

We perform extensive experiments with a public dataset KDDCUP 99, which is the most widely used in the evaluation of various anomaly detection approaches and systems. Table 1 shows the statistics of KDDCUP 99.

Data Pre-processing. Constructing datasets to simulate the private data of the individual clients which fulfill the associated requirements is thus needed for this study. In the FL setting, private datasets stored in clients are Non-IID and unbalanced. In the experiments, the whole KDDCUP 99 dataset is split into two

Table 1. Statistics of KDDCUP 99

# Dimensions	# Instances	Anomaly ratio
120	4898431	19.86%

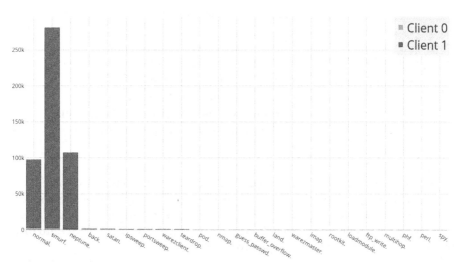

(a) Scenario 1: Client 0 with 1% common and 100% rare attacks.

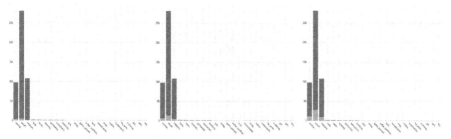

(b) Scenario 2: Client 0 with 2% common and 100% rare attacks. (c) Scenario 3: Client 0 with 5% common and 100% rare attacks. (d) Scenario 4: Client 0 with 10% common and 100% rare attacks.

Fig. 4. Stacked bar-charts of attack types belong to two clients.

sets belonging to *Client 1* and *Client 2* through selecting records in the complete KDDCUP. These two clients play the roles of two companies in which Client 1 with limited data asks for help from Client 2 under the framework of FDAGMM.

Attack instances are separated according to their types. Training sets only include half of the attack samples belonging to its client, while test sets consist of both normal and the other half attack instances. The data belonging to *Client 1* are similar in anomaly ratio as the other client. Since there is a very sharp distinction between rare and common attacks, which is reflected in Fig. 4, *smurf*

and *neptune* are included as **Common Attacks** and the rest comprises the
Rare Attacks. The details are shown in Table 2. The experiments are expected
to prove that *Client 1* can improve its performance with the help of *Client 2*.

Table 2. Common and rare attacks

	Types
Common Attacks	smurf, neptune
Rare Attacks	satan, ipsweep, portsweep, nmap, back,
	warezclient, teardrop, pod, guess_passwd,
	buffer_overflow, land, warezmaster, imap, rootkit,
	loadmodule, ftp_write, multihop, phf, perl, spy

As shown in Fig. 4(a), the scenario with *Client 1* holding 1% common and
100% rare (bars in orange) attack instances and *Client 2* holding the rest, i.e.
99% common attacks (bars in blue), is denoted as **Scenario 1**. Those corresponding to the remaining three figures, (b) to (d) are denoted as **Scenario 2**,
Scenario 3 and **Scenario 4** respectively.

Table 3. Default setting of DAGMM

Notion	Parameter range
$\eta^\star$	0.0001
N^*	1024
$\lambda_1^\dagger$	0.1
$\lambda_2^\dagger$	0.005

$\star$ Learning Rate
$*$ Batch Size
$\dagger$ Involved in Eq. (5)

Parameter Setting on DAGMM. For the purpose of a fair comparison, all
these experiments adopt the default DAGMM setting as the hyper-parameters
of the local models in the proposed FDAGMM. The settings are summarized in
Table 3.

4.2 Experiments on KDDCUP 99

The experiments are designed to evaluate the effectiveness of the proposed
FDAGMM. According to their attack types, i.e. rare or common, instances making up each training set belong to two clients. Four scenarios with distinct combinations of attacks are considered and illustrated in Fig. 4. Three metrics are

adopted to measure the performance of the compared algorithms. They are **Precision**, **Recall**, and **F1-Score**.

All the experiments are run independently for five times; each time, the random seed is fixed. The average (AVG) values of F1-Score are presented in Fig. 5. Tables 4 and 5 show the AVG and standard deviation (STDEV) of precision and recall values respectively. In the tables, the AVG value is listed before the standard deviation in parentheses.

The names of the algorithms compared in Tables 4 and 5 indicate not only the adopted technique but also the involved training set, i.e. Training set from *Client 0* or *Client 1*.

- **DAGMM_C0** employs DAGMM and trains the model with only data from *Client 0*. Test set comprises half of the attack instances belonging to *Client 0* and the corresponding proportion of normal samples.
- **DAGMM_C1** employs DAGMM and trains the model with only data from *Client 1*. Test set comprises half of the attack instances belonging to *Client 1* and the corresponding proportion of normal samples. *DAGMM_C1* thus complements the training on the limited data of *Client 0* to increase the detection performance on *Client 0* without comprising the privacy of its data.
- **DAGMM_C0&C1** employs DAGMM and trains the model with a mixture of the data from two clients. Test set comprises half of the attack instances belonging to *Client 0* and the corresponding proportion of normal samples. *DAGMM_C0&C1 (Ideal Bound)* denotes where the performance limit of FDAGMM is. This is under the ideal scenario where both clients are willing to share their data so that DAGMM can be trained to achieve the best performance.
- **FDAGMM** includes two clients, and each employs a DAGMM and trains the model with only data from itself. Test set comprises half of the attack instances belonging to *Client 0* and the corresponding proportion of normal samples. The performance of FDAGMM reflects how much help *Client 0* can receive from *Client 1* under the FL framework. In other words, how close the presented FDAGMM can approach the ideal performance limit, i.e. DAGMM_C0&C1 (Ideal Bound) where there is an abundance of data for training and clients are willing to contribute their data for collective training of the DAGMM.

Table 4. Comparative studies on FDAGMM: precision.

Scenario	DAGMM_C0	FDAGMM	Ideal Bound	DAGMM_C1
Scenario 1	0.397(0.1304)	**0.4532(0.0317)**	0.5366(0.0215)	0.9881(0.0024)
Scenario 2	0.4793(0.1712)	**0.5116(0.0289)**	0.6266(0.0444)	0.9787(0.0073)
Scenario 3	0.6069(0.0872)	**0.6553(0.0403)**	0.7261(0.0612)	0.9832(0.0129)
Scenario 4	0.6735(0.0883)	**0.7447(0.0369)**	0.7583(0.0302)	0.9813(0.0018)

Table 5. Comparative studies on FDAGMM: recall.

Scenario	DAGMM_C0	FDAGMM	Ideal Bound	DAGMM_C1
Scenario 1	0.9263(0.1434)	**0.9769(0.0459)**	0.9222(0.0422)	0.9952(0.0097)
Scenario 2	0.8166(0.127)	**0.9927(0.0121)**	0.9409(0.0488)	0.9978(0.0044)
Scenario 3	0.8777(0.0698)	**0.9825(0.0284)**	0.9341(0.0482)	0.9984(0.0016)
Scenario 4	0.9237(0.038)	**0.9803(0.0264)**	0.976(0.0334)	0.9987(0.0019)

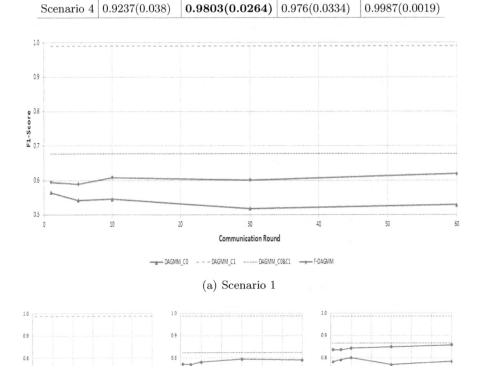

Fig. 5. Comparative studies on FDAGMM on KDDCUP 99.

Based on the results shown in these tables and figures, the following observations can be made:

– The proposed FDAGMM outperforms DAGMM on all metrics, i.e. F1-Score, Precision and Recall for all four scenarios.
– According to its lower STDEV values, FDAGMM's performance is more stable than DAGMM.

– The more Non-IID and unbalanced the data distribution is across clients, the more challenging the scenario tends to be, which is reflected by the blue dotted lines corresponding to DAGMM_C0&C1 in Fig. 5.

5   Conclusion

With the help of other clients holding sufficient feature-similar records under the FL framework, we show that the less than satisfactory performance of DAGMM suffering from limited dataset can be addressed and improved using FDAGMM. Empirical studies comparing the performance of the proposed FDAGMM and DAGMM under four distinct scenarios demonstrate the superiority of the FDAGMM in terms of all the associated performance metrics.

This study follows the assumption that all local models adopt the same neural networks architecture and share the same hyperparameters, which implies all the involved data records share the same feature structure. This renders FDAGMM to be less versatile to be deployed to other application domains. In future research, we are going to develop new federated learning assisted DAGMM address the weakness.

References

1. LeCun, Y., Bengio, Y., Hinton, G.: Deep learning. Nature **521**(7553), 436 (2015)
2. Zhang, J., Yeung, S.H., Shu, Y., He, B., Wang, W.: Efficient memory management for GPU-based deep learning systems. arXiv preprint arXiv:1903.06631 (2019)
3. Chalapathy, R., Chawla, S.: Deep learning for anomaly detection: a survey. arXiv preprint arXiv:1901.03407 (2019)
4. Javaid, A., Niyaz, Q., Sun, W., Alam, M.: A deep learning approach for network intrusion detection system. In: Proceedings of the 9th EAI International Conference on Bio-inspired Information and Communications Technologies (formerly BIONETICS). ICST (Institute for Computer Sciences, Social-Informatics and Telecommunications Engineering), pp. 21–26 (2016)
5. Zong, B., et al.: Deep autoencoding gaussian mixture model for unsupervised anomaly detection (2018)
6. McMahan, H.B., Moore, E., Ramage, D., Hampson, S., et al.: Communication-efficient learning of deep networks from decentralized data. arXiv preprint arXiv:1602.05629 (2016)
7. Edgeworth, F.Y.: Xli. on discordant observations. Lond. Edinb. Dublin Philos. Mag. J. Sc. **23**(143), 364–375 (1887)
8. Tsai, C.F., Hsu, Y.F., Lin, C.Y., Lin, W.Y.: Intrusion detection by machine learning: a review. Expert Syst. Appl. **36**(10), 11994–12000 (2009)
9. M. LLC.: MS Windows NT kdd cup (1999). http://kdd.ics.uci.edu/databases/kddcup99/task.html
10. Stolfo, S.J., Fan, W., Lee, W., Prodromidis, A., Chan, P.K.: Cost-based modeling for fraud and intrusion detection: results from the jam project. In: Proceedings DARPA Information Survivability Conference and Exposition, DISCEX 2000, vol. 2, pp. 130–144. IEEE (2000)

11. Konečný, J., McMahan, B., Ramage, D.: Federated optimization: distributed optimization beyond the datacenter. arXiv Prepr arXiv:1511.03575, no. 1, pp. 1–5 (2015)
12. Konecný, J., McMahan, H.B., Yu, F.X., Richtárik, P., Suresh, A.T., Bacon, D.: Federated learning: strategies for improving communication efficiency. CoRR, vol. abs/1610.0, no. NIPS, pp. 1–5 (2016)
13. Ma, C., et al.: Distributed optimization with arbitrary local solvers. Optim. Methods Softw. **32**(4), 813–848 (2017)
14. Reddi, S.J., Konečnỳ, J., Richtárik, P., Póczós, B., Smola, A.: Aide: fast and communication efficient distributed optimization. arXiv preprint arXiv:1608.06879 (2016)
15. Chen, Y., Sun, X., Jin, Y.: Communication-efficient federated deep learning with asynchronous model update and temporally weighted aggregation. arXiv preprint arXiv:1903.07424 (2019)
16. House, W.: Consumer data privacy in a networked world: a framework for protecting privacy and promoting innovation in the global digital economy. White House, pp. 1–62. Washington, DC (2012)
17. Chen, Y., Sun, X., Hu, Y.: Federated learning assisted interactive EDA with dual probabilistic models for personalized search. In: Tan, Y., Shi, Y., Niu, B. (eds.) ICSI 2019. LNCS, vol. 11655, pp. 374–383. Springer, Cham (2019). https://doi.org/10.1007/978-3-030-26369-0_35
18. McMahan, B., Moore, E., Ramage, D., Hampson, S., Arcas, B.A.: Communication-efficient learning of deep networks from decentralized data. In: Artificial Intelligence and Statistics, pp. 1273–1282 (2017)
19. Candès, E.J., Li, X., Ma, Y., Wright, J.: Robust principal component analysis? J. ACM (JACM) **58**(3), 11 (2011)

Towards a Hierarchical Deep Learning Approach for Intrusion Detection

François Alin, Amine Chemchem[✉], Florent Nolot, Olivier Flauzac, and Michaël Krajecki

CRESTIC - Centre de Recherche en Sciences et Technologies de l'Information et de la Communication - EA 3804, Reims, France
{francois.alin,mohamed-lamine.chemchem,florent.nolot,olivier.flauzac, michael.krajecki}@univ-reims.fr

Abstract. Nowadays, it is almost impossible to imagine our daily life without Internet. This strong dependence requires an effective and rigorous consideration of all the risks related to computer attacks. However traditional methods of protection are not always effective, and usually very expensive in treatment resources. That is why this paper presents a new hierarchical method based on deep learning algorithms to deal with intrusion detection. This method has proven to be very effective across traditional implementation on four public datasets, and meets all the other requirements of an efficient intrusion detection system.

Keywords: Machine learning · Deep learning · Intrusion detection · Artificial intelligence · Cyber security

1 Introduction

Over the last two decades, many solutions have emerged to protect and secure computer systems. They are complementary but not always sufficient. Thus, antivirus software acts on a host computer and protects against viruses and malicious programs. If this solution is effective for isolated machines and viruses already known, it is not recommended to trust them, especially when connected to the Internet.

To solve the problem of network intrusion, firewalls come to the rescue. A firewall is used to control communications between a local network or host machine and the Internet. It filters traffic in both directions and blocks suspicious traffic according to a network security policy. It is therefore the tool that defines the software and the users that have the permission to connect to the Internet or to access the network [1]. With anti-virus, the firewall increases the security of data in a network. However, this combination remains powerless in front of malicious users with knowledge of all the requirements of security policy. Indeed, once a software or user has the permission to connect to the Internet or a network, there is no guarantee that they will not perform illegal operations. Moreover, many studies have shown that 60% to 70% of attacks come from within systems [2].

© IFIP International Federation for Information Processing 2020
Published by Springer Nature Switzerland AG 2020
S. Boumerdassi et al. (Eds.): MLN 2019, LNCS 12081, pp. 15–27, 2020.
https://doi.org/10.1007/978-3-030-45778-5_2

To handle this problem, in addition to antivirus and firewalls, intrusion detection systems (IDS) are used to monitor computer systems for a possible intrusion considered as unauthorized use or misuse of a computer system [3].

IDSs are designed to track intrusions and protect system vulnerabilities. They are very effective at recognizing intrusions for which they are programmed. However, they are less effective if the intruder changes the way he attacks. Whatever the performance of the IDS, they are often limited by the amount of data that the IDS can handle at a time. This limitation does not allow for permanent monitoring and leaves violations for intruders.

This research explores a first implementation of classical machine learning models for intrusion detection. These models take into account the historical IDSs data with their corresponding classes. Then, through a comparative study, we select the best method to use it for the inference in order to detect intrusions in real-time. In addition, we present in this study a hierarchical learning method that is proving to be very effective in comparison with the traditional classification method.

The rest of this paper is organised as follows: next section presents some related works, followed by the proposed strategy in Sect. 3. In Sect. 4 we show the experimental results and give some discussions. Finally, in Sect. 5 we conclude with some perspectives.

2   Related Works

Many recent researches try to handle the intrusion detection problem with the artificial intelligence and machine learning approaches.

The authors of [4] explore the issue of the game theory for modelling the problem of intrusion detection as a game between the intruder and the IDS according to a probabilistic model, the objective of their study is to find a frequency for an IDS verification activities that ensures the best net gain in the worst case. We think that is a good idea, but if the attacker changes his behaviour, the proposed approach will no more be able to intercept him effectively.

In [5] a new agent architecture is proposed. It combines case-based reasoning, reactive behavior and learning. Through this combination, the proposed agent can adapt itself to its environment and identify new intrusions not previously specified in system design. Even if the authors showed that the hybrid agent achieves good results compared to a reactive agent, their experimental study did not include other learning approaches such as support vector machine, K nearest neighbors... In addition, the learning set used in this study is very small, only 1000 records.

The authors in [6] proposed an intrusion detection framework based on an augmented features technique and an SVM algorithm. They validated their method on the NSL-KDD dataset, and stated that their method was superior to other approaches. However, they did not mention which data are used for the test. In addition, the application of features augmentations technique increases the risk of falling into an over fitting case, especially when processing large data, so we believe this is not an ideal choice for analyzing large network traffic for intrusion detection.

In [7], the authors applied a hybrid model of genetic algorithms with SVM and KPCA to intrusion detection. They used the KDD CUP99 dataset to validate their system. However, this dataset contains several redundancies, so the classifier will probably be skewed in favor of more frequent records. With the same way, the authors of [8] combined decision tree with genetic algorithms and features selection for intrusion detection. They used also the KDD dataset which is not really reliable for validating methods, it would have been more interested to take other datasets to confirm the proposed approach. An interesting multi-level hybrid intrusion detection model is presented in [9], it uses support vector machine and extreme learning machine to improve the efficiency of detecting known and unknown attacks. The authors apply this model on the KDD99 dataset, which has the previously mentioned drawbacks. On the same dataset, a new LSTM: long short term memory model is presented in [10] to deal with four classes of attacks, and the results are satisfactory. In [11] the authors present a text mining approach to detect intrusions, in which they present a new distance measure. However, most of logs features are in numerical format, and taking them as text data will considerably increase the complexity of the calculation in terms of memory capacity and also in terms of learning time. This is the biggest flaw in the last two papers mentioned.

A very interesting survey is presented in [12]. The paper describes the literature review of most of machine learning and data mining methods used for cyber security. However, the methods that are the most effective for cyber applications have not been established by this study, the authors affirm that given the richness and complexity of the methods, it is impossible to make one recommendation for each method, based on the type of attack the system is supposed to detect. In our study we draw inspiration from the methods cited by this paper such as decision trees, support vector machine, k nearest neighbors,... for a comparative study established on the most popular datasets. In addition, we enrich our comparative study with several neural networks models, and with a new proposed hierarchical classification method.

3   The Proposed Strategy

Contrary to the idea in [13], in which the authors present an hierarchical classification of the features for intrusion detection. In our study we propose a hierarchical method that starts by detecting malicious connexion with a binary classification, and then, in a second time, the algorithm tries to find the corresponding attack class by a multi-label classification as shown in Fig. 1. The idea is to detect an abnormal connection very quickly and launch a warning to the admin, then try to classify this connection in the corresponding attack class.

For this, we will design a hierarchical approach to the machine learning methods, and compare their performances with the classical algorithms of classification.

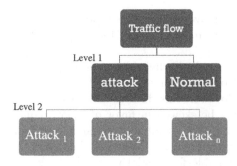

Fig. 1. The proposed hierarchical classification method

In the context of data mining and machine learning: classification is done using a model that is built on historical data. The goal of predictive classification is to accurately predict the target class for each record in new data. A classification task begins with build training data for which the target values (or class assignments) are known. Many classification algorithms use different techniques for finding relations between the predictor attribute's values and the target attribute's values in the build data [14,15]. In the following subsections, a summarised overview of the implemented machine learning algorithms is reported.

3.1   Naïve Bayes Classifier

The naïve Bayes algorithm is based on Bayesian probability theory following assumptions of naive independence [16]. It is one of the most basic classification techniques with various applications, such as email spam detection, personal email sorting, and document categorization.

Even though it is often outperformed by other techniques. The main advantage of the naïve Bayes remains that it is less computationally intensive (in both CPU and memory), and it requires a small amount of training data. Moreover, the training time with Naive Bayes is significantly smaller as opposed to alternative methods [17].

3.2   K-Neighbors Approach

Nearest Neighbors is one of the simplest, and rather trivial classifiers is the rote classifier, which memorizes all training data and performs classification only if the attributes of the test object match one of the training examples exactly [18]. A more known variation, the k-nearest neighbor (kNN) classification [19], finds a group of k objects in the training set that are closest to the test object and bases the assignment of a label on the predominance of a particular class in this neighborhood. There are three key elements of this approach: a set of labeled objects, a distance or similarity metric to compute distance between objects, and the value of parameter k, which represents the number of nearest neighbors.

To classify an unlabeled object, the distance of this object to the labeled objects is computed, its k-nearest neighbours are identified, and the class labels of these nearest neighbours are then used to determine the class label of the object.

Given a training set DR and a test object $z = (x', y')$ the algorithm computes the distance (or similarity) between z and all the training objects $(x, y) \in DR$ to determine its nearest-neighbor list: D_z. (x_i is the training data of $object_i$, while y_i is its class. Likewise, x' the data of the test object and y' is its class.) Once the nearest-neighbors list is obtained, the test object is classified based on the majority class of its nearest neighbors:

$$Majority\_Voting\_y' = argmax_v \sum_{x_i, y_i \in D_z} I(v = y_i).$$

where v is a class label, y_i is the class label for the ith nearest neighbors, and $I()$ is an indicator function that returns the value 1 if its argument is true and 0 otherwise.

3.3   Support Vector Machine

Support vector machines (SVM) have exhibited superb performance in binary classification tasks. Intuitively, SVM aims at searching for a hyperplane that separates the two classes of data with the largest margin (the margin is the distance between the hyperplane and the point closest to it) [20, 21].

Most discriminative classifiers, including SVMs, are essentially two-class classifiers. A standard method of dealing with multi-class problems is to create an ensemble of yes/no binary classifiers, one for each label. This method is called "one-vs-others" [22].

3.4   Random Forests

Random Forests are a part of ensemble learning. Ensemble learning [23] deals with methods which employ multiple learners to solve a problem. The capacity of working with several learners in the same time achieve better results than a single learner. Random forest works by creating various decision trees in the training phase and output class labels those have the majority vote [24]. They achieve high classification accuracy and can handle outliers and noise in the data. Random Forest is implemented in this work because it is less susceptible to over-fitting and it has previously shown good classification results.

3.5   Multilayer Perceptron Neural Networks

The basic unit in a neural network is called "neuron" or "unit". Each neuron receives a set of inputs, which are denoted by the vector $\overline{X}_i$ [25]. In addition, each neuron is associated with a set of weights A, which are used for computing a function f of its inputs. A typical function that is basically used in the neural network is the linear function, it is defined as follows: $p_i = A.\overline{X}_i$. We assume that the class label is denoted by y_i. The goal of this approach is to learn the set of weights A with the use of the training set. The idea is to start off with random weights, and gradually update them when a mistake is done by applying the current function on the training example. The magnitude of the update is regulated by a learning rate μ. This forms the core idea of the perceptron algorithm.

Algorithm 1. Perceptron Algorithm [26]

---

inputs: Learning Rate: μ
        Training rules $(\overline{X}_i, y_i) \forall i \in \{1...n\}$.
Initialize weight vectors in A to 0 or small random numbers.
Repeat
– Apply each training rule to the neural network
– **if** $((A.\overline{X}_i)$ does not matches $y_i)$ **then**
        update weigts A based on learning rate μ.
until weights in A converge.

3.6   Convolutional Neural Network Classifier

A Convolutional Neural Network (CNN) is comprised of one or more convolutional layers, and then followed by one or more fully connected layers as in a standard multilayer neural network. The neurons of a convolutional layer are grouped in feature maps sharing the same weights, so the entire procedure becomes equivalent to convolution [27]. Convolutional layers are usually followed by a non-linear activation-layer, in order to capture more complex properties of the input data. The pooling layers are usually used for subsampling the preceding layer, by aggregating small rectangular values subsets. Maximum or average pooling is often applied by replacing the input values with the maximum or the average value, respectively. Finally, one or more dense layers are put in place, each followed by an activation-layer, which produce the classification result.

The training of CNNs is performed similarly to that of classical Multilayer Perceptron Networks, by minimizing a loss function using gradient descent-based methods and back-propagation of the error.

In this study, our best CNN model is reached after trying many architectures. It is inspired by the contribution of [28] in sentiments detection. Since its model demonstrated well performant results, we adapted it in our study for intrusion detection. It is mainly composed of four hidden layers in addition to the input and the output layer. In the following, we show a summarization of the architecture of our best CNN model:

– The input layer is a convolution of one dimension with a number of neurones equals to the number of the dataset features.

- The second layer is a max pooling 1D with a pool size equals to 4.
- The third layer is a flatten with 512 neurones.
- The fourth is a dense layer with 512 neurones with 'relu' as activation function.
- The output layer is a dense layer with the 'sigmoid' function, and a number of neurones equals to the number of classes.

4   Experimentation Results and Discussion

In this section, we implement various machine learning models to classify the different sets of network traffic of the four well known benchmarks, which are summarised in Table 1:

Table 1. The datasets characteristics

Dataset	Rows	Features	Classes
KDD 99	4,898,430	42	23
NSL-KDD	125,973	42	23
UNSW-NB15	2,540,044	49	10
CIC-IDS 2017	2,832,678	79	14

4.1   Performance Evaluation

For the performance evaluation, we calculate the accuracy with the F-score of each approach. The F-score also called F-measure is based on the two primary metrics: precision and recall. Given a subject and a gold standard, precision is the proportion of cases that the subject classified as positive that were positive in the gold standard. It is equivalent to positive predictive value. Recall is the proportion of positive cases in the gold standard that were classified as positive by the subject. It is equivalent to sensitivity. The two metrics are often combined as their harmonic mean, the formula can be formulated as follows:

$$F = \frac{(1 + \beta^2) \times recall \times precision}{(\beta^2 \times precision) + recall}$$

$$Precision = \frac{TP}{TP + FP}, Recall = \frac{TP}{TP + FN}$$

Where TP is the number of true positives, TN is the number of true negatives, FP is the number of false positives and FN is the number of false negatives. The F-measure can be used to balance the contribution of false negatives by weighting recall through a parameter $\beta \geq 0$. In our case β is set to 1, F1-score is than equal to:

$$F1\_score = \frac{2 \times recall \times precision}{precision + recall}$$

4.2   Environment and Materials

We use in this implementation, Python language, Tensorflow tool, and Keras library. All the algorithms are executed and compared using the **NVIDIA DGX-1**[1]. The DGX1 is an Nvidia server which specializes in using GPGPU to accelerate deep learning applications. The server features 8 GPUs based on the Volta daughter cards with HBM 2 memory, connected by an NVLink mesh network.

4.3   Multi-class Classification Results

In this paper, first, we classify the datasets as they are labelled, without any modification. The obtained results are shown in Table 2, columns of Multi-label classification.

We notice from these results that most of the approaches succeed in obtaining good training and test scores. However, the best of these models does not exceed 71% accuracy on the test benchmark.

Considering the delicacy of the domain, and the dangerousness that can generate an IDS which classifies network traffic as normal when it is an attack. We propose a hierarchical classification strategy to achieve greater accuracy.

4.4   Hierarchical Classification Strategy

Given the main objective of an intrusion detection system, which is to detect potential attacks, we have decided in this strategy to adopt an hierarchical classification.

First, we start with a binary classification, merging all attack classes into one large class and labelling it 'attack', and on the other hand keeping the 'normal' class without any modification.

Then, after separating normal network traffic to that which represents a potential attack, a multi-class classification can be applied within the "attack" class to know what type of attack it is.

Binary Classification (Level 1: Normal/Attack). We start the hierarchical classification strategy by the detection of an abnormal connexion. This is reached by the binary classification task. For this, we have merged all the connections which have a different label from 'normal' into a single class that we have labelled 'attack'. The obtained results are show in Table 2, columns of hierarchical (level 1).

---

[1] https://www.nvidia.fr/data-center/dgx-1/.

Table 2. Classifications reports

Data set	ML approach	Multi-label classification		Hierarchical (level 1)		Hierarchical (level 2)	
		Training	Test score	Training	Test score	Training	Test score
KDD 99	Naive Bayes classifier	0.91	0.30	0.98	0.79	0.84	0.46
	Decision tree classifier	0.99	0.53	0.99	0.69	0.99	0.58
	K neighbors classifier	0.99	0.50	0.99	0.81	0.99	0.58
	Logistic regression classifier	0.99	0.31	0.99	0.83	0.99	0.55
	Support vector classifier	0.99	0.43	0.99	**0.84**	0.99	0.58
	Ada_Boost classifier	0.66	0.34	0.92	0.83	0.92	0.40
	Random forest classifier	0.99	0.53	0.99	0.83	0.99	0.56
	Multilayer perceptron	0.99	0.33	0.99	0.83	0.99	**0.61**
	Best NN model	0.99	0.35	0.99	**0.84**	0.99	**0.61**
NSL-KDD	Naive Bayes classifier	0.87	0.56	0.90	0.77	0.83	0.44
	Decision tree classifier	0.99	0.59	0.99	0.80	0.99	0.42
	K neighbors classifier	0.99	0.69	0.99	0.78	0.99	**0.75**
	Logistic regression classifier	0.97	0.71	0.95	0.78	0.99	**0.75**
	Support vector classifier	0.99	0.70	0.99	**0.82**	0.99	0.73
	Ada_Boost classifier	0.84	0.62	0.98	0.75	0.76	0.23
	Random forest classifier	0.99	0.67	0.99	0.79	0.99	0.55
	Multilayer perceptron	0.99	0.70	0.99	0.81	0.99	0.73
	Best NN model	0.98	0.73	0.99	0.81	0.99	0.73
UNSW-NB	Naive Bayes classifier	0.64	0.56	0.80	0.72	0.60	0.59
	Decision tree classifier	0.80	0.32	0.94	0.70	0.78	0.22
	K neighbors classifier	0.76	0.70	0.93	0.83	0.74	0.74
	Logistic regression classifier	0.75	0.62	0.93	0.78	0.73	0.53
	Support vector classifier	0.78	0.53	0.93	0.80	0.77	0.73
	Ada_Boost classifier	0.59	0.50	0.94	0.77	0.59	0.12
	Random forest classifier	0.81	0.43	0.95	0.76	0.79	0.22
	Multilayer perceptron	0.79	0.71	0.94	0.81	0.78	0.67
	Best NN model	0.74	0.71	0.94	**0.86**	0.78	**0.79**
CIC-IDS	Naive Bayes classifier	0.69	0.63	0.65	0.64	0.81	0.43
	Decision tree classifier	0.99	0.15	0.99	0.30	0.99	0.10
	K neighbors classifier	0.99	0.73	0.99	0.73	0.99	0.91
	Logistic regression classifier	0.94	0.88	0.95	0.90	0.98	0.94
	Support vector classifier	0.96	0.72	0.97	0.67	0.99	0.86
	Ada_Boost classifier	0.65	0.35	0.99	0.53	0.50	0.50
	Random forest classifier	0.99	0.30	0.99	0.32	0.99	0.45
	Multilayer perceptron	0.98	0.87	0.99	0.91	0.99	0.91
	Best NN model	0.96	0.88	0.97	**0.92**	0.99	**0.95**

Classification of Attack Types (Level 2: Multi-class Classification).
After applying a binary classification to detect abnormal connections. We implement a multi-class classification approach on the 'attack' class, to obtain more
details on the type of attack. All the well known machine learning approaches
are implemented and compared in this way. The results are noted in Table 2,

columns of hierarchical classification (level2). We can imagine a third level, if we have subclasses in an attack class.

4.5   Discussion

From these results, we note that the proposed hierarchical approach has considerably improved the effectiveness of the classical classification approach on all the benchmarks studied.

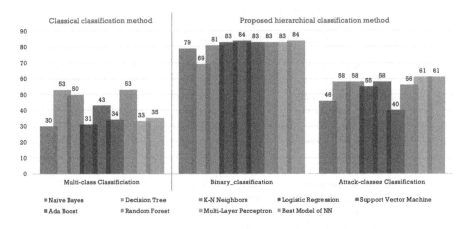

Fig. 2. Classification methods comparison on KDD Dataset

For instance, on the KDD99 dataset, the proposed hierarchical classification approach surpassed the traditional multi-class approach. Like demonstrated on Fig. 2. While the best approach obtained an accuracy of 53% in a multi-class classification, the proposed approach allows to detect an abnormal connection with a rate of 84%, and to predict the attack type with a success rate of 61%.

Also on the NSLKDD dataset, the proposed hierarchical classification approach surpassed the traditional multi-class approach. We can note on Fig. 3 that, while the best approach obtained an accuracy of 73% in a multi-class classification, the proposed approach allows to detect an abnormal connection with a rate of 82%, and to predict the attack type with a success rate of 75%.

In Fig. 4, we note that the proposed approach of hierarchical classification has increased the accuracy rate. While, the best approach obtained an accuracy of 71% in a multi-class classification, the proposed approach allows to detect an abnormal connection with a rate of 86%, and to predict the attack type with a success rate of 79%.

We valid our hypothesis also on the CIC-IDS 2017 dataset. In Fig. 5, we note that the proposed approach of hierarchical classification has increased the accuracy rate. While, the best approach obtained an accuracy of 88% in a multi-class classification, the proposed approach allows to detect an abnormal connection with a rate of 92%, and to predict the attack type with a success rate of 95%.

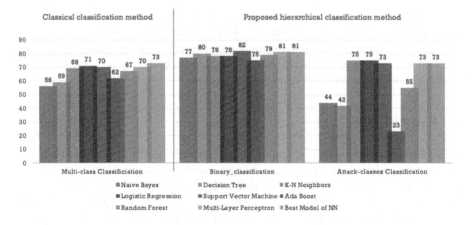

Fig. 3. Classification methods comparison on NSLKDD Dataset

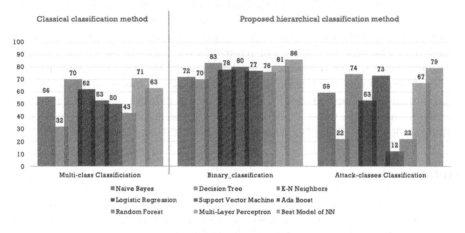

Fig. 4. Classification methods comparison on UNSW15 Dataset

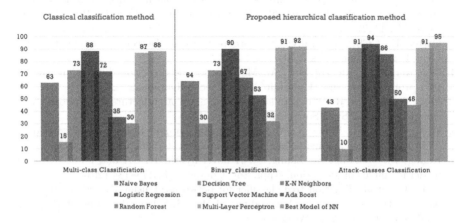

Fig. 5. Classification methods comparison on CIC_IDS 2017 Dataset

According to the results of our comparative study, in general we can validate the hypothesis that to achieve effective intrusion detection, we must start with a binary classification (attack/normal) using our best neural network model, followed by the application of the KNN algorithm or the best neural network model to find out what type of attack is involved.

5   Conclusion

The study presented here leads us to make two conclusions. It appears first that the learning model best suited to the intrusion detection problem is that based on convolutional neural networks. Moreover, by comparing different learning strategies, the approach based on a hierarchical detection of the attacks (starting with a first level of binary classification discriminating only the compliant traffic of the nonconforming traffic) presents the best performances, well in front of the methods of multi-label classification. The system thus obtained has an intrusion detection rate of 95%. These results allow us to consider the implementation of a real-time intrusion detection system based on our CNN model and binary classification. This will require larger datasets and more powerful training infrastructure solutions to further improve the detection rate. Finally, one of the challenges of intrusion detection remains zero-day attack detection. It turns out that the method used to train our neural network gives him the ability to identify as invalid data he has never met during his training. The next task will be to develop this capacity and especially to measure its effectiveness.

References

1. da Costa Júnior, E.P., da Silva, C.E., Pinheiro, M., Sampaio, S.: A new approach to deploy a self-adaptive distributed firewall. J. Internet Serv. Appl. 9(1), 1–21 (2018). https://doi.org/10.1186/s13174-018-0083-6
2. Stolfo, S.J., Salem, M.B., Keromytis, A.D.: Fog computing: mitigating insider data theft attacks in the cloud. In: 2012 IEEE Symposium on Security Privacy Workshops. IEEE (2012)
3. Carter, E.: CCSP Self-study: Cisco Secure Intrusion Detection System (CSIDS). Cisco Press, Indianapolis (2004)
4. Ouharoun, M., Adi, K., Pelc, A.: Modélisation de détection d'intrusion par des jeux probabilistes. Diss. Université du Québec en Outaouais (2010)
5. Leite, A., Girardi, R.: A hybrid and learning agent architecture for network intrusion detection. J. Syst. Softw. 130, 59–80 (2017)
6. Wang, H., Jie, G., Wang, S.: An effective intrusion detection framework based on SVM with feature augmentation. Knowl.-Based Syst. 136, 130–139 (2017)
7. Kuang, F., Weihong, X., Zhang, S.: A novel hybrid KPCA and SVM with GA model for intrusion detection. Appl. Soft Comput. 18, 178–184 (2014)
8. Stein, G., et al.: Decision tree classifier for network intrusion detection with GA-based feature selection. In: Proceedings of the 43rd Annual Southeast Regional Conference-Volume 2. ACM (2005)

9. Al-Yaseen, W.L., Othman, Z.A., Nazri, M.Z.A.: Multi-level hybrid support vector machine and extreme learning machine based on modified K-means for intrusion detection system. Expert Syst. Appl. **67**, 296–303 (2017)
10. Kim, J., et al.: Long short term memory recurrent neural network classifier for intrusion detection. In: 2016 International Conference on Platform Technology and Service (PlatCon). IEEE (2016)
11. RajeshKumar, G., Mangathayaru, N., Narsimha, G.: Intrusion detection a text mining based approach. arXiv preprint arXiv:1603.03837 (2016)
12. Buczak, A.L., Guven, E.: A survey of data mining and machine learning methods for cyber security intrusion detection. IEEE Commun. Surv. Tutorials **18**(2), 1153–1176 (2016)
13. Wang, W., et al.: HAST-IDS: learning hierarchical spatial-temporal features using deep neural networks to improve intrusion detection. IEEE Access **6**, 1792–1806 (2018)
14. Chemchem, A., Alin, F., Krajecki, M.: Improving the cognitive agent intelligence by deep knowledge classification. Int. J. Comput. Intell. Appl. **18**, 1950005 (2019)
15. Chemchem, A., Alin, F., Krajecki, M.: Combining SMOTE sampling and machine learning for forecasting wheat yields in France. In: 2019 IEEE Second International Conference on Artificial Intelligence and Knowledge Engineering (AIKE). IEEE (2019)
16. Jain, A., Mandowara, J.: Text classification by combining text classifiers to improve the efficiency of classification. Int. J. Comput. Appl. (2250–1797) **6**(2) (2016)
17. Huang, J., Lu, J., Ling, C.X.: Comparing naive Bayes, decision trees, and SVM with AUC and accuracy. In: Third IEEE International Conference on Data Mining (ICDM), p. 553 (2003)
18. Adeniyi, D., Wei, Z., Yongquan, Y.: Automated web usage data mining and recommendation system using K-nearest neighbor (KNN) classification method. Appl. Comput. Inform. **12**(1), 90–108 (2016)
19. Wu, X.: Top 10 algorithms in data mining. Knowl. Inf. Syst. **14**(1), 1–37 (2008)
20. Guyon, I., Weston, J., Barnhill, S., Vapnik, V.: Gene selection for cancer classification using support vector machines. Mach. Learn. **46**(1), 389–422 (2002)
21. Vapnik, V.: The Nature of Statistical Learning Theory. Springer, Heidelberg (2013)
22. Hsu, C.-W., Lin, C.-J.: A comparison of methods for multiclass support vector machines. IEEE Trans. Neural Netw. **13**(2), 415–425 (2002)
23. Zhang, C., Ma, Y. (eds.): Ensemble Machine Learning: Methods and Applications. Springer, Heidelberg (2012). https://doi.org/10.1007/978-1-4419-9326-7
24. Dietterich, T.G.: Machine learning: four current directions. AI Mag. **18**(4), 97–136 (1997)
25. Liu, B., Zhang, L.: A survey of opinion mining and sentiment analysis. In: Aggarwal, C., Zhai, C. (eds.) Mining Text Data, pp. 415–463. Springer, Heidelberg (2012). https://doi.org/10.1007/978-1-4614-3223-4_13
26. Aggarwal, C.C.: Data Classification: Algorithms and Applications. CRC Press, Boca Raton (2014)
27. Anthimopoulos, M., Christodoulidis, S., Ebner, L., Christe, A., Mougiakakou, S.: Lung pattern classification for interstitial lung diseases using a deep convolutional neural network. IEEE Trans. Med. Imaging **35**(5), 1207–1216 (2016)
28. Kim, Y.: Convolutional neural networks for sentence classification, CoRR abs/1408.5882 (2014)

Network Traffic Classification
Using Machine Learning for Software
Defined Networks

Menuka Perera Jayasuriya Kuranage, Kandaraj Piamrat[✉],
and Salima Hamma

LS2N/University of Nantes, 2 Chemin de la Houssiniere,
BP 92208, 44322 Nantes Cedex 3, France
jkmenukaperera@gmail.com, {kandaraj.piamrat,salima.hamma}@univ-nantes.fr

Abstract. The recent development in industry automation and connected devices made a huge demand for network resources. Traditional networks are becoming less effective to handle this large number of traffic generated by these technologies. At the same time, Software defined networking (SDN) introduced a programmable and scalable networking solution that enables Machine Learning (ML) applications to automate networks. Issues with traditional methods to classify network traffic and allocate resources can be solved by this SDN solution. Network data gathered by the SDN controller will allow data analytics methods to analyze and apply machine learning models to customize the network management. This paper has focused on analyzing network data and implement a network traffic classification solution using machine learning and integrate the model in software-defined networking platform.

Keywords: Machine learning · Classification · Network traffic · Software defined networking

1 Introduction

Recent advances in software defined networking and machine learning techniques have created a new era of network management. This new concept has combined network intelligence and network programmability to create autonomous high performing networking, which will expand 5G (5th Generation) capabilities. With the recent improvements in Internet of Things (IoT), cloud computing self-driving vehicles, etc., the demand for bandwidth consumption has increased exponentially and pushed network operators the ability to search for new concepts of network management.

Software defined networks provide a programmable, scalable and highly available network solution. This solution separates the control plane and the data

S. Boumerdassi et al. (Eds.): MLN 2019, LNCS 12081, pp. 28–39, 2020.
https://doi.org/10.1007/978-3-030-45778-5_3

plane from the network devices and logically centralized the controlling component. The centralized controller has a global view of the network and enables the network operator to program their policies rather than depending on network equipment vendors.

For the past decades, Artificial Intelligence (AI) and Machine Learning (ML) concepts were developed for different use cases with different approaches. The latest concept of AI/ML technologies are developed based on statistics. Integrating these tools into the networking industry will enable network operators to implement self-configuring, self-healing, and self-optimizing networks. We can name this type of network as Knowledge Defined Networks (KDN) as mentioned in [1].

This new concept of intelligent and programmable network is an end-to-end network management solution. It is important to manage existing network resources efficiently. Even the number of users connected to the network is increasing, not all users required the same amount of network resources. Identifying each user's demand and behavior on the network will enable the operator to manage network resources much more efficiently.

In a network, there are two basic types of traffic flows: elephant flows and mice flows. Elephant flows are referred to as heavy traffic flows and mice flows are referred to as light traffic flows. And typically the resource allocation process for these flows are standard. This approach of resource allocation is a waste of network resources and allocating the same amount of resources for both flows is not an optimum solution. There are currently few methods to identify network traffic but the recent technological advancements made these concepts inefficient. Port-based classification is one of the methods that classifies network traffic based on port numbers extracted from packet header, which allow to understanding the traffic behavior and the type of applications having been used. But nowadays, modern applications use dynamic ports or tunneling, which makes this method ineffective. In Payload-based classification method, network traffic is classifying by inspecting packet payload. But this method requires a high level of computing power and storage, which will increase the cost. Another issue with this method is the privacy laws and data encryption.

When it comes to network traffic classification, ML algorithms depend on a large number of network features. And software defined networking will enable ML algorithms to control the network and can become automatic resource allocation process. Therefore, in this study, ML-based traffic classification solution was introduced for SDN. The proposed architecture uses existing network statistics and an offline process for understanding network traffic patterns with a clustering algorithm. For the online process, a classification model is used to classify incoming network traffic in real-time.

The rest of the paper will be presented as follows: Sect. 2 discusses on related work of similar researches on network traffic classification. Section 3 describes the proposed system architecture. Section 4 presents the experimental result of the system and Sect. 5 concludes this paper.

2   Related Work

In the paper [2], the authors have used the ML algorithm for classifying network traffic by application. They have trained few ML models using labeled data by applications such as Post Office Protocol 3 (POP3), Skype, Torrent, Domain Name System (DNS), Telnet were recognized by the classifier. For this experiment, they have tested six different classification models and compared accuracy. AdaBoost, C4.5, Random Forest, Multi-layer Perceptron (MLP), Radial Biased Function (RBF), Support Vector Machine (SVM) are the classifiers used for this research. They have concluded that Random Forest and C4.5 classifiers give better accuracy than the other models.

Authors of [3] have experimented with mobile network traffic classification ML models. In their project, there are three main objectives. Comparing the accuracy of three classification models [SVM, Multi-Layer Perceptron with Weight Decay (MLPWD), MLP]. Analyzing the effect on accuracy by varying the size of the sliding window. Comparing the accuracy of predictions of the models for unidimensional /multi-dimensional datasets. In their project, they have selected 24 features and selected one of the feature as the target to predict. In terms of accuracy, the paper has concluded that in multi-dimensional data sets SVM performs better and in unidimensional data sets, the MLPWD model performs better.

In the paper [4] they have experimented with the data collection and traffic classification process in software defined networks. In their work, they have developed a network application to collect OpenFlow data in a controlled environment. Only Transmission Control Protocol (TCP) traffic was considered for this project. Several packets of information were gathered using different methods. For example, Packet_IN messages were used to extract source/destination IP addresses and port addresses. First five packet sizes and timestamps were collected from the controller since in this experiment the next five packets after the initial handshake between server and client flow through the controller. Flow duration was collected by subtracting the timestamp of the initial packet and the time stamp of the message received by the controller regarding the removal of the flow entry. To avoid the high variance of the data set, they have used a scaling process named standard score. They have also mentioned that highly correlated features are not contributing much to the algorithm but increase the complexity in computation. They have used the Principle Component Analysis (PCA) algorithm to remove these high correlated factors. Random Forest, Stochastic Gradient Boost, Extreme Gradient Boost are the classifiers used in their research. The results were compared by evaluating the accuracy of each label.

In the paper [5] discussed ML-based network traffic classification. Their motivation for this project is to optimize resource allocation and network management using ML based solution. According to the paper, there are four levels of resource allocation, which are spectrum level, network level, infrastructure level, and flow level. In their paper, they have tested classifying network traffic by applications and they have used support vector machine and Kmeans clustering algorithm. The data set contains 248 features and manually labeled. The traffic labels were www, mail, bulk, service, p2p, database, multimedia, attack, interactive and games. In the SVM model, they have used four kernels namely linear, polynomial, RBF and sigmoid. And evaluated its performance using the following parameters: accuracy, recall, precision. Considering overall accuracy, the RBF kernel of SVM outperforms other kernels. They have also tested the classification accuracy by varying the number of features. And accuracy is higher with a maximum 13 selected features. In the Kmeans clustering algorithm, they have used the unlabeled data with a predetermined number of clusters. They have compared results with supervised and unsupervised models and according to the paper, SVM has the highest precision and overall accuracy.

Authors of [6] have discussed and concepts of SDN, Network Function Virtualization(NFV), Machine learning, and big data driven network slicing for 5G. In their work, they have proposed an architecture to classify network traffic and used those decisions for network slicing. According to the paper, with the exponentially increasing number of applications entering the network is impossible to classify traffic by a single classification model. So they have used the Kmean clustering algorithm to cop this issue. By using this unsupervised algorithm, they have grouped the data set and labeled them. They have set the number of clusters k=3 associating three bandwidths. With this grouping and labeling, they have trained five classification models: Navie Bayes, SVM, Neural networks, Tree ensemble, Random Forest. And compared its accuracies. The results show that Tree ensemble and Random forest perform with the same accuracy. Depend on the ML output, bandwidth was assigned in the SDN network applications. They have ed this system by streaming YouTube a video before the classification process and check the quality of the video. And compared it with the quality of the video after the classification and bandwidth allocation.

In this study, the number of features was selected based on keeping the compatibility with the implementation (SDN controller) and avoid complexity and heavy computations in the network application. An unsupervised learning algorithm was used to identify the optimum number of network traffic classes rather than selecting a predefined number of network traffic classes, which makes this method a more customized network traffic classification solution for network operators.

3   Proposed Solution

This proposed solution was divided into two sections. One of the sections was to train the machine learning algorithm and the other section was to create a network experiment to run the trained ML model on an SDN platform as a proof of concept. In the first section, a related dataset was selected, cleaned and prepared for ML models. An unsupervised ML algorithm is applied to cluster and label the dataset then we used that dataset to trained multiple classification models. In the second section, the SDN bed was implemented, a network application containing the trained ML model was created and deployed to the network for real-time classification.

3.1   ML Model Training

For this paper, "IP Network Traffic Flows, Labeled with 75 Apps" dataset from Kaggle [7] database was used. This dataset was a perfect match for our objectives and satisfy all the three main components of a good dataset, which are real-world, substantial and diverse. This dataset was created by collecting network data from Universidad Del Cauca, Popayán, Colombia using multiple packet capturing tools and data extracting tools. This dataset is consisting of 3,577,296 instances and 87 features and originally designed for application classification. But for this work, only a fraction of this dataset is needed. Each row represents a traffic flow from a source to a destination and each column represents features of the traffic data.

(1) **Data Preparation.** As mentioned above only a few features were used for this research. The most important factors that concerned when selecting features were relatability to the research objective and easily accessed by the controller without using tools or other network applications to reduce high computations. Selected features as follows: Source and destination MAC addresses and port addresses, flow duration, flow byte count, flow packet count, and average packet size. In the data cleaning process, several operations need to be done before it is ready for machine learning model training. If there are duplicate instances in the dataset, it will cause bias in the machine learning algorithm. So to avoid the biasing, those duplicates need to be identified and remove from the dataset. Moreover, some ML models cannot handle missing data entries. In that case, rows with missing data have to remove from the dataset or fill them with the values close to the mean of that feature. In this dataset, there are several features contains different data types. But some ML models can only work with numeric values. To use those data types for the ML model training, it is necessary to convert or reassign numeric values to represent its correlations with other features. Next, Min/Max normalization was used to normalize features with high variance.

(2) **Data Clustering.** Even though the data was clean enough to train ML models, data was not labeled. Classification process is a supervised learning algorithm that need labeled data for the training process. Understanding the traffic patterns in the dataset is a complicated and time-consuming task. Since the dataset is very large, it is very hard to label traffic flows manually. To avoid manual labeling, an unsupervised learning model can be used. By using an unsupervised learning algorithm, network traffic data will be clustered based on all the possible correlations of network traffic data. For this process, Kmeans unsupervised learning model was used as shown in Fig. 1. It is a high accuracy, fast learning model ideal for large datasets. The number of clusters will be selected using the Davies-Bouldin algorithm [8]. This method is calculating distances of clusters by using Euclidean distances and lower the score better the cluster

in terms of similarity ratio of within-cluster and between cluster distances. By selecting k value with the lowest Davies-Bouldin score, Dataset was clustered and labeled.

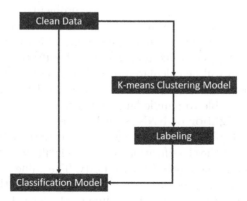

Fig. 1. Labeling dataset using Kmeans clustering algorithm

(3) Classification. Next, labeled data was used to train classification models. There are multiple classification models available and each and every model classify data with different mathematical models. Therefore, results of each model could be different from each other. Some models could perform better and some models perform poorly. In other word, it is better to train and test multiple classification models to find out which model fit better for the project. The tested models are briefly described below.

– **Support Vector Machine (SVM)** algorithm is a supervised learning algorithm that uses labeled data to train the model. SVM model will calculate decision boundaries between labeled data also known as hyper planes. And points near these hyper-planes are called extreme points. The algorithm will optimize these decision boundaries by setting up margins that separate hyperplanes. Several kernels that uses to optimize these decision boundaries. Linear, RBF, Polynomial and Sigmoid are the most commonly used kernels. Real-world data can be one dimensional or multidimensional. And these data sets are not always linear separable. The linear kernel can handle datasets that can linear separable and for nonlinear datasets, can use other kernels that can transform nonlinear datasets into linear datasets and classify. SVM is effective in multi-dimensional datasets and it is a memory-efficient model.
– **Decision Tree** is another supervised learning model that classifies data based on information gains by calculating the entropy of the dataset. It is a graphical representation of all the conditions and decisions of the dataset. The root node will be calculated using entropy with the highest information gain among the dataset. This process will continue to split branches and complete the tree. Each internal node is a test on attribute and branches represent the outcome.

Leaf represents a class label. The decision tree can use numeric and categorical data for the classification problems. It also supports nonlinear relationships between features.

– **Random Forest** is one of the powerful supervised learning algorithm, which can perform both regression and classification problems. This is a combination of multiple decision tree algorithms and higher the number of trees, higher the accuracy. It works as same as the decision tree, which based on information gain. In classification, each decision tree will classify the same problem and the overall decision will be calculated by considering the majority vote of the results. The most important advantage of this model is that it can handle missing values and able to handle large datasets.

– *Kth* **Nearest Neighbor** or KNN is an instance based supervised learning algorithm. In the KNN model, the value k represents the number of neighbors needs to consider for the classification. The model will check the labels of those neighbors and select the label of the majority. The value k should be an odd number to avoid drawing the decision. It is a robust model that can work with noisy data and perform better if the training data set is large. However, it is not performing well in multidimensional datasets and could reduce efficiency, accuracy, etc.

3.2   Network Application Development

For the simulation testbed, a simple virtual network was created on Mininet [9] network emulator with five hosts, one OpenFlow [10] enabled open vSwitch and one SDN controller (RYU) [11]. For the simplicity of this research, tree topology was used as shown in Fig. 2. There are two other network applications that need to be installed, which are simple_switch and ofctlrest. These applications will allow the controller to switch packets within the network and enable REST API calls. This switching application manages to install flow rules on the flow tables based on source, destination and flow information. These flow tables are the source of information for the classification application (Table 1).

Table 1. System configurations

System OS	Ubuntu(18.10)
SDN controller	RYU(4.30)
Switches	Open vSwitch(2.11)
Network emulator	Mininet(2.2.2)

This network traffic classification application is the program that contains the trained machine learning model. It is a python based program and communicates with the SDN controller via REST API calls. It is also responsible for extracting data from the controller, cleans, normalize and feed the ML model. The model will classify traffic flows each time when the program runs.

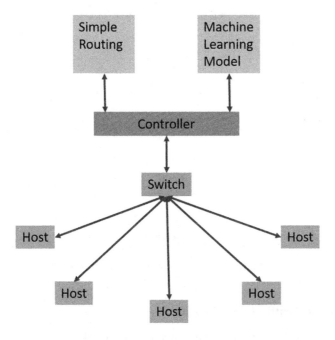

Fig. 2. SDN testing platform

In this paper, traffic has to be generated artificially. To generate traffic, the tool D-ITG [12] as been used. In this tool, various parameters can be modified to mimic real-world network traffic. Bandwidth, window size, packet size are some of them. There are also CBR (Constant Bit Rate) and VBR (Variable Bit Rate) options available within this tool. For this experiment, multiple traffic flows were generated between hosts to evaluate the machine learning output and compare it with its traffic flow characteristics.

4   Performance Evaluation

4.1   Kmeans Clustering

In the Kmeans clustering results, the number of clusters (k value) will be varied from 2 to 15 and calculate the Davies-Bouldin score for each k value. From Fig. 3, k = 4 has the lowest Davies-Bouldin score, which reflects that there are four types of traffic behaviors that can be identified from this dataset.

The four types of network traffic behaviors recognized by the Kmeans algorithm were analyzed for understanding their characteristics. However, they are not clearly specific to typical traffic classes that we encounter on the internet. Therefore, in order to better define each cluster, more features need to be added to refine the clusters. This needs to be done in the future work. Nevertheless, for this research, ranges from features of each cluster are sufficient to continue with the classification process.

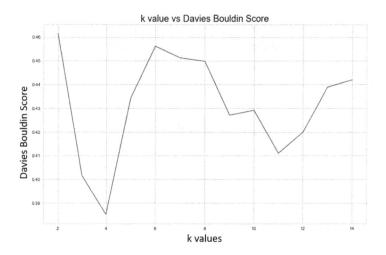

Fig. 3. k value vs Davies Bouldin Score

4.2 Network Traffic Classification

Using the labeled dataset from the above clustering algorithm, five supervised learning models were trained and evaluated. The labeled dataset was divided into two parts as training dataset and testing dataset with 70% to 30% ratio. All the models were trained using the training dataset separately and as shown in Table 2, model accuracies were calculated using the testing dataset. All the classification models were further analyzed using confusion matrices to checking the cluster accuracies and Fig. 4 shows the results for each model. From the confusion matrices, it is clear that SVM linear model has the most accurate clusters. Decision Tree and Random Forest models have failed to classify cluster No.2 correctly even though those have classified other clusters correctly. With the highest overall accuracies and high cluster accuracies, SVM linear model was selected for the network application.

Table 2. Classification model accuracies

Model	Accuracy
SVM (Linear)	96.37%
SVM (RBF)	70.40%
Decition Tree	95.76%
Random Forest	94.92%
KNN	71.47%

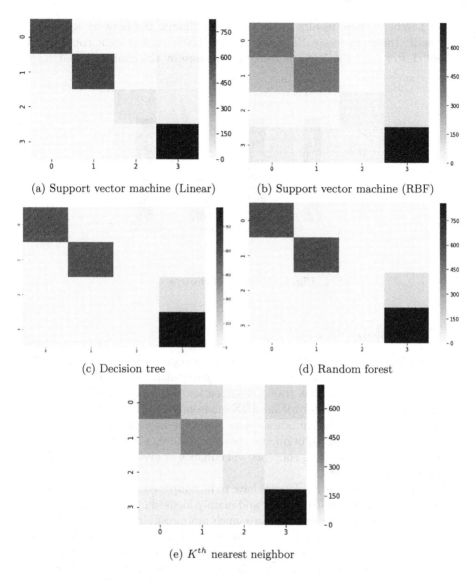

Fig. 4. Confusion matrices of classification models

4.3 Network Performance

The trained classification model was integrated with the network application and evaluated the real-time network traffic classification by generating network traffic in the testbed using D-ITG tool. For this evaluation, 50 traffic flows were generated considering cluster characteristics identified by the clustering algorithm. Generated traffic were compared with its characteristics and classification outputs. Figure 5 shows the percentages of accurate classifications by cluster number. These results shows that even though the network application can classify three clusters with high accuracy (100%), it has some confusions to classify cluster No.2 (96.50%) as recognized before by the confusion matrix.

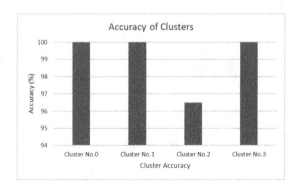

Fig. 5. Accuracy of clusters

5   Conclusion

This work has been carried out as a proof of concept while combining machine learning with software defined networking, in particular, for network traffic classification. It can be seen that traffic classification using machine learning algorithms provides good results within SDN environment. This is possible thanks to the ability of collecting information in this type of architecture. It is clear that this is a promising solution. In the near future, these high performing, intelligence-based networking concepts will enhance or even replace conventional networking management.

For the future work, several issues have to be addressed. First, the proposition was tested only on a simple topology and mainly focused on ML model accuracy. But in the real world, the networks are much more complicated and accuracy is not enough. There are other factors such as scalability, availability, etc., which directly effect the performance of a real-world network. Furthermore, the four traffic pattern detected by the clustering algorithm needs to be refined while keeping complexity reasonable when increasing number of features. This result is also context-dependent because user behavior patterns are different from network to another. For example, the number of clusters in a data center dataset would be different from the number of clusters in a sensor network dataset. Finally, for the classification, only five models were trained and compared. However, there might be another classification model that can be a better fit for this type of classification problem.

Acknowledgement. This work is a part of CloudIoT project, funded by Atlanstic 2020 programme, which is supported by the Pays de la Loire region and the cities of Nantes, Angers and Le Mans.

References

1. Mestres, A., et al.: Knowledge-defined networking. SIGCOMM Comput. Commun. Rev. **47**(3), 2–10 (2017). https://doi.org/10.1145/3138808.3138810
2. Jaiswal, R.C., Lokhande, S.D.: Machine learning based internet traffic recognition with statistical approach. In: 2013 Annual IEEE India Conference (INDICON), Mumbai, pp. 1–6 (2013). https://doi.org/10.1109/INDCON.2013. 6726074. http://ieeexplore.ieee.org/stamp/stamp.jsp?tp=&arnumber=6726074& isnumber=6725842
3. Nikravesh, A.Y., Ajila, S.A., Lung, C., Ding, W.: Mobile network traffic prediction using MLP, MLPWD, and SVM. In: 2016 IEEE International Congress on Big Data (BigData Congress), San Francisco, CA, pp. 402–409 (2016). https://doi.org/10.1109/BigDataCongress.2016.63. http://ieeexplore.ieee. org/stamp/stamp.jsp?tp=&arnumber=7584969&isnumber=758490
4. Amaral, P., Dinis, J., Pinto, P., Bernardo, L., Tavares, J., Mamede, H.S.: Machine learning in software defined networks: data collection and traffic classification. In: 2016 IEEE 24th International Conference on Network Protocols (ICNP), Singapore, pp. 1–5 (2016). https://doi.org/10.1109/ICNP.2016. 7785327. http://ieeexplore.ieee.org/stamp/stamp.jsp?tp=&arnumber=7785327& isnumber=7784399
5. Fan, Z., Liu, R.: Investigation of machine learning based network traffic classification. In: 2017 International Symposium on Wireless Communication Systems (ISWCS), Bologna, pp. 1–6 (2017). https://doi.org/10.1109/ISWCS.2017. 8108090. http://ieeexplore.ieee.org/stamp/stamp.jsp?tp=&arnumber=8108090& isnumber=8108089
6. Le, L., Lin, B.P., Tung, L., Sinh, D.: SDN/NFV, machine learning, and big data driven network slicing for 5G. In: 2018 IEEE 5G World Forum (5GWF), Silicon Valley, CA, pp. 20–25 (2018). https://doi.org/10.1109/5GWF.2018. 8516953. http://ieeexplore.ieee.org/stamp/stamp.jsp?tp=&arnumber=8516953& isnumber=8516707
7. Rojas, J.S., Gallón, Á.R., Corrales, J.C.: Personalized service degradation policies on OTT applications based on the consumption behavior of users. In: Gervasi, O., et al. (eds.) ICCSA 2018. LNCS, vol. 10962, pp. 543–557. Springer, Cham (2018). https://doi.org/10.1007/978-3-319-95168-3_37
8. Davies, D.L., Bouldin, D.W.: A cluster separation measure. IEEE Trans. Pattern Anal. Mach. Intell. **PAMI-1**(2), 224–227 (1979). https://doi.org/ 10.1109/TPAMI.1979.4766909. http://ieeexplore.ieee.org/stamp/stamp.jsp?tp=& arnumber=4766909&isnumber=4766893
9. http://mininet.org/
10. https://www.opennetworking.org/
11. https://osrg.github.io/ryu/
12. http://www.grid.unina.it/software/ITG/

A Comprehensive Analysis of Accuracies of Machine Learning Algorithms for Network Intrusion Detection

Anurag Das, Samuel A. Ajila, and Chung-Horng Lung[✉]

Department of Systems and Computer Engineering, Carleton University,
1125 Colonel By Drive, Ottawa, ON K1S 5B6, Canada
anuragdas@cmail.carleton.ca,
{ajila, chung}@sce.carleton.ca

Abstract. Intrusion and anomaly detection are particularly important in the time of increased vulnerability in computer networks and communication. Therefore, this research aims to detect network intrusion with the highest accuracy and fastest time. To achieve this, nine supervised machine learning algorithms were first applied to the UNSW-NB15 dataset for network anomaly detection. In addition, different attacks are investigated with different mitigation techniques that help determine the types of attacks. Once detection was done, the feature set was reduced according to existing research work to increase the speed of the model without compromising accuracy. Furthermore, seven supervised machine learning algorithms were also applied to the newly released BoT-IoT dataset with around three million network flows. The results show that the Random Forest is the best in terms of accuracy (97.9121%) and Naïve Bayes the fastest algorithm with 0.69 s for the UNSW-NB15 dataset. C4.5 is the most accurate one (87.66%), with all the features considered to identify the types of anomalies. For BoT-IoT, six of the seven algorithms have a close to 100% detection rate, except Naïve Bayes.

Keywords: Network intrusion detection · Supervised learning · UNSW-NB15 dataset · BoT-IoT dataset

1 Introduction

Due to the massive growth of computer networks and its many applications, the number of network flows has increased tremendously. The considerable large number of traffic flows leads to a massive amount of data, which eventually leads to the vulnerability of the data and network as a whole. One of the many challenges of cybersecurity research is to identify intrusion/anomaly in the traffic flow. A network Intrusion Detection System (IDS) is one of the solutions to detect such attacks before they compromise the network. An IDS monitors the normal traffic flows and identifies its characteristics or patterns. If a new flow does not follow the same characteristics, it might be an anomaly. Hence, an IDS may help identify even detect unknown attacks. Note that this paper uses intrusion and anomaly interchangeably.

© IFIP International Federation for Information Processing 2020
Published by Springer Nature Switzerland AG 2020
S. Boumerdassi et al. (Eds.): MLN 2019, LNCS 12081, pp. 40–57, 2020.
https://doi.org/10.1007/978-3-030-45778-5_4

This research is an experimental investigation of nine machine learning algorithms on the dataset UNSW-NB15 (released in November 2015) [1] and seven machine algorithms on the recently released BoT-IoT dataset (released in November 2018) [2]. This research intends to discuss the following questions:

1. *Network Intrusion Detection:* How effective is it to detect network intrusion based on the traffic flow features present in datasets using different machine learning techniques?
2. *Types of Intrusion Classification:* Different cyberattacks can be stopped by different mitigation techniques. Hence classification of attack is as important as the detection of attacks. How effective can the classification of the types of attacks be from different features of network traffic flows present in datasets?
3. *Accuracy of models:* Which machine learning model has the highest accuracy for classifying the network anomalies for the selected datasets?
4. *Efficiency of models:* Which machine learning model is efficient for detecting network intrusion without compromising on accuracy? The earlier an attack is detected, the less harm it can generate on the network. Furthermore, by selecting a fewer number of features from the complete dataset, we can reduce the computation time a machine learning model takes to build.

The main contributions of the paper are: Firstly, comparing the accuracy and the time to build in the evaluation of network intrusion detection of the UNSW-NB15 dataset using nine machine learning techniques. Secondly, using the same nine machine learning techniques and nine different features selections, we compared and evaluated the performance of the various methods to identify the types of network intrusions in the UNSW-NB15 dataset. Thirdly, we analysed and evaluated the accuracy of and time to build seven machine learning techniques on the newly released BoT-IoT dataset. The premise upon which this research is based is to synthesis the previous research works on the UNSW-NB15 dataset [1–9]. Some (if not all) of related research works only used one or two machine learning algorithms to analyse the dataset, and in some cases do not even identify the different anomalies.

The rest of this paper is structured as follows. Section 2 presents background information about different supervised learning algorithms. Section 3 gives a literature review of previous research works and the different feature selection methods used in this paper. Section 4 presents the two datasets used in this research. Section 5 describes the methodology and the three sets of experiments. Section 6 provides the results and discussion and the conclusion is given in Sect. 7.

2 Background

2.1 Supervised Learning Algorithms

Machine learning is the study of algorithms that can learn complex relationships or patterns from empirical data and make accurate decisions [10]. Machine learning can be classified into supervised learning, semi-supervised learning, and unsupervised learning. Supervised learning deduces a functional relationship from training data that generalizes well to the whole dataset. In contrast, unsupervised learning has no training dataset and the goal is to discover relationships between samples or reveal the latent variables behind the observations [11]. Semi-supervised learning falls between supervised and unsupervised learning by utilizing both labeled and unlabeled data during the training phase [10]. Among the three categories of machine learning, supervised learning is the best fit to solve the prediction problem in the auto-scaling area [11]. Therefore, this research focuses on supervised learning.

After conducting an in-depth search and review of research papers that have previously used the UNSW NB15 dataset, we selected nine machine learning algorithms that appear frequently in different papers [1, 3, 7], and [8].

Random Tree is an algorithm with a random number of features at each node, and it is used for classification [12]. This algorithm is very fast, but it suffers from overfitting. To overcome overfitting, Random Forest is used with this algorithm. We used the WEKA [13] implementation of this algorithm in which the Random Tree classifier constructs a tree that considers K random chosen attributes at each tree node. There is no pruning and has an option to estimate classifier probabilities (or target mean in the case of regression) based on a hold-out set (i.e. back fitting). We set the seed to be 1, that is, the random number seed used for selecting attributes.

Random Forest is an ensemble learning algorithm which can be applied on classification as well as a regression problem [12]. In this technique, lots of decision trees are produced at training time. For a regression problem, the mean is considered, and for the classification problem, the mode is used. Random Forest was designed to combat the overfitting problem in the random tree. Random Forest is a classifier for constructing a "forest" of random trees.

Bayesian Networks (WEKA Bayes Net) [13] - These networks show the probabilistic relations between different features with the target attribute (one which is to be classified) [12]. In this research, this algorithm is used to calculate the probability of different features with an impact on the anomaly. The dual nature of a Bayesian network makes learning a Bayesian network a two stage processes: first learn a network structure, then learn the probability tables. All Bayes network algorithms implemented in Weka assume that all variables are discrete finite and no instances have missing value [14]. In general, Bayes Network learning uses various search algorithms and quality measures [13]. In our case, we used the K2 search algorithm and the SimpleEstimator for the estimate function [13].

Naïve Bayes - These are traditional classifiers and they are based on the Bayes theorem of independent relation between the features [12]. Although it is an old technique, this algorithm is still highly scalable and it can be used to build the fastest model for large dataset such as UNSW NB15. These classifiers are family of simple probabilistic classifiers with strong (naïve) independence. The assumption here is that the features of measurement are independent of each other.

k-Nearest Neighbours (k-NN) - k-NN is an algorithm which can be used for both regression and classification [12]. The model consists of training k closest samples in the feature space. In classification, the class having the maximum number of k nearest neighbours is chosen. Weights are assigned to nearer neighbours that contribute more to the result compared to the ones that are farther away. It is an instance-based learning algorithm where all the calculations are deferred until the final regression/classification.

Hence it is known as a lazy algorithm. This algorithm is called "IBk" in Weka [13]. It selects an appropriate value of k based on cross-validation and can also do distance weighting.

C4.5 - It is a decision tree-based classifier [12]. It is an extension of the classical ID3 algorithm from the same author - Ross Quinlan [15]. It uses information entropy and gain for decision making. On the Weka platform [13], it is called J.48, which is an implementation of C4.5 in Java. J.48 generates a pruned (or unpruned) C4.5 decision tree. The seed in this classifier is used for randomizing the data when reduced error pruning is used.

REPT - Reduced Error Pruning Tree (REPT) is a fast decision tree based on the C4.5 algorithm and it can produce classification (for discrete outcome) or regression trees (for continuous outcome). It builds a regression/decision tree using information gain/variance and prunes it using reduced-error pruning with back-fitting) [12, 13]. Missing values are replaced by breaking down the corresponding instances.

RIPPER - Repeated incremental pruning to produce error reduction (RIPPER) is an inductive rule-based classifier which is the optimized version of incremental reduced error pruning [12]. It uses incremental reduced-error pruning and a somewhat complicated global optimization step. It makes rules for all features. Depending on the satisfaction of those rules, a network flow is classified as normal or an anomaly. On the Weka platform, it is called Jrip [13]. Generally, the Weka Jrip implements a propositional rule learner.

PART (Partial Decision Tree) - Here rules are made according to the features of the past observations and classification of whether the data is an anomaly or normal is done according to the rules [12]. The Weka [13] implementation builds a partial C4.5 decision tree in each iteration and makes the "best" leaf into a rule.

These nine algorithms are either tree-based or partial tree or forest (a collection of trees) or networks (a form of tree). We have set the "seed" to 1 and the batch size to 100 where needed.

3 Literature Review

A number of research efforts [1–8, 12] have been conducted for network anomaly or intrusion detection using the UNSW-NB15 dataset. These approaches have certain limitations. Some research papers considered one or two machine learning algorithms. For instance, only the research works in [4] and [12] use more than one machine learning technique on the UNSW-NB15 dataset. Furthermore, the following research works: [4–6] and [12] do not identify the types of attacks. They only detect if a flow is normal or an anomaly. In addition, the research work in [12] does not adopt a feature selection method. Research works in [1–3] and [7, 8] do classify the attack types, but they only investigate a single machine learning technique.

In this paper, we investigate the detection, the types of attacks, and make a comparison between the effectiveness of nine different machine learning techniques as well as the impact of different feature selection techniques on those machine learning algorithms. Our research, like [1] and [5], also does a benchmark of time taken for various machine learning techniques when applied together with specific feature selection methods.

The authors of [1] divided their network intrusion detection model into 3 stages. In the first stage, they applied Correlation-based feature selection on all the 45 features (as shown in Table 1) along with the genetic search. They used a statistical filter-based feature selection method on the complete dataset. Once they obtained the best features from stage 1, they applied a wrapper-based filter on those selected features only. The machine learning algorithm used in the wrapper-based filter was Random Forest. At the end of stage 2, the authors identified five best features, namely, *sbytes, tcprtt, synack, dmean and response_body_len*. In stage 3, they used the Random Forest classifier to detect the anomaly. They were able to improve the accuracy of the model from 94.70% to 99.63%. One problem in this approach is that only 5 out of 45 features were finally considered.

Table 1. Features for UNSW-NB15

Feature number	Feature name	Feature number	Feature name
1	id	23	dtcpb
2	dur	24	dwin
3	proto	25	tcprtt
4	service	26	synack
5	state	27	ackdat
6	spkts	28	Smean
7	dpkts	29	dmean
8	sbytes	30	trans_depth
9	dbytes	31	response_body_len
10	rate	32	ct_srv_src
11	sttl	33	ct_state_ttl
12	dttl	34	ct_dst_ltm
13	sload	35	ct_dst_dport_ltm
14	dload	36	ct_dst_sport_ltm
15	sloss	37	ct_dst_src_ltm
16	dloss	38	is_ftp_login
17	sinpkt	39	ct_ftp_cmd
18	dinpkt	40	ct_flw_http_mthd
19	sjit	41	ct_srv_ltm
20	djit	42	ct_srv_dst
21	swin	43	is_sms_ips_ports
22	stcpb	44	attack_cat
		45	label

The authors of [2] were the original authors of UNSW-NB15. Their method for feature selection has three parts: feature conversion, feature reduction and feature normalization. After all the three steps, they selected the best features which are *Dttl, synack, swin, dwin, ct_state_ttl, ct_src_ltm, ct_srv_dst, Sttl, et_dst_sport_ltm,* and *Djit.* Association rule mining was also used to find features which are not correlated to each other, but highly correlated to the target attribute that the authors wanted to predict. They ranked all the 43 features (excluding id). From the ranking, they selected the top 25% (i.e., top 11 out of 43) features. Furthermore, Independent Component Analysis (ICA) was used to find the best features in [7].

The authors of [3] used the Weka tool [14] to select the optimal features. They used *CfsSubsetEval* (attribute evaluator) + *GreedyStepwise* method and *InfoGainAttibuteEval* (attribute evaluator) + *Ranker* method. The classifier Random Forest was used to evaluate the accuracy. The combination of features which generated the highest accuracy was selected. The five selected features are *service, sbytes, sttl, smean, ct_dst_sport_ltm.* Their test accuracy was around 83% for anomaly type classification.

The authors of [5] used a pure statistical filter-based subset evaluation (FBSE) method of correlation-based feature selection to detect Denial of Services (DoS) attacks. The final features selected are F7, F10, F11, F12, F18 and F32 (see Table 1). They used Artificial Neural Network to detect the attacks and obtained an accuracy of 81.34%. Their false alarm rate was quite high with 21.13%.

The authors of [6] trained a deep learning model on the entire dataset for 10-fold cross-validation. The most important features were then selected using the Gedeon method [9]. Gedeon method selects features which are unique from one another even if the information they provide is minor. They discarded the features which generate huge amount of redundant information. The accuracy obtained from the proposed model was 98.99% with a low false alarm rate of 0.56%.

The authors of [8] used Genetic Algorithm to find the best features. They used Support Vector Machine (SVM) to check the accuracy of the selected features.

4 Datasets

Two datasets have been selected for our experimental validation. They are UNSW-NB15 and BoT-IoT, which are described in the following subsections.

4.1 UNSW-NB15

This dataset was created by Moustafa and Slay [10]. The UNSW-NB15 dataset is a mixture of real normal traffic flow and synthetic attacks. The types of attacks and the number of each attack in the testing dataset are shown in Fig. 1. The testing dataset has 82,332 records (37,000 normal and 45,332 anomalies) and 45 attributes or features (see Table 1).

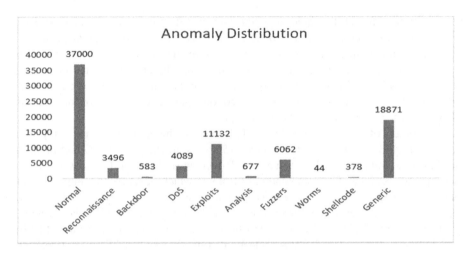

Fig. 1. Distribution of anomalies (attack types) in the UNSW-NB15 training dataset

4.2 The BoT-IoT Dataset

One of the original authors of UNSW-NB15 was also involved in creating the BoT-IoT dataset [11], as depicted in Fig. 2, in the Cyber Range Lab of UNSW Canberra Cyber Center. This dataset is a combination of standard and botnet traffic (hence the name). Attack distribution in the training dataset is depicted in Fig. 2. The training dataset has 2,934,817 records. The features used for the experiments were the top 10 features [11] selected by the creators of this dataset.

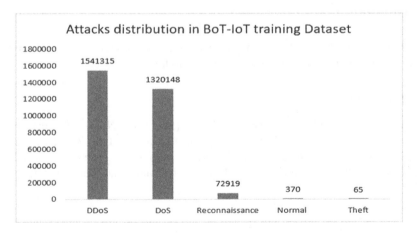

Fig. 2. Distribution of anomalies (attack types) in BoT-IoT training dataset

5 Methodology and Experiments

The specifications of the system environment for our experiments are shown in Table 2. Weka [14] tool was used for detection (training and testing), and the RStudio for data preprocessing with the R programming language. Microsoft Excel is used for data visualization. Three different experiment sets were conducted, which are described as follows.

Table 2. Hardware specifications

Processor	Intel(R) Xeon(R) @ 2.50 GHz (2 processors)
RAM	32.0 GB
Operating System	Windows 7 - 64 bit OS
Architecture	Microarchitecture - Ivy Bridge, Multiprocessor (2 Processors)

Experiment set 1: This set of experiments is to detect if a network flow is normal or an anomaly using supervised learning techniques. The nine machine learning algorithms described above have been evaluated and compared for accuracy (%) and the time taken to build the model (in seconds). The dataset used for this experiment set is the UNSW-NB15 training dataset.

Experiment set 2: In this set of experiments, the type of network attacks is also identified using the nine supervised learning algorithms for validation. Further, eight different feature selection techniques and the complete dataset (making a total of nine different feature sets) together with the nine different machine learning algorithms making 81 different combinations of feature selection methods and machine learning techniques to identify the types of attack. The dataset used is the UNSW-NB15 training dataset.

Experiment set 3: The types of network attacks are identified using seven supervised machine learning algorithms on the ten best features pre-selected by the authors of the BoT-IoT training dataset [11].

In addition, 10-fold cross-validation has been adopted [13] in our experiments. The standard way of predicting the error rate of a learning technique given a single, fixed sample of data is to use stratified 10-fold cross-validation. The dataset is divided randomly into 10 parts in which the class is represented in approximately the same proportions as the full dataset. Each part is held out in turn and the learning scheme trained on the remaining nine-tenths; then its error rate is calculated on the holdout set. In the end, the average of all the iterations is calculated. As all the values have been tested at least once, this step helps in avoiding overfitting. Why 10? Previous extensive works in the domain have shown that the number 10 is about the right number of folds to get the best estimate of error.

5.1   Experiment Set 1 - Supervised Learning on UNSW-NB15 to Detect the Anomaly

Our first experiment-set used nine supervised learning algorithms on the UNSW-NB15 training dataset to detect the anomaly. The methodology in this experiment is that all the attributes (i.e. features) are considered and all nine machine learning algorithms are used to build anomaly detection models. The machine learning algorithms are Random Forest, Random Tree, Bayes Network, Naive Bayes, k-NN, C4.5, Reduced Error Pruning Tree, RIPPER and PART. The experimental results are presented in Table 3.

Table 3. Experiment I results

Machine learning algorithms	Accuracy (%)	False positive rate (%)	Precision (%)	Recall (%)	Time to build the model (s)
Random Forest	97.9121	2	97.9	97.9	57.25
Random Tree	96.1036	4	96.1	96.1	0.93
Bayes Network	81.6961	17.2	82.7	81.7	4.93
Naive Bayes	76.1952	21.4	79.1	76.2	0.69
k-nearest neighbours	93.4691	6.5	93.5	93.5	1.51
C4.5	97.3194	2.7	97.3	97.3	15.63
REPT	97.068	2.9	97.1	97.1	3.43
RIPPER	96.7582	3.2	96.8	96.8	185.36
PART	97.3109	2.7	97.3	97.3	53.69

As presented in Table 3, in terms of accuracy, Random Forest is the most accurate anomaly detection model with 97.91% closely followed by C4.5 (97.3194%), then PART (97.3109). Naïve Bayes is the least accurate considering the nine algorithms. Judging by the false positive rates and the precision (Table 3), there are little or no significant statistical differences between the following algorithms in terms of accuracy – Random Forest, Random Tree, C4.5, REPT, RIPPER, and PART. In terms of speed, that is, time to build, Naive Bayes is the fastest model with 0.69 s although the least accurate. Random Tree is equally fast with a build time of just 0.93 s and gives a high accuracy of 96.10%. Table 4 shows the confusion matrix of the Random Forest.

Table 4. Confusion matrix of the Random Forest algorithm

Classified as		
Normal	Anomaly	
36354	646	Normal
1073	44259	Anomaly

The diagonal in green indicates the correct classification. There are total of 37,000 normal flows and 45,332 anomalies in the UNSW-NB15 training dataset. Out of 37,000 normal, 36,354 were classified normal (98.25%) correctly, but 646 observations were classified anomaly incorrectly. Out of 45,332 anomalies, 1073 were classified normal incorrectly, but 44,259 were classified anomaly (97.63%) correctly.

5.2   Experiment Set 2 - Supervised Learning on UNSW-NB15 to Detect the Anomaly Type

Experiment set 2 was conducted to identify not only if a network traffic flow is normal or an anomaly, but also the types of anomaly. The objective is to support an appropriate action to mitigate it. The methodology for Experiment-set 2 is depicted in Fig. 3.

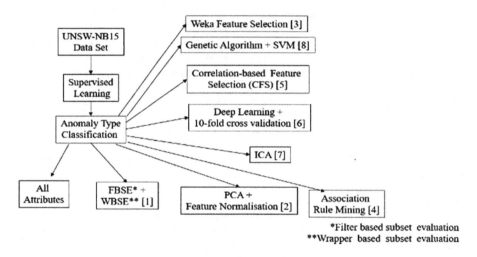

Fig. 3. Experiment 2 methodology

This experiment set 2 has nine sub-experiments with different feature sets. The experiments are based on methods previously published in the literature [1–8]. The features for the nine sub-experiments are described as follows

1. **All the features** of the UNSW-NB15 training dataset are considered.
2. **FBSE+WBSE** - Features selected by FBSE and wrapper WBSE. This experiment uses features suggested in [1].
3. **PCA + Feature Normalization** - Features selected by principal component analysis (PCA) and feature normalization are considered. This experiment is based on the features selected by [2].
4. **Weka Feature Selection** - Optimal features of UNSW-NB15 were selected using the Weka tool. This method was proposed in [3].
5. **Association Rule Mining** - Feature selection based on association rule mining. This experiment is based on work done in [4].

6. **Correlation based Feature Selection** - Pure statistical feature selection method based on correlation was used to select features. This experiment is based on [5], [16].
7. **Deep Learning with 10-fold CV** - The authors of [6] selected the best features which generated the highest accuracy for their deep learning model after 10-fold cross-validation [17–19]. These features were used in this sub-experiment.
8. **ICA** - Features used in this experiment were based on ICA proposed in [7].
9. **Genetic Algorithm with SVM** - Feature selection was done using the genetic algorithm. The classifier used in the genetic algorithm to check the highest accuracy was SVM. This methodology was proposed in [8].

Table 5. Accuracies of different algorithms for different feature sets for UNSW-NB15

Feature sets	Random Forest	Random Tree	Bayes Network	Naive Bayes	k-NN	C4.5	REPT	RIPPER	PART
All Features	87.08	84.17	65.28	46.16	80.62	**87.66**	86.62	80.24	87.05
FBSE + WBSE	82.85	80.92	74.25	17.95	76.59	82.56	82.33	76.67	82.30
PCA + Feature Normalization	85.85	83.48	71.55	41.87	81.27	85.78	85.13	79.58	85.65
Weka Feature Selection	82.99	82.85	74.55	57.57	82.5	83	82.8	76.44	82.9
Association Rule Mining	77.70	75.18	62.72	51.07	77.04	78.22	77.99	72.69	77.83
Correlation based Feature Selection	74.31	71.93	60.32	43.05	73.59	74.58	74.28	71.15	74.3
Deep Learning with 10-fold CV	85.57	83.8	66.55	56.59	84.17	86.16	85.10	79.24	85.91
ICA	85.68	83.59	74.27	37.87	80.12	85.31	84.86	77.6	85.03
Genetic Algorithm with SVM	77.35	74.18	69.68	35.83	71.67	76.05	76.80	71.86	75.78

Table 5 shows the experimental results. C4.5 generates the best accuracy for *All Features* (87.66%) followed by Random Forest (87.08%) then comes PART (87.05%) and the rest are REPT (86.62%), Random Tree (84.18%), k-NN (80.62%), Bayes Network (65.28%) and finally Naïve Bayes (46.16%)

Four algorithms (Random Forest, C4.5, REPT, and PART) are on par (roughly 83%) for FBSE + WBSE [2] and the rest are below 80% with Naïve Bayes as the worst (17.955%).

In the case of PCA + Feature Normalization [2], Random Forest produces the best accuracy (85.85%), followed by C4.5 (85.78%), PART (85.65%), REPT (85.13%), then Random Tree, k-NN, RIPPER and finally Bayes Network.

For Weka Feature Selection [3], six algorithms have roughly the same accuracies of around 83%. The maximum accuracy for Association Rule Mining [4] is 78.22% for C4.5 and the minimum is 51.07% for Naive Bayes.

The accuracy results for Correlation based Feature Selection [5] ranges from 74.58% for C4.5 to 43.05% for Naïve Bayes. Random Forest has the best accuracy for ICA [7].

All the accuracy measurements for the Genetic Algorithm with SVM [8] are below 80% with the highest as 77.35% for Random Forest and the minimum is 35.83% for Naïve Bayes.

In general, Random Forest came out to be the top in terms of accuracy for the feature sets for UNSW-NB15. This is closely followed by C4.5.

5.3   Experiment Set 3 - Supervised Learning on BoT-IoT to Detect the Anomaly Type

Bot-IoT training dataset [11] was used for experiment-set 3. It has around 3 million values. Only the top 10 features according to the original dataset authors were selected for this experiment set. Unfortunately, the Weka tool crashed for k-NN and RIPPER algorithms. It was unable to build models for these algorithms. This probably is due to the size of the dataset. Table 6 shows the results for the seven remaining algorithms and almost all the algorithms have around a 100% detection rate, except Naïve Bayes. These results are like that of the authors of [10]. Random Tree has high accuracy with a minimal build time of 96.98 s. Table 7 shows the confusion matrix of Random Tree for the BoT-IoT training dataset with 10-fold Cross-Validation.

Table 6. Accuracies of different algorithms for different feature sets for BoT-IoT

Algorithms	Accuracy %	False positive %	Precision %	Recall %	Time to build (seconds)
Random Forest	99.99	0	100	100	5628.96
Random Tree	**99.9937**	0	100	100	**96.98**
Naïve Bayes	**73.4121**	2.8	73.4	71.1	**11.46**
C4.5	99.99	0	100	100	448.57
REPT	99.99	0	100	100	205.96
Bayes Network	99.6	0.2	99.6	99.6	228.6
PART	99.99	0	100	100	1210.6

Table 7. Confusion matrix for Bot-IoT

Classified as					
Normal	DDoS	DoS	Reconnaissance	Theft	
349	1	6	12	2	Normal
2	1541262	48	3	0	DDoS
4	54	1320083	6	1	DoS
14	13	12	72878	2	Reconnaissance
1	0	0	3	61	Theft

The green diagonal shows the correct classification. As illustrated in Fig. 2, there are total of 370 normal flows, 1,541,315 DDoS, 1,320,148 DoS, 72,919 Reconnaissance, and 65 Theft in the BoT-IoT dataset. From the 370 normal flows, 349 (94.32%) were classified correctly. Out of 1,541,315 DDoS, 1,541,262 (99.99%) were classified correctly. For DoS 1,320,083 (99.9%) of 1,320,148 were classified correctly. 61 (93.85%) out of 65 theft were identified successfully.

6   Results and Discussions

In Experiment-set 1, nine machine learning algorithms have been evaluated and compared on UNSW-NB15 for network intrusion detection. In addition, the types of network intrusion are identified as well.

In Experiment-set 2, 81 (nine machine learning algorithms with eight different feature selection methods and all features together) different techniques have been compared. To the best of our knowledge, this is the first time all these nine methods are compared for the same dataset.

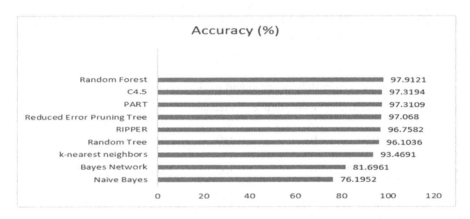

Fig. 4. Comparison of accuracy for different machine learning algorithms

Benchmarking has also been conducted on time taken to build each model [20–25].

Figure 4 displays the results of the comparison of different machine learning algorithms for anomaly detection on all the UNSW-NB15 dataset features. Random Forest provides the best accuracy (97.91%) but very costly in terms of computational time (57.25 s).

According to Fig. 5, Naive Bayes is the fastest algorithm for all the features. Random Tree is the most optimal algorithm with a high accuracy of 96.10% and second fastest with a build time of only 0.93 s.

Figure 6 is the comparison of all the different feature selection methods used from the existing literature that are applied to Random Forest to detect the anomaly.

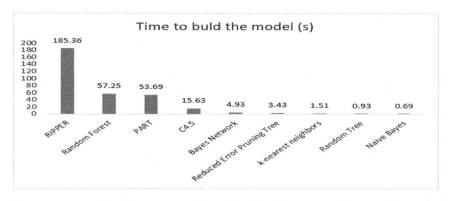

Fig. 5. Comparison of build time for different machine learning algorithms

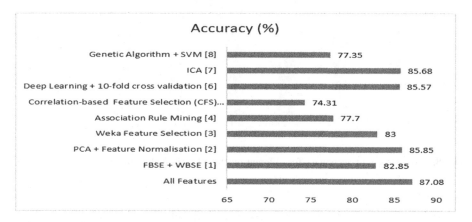

Fig. 6. Comparison of accuracies for feature selection methods on the Random Forest algorithm

When all features are used, Random Forest generates the best accuracy. For feature selection, PCA + Feature Normalisation, Deep Learning + 10-fold cross-validation and ICA all are valid regarding high accuracy.

According to Fig. 7, FBSE + WBSE, Weka feature selection is the fastest with around 19 s to build using the Random Forest algorithm. PCA + Feature Normalisation is a balanced approach for feature selection. It has the second highest accuracy with 85.85% and third fastest with 26.32 s.

According to Fig. 8, almost all the algorithms have around 100% detection accuracy for the BoT-IoT. The exception is Naïve Bayes with 73.4121%.

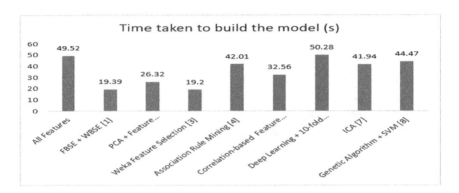

Fig. 7. Comparison of build time for feature selection methods on the Random Forest algorithm

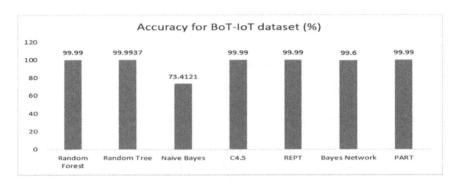

Fig. 8. Comparison of accuracies for different machine learning algorithms on BoT-IoT dataset

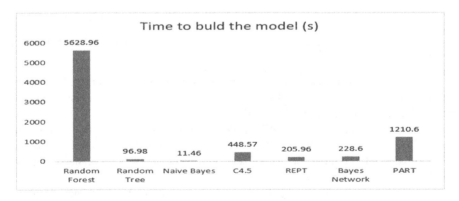

Fig. 9. Comparison of computational time for different machine learning algorithms on BoT-IoT dataset

Figure 9 shows that the time taken to build the model is very high for Random Forest. Naïve Bayes has the shortest time.

7 Conclusion and Future Work

We asked four questions at the beginning of this paper. The answers or the explanation to these questions are examined below.

Question 1 - *How effective is it to detect network intrusion based on the traffic flow features present in datasets using different machine learning techniques?*
As shown in Table 3 and Fig. 4, different machine learning algorithms give different accuracies for the UNSW-NB15 dataset. Firstly, this shows that the choice of a machine learning technique for anomaly detection is crucial based on the dataset. Secondly, there is a set of machine learning algorithms that have similar "accuracies" judging by the false positive rates and precisions which means that if speed is important one can choose a machine learning technique among others based on time to build the model.

Question 2 - *How effective can the classification of the types of attacks be from different features of network traffic flows present in datasets?* Table 5 shows the results of using the nine algorithms for nine different feature selection sets (81 different sub experiments). From this table, we can see that the set "All Features" shows "better accuracy" compared to other feature selections supposedly contains the best features. These results are somewhat different from previous research works (for example in papers [2] to [8]) that used feature selections. The only conclusion we can make here is that if timing (i.e., speed) is not important, then it may be better to use all the features if an efficient algorithm is used.

Question 3 - *Which machine learning model has the highest accuracy for classifying the network anomalies for the selected datasets?* To answer this question, there is no single machine learning algorithm that can be tagged as "the best" for classifying anomalies. However, based on the "nature" of the data and, these two datasets (UNSW-NB15 and BoT-IoT) the tree-based algorithms appear to perform better than non-tree based. In Table 3, other than k-NN; Bayes Network and Naïve Bayes are less accurate compared to the rest of the algorithms.

Question 4 - *Which machine learning model is efficient for detecting network intrusion without compromising on accuracy?* So far as this research is concerned and based on the dataset (i.e., UNSW-NB15), the best algorithm that does not compromise on accuracy is Random Tree (Table 3) with 96.9121% accuracy and a time of 0.93 s. The second is REPT with 97.3109% and 3.43 s. Other algorithms (Random Forest, PART, RIPPER, etc.) have higher accuracies but not that efficient in terms of build time.

In conclusion, in this research, supervised learning techniques were used to detect anomaly in the UNSW-NB15 dataset. Although Random Forest has the best accuracy (97.9121%) and Naïve Bayes is the fastest, Random Tree is the most optimal algorithm with an accuracy of 96.10% and the second fastest with a build time of only 0.93 s. Furthermore, supervised learning techniques were used to classify the types of attacks as well. C4.5 is the most accurate one (87.66%) with all the features considered. Eight different types of feature selection methods were used from existing literature to investigate the accuracy and timing of each model. PCA + Feature Normalisation [2] is a balanced approach for feature selection; it has the second highest accuracy with

85.85% and the third fastest with 26.32 s. Supervised Learning techniques were used on the BoT-IoT dataset to classify anomaly types. Random Tree is an optimal algorithm with almost perfect accuracy, and it is the second fastest one with just 96.98 s taken to build a model for 3 million network flows.

For future work, various feature selection methods can be applied to supervised learning algorithms. Weka can also be used for more feature selection methods. GPUs or distributed systems can be used to ease the computation burden. Unsupervised Learning algorithms can be applied to the BoT-IoT datasets. Deep learning models like Convolutional Neural Networks and Recurrent neural networks can be trained and compared to traditional machine learning algorithms regarding accuracy and speed.

References

1. Moustafa, N., Slay, J., Creech, G.: Novel geometric area analysis technique for anomaly detection using trapezoidal area estimation on large-scale networks. In: IEEE Transactions on Big Data, p. 1 (2017)
2. Koroniotis, N., Moustafa, E., Sitnikova, B., Turnbull, B.: Towards the Development of Realistic Botnet Dataset in the Internet of Things for Network Forensic Analytics: Bot-IoT Dataset (2018). https://arxiv.org/abs/1811.00701
3. Janarthanan, T., Zargari, S.: Feature selection in UNSW-NB15 and KDDCUP'99 datasets. In: Proceedings of the IEEE 26th International Symposium on Industrial Electronics (2017)
4. Moustafa, N., Slay, J.: A hybrid feature selection for network intrusion detection systems: central points. In: Proceedings of the 16th Australian Information Warfare Conference, November 2015
5. Idhammad, M., Afdel, K., Belouch, M.: DoS detection method based on artificial neural networks. Int. J. Adv. Comput. Sci. Appl. 8(4), 465–471 (2017)
6. Al-Zewairi, M., Almajali, S., Awajan, A.: Experimental evaluation of a multi-layer feedforward artificial neural network classifier for network intrusion detection system. In: Proceedings of the International Conference on New Trends in Computing Sciences, pp. 167–172 (2017)
7. Gharaee, H., Hosseinvand, H.: A new feature selection IDS based on genetic algorithm and SVM. In: Proceedings of the 8th International Symposium on Telecommunications, pp. 139–144 (2016)
8. Moustafa, N., Slay, J.: UNSW-NB15: a comprehensive data set for network intrusion detection systems (UNSW-NB15 network data set). In: Proceedings of the IEEE Military Communications and Information Systems Conference (MilCIS) (2015)
9. Moustafa, N., Slay, J.: The evaluation of network anomaly detection systems: Statistical analysis of the UNSW-NB15 data set and the comparison with the KDD99 data set. Inf. Secur. J. 25(1), 18–31 (2016)
10. Nikravesh, A.Y., Ajila, S.A., Lung, C.-H.: An autonomic prediction suite for cloud resource provisioning. J. Cloud Comput. 6(3), 1–20 (2017). https://doi.org/10.1186/s13677-017-0073-4
11. Ajila, S.A., Bankole, A.A.: Using machine learning algorithms for cloud prediction models in a Web VM resource provisioning environment. Trans. Mach. Learn. Artif. Intell. 4(1), 29–51 (2016)
12. Marsland, S.: Machine Learning: An Algorithmic Perspective, 2nd edn. Chapman and Hall/CRC, Boca Raton (2014)

13. Weka. https://www.cs.waikato.ac.nz/ml/weka/. Accessed June 2019
14. Bouckaert, R.R.: Bayesian Network Classifiers in Weka for Version 3-5-7, 12 May 2008. https://www.cs.waikato.ac.nz/ ~ remco/weka.bn.pdf. Accessed June 2019
15. Quinlan, J.R.: Induction of decision trees. Mach. Learn. **1**, 81–106 (1986). https://doi.org/10.1007/BF00116251
16. Nguyen, H., Franke, K., Petrovic, S.: Improving effectiveness of intrusion detection by correlation feature selection. In: Proceedings of the International Conference on Availability, Reliability and Security, pp. 17–24 (2010)
17. Shone, N., Ngoc, T.N., Phai, V.D., Shi, Q.: A deep learning approach to network intrusion detection. IEEE Trans. Emerg. Top. Comput. Intell. **2**(1), 41–50 (2018)
18. Pervez, M.S., Farid, D.M.: Feature selection and intrusion classification in NSL-KDD cup 99 dataset employing SVMs. In: Proceedings of the International Conference on Software, Knowledge, Information Management and Applications, pp. 1–6 (2014)
19. Nguyen, H.T., Franke, K., Petrovic, S.: Towards a generic feature-selection measure for intrusion detection. In: Proceedings of the International Conference on Pattern Recognition, pp. 1529–1532 (2010)
20. Zainal, A., Maarof, M.A., Shamsuddin, S.M.: Feature selection using rough set in intrusion detection. In: TENCON IEEE Region 10 Conference, Hong Kong, pp. 1–4 (2006)
21. Muda, Z., Yassin, W., Sulaiman, M.N., Udzir, N.I.: Intrusion detection based on K-Means clustering and Naïve Bayes classification. In: Proceedings of the 7th International Conference on Information Technology in Asia, pp. 1–6 (2011)
22. Kumar, S., Yadav, A.: Increasing performance of intrusion detection system using neural network. In: Proceedings of the IEEE International Conference on Advanced Communications, Control and Computing Technologies, pp. 546–550 (2014)
23. Ingre, B., Yadav, A.: Performance analysis of NSL-KDD dataset using ANN. In: Proceedings of the International Conference on Processing and Communication Engineering Systems, Guntur, pp. 92–96 (2015)
24. Garg, T., Khurana, S.S.: Comparison of classification techniques for intrusion detection dataset using WEKA. In: Proceedings of the International Conference on Recent Advances and Innovations in Engineering, pp. 1–5 (2014)
25. Paulauskas, N., Auskalnis, J.: Analysis of data pre-processing influence on intrusion detection using NSL-KDD dataset. In: Proceedings of the Open Conference of Electrical, Electronic and Information Sciences (eStream), pp. 1–5 (2017)

Q-routing: From the Algorithm to the Routing Protocol

Alexis Bitaillou[1]([✉]), Benoît Parrein[1], and Guillaume Andrieux[2]

[1] University of Nantes, LS2N, Polytech Nantes, Nantes, France
`alexis.bitaillou@univ-nantes.fr`
[2] University of Nantes, IETR, IUT La Roche-sur-Yon, La Roche-sur-Yon, France

Abstract. Routing is a complex task in computer network. This function is mainly devoted to the layer 3 in the Open Standard Interconnection (OSI) model. In the 90s, routing protocols assisted by reinforcement learning were created. To illustrate the performance, most of the literature use centralized algorithms and "home-made" simulators that make difficult *(i)* the transposition to real networks; *(ii)* the reproducibility. The goal of this work is to address those 2 points. In this paper, we propose a complete distributed protocol implementation. We deployed the routing algorithm proposed by Boyan and Littman in 1994 based on Q-learning on the network simulator Qualnet. Twenty-five years later, we conclude that a more realistic implementation in more realistic network environment does not give always better Quality of Service than the historical Bellman-Ford protocol. We provide all the materials to conduct reproducible research.

Keywords: Routing protocol · Q-learning · Quality of Service · Qualnet · Reproducible research

1 Introduction

Routing is a complex task in computer networks. A common solution uses the shortest path algorithm as the Bellman-Ford routing protocol [2]. But the shortest path is not necessarily the one that maximizes the Quality of Service (QoS), especially in wireless networks. In order to solve routing problem, two original approaches appeared in the 90s: *(i)* bio-inspired algorithm and *(ii)* Q-routing. In bio-inspired approaches, the idea is to model ant colonies as routing algorithm [11]. In 1992, a new reinforcement learning algorithm was created by Watkins and Dayan: Q-learning [12]. Two years later, Boyan and Littman [3] experience Q-Learning in routing algorithm named Q-routing. On their personal simulator, Q-routing offers a better average end-to-end delay than the Bellman-Ford protocol in high load condition. In fact, in congestion state, the Q-routing

Supported by the COWIN project from the RFI Wize and Atlantic 2020, Région Pays de la Loire.

S. Boumerdassi et al. (Eds.): MLN 2019, LNCS 12081, pp. 58–69, 2020.
https://doi.org/10.1007/978-3-030-45778-5_5

proposes alternative route based on the end-to-end delay while Bellman-Ford protocol is focused on the shortest path in terms of hops count. Those results have many potential applications especially for mesh and mobile ad hoc networks (MANET). But the work of Boyan and Littman is not complete. First, they do not supervise other QoS metric like the Packet Delivery Ratio (PDR). Second, their implementation is not totally specified. Even if their algorithm is distributable, we don't know if their implementation is really distributed. Third, their simulator is "home-made" and the simulation parameter is not highly depicted that makes this work hard to reproduce.

In this paper, we propose to evaluate the performances of Q-routing in a more realistic environment provided by a reference discrete event simulator like Qualnet. All of our experiences are available in a public git repository that makes this research reproducible and upgradable[1]. Furthermore, our implementation is fully distributed that enables to consider deployments in MANET. In such realistic conditions, we highlight that our Q-routing implementation over Qualnet simulator experiences routing message flooding and routing loops that leads to high packet loss rate in the original grid used by Boyan and Littman. We propose some counter-measures at the end of the paper.

The organisation of the paper is the following. In Sect. 2, we summarize some previous works about reinforced routing. In Sect. 3, we detail the implementation of our distributed Q-routing protocol. Section 4 provides results in terms of QoS and a discussion. The last section concludes the work and draws some perspectives.

2   Related Works

In this section, we provide more details on Q-routing in the related works.

2.1   Q-routing

Watkins and Dayan [12] created Q-Learning, a reinforcement learning algorithm in 1994. Two years later, Boyan and Littman proposed to integrate Q-Learning in routing algorithm. They name their algorithm Q-routing in reference to Q-Learning. In this algorithm, each node x looks for the lowest Q-value, defined using the Q function. The estimated delivery time from node x to node d by node y is noted: $Q_x(d, y)$. They define Q-value of function Q as:

$$Q_x(d, y) = Q_x(d, y) + \eta(q + s + t - Q_x(d, y)) \tag{1}$$

where η is the step size α in Q-Learning (usually 0.5 in [3]) q the unit of time spent in node x's queue, s the unit of time spent during the transmission between x and y and t as $t = \min\limits_{z \in neighbour \ of \ y} Q_y(d, z)$. In this case, the effective delivery time is the reward R and defined as: $R = q + s + t$.

---

[1] https://gitlab.univ-nantes.fr/ls2n-rio/qrouting-qualnet, it assume a valid Qualnet license.

The Q-value is initialized with 0. Q-routing is greedy. It always chooses the lowest Q-value. Several networks topologies are tested in the work of [3]: 7-hypercube, the 116-nodes LATA telephone network and a 6×6 irregular grid. The authors argue that only local information is used to proceed. The presented results of [3] concern only the 6×6 irregular grid. Q-routing is compared to Bellman-Ford shortest path algorithm. The average latency is higher in the exploration phase because the packets are randomly sent. Then, the latency is similar to the shortest path in low load condition. Q-routing is not always able to find the shortest path under low network load. But Q-routing clearly outperforms the shortest path in high load condition (even if the high load condition is not clearly defined). When the traffic load decreases, Q-routing keeps the high load policy. The original approach is thus not adapted to dynamic changes.

2.2 Predictive Q-routing

Choi and Yeung [4] proposed an improvement for Q-routing in 1996. Their algorithm is Predictive Q-routing (PQ-routing). It corrects the problem of suboptimal policy after a high network load. Unlike Q-routing, PQ-routing is not memory-less. It keeps track of the best Q-value. Under low network load, PQ-routing uses the shortest path algorithm to get the optimal policy. Under high network load, PQ-routing uses the latency as main metric. Thanks to its memory, it can come back to the optimal policy when the network load decreases. PQ-routing is composed of 4 tables: Q (estimation), B (best Q-value), R (recovery), U (last update). Q is defined like in [3] (cf. Eq. 1). Each table can be finely tuned by parameters. These tables are updated at each packet reception.

PQ-routing is compared to Q-routing. Two network topologies are tested: a 15-nodes network and the 6×6 irregular grid from [3]. PQ-routing performs better than Q-routing independently of the network load. Under high network load, PQ-routing is quite similar to Q-routing. These results are also obtained on a "home-made" network simulator. The average delivery time is the only metric provided. PQ-routing has higher memory requirements due to additional tables and higher computational cost because the 4 tables need to be updated.

2.3 Dual Reinforcement Q-routing

In 1997, Kumar and Miikkulainen [8] proposed to add backward exploration to Q-routing. As there is forward and backward, they name their algorithm Dual Reinforcement Q-Routing (DRQ-Routing). The evaluated network is the 6×6 irregular grid from [3]. They use once again a "home-made" simulator. They define low network load as 0.5 to 1.5 packets per simulation step, medium as 1.75 to 2.25 and high as 2.5 or more. DRQ-Routing is compared to Q-routing and shortest path. The average delivery time is the unique metric of comparison. According to their results, DRQ-Routing outperforms Q-routing in low network load. It outperforms Q-routing and shortest path in medium and high network load. Moreover, DRQ-Routing learns twice faster than Q-routing. They use unbounded FIFO queues. This means that a packet cannot be dropped by queue overflow. This simplifies the simulation but it cannot be applied in realistic network.

2.4   Other Related Works

There are several other related works about Q-routing. Many other extensions of Q-routing have been proposed: Gradient Ascent Q-routing [9], Enhanced Confidence-Based Q-routing [14], K-Shortest Paths Q-routing [7], Credence based Q-routing [6]. There are also extensions for wireless networks for Q-probabilistic routing [1] and for the Mobile Ad-hoc Networks (MANETs) [10]. Xia *et al.* [13] propose to use Q-routing in cognitive radio context. The average delivery time is the only metric used for most of those papers. Arroyo-Valles *et al.* do not use average delivery time. Instead, they prefer to use packet delivery ratio. Except [13] on OMNET++, those related works use their own simulator. Unfortunately [13] do not give any details about their implementation.

3   A Distributed Q-routing Implementation

In this section, we describe our implementation of Q-routing and the complete experimentation set-up. Our experimental plan concerns two topologies: one simple with two main paths and the 6×6 grid of [3].

3.1   Implementation

We have implemented Q-routing based on the Bellman-Ford implementation of Qualnet. The routing table has been replaced by the function Q (see Eq. 1). We reuse some parameters from Bellman-Ford implementation such as the maximum route length (16 hops), the timeout delay (120 s), the maximum number of routes per packet (32 routes per packets), and the periodic update delay (10 s). Our protocol is totally distributed. As Bellman-Ford protocol, nodes have access to local information only. Every 10 s, nodes broadcast their routing to their 1-hop neighbourhood. During a periodic updates, all routes are broadcast. Additionally, there are aperiodic updates. Aperiodic updates help to broadcast more quickly new route for example. There are also triggered to broadcast new latency value. During aperiodic updates, only the new or modified routes are sent.

The header of the routing packet contains a timestamp. Thanks to this information, the receiver can know the delay thanks to this timestamp. This method limits the network overhead but nodes have to use only one queue. The clock of the nodes needs to be synchronized. The level of synchronization determines the accuracy. The function Q is updated when the routes are updated, at least every 10 s. We define a route as a destination, a subnet mask, a next hop, a distance, a latency, two timestamps (local and from the original node), the incoming and outgoing interfaces. The first timestamp is defined when a node gets the latency from its 1-hop neighbour. This timestamp is not modified when the information is broadcast. The second timestamp is local and is used for checking the time out. This timestamp is updated when the node receives an update for this route. Q-routing has an exploration phase during 2 s. During the exploration phase, the Q-values are not updated.

3.2   Simulation Parameters

We use Qualnet 8.2 (from Scalable Networks Technology) as network simulator. The networks are composed of symmetric 10 Mb/s wired links. In order to prevent side effect, we used an abstract link layer. All links propagation delays are set to 1 ms that defines the latency of one hop. Unlike Kumar [8], each node has a finite FIFO queue of 150k packets. With Qualnet (and other discrete event simulators), the seed of the pseudo-random generator has a high impact on the results. For the same seed, the number of trials doesn't have a significant importance. Both foreground traffic and background traffic are constant bit rate (CBR) traffic flow. All CBR messages are 512 bytes long. CBR messages are sent in UDP packets. The CBR receiver drops disordered messages. We compare Q-routing to Bellman-Ford protocol because it uses the shortest path. The Table 1 sums up the simulation parameters.

Table 1. Simulation parameters

Element	Parameter	Value
Network	Link	Symmetric 10 Mb/s wired link
	Link propagation	1 ms
	Link layer	Abstract MAC
Node	Number of queue	1 FIFO queue
	Queue size	150k packets
CBR	Message size	512 bytes

A Toy Example. Before evaluating Q-routing on the topology of [3], we test it first on a simple topology as depicted on Fig. 1. Our test CBR is between node 1 and node 4 which are the source and the destination respectively. In this simple network, a large background traffic appears on the shortest path between node 2 and node 3 as shown Fig. 2. The goal is simply to verify that our Q-routing implementation prefers the longer path (through node 5) as soon as congestion occurs. The CBR source starts sending at 1 s and stop at the end of the simulation. The interval between two messages is 1 s. Background traffic appears at 15 min. The simulation time is only 60 min for this toy example. We test 10 different and arbitrary seeds.

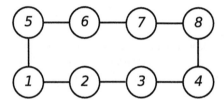

Fig. 1. Simple topology to test our implementation. Numbers correspond to the node ID.

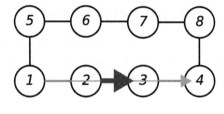

Fig. 2. CBR traffic flow location on the simple topology at 15 min. Numbers correspond to the node ID. (Color figure online)

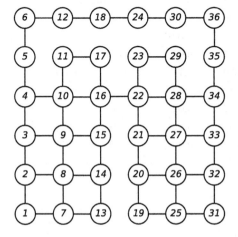

Fig. 3. The 6 × 6 irregular grid used by Boyan and Littman [3]

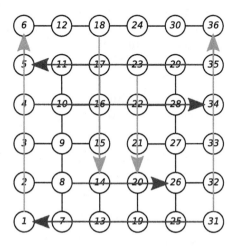

Fig. 4. CBR location on the 6 × 6 irregular grid (Color figure online)

The Irregular Grid. We evaluate then our implementation on the original 6 × 6 irregular grid from [3]. The Fig. 3 illustrates this particular grid and the Fig. 4 shows the location of the CBR couples. The location of the CBR flow is arbitrary because Boyan and Littman [3] don't give so many details in their works. However, we experience that the location of the couples has a great impact on the results. As in [3], the CBR traffic flow will be alternatively "horizontal" and "vertical" every 30 min. The "vertical" CBR traffic flow (in red) will be active then it will be the "horizontal" CBR traffic flow (in blue). All CBR sources have the same throughput for a given simulation as depicted on the 2nd column of the Table 3. We used 36 different and arbitrary seeds in order to validate our simulation. The simulation time is 180 min.

4   Results and Discussion

In this section, we present the results of the experimentation. We focus on two metrics: the average end-to-end delay (or average delivery time) and the packet delivery rate (PDR). Both are measured at the application layer (layer 7).

Table 2. Comparison between Q-routing (Q-r) and Bellman-Ford protocol (B-F) with background traffic on the toy example, 10 seeds.

Protocol	Throughput of bg traffic	Avg. EtE delay (ms)		PDR (%)	
		Average	SD	Average	SD
B-F	8.2 Mb/s	4.36	<0.01	100	0
B-F	10.2 Mb/s	82.9	<0.01	66.6	0.07
B-F	11.7 Mb/s	67.7	0.01	50.3	0.08
B-F	13.7 Mb/s	67.9	0.11	50.1	0.09
Q-r	8.2 Mb/s	4.37	<0.01	100	0
Q-r	10.2 Mb/s	6.57	0.02	100	0.04
Q-r	11.7 Mb/s	6.57	0.02	100	0.04
Q-r	13.7 Mb/s	6.58	0.05	100	0.07

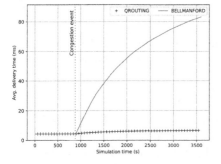

Fig. 5. Average End-to-End delay over the simulation, measured by the CBR between node 1 and node 4.

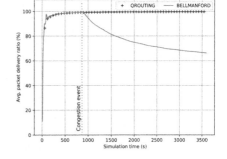

Fig. 6. PDR over the simulation, measured by the CBR between node 1 and node 4.

Results on the Toy Example. The Table 2 sums up the results with different throughput of background traffic for the simple grid of Fig. 1. On average we experience low delay and the highest PDR for Q-routing for the considered traffic pattern. The singularity at 10.2 Mb/s for BF protocol means that we are in a congested status with relative good PDR but a degraded end-to-end delay.

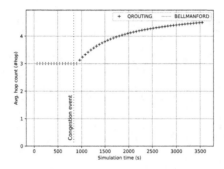

Fig. 7. Average hop count over the simulation, measured by the CBR between node 1 and node 4.

Table 3. Comparison between Q-routing and Bellman-Ford on the irregular grid, 36 seeds.

Protocol	Throughput	Avg. EtE delay (ms)			PDR (%)			Avg. drop of IP packets (number of packets)		
		Average	Median	SD	Average	Median	SD	No route to host	Expired TTL	Queue overflow
B-F	4.1 kb/s	9.27	9.27	<0.01	100	100	<0.01	2	0	0
B-F	41 kb/s	9.27	9.27	<0.01	100	100	<0.01	3	0	0
B-F	410 kb/s	9.27	9.27	<0.01	100	100	<0.01	18	0	0
B-F	4.1 Mb/s	9.27	9.27	<0.01	100	100	<0.01	164	0	0
B-F	8.2 Mb/s	51.0	51.4	7.70	88.0	88.5	0.74	49.0×10^3	0	10.4×10^6
B-F	10.2 Mb/s	242	242	12.9	75.3	75.4	1.17	295×10^3	14	26.8×10^6
B-F	13.7 Mb/s	254	255	14.1	56.5	56.5	0.96	398×10^3	3	62.8×10^6
Q-r	4.1 kb/s	9.33	9.33	<0.01	99.9	99.9	<0.01	14	32	0
Q-r	41 kb/s	9.33	9.33	<0.01	99.9	99.9	<0.01	102	318	0
Q-r	410 kb/s	9.79	9.70	0.38	99.9	99.9	<0.01	980	2776	0
Q-r	4.1 Mb/s	11.8	11.7	0.84	99.7	99.8	0.20	10×10^3	10×10^3	140×10^3
Q-r	8.2 Mb/s	129	135	91.6	82.7	79.7	9.74	21×10^3	54×10^3	17×10^6
Q-r	10.2 Mb/s	680	674	32.6	65.1	66.7	8.12	115×10^3	258×10^3	45×10^6
Q-r	13.7 Mb/s	673	675	42.4	47.1	48.2	5.73	141×10^3	280×10^3	85×10^6

This singularity disappears for Q-routing. Figures 5, 6 and 7 give a temporal representation of the simulation. The background traffic is 10.2 Mb/s. The dotted line represents the appearing of the congestion. With Q-routing, the average delivery time stays low compared to Bellman-Ford protocol (Fig. 5) when the congestion occurs at 15 min (900 s). Moreover, Q-routing drops only few packets as depicted on Fig. 6. Finally, the average hop count (Fig. 7) shows that Q-routing bypass the congested path through a longer way. There is no more than 1 packet lost. The throughput is 1 packet per second. The results do not show clearly the convergence time. If we consider that dropped messages are due to congestion, we estimate it no higher than 2 s. Q-routing performs pretty well on our simple test. But Q-routing uses a greedy strategy. Even if the congestion disappears, the packets will continue to pass by the longest route.

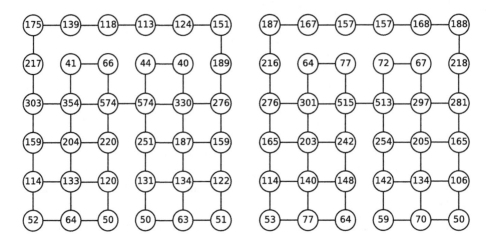

Fig. 8. Policy summary: Bellman-Ford under 4.1 Mb/s (medium load).

Fig. 9. Policy summary: Q-routing under 4.1 Mb/s (medium load).

Results on the Irregular Grid. Boyan and Littman [3] expressed their results in the unit of their "home-made" simulator. The time is in simulator time. The network load and its unit are not defined. We express the duration in seconds and the network load in bytes per second. Thus, we cannot compare directly their results to ours. The Table 3 sums up the results for the irregular grid. We experience many packet drops. We add to this table the origin of those drops: (1) route starvation i.e. the "no route to host" message; (2) abnormal route length (monitored by the Time-To-Leave parameter); (3) queue overflow. For Q-routing, the number of packets dropped due to expired TTL is relatively high even at low throughput. Routing loops can only explain the number of packets dropped due to expired TTL. Performance under low network load is similar to Bellman-Ford protocol. The behaviour of the Q-routing protocol changes between 410 kb/s and 4.1 Mb/s values. The average delivery time increases and the PDR decreases up to reach 47.1% at 13.7 Mb/s. The number of dropped packets by queue overflow reaches 140×10^3 packets at 4.1 Mb/s. At 13.7 Mb/s, the number of dropping by queue overflow is very high, around 85×10^6 dropped packets for 144×10^6 messages sent. From a topological point of view, the Figs. 8 and 9 resume the routing policies by showing the average number of routes passing by each node under medium load at the end of the simulation. For Bellman-Ford protocol, the distribution of routes is very similar to [3]. However, the Fig. 9 gives a different result compared to [3]. With Q-routing, nodes 16 and 22 are less solicitate but the number of routes is not so balanced as in [3].

We extract the number of routing packets sent and received. With the Bellman-Ford protocol, there are not so much aperiodic updates because the minimal distance to a destination is constant in our scenario. For example, under low network load, aperiodic updates represent 7 packets per hour and per node. The situation is different with Q-routing. The average number of packets sent for periodic updates is around 15×10^3 packets over the simulation. The average number of packets sent for aperiodic updates is around 661×10^3 packets. Aperiodic updates represent about 90% of routing packets. At 4.1 kb/s, on the irregular grid, there are about 190 routing packets per seconds for 4 data packets per seconds.

4.1 Discussion

Although the test over a simple topology is very encouraging, our test on the irregular grid doesn't give a similar result to [3]. As the authors of this work, we expected Q-routing to outperform Bellman-Ford protocol under high network load condition. The implementation of the Q-routing in a real packet simulator in a distributed manner is the main reason from our point. Traffic that needs to pass through those links will be penalized. The quantity of packets dropped by queue overflow is really important. This is not considered in [3]. We made some additional tests with a larger queue in order to give more chance to have a higher latency but the results have not changed significantly.

Packets dropped by expired TTL are the main cause of dropping under low network load. An additional mechanism is needed in order to prevent routing loops. For example, source routing or tracking packet's ID can be used periodically to check the route. Another possibility is to change the reward in the Q function. Indeed, the distance (in hop count) could be considered in order to help the choice of the best route. The reward can also take account of the number of dropped packets.

Moreover, Q-routing has a higher memory requirement than Bellman-Ford protocol. Indeed, Q-routing needs memory to store all destinations by all the next hops whereas Bellman-Ford protocol stores also all destinations only once. Furthermore, computational costs are higher with Q-routing due to data structure and the quantity of data. For example, on the grid, a node with 4 neighbours like node 16 has to store 144 routes. When it broadcasts all routes, it needs 5 packets. With the Bellman-Ford protocol, nodes have 36 routes. They need to send just 2 packets to broadcast all routes. The network overhead is also higher with Q-routing. The value of the latency changes a lot so the number of aperiodic updates can be very high. In order to limit the number of aperiodic updates, several solutions are possible. For example, the aperiodic update can be schedule only if the difference between the old and the new value is greater than a threshold. Moreover, we can introduce partial flooding mechanism as it is proposed in MANET protocol as OLSR [5]. In this protocol, Multi-Point Relays reduce drastically routing message flooding.

5 Conclusion

In this paper, we presented a distributed implementation of the Q-routing algorithm. We experienced it on the professional packet driven simulator Qualnet. Q-routing works well on a simple topology composed of 8 nodes. However, the

Quality of Service parameter as the end-to-end delay and the packet delivery ratio are degraded as soon as Q-routing protocol is deployed on the irregular grid proposed in [3]. High network overhead and routing loops are the main explanation in real networks conditions. They also explain routing starvation. We provide all the materials to conduct reproducible research. Auxiliary functions to prevent flooding, the integration to existing MANET protocols (as OLSR) and the extension of Q function are the perspective of this work.

References

1. Arroyo-Valles, R., Alaiz-Rodriguez, R., Guerrero-Curieses, A., Cid-Sueiro, J.: Q-probabilistic routing in wireless sensor networks. In: Sensor Networks and Information 2007 3rd International Conference on Intelligent Sensors, pp. 1–6, December 2007. https://doi.org/10.1109/ISSNIP.2007.4496810
2. Bellman, R.: On a routing problem. Q. Appl. Math. **16**(1), 87–90 (1958). https://doi.org/10.1090/qam/102435
3. Boyan, J.A., Littman, M.L.: Packet routing in dynamically changing networks: a reinforcement learning approach. In: Advances in Neural Information Processing Systems, pp. 671–678 (1994)
4. Choi, S.P., Yeung, D.Y.: Predictive Q-routing: a memory-based reinforcement learning approach to adaptive traffic control. In: Advances in Neural Information Processing Systems, pp. 945–951 (1996)
5. Clausen, T.H., Jacquet, P.: Optimized Link State Routing Protocol (OLSR). RFC 3626, October 2003. https://doi.org/10.17487/RFC3626. https://rfc-editor.org/rfc/rfc3626.txt
6. Gupta, N., et al.: Improved route selection approaches using Q-learning framework for 2D NoCs. In: Proceedings of the 3rd International Workshop on Many-core Embedded Systems, MES 2015, Portland, OR, USA, pp. 33–40. ACM, New York (2015). https://doi.org/10.1145/2768177.2768180
7. Hoceini, S., Mellouk, A., Amirat, Y.: K-shortest paths Q-routing: a new QoS routing algorithm in telecommunication networks. In: Lorenz, P., Dini, P. (eds.) ICN 2005. LNCS, vol. 3421, pp. 164–172. Springer, Heidelberg (2005). https://doi.org/10.1007/978-3-540-31957-3_21
8. Kumar, S.: Dual reinforcement Q-routing: an on-line adaptive routing algorithm. In: 1997 Proceedings of the Artificial Neural Networks in Engineering Conference (1997)
9. Peshkin, L., Savova, V.: Reinforcement learning for adaptive routing. In: Proceedings of the 2002 International Joint Conference on Neural Networks, IJCNN 2002 (Cat. No. 02CH37290), vol. 2, pp. 1825–1830, May 2002. https://doi.org/10.1109/IJCNN.2002.1007796
10. Santhi, G., Nachiappan, A., Ibrahime, M.Z., Raghunadhane, R., Favas, M.K.: Q-learning based adaptive QoS routing protocol for MANETs. In: 2011 International Conference on Recent Trends in Information Technology (ICRTIT), pp. 1233–1238, June 2011. https://doi.org/10.1109/ICRTIT.2011.5972411
11. Schooenderwoerd, R., Holland, O., Bruten, J., Rosenkrantz, L.: Ants for load balancing in telecommunication networks. Technical report 96-35, HP Labs, Bristol (1996)
12. Watkins, C.J.C.H., Dayan, P.: Q-learning. Mach. Learn. **8**(3), 279–292 (1992). https://doi.org/10.1007/BF00992698

13. Xia, B., Wahab, M.H., Yang, Y., Fan, Z., Sooriyabandara, M.: Reinforcement learning based spectrum-aware routing in multi-hop cognitive radio networks. In: 2009 4th International Conference on Cognitive Radio Oriented Wireless Networks and Communications, pp. 1–5, June 2009. https://doi.org/10.1109/CROWNCOM. 2009.5189189
14. Yap, S.T., Othman, M.: An adaptive routing algorithm: enhanced confidence-based Q routing algorithm in network traffic. Malays. J. Comput. Sci. **17**(2), 21–29 (2004)

Language Model Co-occurrence Linking for Interleaved Activity Discovery

Eoin Rogers[✉], John D. Kelleher, and Robert J. Ross

Applied Intelligence Research Centre, Technological University Dublin,
Dublin, Ireland
{eoin.rogers,john.d.kelleher,robert.ross}@tudublin.ie

Abstract. As ubiquitous computer and sensor systems become abundant, the potential for automatic identification and tracking of human behaviours becomes all the more evident. Annotating complex human behaviour datasets to achieve ground truth for supervised training can however be extremely labour-intensive, and error prone. One possible solution to this problem is activity discovery: the identification of activities in an unlabelled dataset by means of an unsupervised algorithm. This paper presents a novel approach to activity discovery that utilises deep learning based language production models to construct a hierarchical, tree-like structure over a sequential vector of sensor events. Our approach differs from previous work in that it explicitly aims to deal with interleaving (switching back and forth between activities) in a principled manner, by utilising the long-term memory capabilities of a recurrent neural network cell. We present our approach and test it on a realistic dataset to evaluate its performance. Our results show the viability of the approach and that it shows promise for further investigation. We believe this is a useful direction to consider in accounting for the continually changing nature of behaviours.

1 Introduction

Activity discovery (AD) refers to the automated and unsupervised extraction of *activities* (recurrent patterns of behaviour) from a given dataset of sequential sensor events [6]. These sensor events could come from any one of a wide range of sensors installed in a real-world environment (such as a house or office), or could be events logged by a server or other computer monitoring a virtual environment. Activity discovery is part of the wider field of *activity recognition*, which refers to the use of machine learning in the identification and classification of activities. AD has a wide range of applications, from the automatic labelling of datasets for more general activity recognition [17], to more complex end-to-end systems which combine elements of activity discovery and recognition [9]. The wider problem of identifying usable sub-sequences in larger sequences that have semantic meaning also has wider applicability for areas like anomaly and crime detection [10,28] and network intrusion detection [16].

© IFIP International Federation for Information Processing 2020
Published by Springer Nature Switzerland AG 2020
S. Boumerdassi et al. (Eds.): MLN 2019, LNCS 12081, pp. 70–84, 2020.
https://doi.org/10.1007/978-3-030-45778-5_6

One issue that poses a challenge for many existing activity discovery systems is *interleaving*, where two or more activities are being carried out by a single actor at the same time. Usually, this appears on the dataset as if the actor is switching back and forth between activities, much like how a compute processor switches back and forth between processes when the operating system performs a context switch. Equivalently, multiple individuals might be carrying out tasks in parallel, which would register on the sensors in a manner that closely resembles interleaving.

With the goal of improving the performance of such systems on interleaved data, we present here a novel system for activity discovery which is explicitly designed to account for the interleaved data of modern datasets. Like most activity discovery systems, we base our work on the intuition that activities appear in datasets as *sub-sequences* that repeat multiple times throughout the dataset. Unlike most other AD systems, however, we do not require that our activities consist of contiguous substrings of the dataset, but rather can be interrupted with sensor events which may be parts of other activities, or may not belong to other activities. This allows activities to be interleaved, and allows the model to explicitly model and *disentangle* the interleaving.

On top of this, we build activities that are hierarchical, tree-like structures. Thus, discovered activities can contain other activities as a subset. This models the hierarchical nature of real-world activities: a large, complex activity such as *washing* could consist of smaller sub-activities such as *using the shower* or *brushing teeth*. This allows our system to take sequential input data and convert it into a rich, complex structure, much like how a language parser can convert a sequential string of symbols into a parse tree, thus exposing non-obvious structure contained within the original sequence.

The remainder of this paper is structured as follows: in Sect. 2 we review prior and related work in activity discovery and sequential pattern discovery more generally, some of which has been an important influence on the work presented here. In Sect. 3 we briefly introduce the formal notation that we use in this paper, before introducing our activity discovery model in Sect. 4. We present our evaluation study design and the results of that study in Sects. 5 and 6 respectively. Before concluding we provide a brief discussion in Sect. 7.

2 Prior Work

A number of techniques have been proposed to tackle the activity discovery problem. A good general introduction to the field is Cook et al. [6], which introduces an activity discovery system that applies a beam search algorithm using an operator called *ExtendSequence* to discover activities in an unlabelled dataset. Like a number of other systems in the field, this algorithm utilises the *minimum description length* (MDL) principle [23,24], which proposes evaluating machine learning models by measuring the degree to which they *compress* their input dataset. This is an important principle, and is one that we shall return to later in this paper.

Activity discovery can also be carried out by relatively simple systems that utilise topic models [17]. Here, the *latent Dirichlet allocation* (LDA) topic model [3] is used to model the relationship between sensor events and latent variables which are presumed to represent activities. The model is shown to have good performance, even on a complex dataset. More recently, other models based on statistical models have been proposed: for example, Fang et al. [11] proposes activity discovery by means of a hierarchical mixture model. [25] propose using activity discovery to build a model of normal behaviour patterns of a person in order to detect anomalous behaviours that may be of interest to medical professionals.

Related fields also provide an important source of ideas. Grammar induction is a concept from computational linguistics which refers to the derivation of grammar productions for a language given only a dataset. Some forms of grammar induction require labelled input, distinguishing positive and negative examples, but others require only positive examples. In the general case, grammar induction is not a tractable problem, regardless of whether the dataset is labelled or not [13,14], but tractable approximations have been demonstrated which solve the problem to a degree [8]. While the problems are related, grammar induction is by no means equivalent to activity discovery (and is in fact in many ways harder), but it does involve the induction of structure from a one-dimensional input vector. *Adios* [27] induces a grammar by loading a dataset into memory as a graph, with words represented as vertexes and sentences represented as directed edges between these. This representation allows for the identification of *equivalence classes* between words and phrases which share the same input and output edges, which can then be added to the graph as nonterminals. A variant of the Adios approach, which supplements the basic grammar induction algorithm with logical predicates to allow for more accurate induction in a limited linguistic domain is presented by [12].

Remaining on the theme of grammar induction, the *eGrids* grammar induction algorithm [22] bears a resemblance to the beam search-based system mentioned previously from [6]. This approach also uses an MDL-based objective function to guide the search. More recently, an interesting deep learning-based grammar induction model using convolutional networks to determine *syntactic distance* (the degree to which two neighbouring words or symbols belong to the same POS phrase) has been proposed by [26]; this approach does have some similarities to the approach we take later in this paper.

More generally, it should be noted that the activity discovery problem as presented by us can also be understood as a non-local variant of the tree structure induction algorithm *Sequitur* [21], which groups input symbols together even if they do not appear contiguously.

However, none of the approaches discussed above were explicitly developed with the goal of dealing with interleaving. Some solve the interleaving problem better than others, but the system we propose in this paper has the advantage of explicitly disentangling interleaved activities from each other, identifying where one activity switches to another and when it switches back. Thus, it solves a problem not tackled by most existing activity discovery systems.

3   Activity Discovery Process Notation

In order to have a clean model description, it is necessary for us to briefly introduce the terminology that we will use later. Formally, an activity discovery system can be modelled as a 5-tuple (Σ, D, A, f, g), where:

- Σ is a set of *event types*;
- D is an ordered sequence of events, $D = \langle d_1, d_2, \ldots, d_L \rangle$ of length L, such that each $d_i \in D$ is drawn from the set Σ. We call this the *dataset*;
- A is a set of *activity types*;
- f is a mapping $f : D \to X^*$, which takes a sequence of events D as input, and returns a set of (possibly non-contiguous) sub-sequences of D as output; and
- g is a mapping $g : X \to A$, where $X \subset D^*$, which takes a sub-sequence produced by f as input, and returns an activity type $a \in A$ as output.

This definition can be made clearer with a concrete example. Supposing we have a dataset $D = \langle d_1, d_2, \ldots, d_N \rangle$. Each $d_i \in \Sigma$ is a sensor event drawn from Σ, our full set of sensor events. In an environment where sensors have been set up in a home, for instance, Σ could consist of events like *open front door, turn oven on, flush toilet* and similar domestic events. An *activity*, then, is simply a sub-sequence of D consisting of events that appear to the activity discovery system to be semantically related. For instance, we would expect that events like *turn oven on, open kitchen cupboard, open refrigerator* might occur in an activity together, since they tend to occur together temporally. It should be noted that D is not a set of sequences as might be the case in a supervised learning setting; D is a single large dataset from which we extract activities.

Multiple similar activities can then be clustered or lifted into one *type*. The activity discovery system might notice that an activity similar to the one mentioned in the previous paragraph seems to occur nightly, and may cluster them all into a single *making dinner* activity type. The concrete sub-sequences of D are referred to as the *instances* of the *making dinner* activity.

Note that we don't generally expect an activity discovery system to operate with human-like semantic knowledge or expectations in the basic case. Thus, it would not be expected to be able to name the new activity type as *making dinner*, only to identify that the instances involved can be sensibly clustered together. A commercial activity discovery system might well be supplemented with real-world knowledge, with the intention of biasing towards the sort of activities we would expect to find in the environment in which it operates. For instance, knowledge that events relating to a fridge or oven indicates activities relating to food preparation such as *making dinner* are taking place. In many ways this would stray over into being a form of activity recognition as well as discovery. For this reason, we stick to a pure form of activity discovery without any real-world knowledge. We do still expect it to be able to discover *making dinner* as an activity, just not to be able to give it a label (*making dinner*) that would be semantically meaningful to a human observer.

4   Model

We now proceed to outline our approach to the activity discovery problem. At a bird's-eye view, one can conceptualise our model as being composed of the following four elements:

- A *neural language model* to analyse the input dataset, and build probability distributions over future events given past events;
- A *linking component* to link events into activity instances using the probability distributions;
- A *clustering component* to cluster activity instances into activity types in a principled way;
- A *hierarchy construction component* to remove discovered activities and replace them with new synthetic events, thus allowing the above steps to be repeated, and a hierarchy to be iteratively built up.

In the following we take each of these components and describe them in detail.

4.1   Association Estimation

All approaches to activity discovery depend upon the assumption that the activities present in a dataset will be composed of events, which are assumed to be associated with each other in some predictable way. For example, if we want an activity like *cooking dinner* to be found in a dataset, we would need to see that a set of events such as *turn on oven, open refrigerator, open cupboard* and the like occur close to one another in the dataset on a daily basis. Intuitively, our approach is based on the insight that if observing the *turn on oven* event allows me to predict that the *open refrigerator* event will soon follow, I can use this fact to infer the existence of the *cooking dinner* activity.

To detect these associations, and to estimate how strong they are, we turn to *language modelling*, a concept originating from the natural language processing (NLP) community. Given a sentence of words $W = \langle w_1, w_2, \ldots w_Q \rangle$, a language model estimates the probability of a word w_i given the previous n words, which can be written $p(w_i | w_{i-1}, w_{i-2}, w_{i-n})$. Equivalently, we can view this as assigning a probability to a given sentence or sentence fragment, since the probability of the entire sentence W must be:

$$p(W) = p(w_1)p(w_2|w_1) \ldots p(w_N | w_{N-1}, w_{N-2}, \ldots w_{N-n}) \tag{1}$$

A typical language model attempts to predict the *next* word of the sequence, but we need to take into account the possible presence of interleaving. In other words, *open refrigerator* may occur a number of events *after* the *turn on oven* event. Keep in mind, therefore, that we build a probability distribution not over the next event, but rather over the next m events.

Traditionally the training of language models entails the collection of statistics from a dataset. Such systems have been used for many decades within the NLP community.

Rather than relying on statistical language models, we apply a so-called *neural language model* (NLM) [2,20] to the association estimation task. In recent years, neural language modelling systems have achieved parity with, and subsequently overtaken, their more classical (statistical) counterparts. These models build on general trends in *deep learning* [19] by applying large artificial neural network architectures and taking advantage of recent advances in hardware and training algorithms. Yoshua Bengio [2] proposed the first NLM system in 2003, with more recent systems adopting sophisticated additions such as recurrent models, attention [4], internal memory [29] and levels of representation other than individual words [20].

Our modelling approach builds on the LSTM network architecture [15]. This architecture is a form of recurrent neural network that is particularly good at encoding long range dependencies into a decision process. An LSTM unit consists of a single memory unit called a *cell*. Computational units called *gates* determine the contents of the cell and the behaviour of the unit by controlling the movement of data in and out of the cell. Typically, three gates are used: an *input gate*, which controls the extent to which the current input to the unit influences the cell, an *output gate*, which controls the extent to which the current cell contents influence the unit's output, and a *forget gate*, which controls the extent to which the current value of the cell will be retained for the next iteration of the LSTM unit. A given LSTM can be trained using labelled data and the backpropagation algorithm in the normal way. Moreover, rather than using a single LSTM network, our approach makes use of a collection of LSTM estimators. Specifically, we train m LSTM networks, one to predict the *next event*, one to predict the event immediately after it, one to predict the event after that one and so on. Each network net_j thus predicts the distribution:

$$p_j(d_{i+j}|d_i, d_{i-1}, \ldots, d_{i-n}) \tag{2}$$

For each $j \in \langle 1, 2, \ldots m \rangle$, where we refer to each j as a *lookahead offset*. This builds distributions over the next m events, motivated above for allowing us to detect interleaving. Since some events might be very common in the dataset, we actually use the difference between the probability distribution in Eq. 2 and the distribution already computed for the previous position of the sliding window $i-1$. In a slight abuse of notation, therefore, we actually use p_j in Eq. 2 to refer to *relative probability*, defined as the following:

$$p_j(d_{i+j}|d_i, d_{i-1}, \ldots, d_{i-n}) - p_j(d_{(i-1)+j}|d_{i-1}, d_{i-2}, \ldots, d_{(i-1)-n}) \tag{3}$$

The equation above gives us the relative probability distribution for the $i+j$th slot in the dataset, where $1 \leq j \leq m$. This process is illustrated in Fig. 1.

In Fig. 1(a), we see a short dataset of events A, B, C, D, E, F, G, H. A sliding window of length 2 ($n = 2$) is placed over events C and D of the dataset, and a lookahead window of length 3 ($m = 3$) is placed over events E to G in the dataset. The sliding window is fed as input into each of the m networks (there would be three in this case, one for each element of the lookahead window). Thus each network is trained to output a probability distribution over the event it expects to observe at a particular offset into the lookahead window.

To illustrate the example further, the network net_1 might predict a 20% higher relative probability that event E would be observed at an offset of 1 compared to when the sliding window ends at event C instead of D. net_2 might predict an 80% higher relative probability that event F would be observed at an offset of 2, and net_3 might predict a 40% higher relative probability of event G being observed at an offset of 3. Note that each network outputs a probability distribution over the entire set of events Σ: we only show the probabilities of the event types that actually occur in Fig. 1(b). Thus, the output from this stage can be viewed as P, an $n \times m \times |\Sigma|$ matrix, where each P_{ijk} is the relative probability that the $i + j$th event in the dataset is predicted to be the kth event type in the set of events Σ.

4.2   Linking Events into Activities

Given a sliding window across the dataset, the association estimator will provide likelihoods of particular events occurring at given offsets into the look-ahead window – these are encoded as the association matrix P. Based on that information we next establish links between strongly associated events.

There are a number of ways in which the linking process could be achieved. For the experiments presented in this paper, we begin by iterating over each $i \in \langle 1, 2, \ldots, n \rangle$ and each $j \in \langle 1, 2, \ldots, m \rangle$. For each (i, j) pair, we can view P_{ij} as the relative probability distribution that each $d \in \Sigma$ will be the jth event *after* the ith event. In the case of Fig. 1(b), we see that net_2 has assigned a high relative probability to the event type F occurring at an offset of two after D. By contrast, net_1 and net_3 have assigned much lower probabilities to their corresponding values. Thus, we would expect that this means that events D and F are part of the activity, which would justify connecting them via a *link*, as shown in Fig. 1(c).

This is essentially a greedy linking strategy as we are guaranteed that the strongest links only for a given symbol are created. This has advantages over alternatives such as a thresholding-based model in that no threshold parameter needs to be derived from the dataset.

This is almost correct, but naive implementations of the above algorithm for linking result in most links being made with the event of an offset of one into the lookahead window, i.e. the first events in the lookahead window. This is obviously not ideal from the perspective of identifying activities that are interleaved. If we train a system like that described above with a window and lookahead length of 20 (so $n = m = 20$), and evaluate the accuracy of the resulting twenty networks, we observe something interesting. The first network, net_1, is about

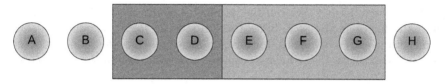

(a) Activities A and B are contained within a sliding window; activities C, D and E are in the lookahead window

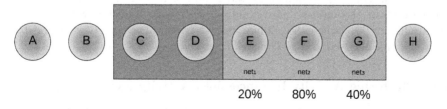

(b) net_1 assigns a 20% relative probability to offset (j) 1 being equal to event type E, net_2 assigns an 80% relative probability of offset 2 being equal to event type F, and net_3 a 40% relative probability of offset 3 being equal to event type G

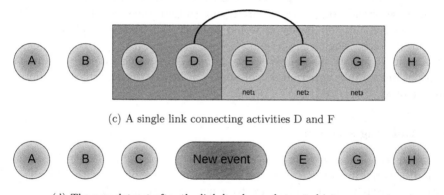

(c) A single link connecting activities D and F

(d) The new dataset after the link has been abstracted into one event

Fig. 1. Using probability distributions to construct links between events

98.8% accurate. The second network's accuracy has dropped down to about 97.9%, which is a trend that continues throughout the lookahead window. By net_{20}, the accuracy is down to about 94%. This makes intuitive sense, since predicting the *next* event will generally be easier than predicting the event that will occur three events from now. Analogously, predicting the next word in a sentence is easier than predicting the word that will come a number of words after. We thus feel that we are justified in modifying the algorithm to explicitly take distance into account, so longer offset networks get a small boost in their probabilities to offset the inherent higher difficulties in what is expected in them.

Thus, we multiply each relative probability by a *correcting factor* that is equal to 1 for an offset of 1, some value larger than one for offsets greater than 1, and which increases linearly. We call the parameter which controls the degree with which the factor increases x. Since this value is no longer a valid probability we refer to it as a *score*. Thus, for formula for computing the score for offset j and window position i is as follows, which is a modification of the probability in Eq. 2:

$$score(d_i, d_{i+j}) = p_j(d_{i+j}|d_i, d_{i-1}, \ldots, d_{i-n}) \times (1 + \frac{j}{x}) \qquad (4)$$

This link will *only* be built if D and F do not link more strongly with some other event in the dataset. For example, if D was predicted with a relative probability of 90% when the sliding window ended at event B, we would have built the link between B and D instead.

4.3   Clustering Activity Instances into Activity Types

We now need to match all the links we have found with links of the same type. We call this step *clustering*. In the case of the example presented in Fig. 1, we would need to cluster all other links between event types D and F (or equivalently between F and D) together. Note that this differs from clustering in the usual sense of the word, since we are trying to find exact matches between link types, not semantic similarity as would be done in a clustering algorithm such as k-means clustering.

At this point, we also apply a threshold factor, which we call y, to remove spurious links. Link types that do not appear at least y times in the entire dataset are removed.

4.4   Building a Hierarchy

The final step in a single iteration of the model is to build a new dataset, where each discovered link of two activities is removed, and replaced with a *new event*, with each activity type giving rise to a new event type. The outcome of this process applied to the small dataset we have used as an example in this section is shown as Fig. 1(d). From here, we can train a new set of m LSTM networks, and repeat the process again. At the end, a tree-like structure will result, showing a hierarchy of (possibly overlapping) activities contained within each other. This is inspired from the way the Sequitur algorithm [21] constructs tree structures from sub-sequences that occur multiple times in a sequence, generalised to allow for non-contiguous sub-sequences. Ideally, the process would be run until a sufficiently high level of abstraction (where the tree-like structures correspond to activities) has been reached. In practice, the process can be stopped early if only a partial result is needed. The new event could be placed into the position formerly occupied by either event D or event F. The choice shouldn't affect the evaluation metrics we are using, so the choice of which position is somewhat arbitrary. In our case, we place the new event in event D's position.

5   Experiment Design

For our experiments, we utilised the Kyoto 3 dataset gathered by the CASAS project [7]. This dataset consists of readings from a range of sensors installed in a small apartment. The dataset was gathered by asking a number of participants to perform *activities of daily living* (ADLs) in a natural manner in the apartment. Most of the sensor readings are either binary (they have a simple on/off state), or can only enter one of a handful of states. This means they can be easily converted to the sequence of events format our system expects by creating event types of the form *SensorName_SensorState*. For example, one of the sensors are refered to as $M17$ in the dataset, and can take the state ON, so $M17\_ON$ becomes an event type in the dataset. For the few sensor types that did have continuous values, we used the *Jenks natural breaks algorithm* [18] to discretise the data. Our system does not take temporal distance into account, so it cannot, for instance, see large gaps between events. This makes the systems task substantially harder, but it allows us to put our system through much more challenging testing than most AD practitioners settle for.

Evaluating activity discovery systems can be a challenge for a number of reasons. Human annotators may not come to an agreement with each other over the start and end points of activities, which makes working from a gold-standard ground truth quite difficult. For example, when does the activity of *Making_Dinner* start? When a person enters the kitchen? When they turn on the oven? In many cases, a ground truth may not even be available (although that isn't an issue for the Kyoto dataset). The output from an AD system may be on a different level of abstraction from the ground truth: for example the system may discover an activity that could be called something like *chop_vegetables*, but the ground truth instead has an activity called *make_dinner*, which *chop_vegetables* would be a constituent of. A good overview of evaluation for activity discovery can be found in [5].

Since we do have access to a ground truth in this experiment, it makes sense to use it, although we must keep the above issues in mind. Because of the abstraction issue mentioned before, we argue that both raw accuracy and F-measures are inappropriate for evaluating this system. Instead, we compare each new event type from the topmost (i.e. most abstract layer) of the hierarchy using the *precision* metric, i.e. the true positives divided by the sum of the true and false positives. Each event type is then matched with the ground truth activity with which it achieves the highest associated precision.

Our system is implemented using Python/TensorFlow [1] running on an Nvidia graphics card. We trained the hierarchy for 5 layers: each layer took roughly an hour to train and cluster. Our LSTMs were two layers deep, with a width of 150 LSTM cells per layer. We used a sliding window length $n = 20$, a lookahead window length $m = 10$, a score factor $x = 400$ and an event type threshold $y = 3$.

We have already mentioned minimum description length (MDL) in Sect. 2 of this paper. This is the second metric that we propose for our system. MDL draws a parallel between machine learning on the one hand, and data compression on the other. If person A wishes to transmit a dataset to person B while minimising the bandwidth, they could do so by encoding the dataset directly according to some optimal encoding scheme and send it. By contrast, person A could also train a machine learning model and send this, since person B could use it to re-create the dataset. Of course, no model will be perfect, so we must also send *data that would be required to correct the model's mistakes.* Since the mistakes will hopefully be small, these corrective measures should not consume many bits of bandwidth. In its purest form, MDL proposes computing a *score* for a machine learning system, which is the length of the machine learning model (in bits), plus the length of all the corrective values (also in bits). The smaller this value, the better the model is taken to be.

However, a pure MDL approach won't work in our case, since the output from the system is actually a compressed form of the input (so the original input can't be recovered from it). For this reason, [5] suggest simply using the *compression ratio* as a metric. In other words, since our model is compressing the input directly, evaluate how well it carries out this compression.

6   Results

We now present the results of the experiment described in Sect. 5.

Table 1. Some event types discovered and associated precision values

Event name	Precision
new_event_10	0.2857142857142857
new_event_11	0.6666666666666666
new_event_12	0.6666666666666666
new_event_13	0.3333333333333333
new_event_14	0.42857142857142855
new_event_160	0.75
new_event_161	0.6666666666666666
new_event_162	0.3333333333333333
new_event_163	0.6666666666666666
new_event_231	0.6666666666666666
new_event_232	0.6666666666666666
new_event_233	0.6666666666666666
new_event_301	0.7142857142857143
new_event_302	0.6666666666666666
new_event_303	0.3333333333333333
new_event_304	0.25
new_event_305	0.3333333333333333

Our system discovers in excess of 500 event types, reproducing the full result of this evaluation here would not be possible. Nonetheless, we present an extract from these results as Table 1. The results are reasonably good: a little over half of the events discovered correlate to a precision of at least 50%, meaning that the results show a correlation (but not a perfect overlap) between the ground truth and the discovered output. Considering both the differing levels of abstraction, and the large amount of interleaving present in the Kyoto 3 dataset, we feel that this is an acceptable initial result. The large number of events suggests that in the future, more needs to be done to combine the discovered event types; see Sect. 7 for more details.

As mentioned earlier, another important evaluation metric is the compression rate. Our system compresses the original input dataset to about 68% of its original input size. This is a good result, albeit one that we hope to improve upon in the future.

Finally, we have produced a visualisation of our system's output to allow us to see that a hierarchy is being built up, and visualising it as a tree-like structure. Because of the length of the input dataset, this is again far too big to show in this paper. However, we present some extracts from it as Fig. 2. The bars at the bottom of the image are the original ground truth, each row represents a certain activity type, and the bar will be present along the row when the activity is active, but not otherwise. The triangles above it represent the discovered events, with the wide bottoms at the bottom of the triangles compressing into the narrower tops. In some places, the hierarchy is quite deep, as visible in Fig. 2(a). Visually, it is noticeable that clusters tend to form around activities. Our previous evaluation methods had no way of picking up on this phenomenon, which we will discuss further in Sect. 7. The events sometimes cross activity boundaries, but these incursions are small. This could be evidence that the human annotator of this dataset and the system are seeing similar activities, but can't agree when they start or end as discussed above.

7 Discussion

Originally, we hypothesised building an NLM that would not predict a distribution over the next event, but rather over the *next m events*. In other words, for each $d \in \Sigma$, it would output $p_{i:i+m}(d|d_{i-1}, d_{i-2}, \ldots, d_{i-n})$, the probability that d would be one of the next m events observed.

The approach shows promise, as shown by our experiment results. Deep learning for activity discovery is in its infancy, so we do not claim that we can outperform other AD techniques, but this is a starting point.

There are a number of ways that we intend to build on this work in the future. As noted in the previous section, visualising the output shows clusters of new events forming around activities. This seems to suggest that the method finds activities, but these aren't being seen or enlarged by subsequent layers of the hierarchy. This could be an artifact of the dataset, or could be evidence that we need to change how we change the LSTM probability distributions into links.

(a) A deep hierarchy of events discovered by the system.

(b) Although events sometimes cross activity boundaries, the incursions are always small, indicating they could still be part of the activity

Fig. 2. A visualisation of the system output, where the red bars at the bottom correspond to ground truth activities, and the triangles correspond to discovered events that can be understood as *compressing* the original dataset.

It could also turn out that we need to find some way to cluster the discovered event types. This could be done based on temporal proximity, for instance. Clustering rare event types into more common ones might also make learning higher layers easier. We *did* attempt to use a more complex clustering method in the past, but this turned out to perform poorly at discriminating between events from different activities. We aim to investigate and correct this issue as future work.

8   Conclusion and Future Work

This paper presented a novel approach to activity discovery (AD), and tested it on a real-world dataset. We have shown that the approach is viable, and appears to show promise for the task of activity discovery. The system has been evaluated on a number of distinct evaluation metrics, which show it to be robust and suggest that the measured performance is not merely a result of picking a favourable evaluation metric. Finally, we have outlined a number of changes we plan to make in the future to further improve the system, and make it capable of discovering activities even in very complex activities.

References

1. Abadi, M., et al.: TensorFlow: large-scale machine learning on heterogeneous systems (2015). https://www.tensorflow.org/. Software available from tensorflow.org
2. Bengio, Y., Ducharme, R., Vincent, P., Jauvin, C.: A neural probabilistic language model. J. Mach. Learn. Res. **3**(Feb), 1137–1155 (2003)
3. Blei, D.M., Ng, A.Y., Jordan, M.I.: Latent Dirichlet allocation. J. Mach. Learn. Res. **3**(Jan), 993–1022 (2003)
4. Chan, W., Jaitly, N., Le, Q.V., Vinyals, O., Shazeer, N.M.: Speech recognition with attention-based recurrent neural networks, US Patent 9,799,327, 24 October 2017
5. Cook, D.J., Krishnan, N.C.: Activity Learning: Discovering, Recognizing, and Predicting Human Behavior from Sensor Data. Wiley, Hoboken (2015)
6. Cook, D.J., Krishnan, N.C., Rashidi, P.: Activity discovery and activity recognition: a new partnership. IEEE Trans. Cybern. **43**(3), 820–828 (2013)
7. Cook, D.J., Schmitter-Edgecombe, M.: Assessing the quality of activities in a smart environment. Methods Inf. Med. **48**(05), 480–485 (2009)
8. Cramer, B.: Limitations of current grammar induction algorithms. In: Proceedings of the 45th Annual Meeting of the ACL: Student Research Workshop, pp. 43–48. Association for Computational Linguistics (2007)
9. Domingo, C., See, S., Legaspi, R.: Unsupervised habitual activity detection in accelerometer data. In: Billingsley, J., Brett, P. (eds.) Mechatronics and Machine Vision in Practice 3, pp. 253–272. Springer, Cham (2018). https://doi.org/10.1007/978-3-319-76947-9_19
10. Emonet, R., Varadarajan, J., Odobez, J.M.: Multi-camera open space human activity discovery for anomaly detection. In: 2011 8th IEEE International Conference on Advanced Video and Signal Based Surveillance (AVSS), pp. 218–223. IEEE (2011)
11. Fang, L., Ye, J., Dobson, S.: Discovery and recognition of emerging human activities using a hierarchical mixture of directional statistical models. IEEE Trans. Knowl. Data Eng. (2019)

12. Gaspers, J., Cimiano, P., Griffiths, S.S., Wrede, B.: An unsupervised algorithm for the induction of constructions. In: 2011 IEEE International Conference on Development and Learning (ICDL), vol. 2, pp. 1–6. IEEE (2011)
13. Gold, E.: Language identification in the limit. Inf. Control **10**, 447–474 (1967)
14. Jain, S.: An infinite class of functions identifiable using minimal programs in all Kolmogorov numberings. Int. J. Found. Comput. Sci. **6**(1), 89–94 (1967)
15. Hochreiter, S., Schmidhuber, J.: Long short-term memory. Neural Comput. **9**(8), 1735–1780 (1997)
16. Huang, J., Kalbarczyk, Z., Nicol, D.M.: Knowledge discovery from big data for intrusion detection using LDA. In: 2014 IEEE International Congress on Big Data, pp. 760–761. IEEE (2014)
17. Huynh, T., Fritz, M., Schiele, B.: Discovery of activity patterns using topic models. In: UbiComp 2008, pp. 10–19 (2008)
18. Jenks, G.F.: The data model concept in statistical mapping. Int. Yearb. Cartography **7**, 186–190 (1967)
19. LeCun, Y., Bengio, Y., Hinton, G.: Deep learning. Nature **521**(7553), 436 (2015)
20. Merity, S., Keskar, N.S., Socher, R.: An analysis of neural language modeling at multiple scales. arXiv preprint arXiv:1803.08240 (2018)
21. Nevill-Manning, C.G., Witten, I.H.: Identifying hierarchical structure in sequences: a linear-time algorithm. J. Artif. Intell. Res. **7**, 67–82 (1997)
22. Petasis, G., Paliouras, G., Karkaletsis, V., Halatsis, C., Spyropoulos, C.D.: e-grids: computationally efficient gramatical inference from positive examples. Grammars **7**, 69–110 (2004)
23. Rissanen, J.: Modeling by shortest data description. Automatica **14**(5), 465–471 (1978)
24. Rissanen, J.: Stochastic Complexity in Statistical Inquiry. World Scientific, Singapore (1989)
25. Saives, J., Pianon, C., Faraut, G.: Activity discovery and detection of behavioral deviations of an inhabitant from binary sensors. IEEE Trans. Autom. Sci. Eng. **12**(4), 1211–1224 (2015)
26. Shen, Y., Lin, Z., Huang, C.W., Courville, A.: Neural language modeling by jointly learning syntax and lexicon. arXiv preprint arXiv:1711.02013 (2017)
27. Solan, Z., Horn, D., Ruppin, E., Edelman, S.: Unsupervised learning of natural languages. Proc. Natl. Acad. Sci. **102**(33), 11629–11634 (2005)
28. Xu, D., Wu, X., Song, D., Li, N., Chen, Y.L.: Hierarchical activity discovery within spatio-temporal context for video anomaly detection. In: 2013 IEEE International Conference on Image Processing, pp. 3597–3601. IEEE (2013)
29. Yogatama, D., et al.: Memory architectures in recurrent neural network language models (2018)

Achieving Proportional Fairness in WiFi Networks via Bandit Convex Optimization

Golshan Famitafreshi[✉] and Cristina Cano

Universitat Oberta de Catalunya, 08860 Castelldefels, Barcelona, Spain
{gfamitafreshi,ccanobs}@uoc.edu

Abstract. We revisit in this paper proportional fair channel allocation in IEEE 802.11 networks. Instead of following traditional approaches based on explicit solution of the optimization problem or iterative solvers, we investigate the use of a bandit convex optimization algorithm. We propose an algorithm which is able to learn the optimal slot transmission probability only by monitoring the throughput of the network. We have evaluated this algorithm both using the true value of the function to optimize, as well as adding estimation errors coming from a network simulator. By means of the proposed algorithm, we provide extensive experimental results which illustrate the sensitivity of the algorithm to different learning parameters and noisy estimates. We believe this is a practical solution to improve the performance of wireless networks that does not require inferring network parameters.

Keywords: Bandit convex optimization · Proportional fairness · WiFi

1 Introduction

Bandit Convex Optimization (BCO) is a type of Online Convex Optimization (OCO) in which we deal with partial information. In BCO, decisions are made between a player and an adversary repeatedly. In each iteration, the player selects a point from a fixed and known convex set. Then the adversary chooses a convex cost function. At the end of the iteration, the only available feedback for the player is the cost of the function at the selected point. In this framework, the player does not have any knowledge about the specific function nor the gradient [6]. The main emphasis of BCO in the machine learning community has been on rigorous theoretical performance analysis of algorithms. However, practical application of BCO algorithms still requires more attention.

We argue that since many wireless network optimization problems can be easily formulated as convex problems; BCO is appealing for the wireless networking community. Some potential applications of the convex optimization formulation

This research is partially funded by the SPOTS project (RTI2018-095438-A-I00) funded by the Spanish Ministry of Science, Innovation and Universities.

S. Boumerdassi et al. (Eds.): MLN 2019, LNCS 12081, pp. 85–98, 2020.
https://doi.org/10.1007/978-3-030-45778-5_7

are pulse shaping filter design, transmit beamforming, network resource allocation, MMSE precoder design for multi-access communication, robust beamforming and optimal linear decentralized estimation.

BCO has two important advantages in wireless network communication. The first advantage is that the player only needs the cost function feedback of a given action, which facilitates practical implementation. Second, in BCO, the adversary is able to choose among a set of convex functions which can capture network dynamics such as changes in the number of nodes and channel conditions.

The main contribution of this article is an investigation of how to apply a bandit convex optimization algorithm to proportional fair resource allocation in wireless networks. This approach can be implemented by the access point allowing learning of the optimal slot transmission probability only by monitoring the throughput of the network. This research can help academia as well as practitioners to assess whether bandit convex optimization algorithms can be a practical solution for commercial use. The ultimate goal is to bring optimal channel allocation in WiFi to practice by addressing the limitations of traditional approaches that need to be fed with as well as track changes in network parameters (such as packet size and data rate used by each node in the network).

This paper is organized as follows. Section 2 describes the main background: the random back-off operation and proportional fair allocation analysis in WiFi networks. Section 3 describes bandit convex optimization and the proposed algorithm. Section 4 presents the evaluation setup and performance results. Section 5 summarizes the related work in the area of WiFi network proportional fair optimization. Finally, in Sect. 6 some final remarks are given.

2   IEEE 802.11 Background

In this section first, we describe the random back-off operation, then summarize the throughput optimization in WiFi networks based on proportional fairness.

2.1   Random Back-Off Operation

The IEEE 802.11 protocol employs the Distributed Coordination Function (DCF) mechanism to access the channel, which is basically a Carrier Sense Multiple Access/Collision Avoidance (CSMA/CA) method with binary exponential back-off. In DCF when a station wants to send a packet, it monitors the channel. If it senses the channel idle for a distributed inter-frame space (DIFS), it will start a back-off countdown timer. Each time the station starts the back-off procedure it initializes CW to CW_{min} and chooses a random number in $(0, CW - 1)$, where CW is the contention window. As long as the channel is sensed idle for a time slot the back-off timer counter decrements. When a transmission is detected on the channel, this timer freezes and is reactivated when the channel is sensed idle for more than DIFS again.

2.2   Throughput Optimization in WiFi Networks

Throughput in the 802.11 standard depends on the number of active stations and the contention window used by each station. In particular, in multi-rate IEEE 802.11 WLANs, stations that use DCF and transmit at lower transmission rates make use of the channel for longer periods of time to transmit the same amount of data compared to stations using higher transmission rates. This reduces the throughput of high-rate stations in the WLAN, since less time is available for transmission in the shared medium. This effect is known as performance anomaly. One solution to approach this problem is proportional fair allocation [10].

As done in [5] we formulate proportional fairness as a convex optimization problem whose objective is the sum of logs of throughput, which will intrinsically capture fairness while trying to achieve maximum performance even in a multi-rate scenario. We assume that all the stations (n) are saturated, (i.e. stations always have a packet to send) and generate uplink traffic to the network. Note that the slot transmission probability (τ) is the reciprocal of the value of CW. Let $S_i(\tau)$ be the throughput of the station i, then:

$$S_i(\tau) = \frac{P_{\text{succ},i} D_i}{\sigma P_{\text{idle}} + T_c(1 - P_{\text{idle}})}. \tag{1}$$

Here two kinds of time slots are considered. The first one is the PHY idle slot duration without any transmission which is of duration σ. The second one is the busy slot which relates to the duration of a packet transmission [7]. The packet transmission duration is denoted by T_c, which is the mean duration of a successful or collided transmission of node i or other stations' packet transmissions. The successful transmission, includes the MAC ACK which is specified by T_{ack}, T_{fra} which defines the duration of a data transmission, a short inter-frame space (SIFS) which represents an amount of time that channel requires for sending the response frame and a DIFS. T_c can be calculated as follows (for clarity of illustration, we consider this duration equal for successful transmissions and collisions but the analysis can be easily extended to consider both as done in [8]):

$$T_c = T_{\text{fra}} + \text{SIFS} + T_{\text{ack}} + \text{DIFS}. \tag{2}$$

The average packet size of the i_{th} station is defined by D_i in bits. $P_{\text{succ},i}$ is the probability of a successful packet transmission of i_{th} station, i.e., that only station i transmits a packet and is given by $P_{\text{succ},i} = \tau_i \prod_{k=1, k\neq i}^{n}(1 - \tau_k)$. The term P_{idle} is the probability that the channel is idle. When none of the stations attempt to transmit a packet, the probability is defined as $P_{\text{idle}} = \prod_{k=1}^{n}(1 - \tau_k)$. The term $1 - P_{\text{idle}}$ is the probability that the channel is busy due to the successful, unsuccessful (collisions) or other stations' packet transmissions and it is defined by $1 - P_{\text{idle}} = 1 - \prod_{k=1}^{n}(1 - \tau_k)$. Therefore, Eq. 1 is the amount of data transmitted per slot when that is successful over the average duration of a slot.

In the following it will be more useful to use the transformed variable $x_i = \frac{\tau_i}{(1-\tau_i)}$ rather than τ_i, $x_i \in [0, \infty)$ for $\tau_i \in [0, 1)$. The optimization problem can then be formulated as [5]:

$$\text{max.} \quad \sum_{i=1}^{n} \tilde{S}_i(x),$$

$$\text{s.t.} \quad \tilde{S}_i(x) \leq \log\frac{x_i D_i}{X(x)T_c}. \tag{3}$$

The constraint certifies that the sum of logs of throughputs is feasible and sits in the rate region. Since the log-transformed rate region $\tilde{Z}$ is strictly convex, there exists a unique solution that satisfies strong duality and Karush-Kuhn-Tucker (KKT) conditions which implies a global maximum [5].

This problem can be solved explicitly but our aim here is to formulate this problem in a bandit framework so that knowing the networks parameters in Eq. 3 is no longer required. This facilitates practical implementation and can help adoption of optimization approaches in commercial WiFi cards, where the selection of the CW used by the stations is generally static.

3   Bandit Convex Optimization

In Bandit Convex Optimization (BCO) three steps are repeated between the player and the adversary. These three steps can be written for iteration t as follows:

- The player chooses a point $x_t \in K \subseteq R^d$.
- The adversary chooses a cost function $f_t \in F \subseteq R^k$.
- The player observes $f_t(x_t)$.

Here, x_t is a point from a fixed and known convex set. K represents a convex subset of a d-dimensional Euclidean space ($K \subseteq R^d$). In addition, all the functions in F are convex [4].

The aim of BCO algorithms is to achieve low regret:

$$R_T = \sum_{t=1}^{T} f_t(x_t) - \min_{x \in K} \sum_{t=1}^{T} f_t(x). \tag{4}$$

This formulation is known as cumulative regret, which measures the difference between the cumulative loss that is revealed to the player and the best-fixed decision in hindsight after T iterations [6]. To achieve low regret most of the BCO algorithms use Online Gradient Descent (OGD) with estimations of the gradient. In fact, the main complication of BCO is to estimate the gradients of the cost functions. Therefore many researchers in BCO have investigated methods for estimating these gradients and used their results in BCO algorithms [6].

Flaxman et al. [2] proposed a scheme that combined the estimated gradients with the OGD algorithm of Zinkevich [11], who showed that a simple gradient descent strategy for the player incurs a $O(\sqrt{T})$ regret bound [3]. The algorithm of Flaxman et al. uses point evaluations of convex functions to approximately estimate the gradient. The regret bound of this algorithm is shown to be $O(T^{3/4})$.

Agarwal et al. showed that in each round knowing the value of each cost function at two points is almost as useful as knowing the value of each function everywhere, therefore their algorithm has a regret bound of $O(T^{2/3})$ improving the $O(T^{3/4})$ bounds achieved by Flaxman et al. However, Flaxman et al. and Agarwal et al. approaches cannot be used in a practical implementation and realistic setting in wireless networks. First, in many settings it is impossible to query cost functions two times in one iteration. Second, the variance of the single point estimators in the approach of Flaxman et al. [2] is large; consequently, speed of convergence is not practical for wireless stations [4,9].

3.1 Sequential Multi-Point Gradient Estimates in WiFi

We use the multi-point BCO algorithm defined in [4] which considers a simpler assumption than that of Agarwal [3]. This algorithm is called Online Gradient Descent with Sequential Multi-Point Gradient Estimates (OGD-SEMP) and to estimate the gradient queries are combined from two consecutive iterations. The algorithm considers a sequence of auxiliary points $y_1, y_2, \ldots$ which are used to keep track of the player's movement by updating gradient descent as follows:

$$y_{k+1} = \prod_k (y_k - \eta_k \tilde{g}_k). \tag{5}$$

Here, $\tilde{g}_k$ is the gradient estimate which is used to update y_{k+1}. The parameter η_k is the gradient descent step size and $\eta = \{\eta_1, \eta_2, \eta_3, \ldots\}$ is a sequence which shrinks over time. This coefficient defines the speed of convergence of gradient descent to the final value y_k. Figure 1 shows a schematic of this algorithm for the k_{th} iteration. Let's consider y_k as the k_{th} point in a one-dimensional convex set with the interval $[y_k - \epsilon_k \delta_k, y_k + \epsilon_k \delta_k]$. Then, the distance between the beginning and the end of the interval is $2\epsilon_k \delta_k$. Here, ϵ_k is a random number that can be either -1 or 1. δ_k is a parameter which shrinks over time and it captures the distance between the selected point and y_k. In the first step, we choose an arbitrary point and obtain its cost function as:

$$\begin{aligned}\bar{x}_t &= y_k + \epsilon_k \delta_k, \\ g_k{}^+ &= f_t(\bar{x}_t).\end{aligned} \tag{6}$$

In the second step we choose another point in time $(t + 1)$ and obtain its cost function as:

$$\begin{aligned}\bar{x}_{t+1} &= y_k - \epsilon_k \delta_k, \\ g_k{}^- &= f_{t+1}(\bar{x}_{t+1}).\end{aligned} \tag{7}$$

In the following step, the gradient can be estimated as follows:

$$\tilde{g}_k = \frac{g_k{}^+ - g_k{}^-}{2\epsilon_k \delta_k}. \tag{8}$$

Here, the numerator is the subtraction of two cost functions evaluations and the denominator is the distance between the beginning and the end of the interval (see Fig. 1), which corresponds to an unbiased estimator of the gradient.

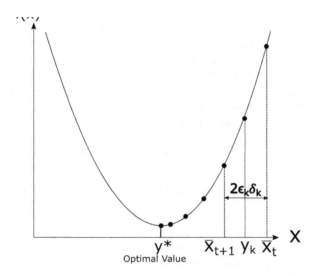

Fig. 1. Sketch of Online Gradient Descent with Sequential Multi-Point Gradient Estimates.

This algorithm was used in [4] for wireless networking optimization. The authors provided the theoretical analysis of this algorithm and evaluated its performance in an unlicensed LTE/WiFi fair coexistence use case. Here we aim to use this method for achieving proportional fair allocation of resources in a IEEE 802.11. Using this method, we only need to know the throughput of each station, regardless of other parameters of Eq. 3, to achieve proportional fairness.

Applied to the WiFi throughput proportional fairness case, the cost function is equal to $f_t = \sum_{i=1}^{n} \tilde{S}_i(x_t)$. Therefore, the cost functions in Eqs. 6 and 7 ($f_t(\bar{x}_t)$, $f_{t+1}(\bar{x}_{t+1})$) define the objective function of our optimization problem at time t and time $t + 1$ respectively, with $\bar{x}_t = \{x_1, x_2, \ldots, x_i\}$ a vector which defines x_i for all the stations at time t,[1] and $\bar{x}_{t+1} = \{x_1, x_2, \ldots, x_i\}$ a vector which defines point x for all the stations at time $t + 1$.

In more detail, consider a repeated game of T rounds. In each round $t = \{1, 2, \ldots, T\}$ the WiFi network selects Eq. 3, which we now denote as f_t with some n, x_i, D_i and T_c. Then, in each round t:

- The WiFi access point chooses $\bar{x}_t$.
- The WiFi network (formed by all WiFi nodes) independently selects $f_t \in F$.
- The WiFi access point observes $f_t(\bar{x}_t)$.

---

[1] Recall that x_i is a transformed variable $x_i = \frac{\tau_i}{(1-\tau_i)}$ rather than τ_i, $x_i \in [0, \infty)$ for $\tau_i \in [0, 1)$. (As seen in Sect. 2).

4   Performance Evaluation

Here we present the evaluation of OGD-SEMP when applied to the WiFi use case by executing an extensive set of simulations in Matlab and in a custom simulator. According to the IEEE 802.11ac standard, we have considered:

$$T_{\text{fra}} = T_{\text{plcp}} + \left\lceil \frac{L_{\text{s}} + n_{\text{agg}}(L_{\text{del}} + L_{\text{mac-h}} + D) + L_{\text{t}}}{n_{\text{sym}}} \right\rceil T_{\text{sym}}, \tag{9}$$

$$T_{\text{ack}} = T_{\text{plcp}} + \left\lceil \frac{L_{\text{s}} + L_{\text{ack}} + L_{\text{t}}}{n_{\text{sym}}} \right\rceil T_{\text{sym}}, \tag{10}$$

with physical and channel access parameters as listed in Table 1.

This custom simulator consists of two main parts, the channel and the node module. The node module includes a network model with physical, MAC, network and application layers. It was previously used for evaluating OGD-SEMP performance in an unlicensed LTE/WiFi fair coexistence use case. In this research we extend the simulator to consider the particularities of the WiFi proportional fair use case gradient descent implementation.

Table 1. Simulation parameters according to IEEE 802.11ac [1].

Parameter	Value
Slot Duration (σ)	9 µs
DIFS	34 µs
SIFS	16 µs
PLCP Preamble + Header Duration (T_{plcp})	40 µs
EIFS	364 µs
TimerACK	314 µs
Propagation Time	1 µs
Tsymbol (T_{sym})	4 µs
PLCP Service Field (L_s)	16 bits
MAC Delimiter Field (L_{del})	32 bits
MAC Header ($L_{\text{mac-h}}$)	288 bits
Tail Bits (L_t)	6 bits
ACK Length (L_{ack})	256 bits
Payload (D)	12000 bits
nsymbol (n_{sym})	1040 bits
number of aggregated packets (n_{agg})	64
MIMO	4

Since the contention window values are discrete while the slot transmission probabilities in our model are continuous, we need to convert discrete values of contention window to the desired continuous one in the simulator. To achieve this we use CW_1 for t_1 seconds and CW_2 for t_2 seconds in a way that the average contention window matches that of the model.[2] Recall that $\tau = 1/CW$. The gradient descent algorithm is executed each T seconds, with $T = c(t_1 + t_2)$, with c a positive integer. The throughput used by the algorithm as feedback is the average throughput during T seconds.

We evaluate the performance of the algorithm using the individual throughput metric (S_t). We use Matlab in order to feed the algorithm with the true values of the individual throughput computed using Eq. 3 for each set of experiments and use the simulator for evaluating the impact of having estimated values instead of the true value of the throughput. Noise in the estimates is caused by the random backoff, collision probabilities and discretization of the slot transmission probability as described above. By comparing the evolution of throughput over time for different settings we evaluate the algorithm's performance regarding the time to convergence.

4.1   Simulation Results

In the simulations first, we have set the gradient descent step size as $\eta_k = \eta/k^{(3/4)}$ and $\delta_k = \omega/h(k)$, with ω as an input parameter and $h(k)$ as some increasing function. We will refer to ω as the exploration parameter and h as the exploration schedule. We vary the number of stations in the network (n) with $n = \{5, 20\}$. Then in order to evaluate the sensitivity of the algorithm to two different exploration schedules, we have changed the exploration schedule to $\eta_k = \eta/k^{(1/2)}$. Note that stations always have a packet to transmit (nodes are saturated) and we consider homogeneous stations (same packet size and transmission probability).

Sensitivity to the Learning Parameters. First, we evaluate the performance of the algorithm by changing the exploration parameter ω -observe that this parameter controls how far from y_k we take the two cost function evaluations at consecutive iterations- and gradient descent step size (η_k). We set $h(k)$ equal to $k^{(3/4)}$ which shrinks the exploration parameter as time goes by.

Figure 2 shows the results of the individual throughput for 5 nodes during 50 iterations. We repeat the same simulations for 30 runs in order to obtain more accurate results. Therefore, in the figures each color represents one simulation run. Optimal results from [5] are shown in Fig. 2 as straight lines. This results are obtained from the implementation of the algorithm in Matlab with cost function computed using Eq. 3 and IEEE 802.11ac parameters from [1]. We show results with different exploration parameter $\omega = \{0.01, 1\}$ and gradient descent step size $\eta = \{0.01, 1\}$.

---

[2] In particular, we set CW_1 and CW_2 to the immediately lower/higher allowed value in IEEE 802.11 and compute t_1 and t_2 accordingly.

Figure 2 shows that by fixing η and increasing parameter ω the rate of convergence increases. As we saw in Eq. 6 increasing the value of exploration parameter (ω), we take bigger steps towards the optimum. By fixing parameter ω and increasing the parameter η for the range of values considered the rate of convergence increases as well since we also make larger steps with bigger gradient descent step sizes. We observe that the increasing trend in the second case is faster than the first case. It can be seen that for exploration parameter equal to $\{0.01, 1\}$ and gradient descent step size equal to 1 the algorithm converges to the optimum value after a few number of iterations (less than 20).

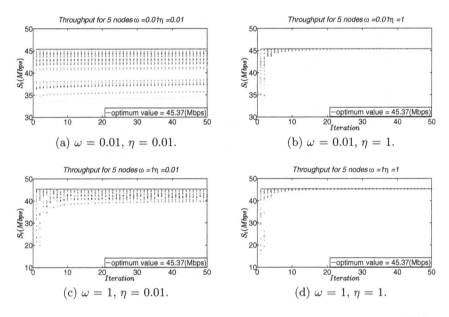

Fig. 2. Individual throughput for $n = 5$ using different ω, η and $h(k) = k^{(3/4)}$.

Here we keep the algorithm setup same as above and increase the number of nodes. Figure 3 illustrates the individual throughput of a network 20 nodes. By comparing Figs. 2 and 3, we observe that for the same value of ω and η by increasing the number of nodes, the algorithm converges to the optimum faster. The reason for this behavior is illustrated in Fig. 4. This figure shows the shape of the objective function for the different number of nodes. As it is shown by increasing the number of the nodes in the network the function becomes steeper, thus the gradients are larger. This means that the algorithm makes larger steps at each iteration and reaches the optimum value faster. Note that the difference between the minimum value of the convex function and its maximum is increased by increasing the number of nodes. For this case the algorithm converges in around 10 iterations or less for $\omega = \{0.01, 1\}$ and $\eta = 1$.

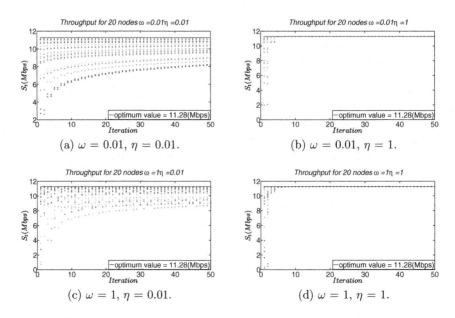

Fig. 3. Individual throughput for $n = 20$ using different ω, η and $h(k) = k^{(3/4)}$.

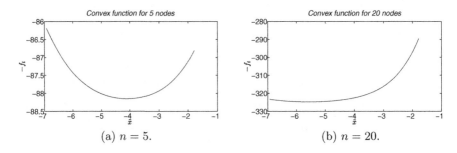

Fig. 4. Shape of the objective functions for $n = 5$ and $n = 20$.

Sensitivity to the Exploration Schedules. In this set of simulations we use the same setup as in the previous subsection but we change the exploration schedule to $h(k) = k^{(1/2)}$. Here we evaluate the sensitivity of the algorithm to two different exploration schedules. Similarly to Figs. 2 and 3, Figs. 5 and 6 show the individual throughput for different values of ω, η and $h(k) = k^{(1/2)}$. We can observe that with $h(k) = k^{(1/2)}$, the convergence speed is almost the same as $h(k) = k^{(3/4)}$. Only for the case $n = 5$, $\omega = 1$ and $\eta = 1$ (Figs. 2(d) and 5(d)), we observe that the convergence speed in Fig. 2(d) with $h(k) = k^{(3/4)}$ is slightly faster than in Fig. 5(d) with $h(k) = k^{(1/2)}$. These results show that the sensitivity of the algorithm to the exploration schedule is negligible for the exploration schedules considered.

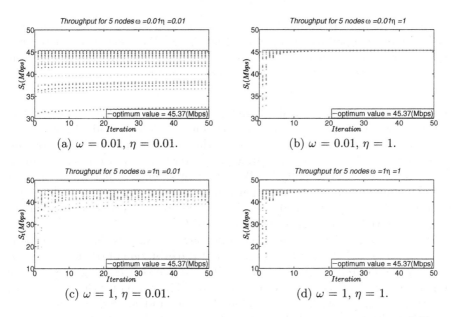

Fig. 5. Individual throughput for $n = 5$ using different ω, η and $h(k) = k^{(1/2)}$.

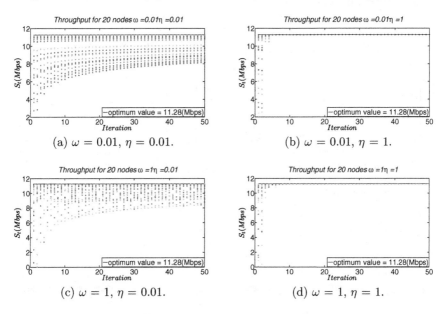

Fig. 6. Individual throughput for $n = 20$ using different ω, η and $h(k) = k^{(1/2)}$.

This means that the impact of $h(k)$ compared to the gradient descent step size η, exploration parameter ω and gradients is low. We also see, as before, that by increasing the number of nodes in the network the individual throughput value converges to its optimum value faster (See Fig. 6).

Sensitivity to Noisy Gradient Estimates. Here we evaluate the performance of the algorithm by having noisy estimates of the individual throughput instead of the true value. In order to achieve this goal we implement the algorithm in the simulator. We set the exploration parameter to $\omega = \{0.01, 1\}$ and gradient descent step size to $\eta = \{0.01, 1\}$. We set the gradient descent timer equal to $T = 100\,\text{s}$ and the value of contention window timer equal to $(t_1 + t_2) = 0.1\,\text{s}$. Each simulation is again run 30 different times in order to achieve more accurate results. Here the exploration schedule is set to $k^{(3/4)}$.

Figure 7 shows that the algorithm still converges in less than 10 iterations for $\omega = 1$ and $\eta = 1$. We see that for $\omega = 0.01$ the evolution of throughput is not following the desired convergence trend. We also see that convergence is worse for smaller values of the exploration parameter. The reason for this behavior is, we argue, the noise: with a small value of ω the gradient estimations are less accurate making more probable for gradient descent to move in the opposite direction of the optimum.

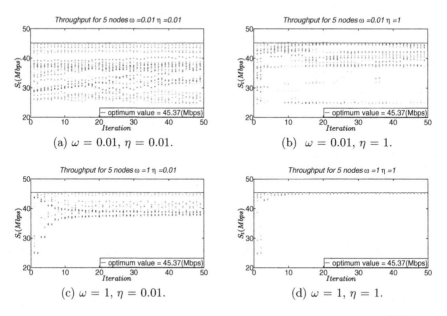

Fig. 7. Individual throughput for $n = 5$ using different ω, η and $h(k) = k^{(3/4)}$ (noisy estimates).

5   Related Work

The most recent works on WiFi network throughput optimization are based on a proportional fairness approach. The reader is referred to those which are presented by Checco et al. [5] and Patras et al. [8], which are similar in nature.

Checco et al. [5] pioneered rigorous analysis of proportional fairness in IEEE 802.11 WLANs. They proved that a unique proportional fair rate allocation exists as the flow total air-time. This algorithm corrects previous works on air-time quantities and uses the IEEE 802.11 rate region as a log-convex. It satisfies per station fairness and per flow fairness. In these approaches [5,8], throughput optimization is achieved by inferring MAC parameters and network metrics such as packet transmission duration, slot transmission probability and average packet size of the stations. These metrics can be estimated but we need to handle estimation errors and network dynamics.

We base our algorithm on these rigorous approaches but without the need to know all parameters of the function to optimize. In this way proportional fairness can be achieved without the need to infer and keep track of network parameters and only by estimating the individual throughput at each station, which can be achieved at the application layer in a commercial access point with minimal changes.

6   Conclusion

The main focus of this article is on achieving proportional fairness in WiFi networks by applying bandit convex optimization. We have applied the OGD-SEMP algorithm based on the BCO algorithm to the WiFi proportional fairness use case. Our results show that, with the appropriate setting of parameters, the algorithm converges to the optimum value in a few number of iterations. However, the parameter of the algorithm that controls the degree of exploration has a significant impact on the algorithm's performance, especially when we are faced with throughput estimation errors. This can be alleviated by increasing the duration of the estimation periods but at the cost of longer convergence times. Other solutions involve using averages of the gradient estimates in different iterations. The evaluation of these approaches to reduce the sensitivity of the algorithm to noise are left as future work. We conclude that the algorithm is a practical solution for wireless network optimization, but that care has to be taken when configuring the algorithm parameters.

References

1. Wireless LAN medium access control (MAC) and physical layer (PHY) specifications. Amendment 4: enhancements for very high throughput for operation in bands below 6 GHz. ANSI/IEEE Standard 802.11ac (2013)

2. Flaxman, A.D., Kalai, A.T., McMahan, H.B.: Online convex optimization in the bandit setting: gradient descent without a gradient. In: Proceedings of the Sixteenth Annual ACM-SIAM Symposium on Discrete Algorithms, SODA 2005, pp. 385–394. Society for Industrial and Applied Mathematics, Philadelphia (2005). http://dl.acm.org/citation.cfm?id=1070432.1070486

3. Agarwal, A., Dekel, O., Xiao, L.: Optimal algorithms for online convex optimization with multi-point bandit feedback. In: COLT, pp. 28–40. Citeseer (2010)

4. Cano, C., Neu, G.: Wireless optimisation via convex bandits: unlicensed LTE/WiFi coexistence. In: Proceedings of the 2018 Workshop on Network Meets AI & ML, NetAI 2018, pp. 41–47. ACM, New York (2018). https://doi.org/10.1145/3229543.3229551

5. Checco, A., Leith, D.J.: Proportional fairness in 802.11 wireless LANs. IEEE Commun. Lett. **15**(8), 807–809 (2011). https://doi.org/10.1109/LCOMM.2011.060811.110502

6. Hazan, E., et al.: Introduction to online convex optimization. Found. Trends® Optim. **2**(3–4), 157–325 (2016)

7. Jiang, L.B., Liew, S.C.: Proportional fairness in wireless LANs and ad hoc networks. In: 2005 IEEE Wireless Communications and Networking Conference, vol. 3, pp. 1551–1556, March 2005. https://doi.org/10.1109/WCNC.2005.1424745

8. Patras, P., Garcia-Saavedra, A., Malone, D., Leith, D.J.: Rigorous and practical proportional-fair allocation for multi-rate Wi-Fi. Ad Hoc Netw. **36**, 21–34 (2016). https://doi.org/10.1016/j.adhoc.2015.06.002. http://www.sciencedirect.com/science/article/pii/S1570870515001262

9. Saha, A., Tewari, A.: Improved regret guarantees for online smooth convex optimization with bandit feedback. In: Proceedings of the Fourteenth International Conference on Artificial Intelligence and Statistics, pp. 636–642 (2011)

10. Siris, V.A., Stamatakis, G.: Optimal CWmin selection for achieving proportional fairness in multi-rate 802.11e WLANs: test-bed implementation and evaluation. In: Proceedings of the 1st International Workshop on Wireless Network Testbeds, Experimental Evaluation & Characterization, WiNTECH 2006, pp. 41–48. ACM, New York (2006). https://doi.org/10.1145/1160987.1160996

11. Zinkevich, M.: Online convex programming and generalized infinitesimal gradient ascent. In: Proceedings of the 20th International Conference on Machine Learning (ICML-2003), pp. 928–936 (2003)

Denoising Adversarial Autoencoder for Obfuscated Traffic Detection and Recovery

Ola Salman[✉], Imad H. Elhajj, Ayman Kayssi, and Ali Chehab

Department of Electrical and Computer Engineering,
American University of Beirut, Beirut 1107 2020, Lebanon
{oms15,ie05,ayman,chehab}@aub.edu.lb

Abstract. Traffic classification is key for managing both QoS and security in the Internet of Things (IoT). However, new traffic obfuscation techniques have been developed to thwart classification. Traffic mutation is one such obfuscation technique, that consists of modifying the flow's statistical characteristics to mislead the traffic classifier. In fact, this same technique can also be used to hide normal traffic characteristics for the sake of privacy. However, the concern is its use by attackers to bypass intrusion detection systems by modifying the attack traffic characteristics. In this paper, we propose an unsupervised Deep Learning (DL)-based model to detect mutated traffic. This model is based on generative DL architectures, namely Autoencoders (AE) and Generative Adversarial Network (GAN). This model consists of a denoising AE to de-anonymize the mutated traffic and a discriminator to detect it. The implementation results show that the traffic can be denoised when different mutation techniques are applied with a reconstruction error less than 10^{-1}. In addition, the detection rate of fake traffic reaches 83.7%.

Keywords: Machine Learning · Network security · Traffic classification · Obfuscation · Deep Learning · IoT · Autoencoder · Generative Adversarial Network

1 Introduction

In the IoT era, billions of things are being connected to the Internet. This results in an unprecedented growth in the amount of generated traffic. This traffic presents different QoS and security challenges. Traffic classification emerges as a key enabling tool to meeting some of these challenges. For example, IoT devices' vulnerabilities have been exploited to perform critical network attacks (e.g. Mirai) [34]. Tracking these devices and detecting abnormal traffic are key to prevent harmful network attacks.

In this context, Machine Learning (ML) based methods were proposed for traffic classification and intrusion detection. Many supervised and unsupervised

© IFIP International Federation for Information Processing 2020
Published by Springer Nature Switzerland AG 2020
S. Boumerdassi et al. (Eds.): MLN 2019, LNCS 12081, pp. 99–116, 2020.
https://doi.org/10.1007/978-3-030-45778-5_8

methods have been employed accordingly. While supervised methods are devoted to classifying the (attack) traffic based on known class labels, the unsupervised ones allow the detection of unknown traffic. The traditional ML methods rely on well-structured and hand-designed features. These features are extracted using statistical traffic measures (e.g. maximum, minimum, standard deviation, etc. of the packet sizes, and packets interarrival times).

However, mutation techniques alter the traffic statistical features, making it very challenging to know the original traffic type. Moreover, the mutation techniques might change the packets characteristics while maintaining the flow statistical features unchanged. In this case, the detection of abnormal traffic can be evaded.

Recently, Deep Learning (DL) has acquired a lot of attention due to its representation learning capabilities. Using a new data representation in this paper, we propose an unsupervised DL model to detect abnormal traffic and de-anonymize the mutated one. Generative DL architectures, namely Autoencoders (AE) and Generative Adversarial Networks (GAN), have been applied mainly in the computer vision domain to detect abnormality and to denoise images. AE is a DL architecture for extracting data representation, and GAN has the capability to generate fake data samples to enhance the discrimination between real and fake data. In this paper, we combine AE and GAN to detect abnormal traffic and de-anonymize the mutated one. The proposed architecture consists of an encoder, a decoder, and a discriminator. The encoder-decoder pair form a denoising AE responsible for learning the original data representation and to denoise the mutated one. In parallel, the discriminator is trained to differentiate between the mutated traffic (abnormal) and the denoised traffic (normal). The training of the proposed model relies on data collected from real IoT devices and IoT attacks. The testing results show the robustness of the proposed method to detect mutated traffic and to recover the original one. Note that the proposed model is not limited to IoT traffic and can be applied to any type of network traffic.

The rest of the paper is organized as follows: in Sect. 2, we present an overview of related concepts. In Sect. 3, we present our proposed model. Section 4 details the implementation and presents the evaluation results. In Sect. 5, we discuss the results and present our future work. Finally, we conclude in Sect. 6.

2   Background and Related Work

In this section, we present the related work including a review of the generative DL architectures, their application in detecting abnormality, traffic classification and intrusion detection methods, along with the obfuscation techniques that can affect their accuracy.

2.1   Unsupervised Deep Learning

In this subsection, we explain the DL architectures that we based our work on. These architectures can learn the input data representation and are therefore called generative DL models.

Generative Adversarial Network: Introduced by Goodfellow et al. in [24], GAN consists of two parts: the generator (G) and the discriminator (D). From a game theoretic perspective, GAN can be interpreted as a zero-sum or min-max game between the generator and the discriminator. The generator tries to learn the input data representation to generate data samples very similar to the real ones. The discriminator tries to maximize the probability of distinguishing between fake and real input. The GAN objective function can be presented as follows: $V(G, D) = E_{x \sim p_{data}(x)}(log(D(x)) + E_{z \sim p_z(z)}(log(1 - D(G(z)))))$ where x is the input data, $p_{data}(x)$ is the data distribution, $D(x)$ is the discriminator output, $p_z(z)$ is the fake data distribution, z is a sample from $p_z(z)$, and $G(z)$ is the generator output. The generator aims at minimizing the probability of fake data detection by the discriminator, which means that the G objective is to find $\min_{G}(E_{z \sim p(z)}(log(1 - D(G(z)))))$. The discriminator aims at maximizing the probability of detecting real data as real and fake data as fake, which means that the D objective is to find $\max_{D}(E_{x \sim p_{data}(x)}(log(D(x))) + E_{z \sim p(z)}(log(1 - D(G(z)))))$. Thus, the GAN objective is to find $\min_{G} \max_{D}(V(G, D))$. Primarily, GAN is applied for synthetic data generation. GAN has been applied also for image anomaly detection [19, 27, 50]. In fact, the adversarial learning permits the discriminator to detect abnormal input data. Furthermore, the generative learning permits the generator to learn the real data representation, which makes GAN suitable for image denoising [40, 47].

Autoencoders: Being a generative model, AE is a type of DL networks that is specialized in extracting the input data representation. The AE consists of two parts: an encoder and a decoder. The encoder extracts a compressed data code by estimating a function f, in such a way $z = f(x)$, where x is the input data, and z is the extracted representation or latent variable. The decoder aims at reconstructing the input data by relying on the extracted representation. In other terms, the decoder tries to infer the inverse function g, in such a way that $y = g(f(x)) = g(z)$. The AE objective function can be represented at the minimization of the difference between $g(f(x))$ and x. In other terms, the AE aims at minimizing the reconstruction error.

A type of AEs is the probabilistic autoencoder, which aims to infer the distribution of x, $p_\theta(x)$ by means of another distribution $q_\phi(z/x)$ (probability of the latent variable z knowing the input x). A well-known type of probabilistic AEs is the Variational Autoencoder (VAE), which imposes a prior restriction on $p(z)$ to be a normal distribution. In this case, the problem reduces to maximizing the Evidence Lower Bound (ELBO) or maximizing the Kullback–Leibler (KL) divergence between $q_\phi(z)$ and $p_{\theta(z/x)}$, represented by $KL(q_\phi(z/x)\|p(z))$. Having the ability to extract the real data distribution, VAEs have been applied for anomaly detection. Indeed, when the reconstruction error is large, anomaly is detected in the input data. Borrowing the adversarial concept from GAN, Adversarial AEs (AAE) were introduced by Makhzani et al. in [30]. Similar to the GAN, AAE includes a discriminator that tries to differentiate between the data sampled from the latent variable prior $p(z)$ and the real data. In this case, the discriminator aims to minimize $L_{dis} = -1/N \sum_{i=0}^{N-1} log(d_x(z_i)) + \sum_{j=N}^{2N} log(1 - d_x(z_j))$, and the generator tries to minimize $L_{prior} = 1/N \sum_{i=0}^{N-1} (log(1 - d_x(z_i))$, where N is the number of samples, and $d_x(z_i)$ is the discriminator output of the latent space variable. In this case, if the total loss function is optimized, $q_\phi(z/x)$ will be very similar to $p(z)$, or in other terms, $KL(q_\phi(z/x)||p(z))$ will be minimized, and thus the log likelihood of the original data distribution will be maximized. AAEs were applied also for anomaly detection in images [5,23,44,46]. Furthermore, a recent work has considered to add the denoising function to the AAEs for image denoising [17]. In this case, the corrupted data $\tilde{x}$ is considered as input and two methods were proposed for model representation. The first consists of matching $\tilde{q}_\phi(z/x)$ to $p(z)$, and the second consists of matching $q_\phi(z/\tilde{x})$ to $p(z)$. However, in our work, we use a sparse AE that aims to minimize the Mean Square Error (MSE) between the reconstructed data and original data. In addition, applying the adversarial concept, we choose to train a discriminator to detect abnormal traffic when the reconstruction error is high. Thus, unlike previous work, the generator part of GAN is omitted [13].

2.2   ML Based Traffic Classification and Intrusion Detection

Traffic classification is an essential network function, which is necessary for traffic engineering, QoS management, and security management. Different methods have been proposed for traffic classification using different sets of features [8,9,18,22,32,33]. More recently, DL has been applied for traffic classification using new features and new data representations [14,21,36,39,45].

On the other hand, intrusion detection is an essential network security component. Several approaches have been adopted to detect and prevent network attacks. Traditional Intrusion Detection Systems (IDSes) are rule-based, where the attack signature is known by identifying some patterns in the packet's fields. However, this method fails to detect unknown attacks (e.g. zero-day attacks) and requires the inspection of the packet header. In addition, traffic anonymization can be used to thwart detection by traditional IDS. ML methods have been proposed to detect network attacks and traffic abnormality by means of statistical features [6]. However, the traditional methods require the extraction of the features by computing statistical measures that might be data dependent. In addition, the detection of abnormal traffic is not a straightforward task, given that the abnormal traffic might present similar statistical behavior to the normal one [4].

Recently, DL has been applied in the network domain for intrusion detection [3,43]. More specifically, IoT, which presents aggravated security challenges, has acquired a special attention from the intrusion detection perspective [10,16]. Convolutional Neural Network (CNN) and Recurrent neural Network (RNN) have been applied for payload-based attack detection in [29]. While supervised learning was mainly considered for identifying specific attacks [48], the detection of unknown and zero-day attacks call for unsupervised-learning-based methods. In fact, the proposed unsupervised [31] or semi-supervised [41] DL methods for intrusion detection use statistical features. To the best of our knowledge, no previous work considered the recovering (denoising) of the mutated traffic. In addition, this is the first work to consider the question of detection of mutated traffic, in addition to the (unknown) attack traffic.

2.3 Traffic Obfuscation Techniques

In the aim of protecting user privacy, a new research direction is considering traffic obfuscation to thwart classification. In this context, many obfuscation techniques have been proposed [15,25]. These methods can be classified in seven categories: steganography, tunneling, anonymization, mutation, morphing, and physical layer obfuscation. While steganography and physical layer obfuscation techniques require specific protocols to recover the original data, the remaining techniques can be applied without imposing changes to the current network protocols. Anonymization and tunneling hide some packet-related information by encryption and establishment of virtual connections (port numbers and internet addresses). However, statistical ML methods still have power to classify the anonymized or tunneled traffic. Mutation and morphing are two techniques that consider modifying the statistical traffic characteristics considering the modification of the packet size and the packet Inter-Arrival Times (IAT) [11]. This has great impact on the accuracy of ML-based classification and intrusion detection. Even though the obfuscation techniques were intended to protect the user privacy, attackers might use them to perform their attacks without being detected. In this context, padding and traffic shaping are proposed to modify the packet size and IAT respectively [7,35].

Recently, the GAN DL architecture has been employed to adapt malware traffic to the normal one [28,37,42,51]. On the contrary, in this paper, we will consider a discriminative denoising DL model to detect abnormality by relying on the trained discriminator and to de-anonymize mutated traffic by means of a denoising AE.

Our contributions in this paper can thus be summarized by the following points:

- The discriminative part of GAN with a denoising AE are combined to allow *mutated traffic detection* and recovery.
- A *new data representation* is used to permit the de-anonymization of mutated traffic by applying DL-based techniques.

3 Proposed Abnormal Traffic Detection

In this section, we detail our proposed DL model, the attack model, and the traffic representation method.

3.1   Attack Model

Our attack model consists of an attacker trying to modify the packet size (padding) and IAT (shaping), in such a way to hide any information that serves for attack detection or traffic classification. The mutations are therefore of two types, padding and shaping [28, 29], as shown in Fig. 1. In the following, we summarized the packet padding techniques listed in [7,12,35], with s being the packet original size and m(s) being the packet size after mutation.

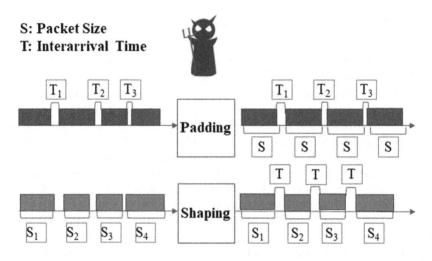

Fig. 1. Attack model

1. **Padding to the Maximum Transmission Unit (MTU):** this technique consists of padding all the flow packets to the same size, which is the MTU. In this case, the mutation can be expressed by $m(s) = MTU$.
2. **Linear padding:** this technique consists of linearly padding the flow packets sizes. In this case, the mutation equation can be expressed as follows: $m(s) = \lceil s/c \rceil * c$, where c is a parameter to choose, and $\lceil s/c \rceil$ is the ceiling of s/c.
3. **Exponential padding:** this technique consists of padding the packet size in an exponential manner. The mutation can be expressed by the following equation: $m(s) = min(2^{log_2(s)}, MTU)$.
4. **Elephants and mice padding:** this technique consists of padding the packet to a certain size c (mice), if the original packet size if less than c. If not, the packet size is padded to the MTU (elephant).
5. **Random padding:** this technique consists of padding the packet randomly to a size chosen randomly from the interval $([s, MTU])$. In this case, the mutation function can be expressed as following: $m(s) = RAND([s, MTU])$.

6. **Probabilistic padding:** this technique assumes that the packet sizes follow a normal distribution. In this case, the mutated size is chosen randomly based on a normal distribution, where the mean (μ) and standard deviation (σ) are computed considering the original traffic packet sizes. In this case, $m(s) = GAUSS(\mu, \sigma)$ [12].

   For the IAT shaping techniques, we list in the following the ones included in [49] in addition to a new technique proposed in [12]:

7. **Constant IAT:** this technique consists of sending the packets at a fixed IAT.

8. **Variable IAT:** this technique consists of sending the packets at a variable interval of time randomly chosen from the interval $[I_1, I_2]$.

9. **Probabilistic IAT shaping:** this technique assumes that the IAT follows a normal distribution. Thus, the packets are transmitted at an interval chosen randomly based on a normal distribution, where the mean and standard deviation are computed based on the original traffic packets IAT.

The last method, described in [20], combines shaping and padding.

10. **Fixed size and fixed IAT:** this method consists of padding the size of all the flow packets to the MTU and sending all the packets at a fixed time interval.

This model considers two types of adversaries, including:

– **Malicious adversary:** this type of attackers aims at mutating the attack traffic to evade intrusion detection.
– **Benign adversary:** this type of attackers aims at hiding the traffic characteristics to protect the user privacy.

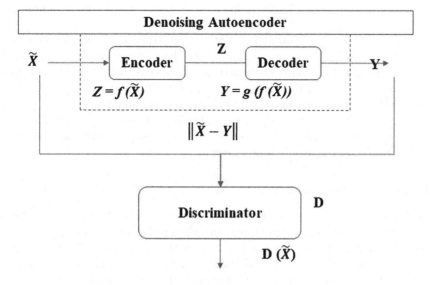

Fig. 2. Proposed model

3.2 Deep Learning Model

Our proposed model, illustrated in Fig. 2, consists of two parts: a denoising AE and a discriminator. The denoising AE consists of an encoder and a decoder. The encoder aims at estimating the function f, that maps the input space of X to the latent space of the latent variable Z, where Z is a compressed representation of X. The denoising AE is fed with a dataset containing the mutated version $\tilde{x}$ of the initial data X. The AE aims at minimizing the reconstruction error $L(X, \tilde{X})$. In our case, the loss function is the MSE function, $L(X, \tilde{X}) = (X - g(f(\tilde{X})))^2$, where X is the original data and $g(f(\tilde{X}))$ is the reconstructed data by considering the mutated data as input. The discriminator aims at differentiating between normal and abnormal traffic. In fact, the discriminator will be trained on classifying the mutated data (i.e. the mutated input data) as abnormal and the reconstructed data (i.e. the decoder output) as normal. The output function of the discriminator is a sigmoid function $p(y = y^{(j)}/x) = 1/(1 + e^{-y^{(j)}})$, where $y^{(j)}$ is the output class that can take the two values: 0 for $j = 0$ and 1 for $j = 1$, and the loss function is a cross entropy function $L_D = 1/m \sum_{j=0}^{1} y^{(j)} log(y^{(j)}) + (1 - y^{(j)}) log(1 - y^{(j)}))$.

3.3 Proposed Model Workflow

After training the model, the proposed scheme workflow, illustrated in Fig. 3, consists of: (1) passing the traffic to the discriminator; (2) if the traffic is normal, it is passed to the classifier of normal traffic; (3) If not, it is passed to the denoiser. (4) After denoising, it is passed again to the discriminator. (5) If a normal traffic is detected, it is passed to the classifier; (6) if not, the traffic is detected as abnormal, so it might be obfuscated or attack traffic. The same workflow is repeated for the abnormal traffic to know the attack type or detect an unknown attack traffic.

In fact, two cases are considered. First, when training the model on normal traffic, the aim is to detect attack traffic and the mutated one as abnormal. In this case, the denoising aims at recovering the normal traffic for classification. However, when training the model with attack traffic, at the testing phase the aim is to detect unknown attacks and to denoise the mutated attack traffic to know the exact attack type. Note that the classifier will be trained in a supervised mode to classify the traffic based on the IoT device type in the normal traffic case and based on the attack type in the attack traffic case.

3.4 Data Representation

In our case, the data consists of collected network traffic. This traffic is filtered by flows, where a flow is defined as being the set of packets having the same: source IP address, source port number, destination IP address, destination port number, and protocol (TCP or UDP). For each packet, we extract three features: size (s), IAT (t), and direction (d). For each flow, we consider the first $4 * 4$ packets in either direction. In fact, as shown in Fig. 3, the extracted features

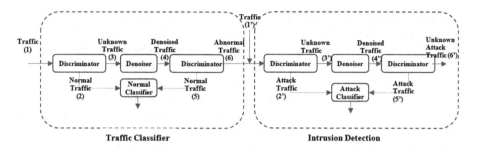

Fig. 3. Proposed scheme workflow

can be visualized in $4 * 4$ RGB images, where the ith pixel RGB coordinates are represented by the ith packet features $[s, t, d]$. Thus, every feature represents a color channel. In our case, R is given for the size, G is given for the interarrival time, and B is given for the direction. This data representation is explained in detail in [38] (Fig. 4).

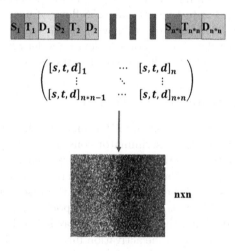

Fig. 4. Data representation

4   Implementation

In this section, we detail the data collection and preprocessing, the model implementation, and the evaluation results.

4.1   Data Collection and Preprocessing

For collecting normal IoT traffic, we used a set of IoT devices, including Wi-Fi-enabled devices and hub-connected devices. The Wi-Fi devices are: D-Link HD

180-Degree Wi-Fi Camera DCS-8200LH, D-Link Wi-Fi Smart Plug DSP-W215, D-Link Wi-Fi Siren DCH-S220, and D-Link Wi-Fi Water Sensor DCH-S160, while the hub-connected devices are components of the Samsung SmartThings Home Monitoring Kit, including a motion sensor, a multi-purpose sensor, and a smart plug. These devices were installed in a home environment and were left to function normally. The Wi-Fi enabled devices are routed through a laptop to the Internet. A bridge is created at this laptop to forward the incoming traffic to the Ethernet interface connected to the home gateway. For the hub-connected devices, the hub is connected to the laptop by Ethernet and this laptop is configured to forward the incoming traffic to the wireless interface connected to the home gateway. In both cases, the traffic is collected at the laptop. One day of traffic for each device was considered for training and one day of traffic was used for testing.

For IoT attack traffic training data, we consider the dataset collected in [26], this dataset consists of multiple PCAP files for each type of attack. We choose one file for each type of attack for training and one file for testing.

To preprocess the data and extract the flows, the dpkt Python library was used [1]. In total, for training, we have 10320 flows of normal traffic and 3308 flows of attack traffic. For testing, we have 258 flows of normal traffic, 350 flows of attack traffic, and 2000 flows of (unknown) attack traffic. The normal traffic is categorized into five classes based on the device model while the attack traffic is categorized into three classes: data theft, denial of service and scanning.

4.2   Model Implementation and Training

The proposed model was implemented in Python using the tensorflow library [2]. The encoder, decoder and the discriminator consist of a 2-layer fully connected neural network with 1000 neurons each. The output layer of the decoder and the discriminator is a sigmoid layer, with the difference that the decoder output is of the same dimension of the input; however, the discriminator output is of dimension one. The model is trained with 100 epochs and a batch size of 100, the learning rate is 10^{-3}, and the momentum decay is 0.9 (beta1). The Rectified Linear Unit (ReLU) was used as the activation function for all hidden layers, and the weights are optimized using the Adam optimizer. For generating mutated data, we implement the mutation techniques listed in section III-B ((1) →(10)). The training data is randomly mutated, where each data sample is uniformly mutated to one of the 10 mutation techniques.

For testing the effectiveness of the denoising process on the classification accuracy, we implement a CNN classifier with three convolutional layers, two max pooling layers, and two fully connected layers with one dropout layer with 50% dropout probability. Similarly, ReLU is applied for activation in the hidden layers, and the Adam algorithm is used for optimization with a learning rate of 10^{-3}. In addition, we apply cross validation for the classifier training with 4 folds. The performance metric used to evaluate the classifier is the accuracy, which is the ratio of the correctly classified samples to the total number of samples.

4.3   Evaluation and Results

For each of the experiments (1) →(10), the corresponding mutation technique is applied to the testing data. First, the traffic is passed to the discriminator. Then, it is passed to the denoising AE. Furthermore, the mutated and denoised traffic are passed to a classifier, that classifies the flow based on the device model for normal traffic and based on the attack type for attack traffic.

Samples of the mutated traffic and resulted images after denoising are included in Fig. 5. It is visually clear from the included images that the denoiser succeeds in learning the original data representation disregarding the mutation level. The difference between the denoised images and the original ones shows that the model does not overfit the data, however it learns the representation and the noise pattern.

For the normal traffic, the denoiser succeeds in recovering flows very similar to the original ones for most of the mutation techniques, except for the mutations (8) and (9). Similarly, for the attack traffic, the denoiser recovers the main flow representation, except for the mutation (2). In Table 1, the MSE between the original data and the mutated one (mutation degradation), the reconstruction loss, and the abnormality detection rate are reported for the normal traffic and the attack traffic.

Table 1. Testing evaluation results

Mutation technique	Autoencoder loss		Mutation loss		Discrimination rate	
	Normal	Attack	Normal	Attack	Normal	Attack
(1)	0.042	0.0357	0.2713	0.1698	100%	100%
(2)	0.03218	0.1569	0.0171	0.082	100%	100%
(3)	0.0149	0.1569	0.0009	0.0065	88.99%	83.13%
(4)	0.0543	0.0463	0.2704	0.1696	50%	49.94%
(5)	0.067	0.0519	0.0171	0.082	100%	100%
(6)	0.0621	0.0703	0.0129	0.0411	49%	49.88%
(7)	0.0378	0.1569	0.2974	0.3245	100%	100%
(8)	0.0595	0.0257	0.09878	0.1016	99.02%	100%
(9)	0.1728	0.0303	0.2825	0.0012	50%	41.59%
(10)	0.0369	0.0519	0.5687	0.4943	100%	100%

The results present a high detection rate for the different mutation techniques, except for the mutation techniques (4), (6), and (9). This can be explained by the fact that (6) and (9) are normal distribution based mutations. In these cases, the main traffic characteristics will remain unchanged and this will harden the differentiation between the original and mutated traffic. (4) is a mice-elephant mutation of the packet sizes. However, in our case (i.e. IoT traffic), most of the packets are of small size and therefore the mutation (4) will

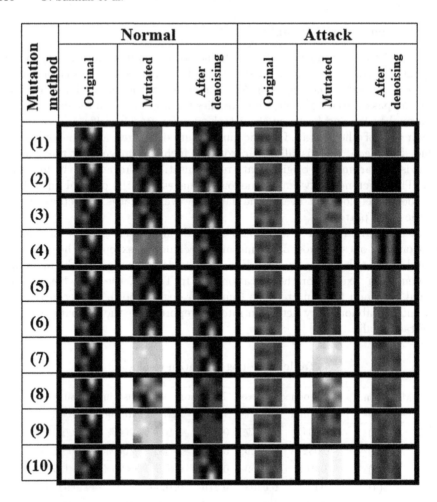

Fig. 5. Visualized images representing network traffic

have limited affect on the traffic characteristics. Moreover, it can be noticed that the MSE between the recovered data and the original one (reconstruction loss) is lower than the MSE between the mutated data and the original one (mutation degradation) in most of the cases. This means that the denoising process decreases the degradation effect by reconstructing a version of the data that is closer to the original one.

Table 2 presents the accuracy of the classification based on the traffic label: device model for normal traffic and attack type for attack traffic. The mutation techniques (1), (4), (7), and (10) affect the accuracy and misleads the classification noticeably in the normal traffic case; however, after denoising, the accuracy increases. However, for the techniques (5), (6), (8), and (9), the denoiser fails to reconstruct the original traffic. This is due to the fact that the randomness will create a denoised traffic of random type. Moreover, for the mutation technique (2), the mutation is linear and this does not affect the CNN classifier accuracy, being immune to the linear degradation of the image. However, for statistical based machine learning methods, the mutation techniques (2), (5), and (6) affect noticeably the classification accuracy. To see the effect of the mutation on the statistical based classification, we include the results of the mutated and denoised traffic statistical based classification in Table 2. The statistical classification uses a subset of the Moore features (see Table 3) and Random Forest (RF) as classifier. It can be noticed that overall our representation method with CNN classifier outperforms the RF method before and after denoising in the normal traffic case. However, in the attack traffic case, our method gives better results after denoising.

Table 2. Classification results

Mutation technique	Before denoising				After denoising			
	Normal		Attack		Normal		Attack	
	CNN	RF	CNN	RF	CNN	RF	CNN	RF
(1)	30.14%	54.47%	51.55%	**80.84%**	**65.19%**	38.78%	53.1%	37.05%
(2)	**93.13%**	25%	**36.27%**	36.24%	79.16%	68.68%	33.41%	33.41%
(3)	**99.01%**	87.19%	76.25%	**90.33%**	87.99%	78.79%	33.41%	33.41%
(4)	29.9%	55.2%	51.55%	**80.1%**	**67.4%**	40.68%	43.79%	42.57%
(5)	**93.13%**	25.06%	36.99%	36.21%	20.09%	20.09%	**63.12%**	38.21%
(6)	**95.09%**	25.06%	56.55%	**62.94%**	56.12%	33.88%	36.99%	42.69%
(7)	75.49%	**85.6%**	33.17%	19.63%	77.69%	55.39%	**33.41%**	**33.41%**
(8)	78.18%	**81.92%**	33.15%	15.96%	62.99%	41.85%	**62.05%**	38.48%
(9)	80.14%	**82.35%**	34.24%	39.64%	20.09%	20.09%	**62.76%**	30.31%
(10)	19.6%	28.18%	33.17%	25.65%	**78.43%**	49.14%	**60.6%**	43.22%

5 Discussion and Future Work

The results show that unsupervised DL architectures are very powerful in learning the data representation. AE is a well-known DL generative model, that is highly effective in extracting compressed representation from image-type data. Consequently, after representing network traffic as images, we applied AE to extract the representation patterns from IoT traffic. Moreover, the capability of AE to denoise images is used to overcome the mutation technique challenges. The results show the effectiveness of the proposed method to recover the original traffic representation for different levels of mutation, some of which are rather severe (e.g. fixed packet size and IAT). In fact, the considered mutation techniques cause any statistical classifier to fail. However, one limitation of this work is that we assume that we know ahead of time the mutation technique used by the attacker, so we can decide to choose to apply the classifier before or after

Table 3. Feature set for statistical classification

Feature	Description
total_fwd_pkt	Total packets in the forward direction
total_fwd_bytes	Total bytes in the forward direction
total_bck_pkt	Total packets in the backward direction
total_bck_bytes	Total bytes in the backward direction
min_pckt_size_fwd	The min packet size in the forward direction
mean_pckt_size_fwd	The mean packet size in the forward direction
max_pckt_size_fwd	The max packet size in the forward direction
std_pckt_size_fwd	The standard deviation packet size in the forward direction
min_pckt_size_bck	The min packet size in the backward direction
mean_pckt_size_bck	The mean size of packets in the backward direction
max_pckt_size_bck	The max packet size in the backward direction
std_pckt_size_bck	The standard deviation packet size in the backward direction
total_size	The total flow size
min_pckt_size	The min packet size in either direction
mean_pckt_size	The mean size of packets in either direction
max_pckt_size	The max packet size in either direction
std_pckt_size	The standard deviation packet size in either direction
min_iat_fwd	The minimum interarrival time in the forward direction
mean_iat_fwd	The mean interarrival time in the forward direction
max_iat_fwd	The maximum interarrival time in the forward direction
std_iat_fwd	The standard deviation interarrival time in the forward direction
min_iat_bck	The minimum interarrival time in the backward direction
mean_iat_bck	The mean interarrival time in the backward direction
max_iat_bck	The maximum interarrival time in the backward direction
std_iat_bck	The standard deviation interarrival time in the backward direction
total_time	The duration of the flow
min_iat	The minimum interarrival time in either direction
mean_iat	The mean interarrival time in either direction
max_iat	The maximum interarrival time in either direction
std_iat	The standard deviation interarrival time in either direction
avg_pckt_fwd	The average number of packets in the forward direction
avg_bytes_fwd	The average number of bytes in the forward direction
avg_pckt_bck	The average number of packets in the backward direction
avg_bytes_bck	The average number of bytes in the backward direction
avg_iat_fwd	The proportion of flow time in the forward direction to the total flow time
avg_iat_bck	The proportion of flow time in the backward direction to the total flow time

denoising. While we considered mutation techniques in this paper, we will study as future work additional obfuscation techniques. In addition to the abnormal traffic detection using the discriminative part of the GAN architecture, in the morphing case for example, we need to consider the generative part also. The generator will be trained to generate morphed traffic similar to the original one. In this case, the model convergence will not be a straightforward task.

6   Conclusion

Applying ML for network management and network security has gained a lot of interest in the last decade. However, new traffic obfuscation techniques have been developed to thwart classification and avoid detection in the sake of user privacy. These techniques have been employed by attackers to mount their attacks without being detected. While the obfuscation techniques might be detected by behavioral or statistical ML techniques, the recovery of the initial traffic is unfeasible. In this paper, inspired from a promising DL application, which is image denoising, we transform the traffic to images then combine two well-established DL architectures (AE and GAN) to reconstruct the original traffic and detect abnormal one. The test results show the effectiveness and robustness of the proposed model to detect abnormal traffic in all its variants.

Acknowledgments. Research funded by the AUB University Research Board, the Lebanese National Council for Scientific Research, and TELUS Corp., Canada.

References

1. dpkt. https://dpkt.readthedocs.io/en/latest/. Accessed 2019
2. tensorflow. https://www.tensorflow.org/. Accessed 2019
3. Alom, M.Z., Taha, T.M.: Network intrusion detection for cyber security using unsupervised deep learning approaches. In: 2017 IEEE National Aerospace and Electronics Conference (NAECON), pp. 63–69. IEEE (2017)
4. Baddar, S.A.H., Merlo, A., Migliardi, M.: Behavioral-anomaly detection in forensics analysis. IEEE Secur. Privacy **17**(1), 55–62 (2019)
5. Beggel, L., Pfeiffer, M., Bischl, B.: Robust anomaly detection in images using adversarial autoencoders. arXiv preprint arXiv:1901.06355 (2019)
6. Buczak, A.L., Guven, E.: A survey of data mining and machine learning methods for cyber security intrusion detection. IEEE Commun. Surv. Tutorials **18**(2), 1153–1176 (2015)
7. Qu, B., Zhang, Z., Guo, L., Zhu, X., Guo, L., Meng, D.: An empirical study of morphing on network traffic classification. In: 7th International Conference on Communications and Networking in China, pp. 227–232, August 2012. https://doi.org/10.1109/ChinaCom.2012.6417481
8. Callado, A.C., et al.: A survey on internet traffic identification. IEEE Commun. Surv. Tutorials **11**(3), 37–52 (2009)
9. Cao, Z., Xiong, G., Zhao, Y., Li, Z., Guo, L.: A survey on encrypted traffic classification. In: Batten, L., Li, G., Niu, W., Warren, M. (eds.) ATIS 2014. CCIS, vol. 490, pp. 73–81. Springer, Heidelberg (2014). https://doi.org/10.1007/978-3-662-45670-5_8

10. Chaabouni, N., Mosbah, M., Zemmari, A., Sauvignac, C., Faruki, P.: Network intrusion detection for IoT security based on learning techniques. IEEE Commun. Surv. Tutorials **21**, 2671–2701 (2019)
11. Chaddad, L., Chehab, A., Elhajj, I.H., Kayssi, A.: App traffic mutation: toward defending against mobile statistical traffic analysis. In: IEEE INFO-COM 2018 - IEEE Conference on Computer Communications Workshops (INFO-COM WKSHPS), pp. 27–32, April 2018. https://doi.org/10.1109/INFCOMW.2018.8406899
12. Chaddad, L., Chehab, A., Elhajj, I.H., Kayssi, A.: Mobile traffic anonymization through probabilistic distribution. In: 2019 22nd Conference on Innovation in Clouds, Internet and Networks and Workshops (ICIN), pp. 242–248, February 2019. https://doi.org/10.1109/ICIN.2019.8685871
13. Chalapathy, R., Chawla, S.: Deep learning for anomaly detection: A survey. arXiv preprint arXiv:1901.03407 (2019)
14. Chen, Z., He, K., Li, J., Geng, Y.: Seq2Img: a sequence-to-image based approach towards IP traffic classification using convolutional neural networks. In: 2017 IEEE International Conference on Big Data (Big Data), pp. 1271–1276. IEEE (2017)
15. Cheng, T., Lin, Y., Lai, Y., Lin, P.: Evasion techniques: sneaking through your intrusion detection/prevention systems. IEEE Commun. Surv. Tutorials **14**(4), 1011–1020 (2012). https://doi.org/10.1109/SURV.2011.092311.00082
16. da Costa, K.A., Papa, J.P., Lisboa, C.O., Munoz, R., de Albuquerque, V.H.C.: Internet of things: a survey on machine learning-based intrusion detection approaches. Comput. Netw. **151**, 147–157 (2019)
17. Creswell, A., Bharath, A.A.: Denoising adversarial autoencoders. IEEE Trans. Neural Netw. Learn. Syst. **30**(4), 968–984 (2018)
18. Dainotti, A., Pescape, A., Claffy, K.C.: Issues and future directions in traffic classification. IEEE Netw. **26**(1), 35–40 (2012)
19. Deecke, L., Vandermeulen, R., Ruff, L., Mandt, S., Kloft, M.: Image anomaly detection with generative adversarial networks. In: Berlingerio, M., Bonchi, F., Gärtner, T., Hurley, N., Ifrim, G. (eds.) ECML PKDD 2018. LNCS (LNAI), vol. 11051, pp. 3–17. Springer, Cham (2019). https://doi.org/10.1007/978-3-030-10925-7_1
20. Dyer, K.P., Coull, S.E., Ristenpart, T., Shrimpton, T.: Peek-a-boo, i still see you: why efficient traffic analysis countermeasures fail. In: Proceedings of the 2012 IEEE Symposium on Security and Privacy, SP 2012, pp. 332–346. IEEE Computer Society, Washington, DC (2012). https://doi.org/10.1109/SP.2012.28
21. Fadlullah, Z.M., et al.: State-of-the-art deep learning: evolving machine intelligence toward tomorrow's intelligent network traffic control systems. IEEE Commun. Surv. Tutorials **19**(4), 2432–2455 (2017)
22. Finsterbusch, M., Richter, C., Rocha, E., Muller, J.A., Hanssgen, K.: A survey of payload-based traffic classification approaches. IEEE Commun. Surv. Tutorials **16**(2), 1135–1156 (2013)
23. Ger, S., Klabjan, D.: Autoencoders and generative adversarial networks for anomaly detection for sequences. arXiv preprint arXiv:1901.02514 (2019)
24. Goodfellow, I., et al.: Generative adversarial nets. In: Advances in Neural Information Processing Systems, pp. 2672–2680 (2014)
25. Khraisat, A., Gondal, I., Vamplew, P., Kamruzzaman, J.: Survey of intrusion detection systems: techniques, datasets and challenges. Cybersecurity **2**(1), 20 (2019). https://doi.org/10.1186/s42400-019-0038-7

26. Koroniotis, N., Moustafa, N., Sitnikova, E., Turnbull, B.: Towards the development of realistic botnet dataset in the internet of things for network forensic analytics: Bot-IoT dataset. Future Gen. Comput. Syst. **100**, 779–796 (2019). https://doi.org/10.1016/j.future.2019.05.041, http://www.sciencedirect.com/science/article/pii/S0167739X18327687

27. Li, D., Chen, D., Goh, J., Ng, S.K.: Anomaly detection with generative adversarial networks for multivariate time series. arXiv preprint arXiv:1809.04758 (2018)

28. Lin, Z., Shi, Y., Xue, Z.: IDSGAN: generative adversarial networks for attack generation against intrusion detection. CoRR abs/1809.02077 (2018). http://arxiv.org/abs/1809.02077

29. Liu, H., Lang, B., Liu, M., Yan, H.: CNN and RNN based payload classification methods for attack detection. Knowl.-Based Syst. **163**, 332–341 (2019)

30. Makhzani, A., Shlens, J., Jaitly, N., Goodfellow, I., Frey, B.: Adversarial autoencoders. arXiv preprint arXiv:1511.05644 (2015)

31. Munir, M., Siddiqui, S.A., Dengel, A., Ahmed, S.: DeepAnT: a deep learning approach for unsupervised anomaly detection in time series. IEEE Access **7**, 1991–2005 (2018)

32. Nguyen, T.T., Armitage, G.: A survey of techniques for internet traffic classification using machine learning. IEEE Commun. Surv. Tutorials **10**(4), 56–76 (2008)

33. Pacheco, F., Exposito, E., Gineste, M., Baudoin, C., Aguilar, J.: Towards the deployment of machine learning solutions in network traffic classification: a systematic survey. IEEE Commun. Surv. Tutorials **21**(2), 1988–2014 (2018)

34. Perrone, G., Vecchio, M., Pecori, R., Giaffreda, R.: The day after mirai: a survey on MQTT security solutions after the largest cyber-attack carried out through an army of IoT devices. In: IoTBDS, pp. 246–253 (2017)

35. Qu, B., Zhang, Z., Zhu, X., Meng, D.: An empirical study of morphing on behavior-based network traffic classification. Secur. Commun. Netw. **8**(1), 68–79 (2015). https://doi.org/10.1002/sec.755

36. Rezaei, S., Liu, X.: Deep learning for encrypted traffic classification: an overview. IEEE Commun. Mag. **57**(5), 76–81 (2019)

37. Rigaki, M., Garcia, S.: Bringing a GAN to a knife-fight: adapting malware communication to avoid detection. In: 2018 IEEE Security and Privacy Workshops (SPW), pp. 70–75, May 2018. https://doi.org/10.1109/SPW.2018.00019

38. Salman, O., Elhajj, I.H., Chehab, A., Kayssi, A.: A multi-level internet traffic classifier using deep learning. In: 2018 9th International Conference on the Network of the Future (NOF), pp. 68–75, November 2018. https://doi.org/10.1109/NOF.2018.8598055

39. Smit, D., Millar, K., Page, C., Cheng, A., Chew, H.G., Lim, C.C.: Looking deeper: using deep learning to identify internet communications traffic. In: 2017 Australasian Conference of Undergraduate Research (ACUR) (2017)

40. Tripathi, S., Lipton, Z.C., Nguyen, T.Q.: Correction by projection: denoising images with generative adversarial networks. arXiv preprint arXiv:1803.04477 (2018)

41. Umer, M.F., Sher, M., Bi, Y.: A two-stage flow-based intrusion detection model for next-generation networks. PLoS One **13**(1), e0180945 (2018)

42. Verma, G., Ciftcioglu, E., Sheatsley, R., Chan, K., Scott, L.: Network traffic obfuscation: an adversarial machine learning approach. In: MILCOM 2018–2018 IEEE Military Communications Conference (MILCOM), pp. 1–6, October 2018. https://doi.org/10.1109/MILCOM.2018.8599680

43. Vinayakumar, R., Alazab, M., Soman, K., Poornachandran, P., Al-Nemrat, A., Venkatraman, S.: Deep learning approach for intelligent intrusion detection system. IEEE Access **7**, 41525–41550 (2019)
44. Vu, H.S., Ueta, D., Hashimoto, K., Maeno, K., Pranata, S., Shen, S.M.: Anomaly detection with adversarial dual autoencoders. arXiv preprint arXiv:1902.06924 (2019)
45. Wang, W., Zhu, M., Wang, J., Zeng, X., Yang, Z.: End-to-end encrypted traffic classification with one-dimensional convolution neural networks. In: 2017 IEEE International Conference on Intelligence and Security Informatics (ISI), pp. 43–48. IEEE (2017)
46. Wang, X., Du, Y., Lin, S., Cui, P., Yang, Y.: Self-adversarial variational autoencoder with Gaussian anomaly prior distribution for anomaly detection. arXiv preprint arXiv:1903.00904 (2019)
47. Warde-Farley, D., Bengio, Y.: Improving generative adversarial networks with denoising feature matching (2016)
48. Xiao, Y., Xing, C., Zhang, T., Zhao, Z.: An intrusion detection model based on feature reduction and convolutional neural networks. IEEE Access **7**, 42210–42219 (2019)
49. Fu, X., Graham, B., Bettati, R., Zhao, W.: On effectiveness of link padding for statistical traffic analysis attacks. In: 2003 Proceedings of 23rd International Conference on Distributed Computing Systems, pp. 340–347, May 2003. https://doi.org/10.1109/ICDCS.2003.1203483
50. Zenati, H., Foo, C.S., Lecouat, B., Manek, G., Chandrasekhar, V.R.: Efficient GAN-based anomaly detection. arXiv preprint arXiv:1802.06222 (2018)
51. Zhang, H., Yu, X., Ren, P., Luo, C., Min, G.: Deep adversarial learning in intrusion detection: A data augmentation enhanced framework. CoRR abs/1901.07949 (2019). http://arxiv.org/abs/1901.07949

Root Cause Analysis of Reduced Accessibility in 4G Networks

Diogo Ferreira[1] , Carlos Senna[2](✉) , Paulo Salvador[1] , Luís Cortesão[3],
Cristina Pires[3], Rui Pedro[3], and Susana Sargento[1,2]

[1] DETI, University of Aveiro, Aveiro, Portugal
{diogodanielsoaresferreira,salvador,susana}@ua.pt
[2] Instituto de Telecomunicações, Aveiro, Portugal
cr.senna@av.it.pt
[3] Altice Labs, Aveiro, Portugal
{luis-a-cortesao,cristina-j-pires,rui-d-pedro}@alticelabs.com

Abstract. The increased programmability of communication networks makes them more autonomous, and with the ability to actuate fast in response to users and networks' events. However, it is usually a difficult task to understand the root cause of the network problems, so that autonomous actuation can be provided in advance.

This paper analyzes the probable root causes of reduced accessibility in 4G networks, taking into account the information of important Key Performance Indicators (KPIs), and considering their evolution in previous time-frames. This approach resorts to interpretable machine learning models to measure the importance of each KPI in the decrease of the network accessibility in a posterior time-frame.

The results show that the main root causes of reduced accessibility in the network are related with the number of failure handovers, the number of phone calls and text messages in the network, the overall download volume and the availability of the cells. However, the main causes of reduced accessibility in each cell are more related to the number of users in each cell and its download volume produced. The results also show the number of PCA components required for a good prediction, as well as the best machine learning approach for this specific use case.

Keywords: Cellular networks · Root cause analysis · Machine learning

1 Introduction

In communication networks, root cause analysis of network problems or failures is essential, so that a fast reaction to these failures, or even an anticipation and prevention of these failures can take place. However, usually it is difficult to assess the cause of reduced network accessibility, since it may happen due to a large number of issues, and impacting in a large number of metrics simultaneously. Knowing the causes that lead to such events can help to prematurely detect

© IFIP International Federation for Information Processing 2020
Published by Springer Nature Switzerland AG 2020
S. Boumerdassi et al. (Eds.): MLN 2019, LNCS 12081, pp. 117–133, 2020.
https://doi.org/10.1007/978-3-030-45778-5_9

them. Besides, when such events do happen, it is easier to know where and how to autonomously actuate in the network to mitigate the failure.

With the increased requirements proposed for the 5G networks, e.g. 1 000 000 devices per km^2, 20 Gbit/s of download peak data rate [4], the new generation of cellular networks promises to handle more traffic than ever before. The incorporation of network slicing, as well as Software-Defined Networking (SDN) and Network Function Virtualization (NFVs) in the 5G architecture, overly increases the management complexity of those networks. With so many metrics to monitor, it is becoming harder to detect the cause of an event due to the complex combinations of various Key Performance Indicators (KPIs). Traditional approaches to detect the root cause of failures, with a knowledge base and a set of rules, is becoming obsolete due to the flexibility of the network. With the advances in machine learning, it is easier to indirectly analyze dependent variables with reduced complexity, but with increased uncertainty.

This work identifies the KPIs that may cause reduced accessibility in 4G networks, using machine learning techniques. Knowing those KPIs helps to create a proactive management of the network, detecting an eventual future drop in network accessibility and having the possibility to avoid it by acting on the network, adjusting the resources that have the most impact on those KPIs. In this work, two different approaches for root cause analysis using machine learning techniques are explained and discussed. The first approach measures the feature importance using internal calculations in the model to determine the importance of each KPI in a reduced accessibility event. However, due to the high number of combinations of the KPIs, it is not feasible to test all possibilities. It is then important to perform feature selection. The second approach proposes a dimensionality reduction algorithm to reduce the number of features and apply the machine learning algorithms. In the evaluation results we present the most important KPIs that are able to predict if the number of E-UTRAN Radio Access Bearer (E-RAB) setup failures is above a specific threshold. Then, we present the most important KPIs that are able to predict if the number of E-RAB establishment failures has high variations, and therefore, are highly correlated to the reduction of the network accessibility.

The remaining of the paper is organized as follows. Section 2 addresses the relevant related work. Section 3 discusses how KPIs impact the network accessibility. Section 4 presents the proposed approaches for root cause analysis, while Sect. 5 discusses the results of the approach to root cause detection. Finally, Sect. 6 presents the conclusion and the future works.

2   Related Work

Understanding the root cause of an observed symptom in a complex system has been a major problem for decades. The main question is how to find appropriate real-time mechanisms to determine root causes [10]. Most root cause analysis approaches for network operators is currently based on Bayesian Networks [7]. They have the capability of representing network metrics and events in nodes, and their relations represent the dependencies, along with a conditional probability. To obtain the most probable cause of an error, a probabilistic inference can be done. In [1,2], the authors argue that a Bayesian network is not suitable for large-scale systems with a large number of components, because the complexity of inference increases exponentially with the number of nodes and dependencies between them. To solve that, they combine the Bayesian network with case-based reasoning techniques to prune the nodes needed to analyze in the network. The results show that the technique used reduces drastically the inference time, as well as the need for human intervention.

In [11], a generic framework for root cause analysis in large IP networks was proposed. To determine the root cause of events, two reasoning engines are included: Bayesian inference and rule-based reasoning. The authors discuss that rule-based logic is often preferred over Bayesian inference, because it is easier to configure, it has an easier interpretation of results and it is effective in most applications. However, Bayesian networks are preferred when the root cause is unobservable (no direct evidence can be collected). In our work, the root cause of reduced accessibility is mostly unobservable.

The proposed solution in [3] computes what are parameters that are most relevant across all different types of failure modes, and use them to build a Bayesian Network to model the cause-effect relationship between the degradation parameters (cause) and failure modes (effect) that occur on the field. Two real life field issues are used as examples to demonstrate the accuracy of the network once it is modelled and built. This paper shows that accurately modelling the hardware system as a Bayesian Network substantially accelerates the process of root cause analysis.

A self-healing method based on network data analysis is proposed to diagnose problems in future 5G radio access networks [8]. The proposed system analyzes the temporal evolution of a plurality of metrics and searches for potential interdependence under the presence of faults. The work in [6] is the only one, that we are aware, that uses the concept of "variable importance" ("feature importance" in our work) to measure how much a feature contributes to predicting an objective variable on a machine learning model. The influence of each variable is then represented in an influence matrix, that represents the influence that each variable has for each event. However, only the Random Forest algorithm is used to create a model. The variable importance is then calculated using the permutation feature importance approach.

3   Problem Statement

The objective of this work is to understand what are the most important KPIs to anticipate reduced accessibility in a 4G network. A 4G network is chosen, since it is a running network with real data. The data available from 5G networks, because they are still not widely implemented, is scarce and can be unbiased by the users (only users with newer phones with 5G available can use it).

In this work, a 31-day data set with 4G network KPIs was used. The interval period between each measurement of a KPI is 1 h, which means that there are

24 values for each KPI for each cell every day. A metric must be chosen to represent reduced accessibility. After that, with time-series analysis, it is possible to calculate the importance of each KPI to the reduced accessibility metric.

The metric used to indicate low accessibility in the network is the number of E-UTRAN Radio Access Bearer (E-RAB) setup failures per hour in the network. The E-RAB setup in a 4G network is a major KPI for accessibility. The E-RAB is a bearer that the User Agents (UEs) need to establish communications in the network. Figure 1 shows the E-RAB setup phase. After the UE has established a connection with the E-UTRAN Node B (eNB), it is needed to setup a context with the Mobility Management Entity (MME), to enable the UE to communicate and send data to the network.

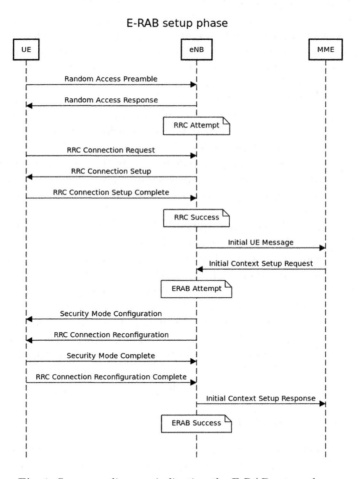

Fig. 1. Sequence diagram indicating the E-RAB setup phase.

When the MME sends a context setup request, it is called an E-RAB setup attempt. After some configuration messages between the UE and the eNB, the context setup response from the eNB to the MME is called an E-RAB setup success. There are more E-RAB setup attempts than E-RAB setup successes. When the network is congested, the difference between the E-RAB setup attempts and the E-RAB setup successes is higher, because some messages after the E-RAB setup attempt are lost due to network problems, such as congestion.

A new accessibility metric is used to measure the accessibility of the network: the number of E-RAB setup failures. If the number of E-RAB setup failures has a high value, the network is congested. The E-RAB setup failure formula is presented in Eq. 1, where esf is the number of E-RAB setup failures, esa is the number of E-RAB setup attempts and ess is the number of E-RAB setup successes.

$$esf(t) = esa(t) - ess(t) \qquad (1)$$

The E-RAB setup failure is depicted in Fig. 2. The biggest congestion happened on the 17th day. The objective of this paper is now to understand which KPIs most contribute to the forecast of this metric.

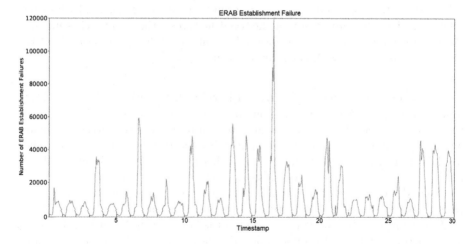

Fig. 2. Number of E-RAB setup failures.

4 Methodology for Root Cause Detection

In this section we describe two different approaches for root cause analysis using machine learning techniques. One approach, that measures feature importance using internal calculations in the model, is used to determine the importance of each KPI in a reduced accessibility event. Our second approach proposes a dimensionality reduction algorithm to decrease the number of features (**feature selection**).

4.1 Approaches to Measure the Feature Importance

A simple test to check if any KPI can accurately forecast the number of E-RAB setup failures is to calculate the Pearson correlation coefficient[1] between the number of E-RAB setup failures and all other KPIs. Since the goal is to understand which KPIs can forecast low accessibility before the number of E-RAB setup failures increases, the KPI values will be shifted (lagged) behind the number of E-RAB setup failures by one hour (the values are sampled hourly; one hour is the minimum time interval).

There are two disadvantages of analyzing the KPIs' importance with the Pearson correlation coefficient. For each KPI, it is only measured its linear contribution to the number of E-RAB setup failures. Besides, combinations of KPIs that can be important are not being taken into account, because it is assumed that the KPIs are independent of each other.

There are other approaches to measure the importance of input features to an output value that takes into account these considerations. These approaches take advantage of machine learning techniques, and they can be divided into two categories: some approaches take into account the error of the model in a test set to calculate the importance of the input features; other approaches measure the feature importance with an internal calculation (algorithm-specific) of the coefficients associated with each input feature.

Approaches Considering the Model Error: One of the approaches that fall into the first category is the drop column feature importance. In this approach, the importance of a feature is measured by comparing the test error of a model when all features are available as input, with the test error of a model when one feature is dropped for training. The higher the error for the model with one feature dropped, the more importance is given to that feature.

A big disadvantage of this approach is that, for each feature, it is needed to train a new model with that feature dropped, which causes the approach to be inefficient for many features, or for models that take a significant training time. In the permutation feature importance approach [5], a similar approach is used without the need to re-train the model for each feature. Instead of dropping the feature, it is applied random shuffling to the test values of that feature among the various examples, to preserve the distribution of that variable. If the model error is almost unchanged, the feature is not much important for the forecast. However, if the model error increases much, it is a sign that the feature is important for the forecast.

Both previous approaches have some advantages. It is possible to apply them to black-box models, given that the feature importance is measured by the model error. They also take into account all interactions between features, which is an advantage when compared with the correlation tests.

---

[1] https://www.statisticssolutions.com/pearsons-correlation-coefficient/.

In the permutation feature importance approach, a disadvantage is that the results are dependent on the shuffling of the features. If the tests are repeated, the results may vary.

Approaches Considering Feature Importance Using Internal Calculations: The approaches that measure the feature importance by inspecting the internals of the models are algorithm-dependent. For some algorithms, like neural networks or Support Vector Machines (SVMs), it is impossible to calculate the importance of each feature, due to the non-linear transformations applied. However, for other algorithms such as Logistic Regression[2], Extra Trees, Random Forest[3], Gradient Boosting[4] or AdaBoost[5], it is possible to estimate the importance of each feature. For linear regression, the importance of each feature can be measured by the absolute value of the coefficient associated with each input (if all features are within the same scale). For the other four tree-based ensemble algorithms (Extra Trees, Random Forest, Gradient Boosting and AdaBoost), the feature importance can be calculated with resort to the mean decrease impurity (or Gini impurity[6]). The importance of a node j in a decision tree is computed as described in Eq. 2, where w_j is the weighted number of samples in node j, C_j is the impurity of this node, and $left(j)$ and $right(j)$ are the respective children nodes.

$$ni_j = w_j C_j - w_{left(j)} C_{left(j)} - w_{right(j)} C_{right(j)} \qquad (2)$$

The feature importance of feature i across all nodes is computed as described in Eq. 3[7].

$$Fi_i = \frac{\sum_{j:\text{node } j \text{ splits on feature } i} ni_j}{\sum_{j \in \text{all nodes}} ni_j} \qquad (3)$$

For this approach, it is important that the features are all normalized within the same scale, and it is recommendable that they are from the same type (continuous/categorical) for better importance estimation.

The advantage of the approaches that measure the feature importance by inspecting the internals of the models is that they do not depend on the test set, only on the model. If the model is accurate and it is not underfitting or overfitting, the feature importance can be calculated more reliably than the previous methods. Otherwise, the feature importance will be highly biased. The biggest challenge is to create accurate models that do not overfit the train set.

---

[2] https://www.statisticssolutions.com/what-is-logistic-regression/.

[3] https://towardsdatascience.com/an-intuitive-explanation-of-random-forest-and-extra-trees-classifiers-8507ac21d54b.

[4] https://towardsdatascience.com/understanding-gradient-boosting-machines-9be756fe76ab.

[5] https://towardsdatascience.com/understanding-adaboost-2f94f22d5bfe.

[6] https://victorzhou.com/blog/gini-impurity/.

[7] https://stats.stackexchange.com/questions/311488/summing-feature-importance-in-scikit-learn-for-a-set-of-features/.

This second category of approaches (measure feature importance with an internal calculation algorithm-specific) will be used to measure the importance of the KPIs.

4.2 Feature Selection

If all KPIs are included as features, the performance of the model will be degraded, because some features are uncorrelated with the output and do not contribute to the output classification. It is then important to perform feature selection. The ideal scenario is to test all combinations of features in the input, and determine the best features by the test error. However, it is not feasible to test all combinations, due to their high number.

The approach chosen was to use a dimensionality reduction algorithm to reduce the number of features. Principal Component Analysis [9] will be used as the algorithm to perform dimensionality reduction. The five algorithms chosen to train the model and to calculate the feature importance were the following: Logistic Regression, Extra Trees, Random Forest, Gradient Boosting and AdaBoost. The feature importance of each model will calculate the importance of each PCA component. Each component importance is then multiplied by the PCA coefficients, to get the KPI importance for each component. Finally, all the importances of the same KPI are added, as described in Eq. 4, where fi_j is the feature importance of the j PCA component and $pca\_coefficient(j, i)$ is the PCA coefficient i for the component j, where i is the number of a KPI and j is a PCA component.

$$kpi\_i_i = \sum_{j=0}^{n\_pca\_components} fi_j * pca\_coefficient(j, i) \qquad (4)$$

To get the best possible model, the number of PCA components cannot be too small, because relevant features can be lost; however, the number cannot be too big either, with the risk of creating overfitted models. The number of PCA components tested will vary from 1 to the number of KPIs, for the five algorithms. The model with a lower test error will be used to calculate the KPI importance.

4.3 Defining the Input and the Output of the Algorithms

The input values will be based on the KPIs. There will be two types of input values: normalized values and normalized variations. For each KPI, the normalized value of the previous hour will be used as input. The number of lags could be increased besides one hour, but in this test it is considered that the low accessibility indicators appear at most one hour before the congestion. For each KPI, the normalized variation will be calculated according to the Eq. 5. The normalized variations are added as features because the network accessibility can depend, not only on the previous values, but also on sudden variations of other KPIs.

$$normalized\_variation(t) = \frac{value(t-1) - value(t-2)}{value(t-2)} \qquad (5)$$

Two types of classification problems will be used to calculate the importance of the KPIs. It is important to understand which KPIs are more important when forecasting the possibility of a low accessibility event in a network. It is also important to understand which KPIs are more important to forecast sudden increases and decreases in the network accessibility. For each case, a different output value will be calculated. For the first case, it will be set a threshold in the 90th percentile of the data, with all the values above the threshold being classified as one, and all other values as zero, turning the problem into a binary classification problem. For the second case, the output will be classified as one if the absolute difference between two consecutive values is higher than the 90th percentile of the data, and zero otherwise. Doing that, it is possible to analyze which KPIs are most important for classifying low accessibility events and also for forecasting bigger increases and decreases in data, which can be important for resource allocation.

The two classification problems will be applied in two different ways, according to the data split. First, the tests will be done with aggregated data. The network KPIs will be added and will be taken into account in the whole network. In this way, it is possible to forecast low accessibility of the network. Another way of performing the tests is to split the data per cell. In this approach, the data will not be aggregated, and 75% of all cells in the network will be used to train, with the other 25% cells to test. The threshold for the tasks will be defined using the specific values of each cell (each cell will have a different threshold, based on its values). With this approach, it is built a model that is capable of detecting low accessibility per cell. This case is expected to perform worse than the aggregated network, because forecasting low accessibility per cell is harder than forecasting low accessibility in the network. However, for a network operator it is very important to forecast low accessibility per cell for various reasons. For a cell in a crowded region, it can be made management adjustments to avoid the low accessibility of the cell, such as the installation of another temporary cell, or the allocation of resources for that cell. Besides, it can be made a time-series analysis about the future accessibility of the cells in a region, for expansion purposes.

4.4   Performance Metric

For both classification problems, a performance metric must be used. Since in the problems previously described the number of positive samples is lower than the number of negative samples, the accuracy performance metric would give similar cost to the false negative and false positive errors. The performance metric used will be the F1-score (Eq. 6). The F1-score is the harmonic mean of precision (Eq. 7) and recall (Eq. 8), and it encompasses the False Negative and False Positive errors, weighted according to the number of samples of each class.

$$F_1 = 2 * \frac{precision * recall}{precision + recall} \tag{6}$$

$$precision = \frac{true\_positives}{true\_positives/false\_positives} \quad (7)$$

$$recall = \frac{true\_positives}{true\_positives/false\_negatives} \quad (8)$$

5   Root Cause Detection Results

Traffic monitoring is essential to provide a good quality do the services in communication networks. Network conditions such as bandwidth, packet loss, delay and jitter are important for traffic engineering to track the quality needs of applications. Therefore, it is important not only to measure the network conditions, but also to analyze them to understand the causes of reduced accessibility in cellular networks. In the following, we show how machine learning techniques can help to understand which network KPIs can indicate a low accessibility event that will happen in the future.

5.1   Aggregated Network Tests

In this subsection, the results for the aggregated network tests will be presented. As explained before, two scenarios are proposed. First, it will be presented (i) the most important KPIs for predicting if the number of E-RAB setup failures is above a threshold. Second, it will be presented (ii) the most important KPIs for predicting if the number of E-RAB establishment failures has high variations.

In Fig. 3 it is shown that the best model for scenario (i) is the Extra Trees algorithm, achieving an F1-score of 86.6%, with 23 PCA components. With a higher number of PCA components, the performance of most of the algorithms starts to deteriorate.

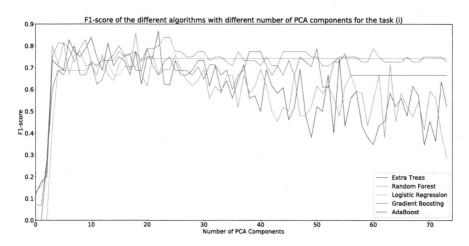

Fig. 3. F1-score with different algorithms varying the number of PCA components for the task (i) in the aggregated network.

Table 1. Ten KPIs that were considered more important for the scenario (i) in the aggregated network.

KPI	Importance
Handover interfreq failure	0.194
Handover intrafreq failure	0.188
CSFB Prep Success	0.181
PDCP Download TX Time	0.169
Variation Handover interfreq failure	0.150
Cell availability	0.131
ERAB Normal Release	0.124
Handover intrafrequency success	0.121
Handover intrafrequency attempt	0.121
PDCP Upload Volume (Mb)	0.115

Table 1 shows the ten KPIs that are considered more important for predicting if the number of E-RAB setup failures is above a threshold (i). The handover failure is considered to be the most important KPI, both inter- and intra-frequency. The third KPI is the Circuit Switched FallBack (CSFB) preparation success. The CSFB is a technology to create circuit-switched calls over a 4G network that does not support LTE voice call standard (VoLTE), which has to fall back on the 3G network. This indicates that the number of phone calls and SMS messages has a high impact on the network accessibility. The Packet Data Convergence Protocol (PDCP) download transmission time, the variation of the inter-frequency handovers and the cell available time are also important KPIs to predict variations in the number of E-RAB setup failures. Finally, the ERAB normal release, handovers intrafrequency success and attempt, and upload traffic volume have also a significant importance.

The best model for scenario (ii) is achieved with the AdaBoost algorithm, achieving an F1-score of 40.0%, with 24 PCA components. The F1-score is lower than in scenario (i) because the task of predicting variations is harder than the one of predicting if the value is above a threshold. The number of PCA components is almost the same as in scenario (i); however in Fig. 4 it can be seen that the algorithms' F1-score is very unstable, and that it is hard to build a model to predict accurately the variations of the number of E-RAB establishment failures in the aggregated network.

Table 2 shows the ten KPIs that are considered more important for predicting if the number of E-RAB establishment failures has high variations (ii). The handover failures are still important, but they are not the most important KPI. In this case, the CSFB preparation success is the most important KPI to predict the variations in the number of E-RAB setup failures. Similar to the task (i),

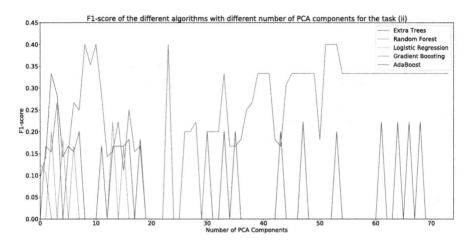

Fig. 4. F1-score with different algorithms varying the number of PCA components for the task (ii) in the aggregated network.

Table 2. Ten KPIs that were considered more important for the scenario (ii) in the aggregated network.

KPI	Importance
CSFB Prep Success	0.188
Handover intrafreq failure	0.170
PDCP Download TX Time	0.157
Handover interfreq failure	0.148
Variation Handover interfreq failure	0.141
Download Active Subscribers (Max)	0.138
Active subscribers (Max)	0.125
RRC Setup Failure	0.121
Cell availability	0.120
Upload active subscribers (Max)	0.119

PDCP download transmission time, the variation of the inter-frequency handovers, the number of active subscribers, download and upload ones, and the cell available time are also important KPIs.

5.2   Individual Cells Tests

In this subsection, the results for the individual cells will be presented. In this case, 75% of the cells in the network will be used for training and the other

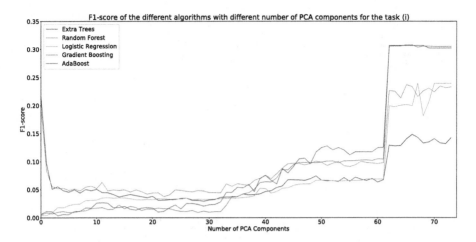

Fig. 5. F1-score with different algorithms varying the number of PCA components for the task (i) in the individual cells tests.

Table 3. Ten KPIs that were considered more important for the scenario (i) in the individual cells tests.

KPI	Importance
RRC Setup Success Rate	0.560
ERAB Setup Success Rate	0.547
Connected subscribers (Max)	0.069
Connected subscribers (Avg)	0.066
Connected active subscribers (Avg)	0.043
Variation Connected Subscribers (Max)	0.034
Variation Connected active Subscribers (Max)	0.032
ERAB normal release	0.031
Variation Radio Bearers (Avg)	0.031
Variation Connected subscribers (Avg)	0.031

25% for testing. Just like in the previous subsection, first it will be presented (i) the most important KPIs for predicting if the number of E-RAB establishment failures is above a threshold, and then it will be presented (ii) the most important KPIs for predicting if the number of E-RAB setup failures has high variations.

The best model for scenario (i) achieved an F1-score of 30.79%, using the Gradient Boosting algorithm, with 66 PCA components. For the cell prediction, more information is needed to obtain the best model when compared with the previous subsection. Figure 5 shows the F1-score varying with the number of components and the different algorithms. In the PCA component 62 there is an increase in the F1-score of all algorithms.

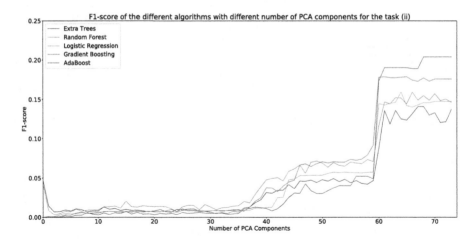

Fig. 6. F1-score with different algorithms varying the number of PCA components for the task (ii) in the individual cells tests.

Table 4. Ten KPIs that were considered more important for the scenario (ii) in the individual cells tests.

KPI	Importance
ERAB Setup Success Rate	0.147
RRC Setup Success Rate	0.147
Variation Cell availablility	0.138
Connected Subscribers (Avg)	0.102
Radio Bearers (Avg)	0.095
Variation RRC Setup Success Rate	0.093
Variation ERAB Setup Success Rate	0.092
ERAB Normal Release	0.088
Connected subscribers (Max)	0.087
Variation Radio Bearers (Avg)	0.081

Table 3 shows the ten KPIs that are considered more important for the task (i). Two KPIs have much more importance than all others: Radio Resource Control (RRC) setup success ratio ($\frac{\text{RRC setup success}}{\text{RRC setup attempt}}$) and E-RAB setup success ratio ($\frac{\text{ERAB setup success}}{\text{ERAB setup attempt}}$), since they are directly correlated with the network accessibility.

The best model for scenario (ii) has an F1-score of 20.39%, using the AdaBoost algorithm with 68 PCA components. The F1-score for different PCA components is similar to the scenario (i), where the F1-score improved its performance significantly after 60 PCA components (Fig. 6).

Table 4 shows the ten KPIs that are considered more important for the scenario (ii). Like in the previous scenario, RRC setup success ratio and E-RAB setup success rate are the most important KPIs for the output. In this task, other KPIs have also similar importance, such as the variation of the cell availability, the average connected subscribers and the average number of radio bearers. A radio bearer is a connection between the eNB and the UE at layer 2, and defines the communication configurations for upper layers.

5.3   Discussion

Aggregated Network Tests: From the results of both aggregated network tests, it can be concluded that most KPIs that cause the number of E-RAB setup failures in a network to be above a threshold are the same that cause it to have high variations. Those KPIs are the number of failed handovers (intra- and inter-frequency), the CSFB preparation success (number of phone calls and SMSs in the network), the PDCP download volume and the cell availability. The maximum number of active subscribers (downloading subscribers and overall subscribers) also causes the number of E-RAB setup failures to vary.

Interpreting the KPIs, the results achieved are according to the intuition about lower network accessibility. When the number of failure handovers is high, the cells are crowded with user sessions and cannot accept any more sessions, which leads to lower network accessibility in the next hour. The high number of CSFB preparation success shows that there is a clear relationship between the high number of phone calls and SMS messages in the network with its lower accessibility. The KPIs of PDCP download volume and the maximum number of active subscribers also show that the number of active subscribers and their download volume influence the network accessibility (more than the number of connected subscribers). Finally, the cell availability indicates that, if many cells are unavailable in the current hour, it is likely that the network accessibility will be lower in the next hour.

Individual Cells Tests: The results of the forecasts of low network accessibility in individual cells had higher error than the results with the aggregated network. In the tests done, the best models needed more data than the aggregated network models to achieve the best result, because more features were needed to be able to generalize the predictions for different cells.

The results show that, just like in the aggregated network tests, the most important KPIs for forecasting the number of E-RAB setup failures in a cell to be above a threshold are the same that cause it to have high variations. However, the most important KPIs that cause lower accessibility in a network are different from the KPIs that cause lower accessibility in a cell. The two most important KPIs are the RRC and the E-RAB setup success rate. These KPIs cannot be understood as the cause for lower network accessibility, but as a consequence. Because they are intrinsically related to the E-RAB setup failure, being themselves accessibility metrics, they can be understood as an indicator of the high autocorrelation between consecutive hours. These results show that the network accessibility per cell is highly dependent on the network accessibility of that cell in the previous hour. If a cell has lower network accessibility in an hour, it is likely that the network accessibility in the next hour for that cell is still low, and vice-versa. It is essential for a network operator to monitor the right KPIs for the different tasks: forecast lower accessibility in a network or forecast lower accessibility per cell.

As opposed to the results for the aggregated network, the important KPIs for low network accessibility are not related with the CSFB preparation success, or with any of the handover metrics. The results by cell show that, besides the RRC and the E-RAB setup success ratio, the most important KPIs are counters related to the number of users in a cell and its utilization: maximum number of connected subscribers, average number of connected subscribers, average number of active subscribers download data, variation of the maximum number connected subscribers, variation of the maximum number of active download subscribers, average number of radio bearers or variation of the average number of radio bearers. It is expected that, as these KPIs have higher values, the accessibility of a cell decreases.

6 Conclusion

Understanding the causes of events in a network, such as low accessibility, helps the network operators to forecast and avoid them to happen, by adjusting network resources that influence their causes. In this work, the goal was to determine the causes of reduced network accessibility in 4G networks, using only historic data.

Two different analysis were made. Besides analyzing the causes of reduced accessibility in the whole network, it was also analyzed the causes of reduced accessibility for each cell. The results showed that the causes of reduced accessibility for each analysis are very different. While for the overall network, the KPIs that most influence the accessibility are the number of failure handovers, the number of phone calls and SMSs in the network, the overall download volume and the availability of the cells, the KPIs that most influence the accessibility of each cell are related with the number of users in a cell and its download volume. For a network operator, it is important to know if it is important to monitor low accessibility in a cell, in a network, or in both, to make the right measurements in the network.

As future work, the next step will be to detect the patterns of those KPIs that indicate future low accessibility, to be able to predict it and adapt the network to prevent it to happen. For example, if the network operator knows that the network accessibility will be lower in the next hour when the number of handover failures intra-frequency and the maximum connected users both exceed a threshold, he can take proactive measures to adapt the network and avoid the low accessibility.

Acknowledgments. This work is supported by the European Regional Development Fund (FEDER), through the Regional Operational Programme of Lisbon (POR LISBOA 2020) and the Competitiveness and Internationalization Operational Programme (COMPETE 2020) of the Portugal 2020 framework [Project 5G with Nr. 024539 (POCI-01-0247-FEDER-024539)], and by FCT/MEC through national funds and when applicable co-funded by FEDER – PT2020 partnership agreement under the project UID/EEA/50008/2019.

References

1. Bennacer, L., Ciavaglia, L., Ghamri-Doudane, S., Chibani, A., Amirat, Y., Mellouk, A.: Scalable and fast root cause analysis using inter cluster inference. In: 2013 IEEE International Conference on Communications (ICC), pp. 3563–3568, June 2013. https://doi.org/10.1109/ICC.2013.6655104

2. Bennacer, L., Amirat, Y., Chibani, A., Mellouk, A., Ciavaglia, L.: Self-diagnosis technique for virtual private networks combining bayesian networks and case-based reasoning. IEEE Trans. Autom. Sci. Eng. **12**(1), 354–366 (2015). https://doi.org/10.1109/tase.2014.2321011

3. Chigurupati, A., Lassar, N.: Root cause analysis using artificial intelligence. In: 2017 Annual Reliability and Maintainability Symposium (RAMS), pp. 1–5, January 2017. https://doi.org/10.1109/RAM.2017.7889651

4. Kennedy, D.: D2.6 final report on programme progress and KPIs (2017). https://5g-ppp.eu/wp-content/uploads/2017/10/Euro-5G-D2.6_Final-report-on-programme-progress-and-KPIs.pdf

5. Fisher, A., Rudin, C., Dominici, F.: All models are wrong but many are useful: variable importance for black-box, proprietary, or misspecified prediction models, using model class reliance (2018)

6. Gonzalez, J.M.N., Jimenez, J.A., Lopez, J.C.D., Hugo, A.P.: Root cause analysis of network failures using machine learning and summarization techniques. IEEE Commun. Mag. **55**(9), 126–131 (2017). https://doi.org/10.1109/mcom.2017.1700066

7. Mfula, H., Nurminen, J.K.: Adaptive root cause analysis for self-healing in 5G networks. In: 2017 International Conference on High Performance Computing Simulation (HPCS), pp. 136–143, July 2017. https://doi.org/10.1109/HPCS.2017.31

8. Muñoz, P., de la Bandera, I., Khatib, E.J., Gómez-Andrades, A., Serrano, I., Barco, R.: Root cause analysis based on temporal analysis of metrics toward self-organizing 5G networks. IEEE Trans. Veh. Technol. **66**(3), 2811–2824 (2017). https://doi.org/10.1109/TVT.2016.2586143

9. Pearson, K.: LIII. On lines and planes of closest fit to systems of points in space. London Edinburgh Dublin Philos. Mag. J. Sci. **2**(11), 559–572 (1901). https://doi.org/10.1080/14786440109462720

10. Solé, M., Muntés-Mulero, V., Rana, A.I., Estrada, G.: Survey on models and techniques for root-cause analysis. CoRR abs/1701.08546 (2017). http://arxiv.org/abs/1701.08546

11. Yan, H., Breslau, L., Ge, Z., Massey, D., Pei, D., Yates, J.: G-RCA: a generic root cause analysis platform for service quality management in large IP networks. IEEE/ACM Trans. Netw. **20**(6), 1734–1747 (2012). https://doi.org/10.1109/tnet.2012.2188837

Space-Time Pattern Extraction in Alarm Logs for Network Diagnosis

Achille Salaün[1,2(✉)], Anne Bouillard[1], and Marc-Olivier Buob[1]

[1] Nokia Bell Labs, Nozay, France
{achille.salaun,anne.bouillard,marc-olivier.buob}@nokia.com
[2] CNRS, Samovar, Télécom SudParis, Institut Polytechnique de Paris,
Evry, France

Abstract. Increasing size and complexity of telecommunication networks make troubleshooting and network management more and more critical. As analyzing a log is cumbersome and time consuming, experts need tools helping them to quickly pinpoint the root cause when a problem arises. A structure called DIG-DAG able to store chain of alarms in a compact manner according to an input log has recently been proposed. Unfortunately, for large logs, this structure may be huge, and thus hardly readable for experts. To circumvent this problem, this paper proposes a framework allowing to query a DIG-DAG in order to extract patterns of interest, and a full methodology for end-to-end analysis of a log.

Keywords: Fault diagnosis · Pattern matching · Online algorithm

1 Introduction

Telecommunication networks management becomes a more and more challenging problem for operators. Indeed, on one hand, their infrastructures involve more and more devices, new technologies and possibly new manufacturers. On the other hand, network providers aim at offering a quality of service according to the Service-Level Agreements (SLAs) established with their clients. Thus, there is a strong need for fault management in order to save money, time, and human resources.

That is why network infrastructures are in general monitored. Monitoring solutions evaluate network performances through measurements. They can also collect *alarms* raised by the equipment involved in the infrastructure. The resulting file storing those messages is called a *log*.

Alarm logs are the raw material used by the expert to understand the cause of outages. Unfortunately, logs are in practice often very verbose and may be noisy. The large number of observed machines and alarm types in the log leads to an important volume of alarms, which complicates the extraction of relevant information, especially when multiple log files are involved. Log analysis is thus a difficult, cumbersome and time-consuming task. Network operators need tools helping them to pinpoint root causes of major incidents and understand the erroneous processes leading to major failures.

S. Boumerdassi et al. (Eds.): MLN 2019, LNCS 12081, pp. 134–153, 2020.
https://doi.org/10.1007/978-3-030-45778-5_10

State of the Art on Root-Cause Analysis. Root cause analysis (RCA) in telecommunication networks has been extensively studied as observed in [13].

Many solutions use neural networks (NN). For instance, [14] investigates the performance of several types of NNs for fault diagnosis of a simulation heat exchanger. In particular, the authors trained a multi-layer perceptron to map symptoms onto causes. Regarding the application, it may be impossible to obtain enough training with ground truth. In this case, supervised learning is not possible. Therefore, the authors also trained self-organizing maps to cluster the observed symptoms. The observation space is mapped onto a 2D grid and clusters are derived a posteriori. However, the interpretation of those clusters remains difficult. [17] considers several time-series and build the correlation matrix of these signals at each instant. The idea is then to train a convolutional and attention-based auto-encoder to predict sequences of correlation matrices. Correlation can be drawn between faulty signals and other ones. Unsupervised, the model takes into account temporal dependencies but is hardly interpretable.

Bayesian networks (BN) are also common in RCA. Thus, in [2], the authors split latent causes and observed symptoms into a bipartite BN. Symptoms are either described by some features or by checking some rules. If the probabilistic framework favors interpretability of the results, BNs generally face scalability issues. Indeed, increasing the complexity of the system dramatically increases the amount of memory to store conditional probabilities. [16] proposes to build BNs for root-cause analysis in an oriented-object fashion. This helps to design proper BNs with regard to prior knowledge about the system structure (thanks to the definition of BN functions), but automation of the model construction is unclear especially when prior knowledge is unavailable.

In order to explain faulty requests in the eBay Internet Service System, [6] trains a decision tree classifying faults and successes. Once trained, the path of a given faulty request is then used as a description of its root cause. The simplicity of decision trees makes them easy to interpret. Nevertheless, increasing the complexity of the system induces instability in the training phase [3]. Interpretability may be unclear in such a situation.

Some other solutions comes from *pattern matching*. [7] introduces a variation of the Smith-Wasserman algorithm [12] which evaluates the similarity between two sequences of events. Indeed, root cause may belong to alarm floods that are similar to a faulty one. [10] splits events streams into chunks that are then compared to a reference database. [15] uses Finite State Machines storing prior faulty patterns. It is possible to update the stored patterns a posteriori.

An other approach, [4], proposes an RCA tool inspired from pattern matching techniques. It builds online an automaton storing space-time causalities between symbols observed in a log. Its construction is unsupervised without losing in interpretability. Moreover, prior knowledge is optional though adding such knowledge makes the resulting structure lighter by discarding irrelevant causalities. Nonetheless, the size of the structure is usually too large for a direct use. For now, we still lack a tool to exploit such a structure and this paper is a first attempt to overcome that lack.

Contributions. Most of works described above try to find the root cause of a failure or to find correlation between alarms.

In this paper, we rather try to find chains of cascading alarms explaining why a given incident has occurred. In our approach, we process an input log with some optional prior knowledge (e.g., the network topology). This information is used to train a data structure, called DIG-DAG [4], which is designed to store every chain of alarm present in the input log.

Such a structure is usually too large to be directly interpretable by an expert. That is why we require a convenient way to extract relevant faulty patterns.

The contributions of this paper are twofold:

1. First, we propose a new query system, allowing to extract small faulty patterns stored in a DIG-DAG and matching the query issued by an expert. Outputs not only contain the possible root causes of a failure, but also the entire chain of alarms leading to the failure.
2. Second, we propose an end-to-end methodology for log analysis. It involves the DIG-DAG and our new query system, but also additional techniques using graph reductions and clustering techniques. We demonstrate the tractability of our framework through the analysis of logs issued by real systems.

Outline of the Paper. The remaining of the paper is organized as follows. Section 2 recalls the DIG-DAG construction from an input log of alarms (and eventual prior knowledge). Section 3 presents the query system built on top of DIG-DAG, which is the core of our contribution. In Sect. 4 details an end-to-end methodology for log analysis and hints to cope with large logs of alarms. Section 5 illustrates our proposal on real datasets. Finally, Sect. 6 concludes the paper.

2   From Log to Space-Time Pattern Storage

In this section, we recall the necessary background related to DIG-DAG, a data structure introduced in [4] able to store space-time patterns. To fix notations, Sect. 2.1 formally defines our representation of an alarm log. Section 2.2 introduces *directed interval graph* (DIG), a graphical representation of the log. Finally, Sect. 2.3 presents the DIG-DAG.

2.1   Alarm Logs

Nowadays, network operators rely on monitoring solutions to manage their infrastructures. Such solutions centralize alarms raised by the equipment into dedicated files called logs.

More formally, an alarm log is a finite sequence of timestamped events. More precisely, we consider that an *event* is a pair $(\sigma, [s, t])$, where σ is a symbol and $[s, t]$ is a non-empty interval of $\mathbb{R}^+$ representing the time interval during which this event is active. The symbol σ can contain any non-temporal information, *e.g.*, the name of the corresponding alarm, its severity, the impacted machine, etc. This constitutes the *space* aspect of the event. We denote by Σ the set of all possible symbols and assume this set finite. For an event $\ell = (\sigma, [s, t])$, we denote its symbol by $\lambda(\ell)$.

A log is then denoted by $L = ((\sigma_i, [s_i, t_i])_{i \in \{1, \dots, n\}})$. We assume, without loss of generality that:

- the values (s_i) and (t_i) are all distinct. Indeed, tie-breaking rules can be used if it is not the case;
- if $\sigma_i = \sigma_j$ for some distinct i and j, then $[s_i, t_i] \cap [s_j, t_j] = \emptyset$: the corresponding events do not temporally overlap. If this is not the case, these two events can be replaced by $(\sigma_i, [s_i, t_i] \cup [s_j, t_j])$.

A log can be processed online. An event is said to be *active at a time* if it corresponds to a pending alarm at this time. More formally, given log L and time τ, an event $(\sigma, [s, t])$ of L is active at time τ if $\tau \in [s, t]$. We note A_τ the set of active events of log L at time τ. The observed log at time τ is defined as $L_\tau \overset{\text{def}}{=} ((\sigma, [s, t]) \in L | s < \tau)$.

Example 1. Consider alphabet $\Sigma = \{a, b, c, d\}$ and log $L = ((a, [1, 4]), (b, [2, 5]), (c, [3, 6]), (a, [7, 10]), (c, [8, 11]), (d, [9, 12]))$. At time $\tau = 5.5$, we have $A_{5.5} = \{(c, [3, 6])\}$ and $L_{5.5} = ((a, [1, 4]), (b, [2, 5]), (c, [3, 6]))$.

2.2   DIG: A Graph-Based Representation of an Alarm Log

In this paragraph, our goal is to represent the log of alarms in a more structured way. Indeed, fault management is based on finding correlation between alarms, hence exhibiting some structure in the log. To do so, we first define the notion of *potential causality* which enables to translate the log into a graph.

Two events share a potential causality if they are *space-related* and if they share a potential time causality. Two events are space-related if their symbol lies in $\mathcal{C} \in \Sigma^2$, which gathers all the relevant pairs of symbols. For example, $\mathcal{C}$ can be tuned to consider the topology of the network (see Sect. 4). Two events share a potential time causality if one of the events occurs before the other, and if their activity period overlap. More formally, we say that event $\ell = (\sigma, [s, t])$ is a potential cause of $\ell' = (\sigma', [s', t'])$ if

- $\sigma\sigma' \in \mathcal{C}$: ℓ and ℓ' are space-related;
- $s' \in [s, t]$: ℓ and ℓ' are co-occurrent and ℓ is active before ℓ'.

This potential causality is denoted by $\ell \to \ell'$.

The *directed interval graph* (DIG) of L with space relation $\mathcal{C}$ is a labeled directed graph $(L, \to, \lambda)$ where:

- L is the set of vertices;
- $\to$ is the set of arcs;
- $\lambda : L \to \Sigma$ is the labeling function inherited form the event: each vertex/event is labeled with its symbol.

This directed graph is acyclic because of the time causality contained in →. As shown in Fig. 1, a DIG can be disconnected.

Example 2. If $\mathcal{C} = \{ab, ac, bc, cd\}$, then $(a, [1,4]) \rightarrow (b, [2,5])$ holds, but $(b, [2,5]) \not\rightarrow (a, [1,4])$ and $(a, [1,4]) \not\rightarrow (c, [8,11])$ as they break the time causality, and $(a, [7,10]) \not\rightarrow (d, [9,12]))$ because $ad \notin \mathcal{C}$.

The DIG corresponding to the log L of Example 1 is represented on Fig. 1.

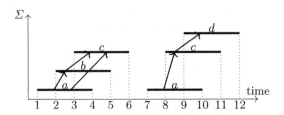

Fig. 1. DIG of L defined in Example 1.

We call *space-time pattern* or *pattern* of the log any word of Σ^* any label of a path of its DIG. The denomination *space-time* comes after potential causalities that can be issued either from topological or from temporal reasons.

2.3   DIG-DAG: A Data Structure for Storing Space-Time Patterns

DIG-DAG is a deterministic automaton-like structure able to store and count every space-time pattern of a log. It is the base of our root-case analysis approach. We use the formal language notations. In particular, the empty word is denoted by ε.

Definition 1 (DIG-DAG). *Let L be an alarm log on alphabet Σ, with spatial relation $\mathcal{C}$ and $A \subseteq L$ a set of active events.*

A DIG-DAG $(V, E, \lambda, \mathcal{A})$ of $(L, \rightarrow, \lambda)$ is a quadruple satisfying:

- *(V, E) is a directed acyclic graph with a unique vertex q_0 with in-degree 0, called the root;*
- *λ is a labeling function $\lambda : V \rightarrow \Sigma \cup \{\varepsilon\}$ with $\lambda(q_0) = \varepsilon$ and $\forall u \in V \backslash \{q_0\}, \lambda(u) \in \Sigma$;*
- *for each vertex $u \in V$ and for each $\sigma \in \Sigma$, vertex u has at most one successor $v \in V$ such that $\lambda(v) = \sigma$;*
- *each path $\ell_1, ..., \ell_k$ of the DIG corresponds to a path $q_0, u_1, ..., u_k \in V^{k+1}$ such that $\lambda(\ell_i) = \lambda(u_i)$ for all $i \in \{1, ..., k\}$ and conversely;*
- *$\mathcal{A} \subseteq V$ is a subset such that $u \in \mathcal{A}$ if and only if for all paths from the root to u there exists a path in $(L, \rightarrow, \lambda)$ ending in an active vertex with the same label. In other words, for each path $q_0, u_1, ..., u$ there exists a path $\ell_1, ..., \ell_u$ in $(L, \rightarrow, \lambda)$ such that $\lambda(\ell_i) = \lambda(u_i)$ and ℓ_u is active.*

One can notice that, given a log L and a spatial relation $\mathcal{C}$, the DIG is unique but not the DIG-DAG. For example, the DIG-DAG can be minimized or not. We assume in the rest of the paper that the DIG-DAG is built according to the deterministic algorithm presented in [4]. This algorithm can be performed online by processing events in their chronological order.

One of its advantages is its capability to count and store the number of occurrences of each space-time pattern occurring in $(L, \rightarrow, \lambda)$. More precisely, $w : E \rightarrow \mathbb{N}$ is a weight function counting the number of occurrences of any pattern ending with that arc: $w((u, v)) = n$ if for any path $q_0, \ldots, u, v$, the pattern $\lambda(q_0) \cdots \lambda(u)\lambda(v)$ corresponds to n paths of $(L, \rightarrow, \lambda)$.

Finally, as already mentioned above, the DIG-DAG can be interpreted as a special case of a deterministic automaton, where q_0 is the initial state and the labels are deported to the targets of the transitions.

Example 3. Figure 2 shows the DIG-DAG built from DIG $(L, \rightarrow, \lambda)$ (cf. Example 1) at time 9.5: the last three events of the log are active. The weights are depicted above the arcs. For example, a and ac occur twice, while abc occurs only once. Active vertices are represented in bold.

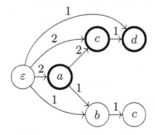

Fig. 2. DIG-DAG built from log L of Example 1; bold vertices are the active states at time $\tau = 9.5$.

To sum up, a DIG-DAG is a graph structure able to store and count any space-time pattern occurring is an alarm log. The size of this structure can grow exponentially with the size of the log. For example, root's out-degree is exactly the size of the alphabet Σ and the depth of the structure is the size of a longest path in the corresponding DIG. The rest of the paper is devoted to the extraction of patterns of interest.

3 Pattern Extraction

In this section, we present a generic solution to extract patterns of interest by queries.

This section presents a new framework able to isolate patterns matching input queries. These patterns are represented as sub-DIG-DAGs:

Definition 2 (sub-DIG-DAG). *Let $\mathcal{D} = (V, E, \lambda, \mathcal{A})$ be a DIG-DAG. For every subgraph (V', E') of (V, E), the 4-tuple $\mathcal{D}' = (V', E', \lambda_{|V'}, \mathcal{A} \cap V')$ is a sub-DIG-DAG of $\mathcal{D}$.*

Remark that a sub-DIG-DAG is not necessarily a DIG-DAG: it may have several roots and be disconnected.

Sub-DIG-DAGs are stable with graph operations like intersection, union, difference.

Definition 3 (query and its resulting sub-DIG-DAG). *A query is a 5-tuple $\mathcal{Q} = (\mathcal{D}, S, T, V_i, E_i)$, where $\mathcal{D} = (V, E, \lambda, \mathcal{A})$ is a DIG-DAG, $S, T, V' \subseteq V$ and $E' \subseteq E$. The result of the query is the largest sub-DIG-DAG $(E', V', \lambda', \mathcal{A}')$ of $\mathcal{D}$ such that:*

- *$V' \subseteq V_i,\ ' \subseteq E_i$;*
- *the vertices with in-degree 0 are in S;*
- *the vertices with out-degree 0 are in T.*

The result of the query is denoted by $\mathcal{D}(\mathcal{Q})$.

Intuitively, the sub-DIG-DAG of $\mathcal{D}$ resulting from a query $\mathcal{Q} = (\mathcal{D}, S, T, V_i, E_i)$ is the subgraph of $\mathcal{D}$ whose maximal paths all start in S and end in T. These paths only traverse vertices of V_i and arcs of E_i.

3.1   Regular Queries

The definition of queries is very broad, and in this paragraph we restrict to queries parametrized by a finite automaton, and local properties on the vertices and arcs. Intuitively, the role of the finite automaton is to extract patterns satisfying some relations between vertices, while local properties select vertices and arcs. These properties do not only depend on the symbols of the nodes (which could otherwise have been done with an automaton), but rely on information that can be attached to the nodes. For example, this can be useful to extract nodes that have been recently active or arcs satisfying some weight-based properties.

Definition 4 (Regular query). *Let $\mathcal{M}$ be a finite automaton, and P_v and P_e be two properties. The regular query $\mathcal{R}(S, T, \mathcal{M}, P_v, P_e) = (S, T, V_i, E_i)$, where:*

- *for all $v \in V_i$, v satisfies P_v;*
- *for all $e \in E_i$, e satisfies E_v;*
- *$\mathcal{D}(\mathcal{R}(S, T, \mathcal{M}, P_v, P_e))$ contains all the paths of $\mathcal{D}(\mathcal{Q})$ labeled by a word recognized by $\mathcal{M}$. Consequently, V_i and E_i are respectively defined by the set of vertices and arcs belonging to one of those paths;*
- *$\mathcal{D}(\mathcal{R}(S, T, \mathcal{M}, P_v, P_e))$ is the minimal sub-DIG-DAG satisfying those properties.*

Algorithm 1. Input query based sub-DIG-DAG extraction

Input: $\mathcal{D}, \mathcal{M} = (Q, \Sigma, \delta, I, F), S, T, P_v, P_e$
Output: a sub-DIG-DAG $\mathcal{D}'$
`// Phase 1: Forward exploration`
foreach $u \in V$ **do** $Q_1(u) \leftarrow \emptyset$;
foreach $u \in V$ *(in the topological order)* **do**
 $\quad$ **if** $u \in S \cap P_v$ **then** $Q_1(u) \leftarrow Q_1(u) \cup I$;
 $\quad$ **foreach** $v \in V \cap P_v$ *such that* $(u, v) \in E \cap P_e$ **do**
 $\quad \quad$ $Q_1(v) \leftarrow Q_1(v) \cup \{q' \in Q \mid \exists q \in Q_1(u),\ \delta(q, q') = \lambda(v)\}$
`// Phase 2: Backward exploration and decision`
$V' \leftarrow \emptyset$; $E' \leftarrow \emptyset$;
foreach $u \in V$ **do** $Q_2(u) \leftarrow \emptyset$;
foreach $u \in V$ *(in the reverse of the topological order)* **do**
 $\quad$ **if** $u \in T \cap P_v$ **then** $Q_2(u) \leftarrow Q_2(u) \cup (Q_1(u) \cap F)$;
 $\quad$ **if** $Q_2(u) \neq \emptyset$ **then**
 $\quad \quad$ $V' \leftarrow V' \cup \{u\}$
 $\quad \quad$ **foreach** v *such that* $(v, u) \in E \cap P_e$ **do**
 $\quad \quad \quad$ **if** $\{q \in Q_1(v) \mid \exists q' \in Q_2(u),\ \delta(q, q') = \lambda(v)\}\} \neq \emptyset$ **then**
 $\quad \quad \quad \quad$ $E' \leftarrow E' \cup \{(v, u)\}$;
 $\quad \quad \quad \quad$ $Q_2(v) \leftarrow Q_2(v) \cup \{q \in Q_1(v) \mid \exists q' \in Q_2(u),\ \delta(q, q') = \lambda(v)\}$
return $\mathcal{D}' = (V', E', \lambda_{|E'}, \mathcal{A} \cap V')$

Algorithm 1 computes the sub-DIG-DAG corresponding to a regular query. We assume that we know a topological order of the vertices. For this one can either use a classical algorithm (see [8] for example), or the topological order can be computed on-the-fly at the DIG-DAG construction.

Algorithm 1 has two phases: the first one identifies, by a forward traversal of the DIG-DAG all the possible paths starting from S, having vertices and arcs satisfying P_v and P_e and whose label are prefixes of words recognized of the automaton. For this, a set $Q_1(u)$ is attached to each node, containing all the states of the automaton that can be reached from a vertex in S.

The second phase performs a backward traversal and identifies the vertices and arcs in the sub-DIG-DAG. For each vertex u, $Q_2(u)$ is the subset of states in $Q_1(u)$ such that there is a path from u to T labeled similarly to a path from a state of $Q_2(u)$ to a final state in the automaton. The vertices and arcs involved in these paths constitute the sub-DIG-DAG.

More formally, let $\mathcal{M} = (Q, \Sigma, \delta, I, F)$ be a finite automaton where Σ denotes its alphabet, Q the its set of states, I its initial states, F its final states and δ its transition map. We use the automaton interpretation of a DIG-DAG: the label of a transition is the label of the extremity of the arc. Thus, the label of a path in the DIG-DAG does not take into account the label of the first node of the path.

We show next that the result of Algorithm 1 is the sub-DIG-DAG corresponding to the regular query as defined in Definition 4.

Let us first state some properties of $Q_2(u)$.

Lemma 1. *With the above notations,*

- $\forall u \in V'$, $p \in Q_2(u)$, $\exists v \in V', q \in Q_2(v)$ *such that* $(u,v) \in E'$ *and* $\lambda(v) \in \delta(p,q)$;
- $\forall v \in V'$, $q \in Q_2(u)$, $\exists u \in V', p \in Q_2(v)$ *such that* $(u,v) \in E'$ *and* $\lambda(v) \in \delta(p,q)$;
- *for all* $v \in V'$, *for all* $q \in Q_2(v)$, *there exists a path from a vertex* $s \in S$ *to* v *corresponding to a path from* $i \in I$ *to* q *in* $\mathcal{M}$;
- *for all* $v \in V'$, *for all* $q \in Q_2(v)$, *there exists a path from* v *to a vertex* $t \in T$ *corresponding to a path from* q *to* $f \in F$ *in* $\mathcal{M}$.

Proof. The first two statements are deduced from lines 5 and 12 of the algorithm, that is the construction of Q_1 and Q_2. The last two statements are obtained by induction from the two firsts.

We prove that the resulting sub-DIG-DAG is indeed the smallest one containing the intersection of $\mathcal{D}$ and $\mathcal{M}$.

Theorem 1. *Consider the regular query* $\mathcal{R}(S,T,\mathcal{M},P_v,P_e)$, *and let* $\mathcal{D}'$ *be the sub-DIG-DAG returned by Algorithm 1. We have* $\mathcal{D}' = \mathcal{D}(\mathcal{R}(S,T,\mathcal{M},P_v,P_e))$.

Proof. We have to check the four properties of Definition 4. The two firsts are straightforward, as all vertices and arcs added to V' and E' respectively check P_v and P_v (lines 9–13).

We now check the third property: let $p = u_1, \ldots, u_f$ be a path in $\mathcal{D}$ labeled by a word accepted by $\mathcal{M}$, with $u_1 \in S$ and $u_f \in T$. Let $q_1, \ldots, q_f$ be a sequence of states visited by $\mathcal{M}$ for accepting this word. For all i, by construction, $q_i \in Q_1(u_i)$ (line 5). As $q_f \in F$ and $u_f \in T$, $q_f \in Q_2(u_f)$ (line 9), and then, $q_i \in Q_2(u_i)$ for all i (line 14): all the arcs of the path are kept.

Finally, we have to check the minimality of the structure, that is that all the arcs of the graph returned by the algorithm belong to a path labeled by a word recognized by $\mathcal{M}$. This is a consequence of Lemma 1: consider an arc (u,v). One can build a path $s \in S \rightsquigarrow u \rightarrow v \rightsquigarrow t \in T$ corresponding to $q_i \in I \cap Q_2(s) \rightsquigarrow p \in Q_2(u) \rightarrow q \in Q_2(v) \rightsquigarrow q_f \in F \cap Q_2(t)$ by applying items 3, 1, 4 of the Lemma 1.

Example 4. Let us extract the patterns satisfying the regular expression $a(\Sigma \setminus \{b\})^*c$ from the DIG-DAG represented in Fig. 3a, that is all the paths starting by a, ending c and not containing any b. We choose $S = \{q_0\}$, $T = V$. Nodes of the DIG-DAG are numbered according a topological order. The regular expression is represented by the automaton shown in Fig. 3b. The sub-DIG-DAG extraction steps of Algorithm 1 are depicted on Fig. 4. Phase 1 is displayed from Fig. 4a to e and phase 2 from Fig. 4f to j. Sets $Q_1(.)$ are reported below each vertex for phase 1, and $Q_2(.)$ are for phase 2. Arcs and vertices that are added to E' and V' are represented in bold, and those non selected are dashed.

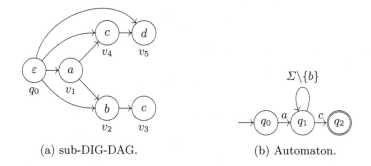

(a) sub-DIG-DAG.   (b) Automaton.

Fig. 3. DIG-DAG and automaton used for Algorithm 1 in Example 4.

3.2 Sub-DIG-DAG Simplification

Queries and in particular regular queries allow to select patterns of interest. Ideally, the extracted sub-DIG-DAG should be easily readable. As we will see in Sect. 5, the number of vertices is often limited, but it happens that the average degree of the sub-DIG-DAG is too high to get a readable graphical representation. In this paragraph, we present three methods to improve and simplify the graph. Note that this simplification cannot be considered as sub-DIG-DAG extraction, as they might change the structure, by merging nodes, erase paths and are not compatible with weights. This is not a big issue since these operations are just for graphical representations.

Transitive Reduction. The aim of the transitive reduction is to decrease the number of arcs. Introduced in [1], this operation removes every arc $e = (u, v)$ whenever there exists another path between nodes u and v. The resulting graph is the minimal subgraph (for the arc-inclusion order) that does not break the connectivity of each connected component of the graph. The transitive reduction removes some paths from the sub-DIG-DAG, but keeps the longest ones.

Minimization. As said above, a DIG-DAG can be seen as a deterministic automaton. Moreover, it is acyclic. As a sub-DIG-DAG is a subgraph of the DIG-DAG, it can also be seen as a deterministic automaton if it has a single source node. Still, we can use Revuz algorithm [11] to minimize it. Dedicated to acyclic deterministic automaton, this algorithm merges equivalent states from the leaves to the root. The resulting sub-DIG-DAG is minimal and recognizes exactly the same patterns. However, some states (resp. arcs) having different weights may be merged in the process.

Source Simplification. As said at the end of Sect. 2, the size of the DIG-DAG grows exponentially with the size of the log. This means that there can be numerous vertices with the same label, especially corresponding to the same

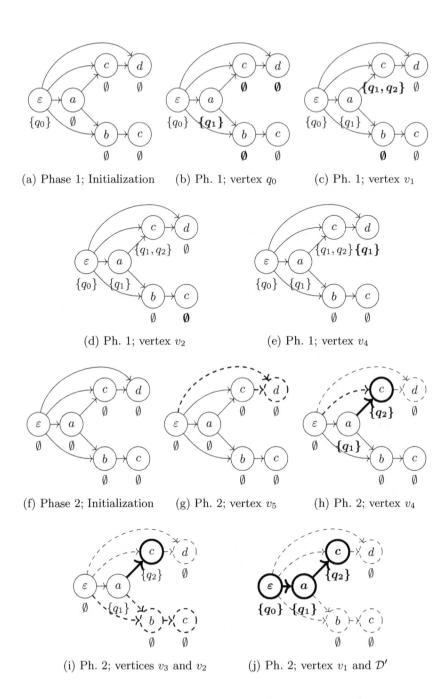

(a) Phase 1; Initialization   (b) Ph. 1; vertex q_0 (c) Ph. 1; vertex v_1

(d) Ph. 1; vertex v_2 (e) Ph. 1; vertex v_4

(f) Phase 2; Initialization   (g) Ph. 2; vertex v_5 (h) Ph. 2; vertex v_4

(i) Ph. 2; vertices v_3 and v_2 (j) Ph. 2; vertex v_1 and $\mathcal{D}'$

Fig. 4. Algorithm 1 applied to sub-DIG-DAG and automaton of Fig. 3.

occurrence of an event. When querying a DIG-DAG, the set S might be described by some property (a given symbol, and set of symbols), and many sources might have the same label. We observe that many of them can have the same sets of successors. Source simplification parses the source nodes and merges those with the same set of successors. Here again, this operation might not be compliant with the arc weights.

4 End-to-End Analysis of an Alarm Log

In this section, we explain how to apply the approach described in Sects. 2 and 3 for the end-to-end analysis of an alarm log in a root-cause analysis context. We assume that the log is given and there is no constraint for performing it online (even if some steps can be performed online). In Sect. 4.1, we explain the general approach, and in Sect. 4.2, we give some solutions when the log is too large for the analysis to be scalable.

4.1   General Approach of the Analysis

Log Parsing. The first step is to transform the raw log into a sequence of events. This includes fixing the alphabet, fixing the activity period of each event, and the time τ when the construction of the log stops.

Alphabet of the Log: The choice of the alphabet is decided based on the features one wants to take into account during the analysis. The most relevant features are:

- the *name of the alarm*: this is generally a short text or a number;
- the emitting *machine*, represented by its identifier. This can be a machine, the port of a switch, a cell in a wireless network, etc.;
- the *severity* of the alarm, represented by a number or a color, that grows or becomes darker with the severity.

Activity Period of an Event: Depending on whether the log has already been pre-processed or not, whether an alarm has a cancel time or not, the activity period of an event must be carefully defined.

The simple case is when an alarm has a emission time and a cancel time, as they respectively define the star time and the end time of the activity period.

When only the emission time is available, this defines only the start time of the activity period. Its end has to be defined.

In some alarm logs, emission times occur at precise dates. This is the case for example for KPIs, when their emission is set a few times per hour. In this case and in order to ensure time causality, a good choice would be to set an activity period a little longer than the periodicity of the measurements.

If there is no such periodicity, then the time interval can be set up by studying the average rate of the events. More details on the choice of the activity period can be found in [4].

Fixing Time τ: Even if a log is given as a file and the analysis can be performed for the whole log, it might be a better idea, sometimes, to stop the construction before. For example, a peak of alarms detected at time τ indicates that a major failure is arising in the network at that time. The origin of the problem occurs before that peak. Moreover, queries of the DIG-DAG can make advantage the active alarms at time τ.

DIG-DAG Construction. Once the log has been well defined, the potential space causality C is required to build the DIG-DAG. We give below some examples:

- if no information is available, then C is simply Σ^2;
- it the geographic location of the network equipment is known (antennas and cells in a wireless network for example), then $\sigma\sigma' \in C$ if and only if the symbols σ and σ' represent machines are distant of no more than a few kilometers.
- if some logical behavior is known, such as a certain type of element or application only communication with some other types of machines, one can build C based on these possible communications.

The DIG-DAG can then be built. Note that additional information can be added to the structure, such as the weight of the arcs, defined in Sect. 2.3, or the last date of activation of each vertex, mentioned in Sect. 3.

Query of the DIG-DAG. Once the DIG-DAG is built, it stores every space-time pattern of the log and one can query it. For a regular query, parameters are $S, T, \mathcal{M}, P_v, P_e$ (see Sect. 3.1). Let us give some examples.

Filtering the Arcs: this is done by defining property P_e. Assume that one wants to extract parts of the DIG-DAG such that there are strong correlations between the nodes. This is done by using the weights of the arcs, and more precisely, the ratio r defined in [4], such that for each arc (u, v) of the DIG-DAG, $r((u,v)) = w((u,v))/|L|_v$, where $|L|_v$ is the number of occurrences of v in log L. This is the ratio between the number of time patterns ending by arc (u, v) has been observed in the log and the total number of occurrences of v in the log. If this ratio approaches 1, this means that v is strongly and positively correlated to these patterns. Property P_e will then select the arcs having a ratio above a given threshold ρ.

Filtering the Nodes: this is done by defining property P_v. Assume that one might want to discard alarms having the lowest severity to focus on the more important messages. This is then equivalent to select a subset of Σ. Assume that one wants to focus on recent alarms only. Then P_v can be set to select nodes that have been recently active.

Sources: We now define S, the possible sources of the paths. By default, one can choose $S = \{q_0\}$ to keep every patterns. On the contrary, one could decide to focus on some types of alarms.

Targets: For the definition of T, one may want to focus on critical events. A good choice, especially if τ has been chosen in a peak of alarm, is to focus on active alarms active at τ that have a high severity level.

More Specific Queries: for more specific queries, the automaton $\mathcal{M}$ can be defined. This can for example be used to follow and check the propagation of faults. For example, to check the propagation of a faulty behavior from a machine m_1 to another machine m_2, one may define an automaton accepting words of the form $\Sigma_1^* \Sigma_2^*$, where Σ_i is the set of alarms emitted from machine m_i.

Once the query has been defined, the DIG-DAG is queried according to Algorithm 1. The readability of resulting sub-DIG-DAGs can be improved by using the techniques described in Sect. 3.2.

4.2   Dealing with Huge Logs

It happens that logs are too huge so that the DIG-DAG can be built in a reasonable time (or online). In this paragraph, we propose several solutions to overcome this difficulty. The first one is based on selecting relevant parts of the log, and the other ones rather modify the alphabet Σ to simplify the log.

Truncation of the Log. A first solution consists in selecting only the most interesting parts of the log. Intuitively, a problem can be detected when the behavior of the alarms emission process deviates from its normal behavior. For this, we are interested in the rate on arrival of the alarms, or of a subset of alarms. Detecting the deviation as soon as possible can help targeting the root cause.

Several techniques have been proposed to detect automatically deviating behaviors. They are all based on tracking the average arrival rate of messages: one can cite the Moving-average model [9] or [5]. In the solution proposed in the next section, we consider the latter solution, as it has been demonstrated to track deviations more precisely. The arrival rate of messages can be computed online. Detecting sudden deviations can then be done using Tchebychev inequality.

Simplification of the Alphabet. The size of the alphabet has a strong impact on the size of the DIG-DAG. Indeed, the out-degree of each vertex is bounded by the size of the alphabet.

Spamming Alarms. The easiest way to simplify a log is to remove some entries. Spamming alarms are frequent alarms that do not provide information: they appear in every log, regularly, whatever the state of the machine. Removing them would not impact the retrieval of cascades of events for root-cause analysis. They can also improve the analysis by discarding irrelevant causalities. Those alarms can be learned through the observation of previous logs, or by pre-processing the log under analysis.

Clustering Co-occurrent Alarms. Another way to simplify the log is to cluster co-occurrent symbols and merge the corresponding events of the log. More precisely, we define a distance between two symbols σ and σ' as the Jaccard distance of their emission intervals: let $I_\sigma = \cup_{\{\ell \in L \mid \lambda(\ell) = \sigma\}} [s_\ell, t_\ell]$ be the union of all the intervals of time where symbol σ is emitted. The Jaccard distance of σ and σ' is

$$d(\sigma, \sigma') = 1 - \frac{|I_\sigma \cap I_{\sigma'}|}{|I_\sigma \cup I_{\sigma'}|},$$

where $|\cdot|$ is the L_1 norm.

This distance can be generalized to the distance between clusters the following way: let C and C' be two sets of symbols. Let $I_C = \cup_{\{\ell \in L \mid \lambda(\ell) \in \sigma\}} [s_\ell, t_\ell]$. The Jaccard distance between C and C' is

$$d(C, C') = 1 - \frac{|I_C \cap I_{C'}|}{|I_C \cup I_{C'}|}.$$

From this distance one can build a clustering of alarms in a bottom-up approach:

– fix a threshold α to merge clusters that have distance less than α;
– while the smallest distance between two any clusters is less than α, merge the two nearest clusters.

When clusters have been computed, the log needs to be simplified. The new set of symbols is the set of clusters obtained, and we replace each event $(\sigma, [s, t])$ by $(C, [s, t])$ is $\sigma \in C$, and merge all the co-occurring events having the same cluster.

As in the previous paragraph, this can be done by pre-processing the log under analysis, but could also be learned from previous logs, and the merging of events be done on-the-fly.

Projection the Alphabet and Two-Step Analysis. A third solution to limit the size of the alphabet, is to project the alphabet. For example, if the alphabet were initially based on both the emitting machine and the name of the alarm, one can build a DIG-DAG by projecting the alphabet on the machines only or the alarm names only. By doing this, events will have to be merged as in the paragraph above. Such a simplification of the log implies some loss of information, but projecting on the machines can help locating the problem, and projecting on the alarm name can help detecting the type of problem that occurred.

This partial information can be used for selecting a sub-set of events of the original log, and perform a second construction of a DIG-DAG: let us assume for example that we used only the information of the machines to build the first DIG-DAG, and performed a query that isolated some parts of the log, hence some machines of interest. Now, consider the original log and keep only the events emitted by the machine of interest. The size of this sub-log and of the alphabet is reduced, so the analysis can be performed by building the DIG-DAG of the sub-log and querying it.

5   Experimental Results

In this section, we apply our methodology (see Sect. 4) to three real datasets using a computer with an Intel i7 microprocessor and 8 GB RAM.

5.1   Datasets

We use three private logs issued by three different GSM network elements. Each line describes a specific event which contains the alarm identifier, the machine name, the emission time of the alarm, and its severity. Note that each row describes a punctual event, $i.e.$ the activity of the event is only characterized by its emission time. Severity is a score indicating the importance of the alarm, equal to 0 for informative messages, 1 for warnings, 2 for mid severe alarms and 3 for major failures. Each triple (`machine`, `alarm-id`, `severity`) corresponds to a symbol as introduced in Sect. 2.1.

Table 1. Experiments summary. First block describes the dimensions of the raw datasets. Second block shows the dimensions of the logs after alphabet simplification. Third block highlights the gains obtained by our query system. When relevant, computation times are mentioned.

	Log 1	Log 2	Log 3
Duration (h)	1042	48	330
Nb. of entries (total)	35,905	6,873	5,591
Nb. of network elements	115	142	41
Nb. of alarms	43	66	39
Size of the alphabet	242	429	113
Nb. of entries after clustering step	1,095	548	480
Nb. of entries after spamming filtering	635	537	226
Nb. of clusters	103	167	69
Clustering execution time	2.5 s	3.7 s	0.4 s
Nb. of DIG-DAG vertices/edges	20,540/443,496	4,364/177,973	190/643
Nb. of query ($\rho = 0.7$) vertices/edges	19/35	142/996	34/54
DIG-DAG construction time	25.1 s	8.9 s	38 ms
Query construction time	0.7 s	1.0 s	42 ms

First block of Table 1 gives for each log its size and its corresponding alphabet. Networks topologies are unknown.

5.2   Simplification of the Alphabet and DIG-DAG Construction

First of all, a potential causality relationship is defined with regard to punctual events. As logs may last several days, we consider that an event could cause another event if it occurred at most one hour earlier. This is enough to catch relevant causalities.

To limit the size of the DIG-DAG, we cluster co-occurring alarms as described in Sect. 4 with $\alpha = 0.3$. Furthermore, we discard spamming clusters, that is clusters whose total activity period is more than 24 h.

The activity in Log 1, before and after alphabet simplification, is represented in Fig. 5. One can check that such simplifications did not alter the behavior observed in the original log.

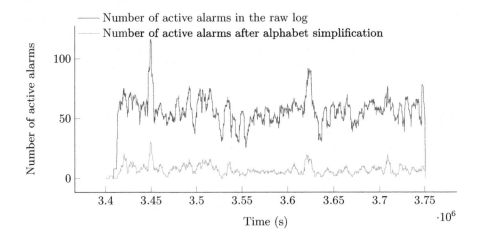

Fig. 5. Number of active alarms over time (raw and simplified logs) in Log 1.

Log sizes after simplifications are shown in the second block of Table 1. Note that skipping the alphabet simplification in the RCA process leads to larger structures. For instance, Log 1 is too big to be built quick.

5.3 DIG-DAG Queries

Despite their quick computation, the DIG-DAGs are hardly human-readable (see third block of Table 1). We now extract patterns leading to critical failures.

For each dataset, we query the DIG-DAG with the following parameters:

- S is restricted to *relevant vertices* that correspond to clusters containing at least one event of severity greater or equal to 1;
- T is restricted to *critical vertices* that correspond to clusters containing at least one event of severity 2 or 3;
- P_v is defined to only consider relevant vertices;
- P_e is defined to only consider arcs of ratio greater than $\rho \in [0, 1]$;
- $\mathcal{M}$ accepts any pattern.

As we have neither prior knowledge nor expertise about the logs we studied, the described patterns are very generic. We will see however in the two next sections that these naive queries still dramatically improve DIG-DAG readability and contain relevant information about failures.

5.4   Results

Evaluating Filtering of Queries. This section presents results for different values of ρ. The greater this parameter, the stricter the filtering.

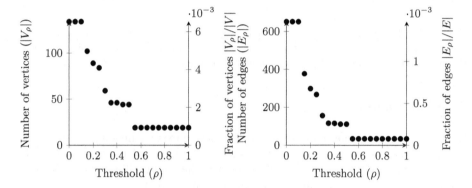

Fig. 6. Number of vertices $|V_\rho|$ (resp. edges $|E_\rho|$) of queries in Log 1 regarding threshold ρ (left axis). This number is divided by the number of vertices $|V|$ (resp. edges $|E|$) in the whole DIG-DAG (right axis).

In Fig. 6, for any ρ, queries select a small fraction of the DIG-DAG built from Log 1. This highlights the efficiency of the queries to extract patterns from the DIG-DAG. For $\rho \geq 0.7$, the corresponding sub-DIG-DAG has less than 19 vertices and 35 arcs, and hence becomes small enough to be human-readable. The choice of ρ is a compromise between readability and quantity of information.

Accuracy of the Queries. For each dataset, the root cause has been provided by experts. These root causes have been highlighted by experts independently of our work. When $\rho = 0.7$, the resulting sub-DIG-DAGs contains all the provided root causes. Moreover, the sub-DIG-DAG provides richer information than just a root cause. For example, Fig. 7 depicts the sub-DIG-DAG obtained for Log 1 and contains the root cause (77397912,4004,2) provided by experts. Unfortunately, we did not get further expert feedback regarding the quality of the sub-DIG-DAGs.

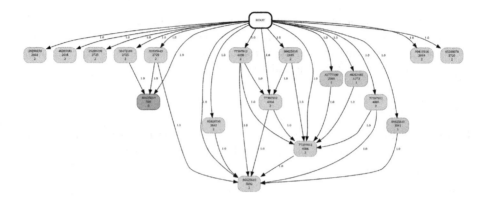

Fig. 7. Sub-DIG-DAG issued by the query of Log 1 ($\rho = 0.7$). Labels (`machine,`
`alarm-id, severity`) are indicated for each vertex. Colors depend on the severity of
the cluster representative. Ratios are indicated on edges.

6   Conclusion

In this article, we proposed a system of queries to extract meaningful and synthetic explanations from causal graph structures. The tool provided is flexible
enough to allow experts to search for specific behaviors within the log. We also
demonstrated how to use this tool for an end-to-end analysis.

However, human expertise is still a necessary component in the RCA process,
and our goal is oriented towards self-diagnosing and self-repairing networks.

Therefore, our future works aim at automatizing the exploitation of the DIG-
DAG structure. Additionally, being able to combine knowledge learnt from similar networks would help us to design a fully automatic RCA solution.

References

1. Aho, A.V., Garey, M.R., Ullman, J.D.: The transitive reduction of a directed graph.
   SIAM J. Comput. **1**(2), 131–137 (1972)
2. Alaeddini, A., Dogan, I.: Using Bayesian networks for root cause analysis in statistical process control. Expert Syst. Appl. **38**(9), 11230–11243 (2011)
3. Aluja-Banet, T., Nafria, E.: Stability and scalability in decision trees. Comput.
   Stat. **18**(3), 505–520 (2003)
4. Bouillard, A., Buob, M.O., Raynal, M., Salaün, A.: Log analysis via space-time
   pattern matching. In: 2018 14th International Conference on Network and Service
   Management (CNSM), pp. 303–307. IEEE (2018)
5. Bouillard, A., Junier, A., Ronot, B.: Hidden anomaly detection in telecommunication networks. In: 2012 8th International Conference on Network and Service
   Management (CNSM) and 2012 Workshop on Systems Virtualiztion Management
   (SVM), pp. 82–90. IEEE (2012)
6. Chen, M., Zheng, A.X., Lloyd, J., Jordan, M.I., Brinewer, E.: Failure diagnosis
   using decision trees. In: Proceedings of the International Conference on Autonomic
   Computing, pp. 36–43. IEEE (2004)

7. Cheng, Y., Izadi, I., Chen, T.: Pattern matching of alarm flood sequences by a modified Smith-Waterman algorithm. Chem. Eng. Res. Des. **91**(6), 1085–1094 (2013)
8. Cormen, T.H., Leiserson, C.E., Rivest, R.L., Stein, C.: Introduction to Algorithms, 3rd edn. MIT Press, Cambridge (2009)
9. Enders, W.: Stationary Time-Series Models. Wiley, New York (2004)
10. Johannesmeyer, M.C., Singhal, A., Seborg, D.E.: Pattern matching in historical data. AIChE J. **48**(9), 2022–2038 (2002)
11. Revuz, D.: Minimisation of acyclic deterministic automata in linear time. Theoret. Comput. Sci. **92**(1), 181–189 (1992)
12. Smith, T.F., Waterman, M.S., et al.: Identification of common molecular subsequences. J. Mol. Biol. **147**(1), 195–197 (1981)
13. Solé, M., Muntés-Mulero, V., Rana, A.I., Estrada, G.: Survey on models and techniques for root-cause analysis. arXiv preprint arXiv:1701.08546 (2017)
14. Sorsa, T., Koivo, H.N.: Application of artificial neural networks in process fault diagnosis. Automatica **29**(4), 843–849 (1993)
15. Van Lunteren, J.: High-performance pattern-matching for intrusion detection. In: Proceedings of the 25th IEEE International Conference on Computer Communications, IEEE INFOCOM 2006, pp. 1–13. Citeseer (2006)
16. Weidl, G., Madsen, A.L., Israelson, S.: Applications of object-oriented Bayesian networks for condition monitoring, root cause analysis and decision support on operation of complex continuous processes. Comput. Chem. Eng. **29**(9), 1996–2009 (2005)
17. Zhang, C., et al.: A deep neural network for unsupervised anomaly detection and diagnosis in multivariate time series data. In: Proceedings of the AAAI Conference on Artificial Intelligence, vol. 33, pp. 1409–1416 (2019)

Machine Learning Methods for Connection RTT and Loss Rate Estimation Using MPI Measurements Under Random Losses

Nageswara S. V. Rao[1]([⊠]), Neena Imam[1], Zhengchun Liu[2],
Rajkumar Kettimuthu[2], and Ian Foster[2]

[1] Oak Ridge National Laboratory, Oak Ridge, TN, USA
`{raons,nimam}@ornl.gov`
[2] Argonne National Laboratory, Argonne, IL, USA
`{zhengchun.liu,kettimut,foster}@anl.gov`

Abstract. Scientific computations are expected to be increasingly distributed across wide-area networks, and Message Passing Interface (MPI) has been shown to scale to support their communications over long distances. Application-level measurements of MPI operations reflect the connection Round-Trip Time (RTT) and loss rate, and machine learning methods have been previously developed to estimate them under deterministic periodic losses. In this paper, we consider more complex, random losses with uniform, Poisson and Gaussian distributions. We study five disparate machine leaning methods, with linear and non-linear, and smooth and non-smooth properties, to estimate RTT and loss rate over 10 Gbps connections with 0–366 ms RTT. The diversity and complexity of these estimators combined with the randomness of losses and TCP's non-linear response together rule out the selection of a single best among them; instead, we fuse them to retain their design diversity. Overall, the results show that accurate estimates can be generated at low loss rates but become inaccurate at loss rates 10% and higher, thereby illustrating both their strengths and limitations.

Keywords: Round Trip Time · Loss rate · Message Passing Interface · Machine Learning · Generalization bounds · Regression · Information fusion

1 Introduction

Computations distributed across geographically dispersed facilities, such as multiple supercomputer sites connected over a wide-area network, are of increasing

This work is funded by RAMSES project and Applied Mathematics program, Office of Advanced Computing Research, U.S. Department of Energy, and by Extreme Scale Systems Center, sponsored by U.S. Department of Defense, and performed at Oak Ridge National Laboratory managed by UT-Battelle, LLC for U.S. Department of Energy under Contract No. DE-AC05-00OR22725.

S. Boumerdassi et al. (Eds.): MLN 2019, LNCS 12081, pp. 154–174, 2020.
https://doi.org/10.1007/978-3-030-45778-5_11

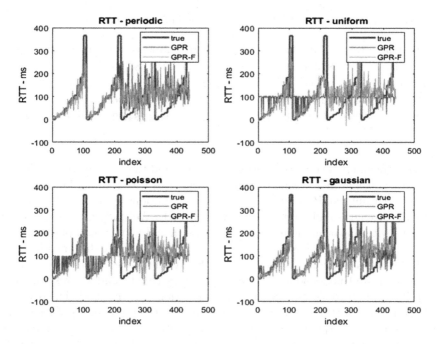

Fig. 1. RTT estimates with lowest RMS error of individual and fuser methods.

interest in science applications. Their execution times are effected by network latencies and loss processes, often in a complex way, due to the close coupling between computations and communications in these applications. Recently, Message Passing Interface (MPI) has been shown to be effective in supporting communications over wide-area connections, including ones long enough to span the globe, under external packet loss rates up to 20% [14]. In contrast with executions at a single facility, these distributed computations need to account for the longer and more varied executions times of MPI operations to avoid inefficiencies due to unbalanced computing and networking operations; for example, MPI join operation over connections with wide ranging latencies will be delayed by the longest. Motivated by such considerations, Round Trip Times (RTT) and loss rates of wide-area connections are estimated using execution time measurements of MPI primitives in distributed computations [15]. A main contributor to these execution times is the Transmission Control Protocol (TCP) which is a dominant underlying transport mechanism of MPI for wide-area connections. In particular, at increased loss rates and randomness, the execution time variations are dominated by TCP's highly non-linear response dynamics [7,9].

Machine Learning (ML) methods have been developed for a number of networking tasks for science data flows, for example, detecting flow anomalies [6] and classifying elephant and mice flows [4]. In particular, ML methods are developed to estimate the connection RTT and loss rate under deterministic periodic losses in [15] for 10 Gbps emulated connections with 0–366 ms RTT; these connections

Table 1. RMS errors of RTT estimation by individual and fuser methods.

Loss type	EOT	GPR	LR	RT	SVM	LR-F	GPR-F	$\tilde{\Delta}_{\text{LR-F}}$	$\tilde{\Delta}_{\text{GPR-F}}$
Periodic	109.20	**87.78**	102.27	95.31	103.49	92.94	**85.50**	−1.63	2.28
Poisson	101.89	**91.90**	104.32	104.31	120.35	89.85	**85.69**	2.05	6.21
Gaussian	84.33	**73.13**	2020.55	73.28	137.29	88.22	**87.06**	−15.06	−13.92
Uniform	109.11	**99.59**	106.47	107.62	121.81	90.12	**84.90**	9.46	14.69

represent local, cross country, continental and round the earth distances. In this paper, we consider more realistic, complex scenarios with random losses, in particular under unform, Poisson and Gaussian distributions up to 20% loss rates, to study the strengths and limitations of ML methods. We study five disparate ML methods, with linear and non-linear, and smooth and non-smooth properties, to estimate connection RTT and loss rate. They include four non-linear estimators, namely, smooth Support Vector Machine (SVM) and Gaussian Process Regression (GPR), and non-smooth Ensemble of Trees (EOT) and Regression Trees (RT), in addition to the baseline Linear Regression (LR) method. The diversity and complexity of these estimators combined with the randomness of losses and TCP's non-linear response rule out the identification of a single best among the estimators. Analytical results establish the finite-sample limits in asserting the performance superiority of any such method based on samples [5]. In particular, the training error is an insufficient indicator of estimator's performance due to potential over-fitting that leads to poor generalization performance on future datasets.

Over-fitting is often specific to an estimator method and is less likely to occur across estimators of radically different designs. In several cases, by fusing diverse estimators both the performance and diversity of design are preserved [10]. However, the fused estimators are also subject to finite sample limits since they are also estimators. We study linear regression fusion (LR-F) and GPR fusion (GPR-F) methods, and our results show that the latter achieves lowest Root Mean Square Error (RMSE) among all estimators for RTT in three out of four scenarios. We develop analytical characterization of the performance improvements of fused estimates over individual RTT estimates under finite sample, distribution-free framework [17].

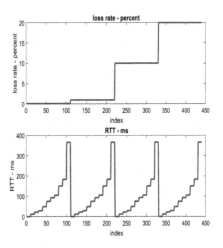

Fig. 2. Index representing increase of loss rate and RTT from left to right.

By using MPI execution times as the independent variable, we formulate the problem of estimating RTT and loss rate as a regression estimation problem. The overall results are illustrated using RTT estimates with smallest RMSE among individual methods and fusers in Fig. 1 for four loss rates. In each plot, datasets are concatenated at four loss rates in increasing order and at each loss rate RTT is increased left to right, and measurements are repeated 10 times at each RTT value as shown in Fig. 2. Among individual RTT estimates, GPR has the lowest RMSE in all four scenarios, and GPR-F fuser achieved even lower RMSE in three out of four loss scenarios, as shown in Table 1 while encompassing the design diversity of individual methods. Overall, our results show that accurate estimates can be generated at low loss rates but become inaccurate at loss rates 10% and higher, wherein the datasets appear much too complex for these methods (as in the case of deterministic periodic losses [15]). In addition, they reveal some subtle performance effects including over-smoothing by some estimators in achieving lower RMSE, and bleeding effects of RTT in loss rate estimates.

The organization of this paper is as follows. The testbed used in collecting MPI execution time measurements is described in Sect. 2. An analytical formulation of the underlying regression problem is presented in Sect. 3. Various datasets of execution time measurements are qualitatively described in Sect. 4. RTT estimators are described in Sect. 5, wherein five different ML methods are described in Sect. 5.1 and two fusers are described in Sect. 5.2. Generalization equations of the fusers for RTT estimation are described in Sect. 6. Loss rate estimators are described in Sect. 7. The performance of the estimators is qualitatively interpreted in the context of datasets at lower and higher loss rates in Sect. 8. A summary of results and directions for future work are described in Sect. 9.

2   Test Configuration

A computing cluster with InifiniBand (IB) interconnect is expanded to constitute a testbed to run MPI codes across the wide-area Ethernet connections. Additional Ethernet Network Interface Cards (NIC) are installed in two cluster computing nodes (tait1 and tait2), which are connected to Ethernet switches and a hardware-based Ethernet emulator in the configuration shown in Fig. 3. The IB connections of the cluster are subject to 2.5 ms latency limit, and hence MPI measurements over IB are not indicative of the performance over long distance connections. Specifically, the shorter distances combined with credit-based IB protocol flow control do not adequately reflect the complex variations of TCP over wide-area connections, particularly under packet losses. Furthermore, due to their latencies, wide-area networks are more prone to more losses compared to IB networks.

Typical wide-area connections consist of a number of switches and routers whereas IB connections have fewer IB switches. This testbed connection consists of two Ethernet switches between the source computing node and a port of the emulator, which reflects a site connection. Similarly at the other end, the connection consists of two Ethernet switches between the second port of the emulator and destination computing node. Thus, this symmetric end-to-end connection consists of four Ethernet internal cross-connections, six short Ethernet segments and one emulated long distance Ethernet connection with variable latency, loss rate and type of loss distribution.

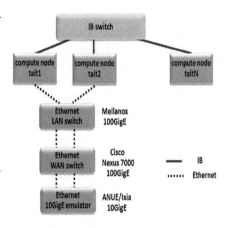

Fig. 3. Configuration for long Ethernet connection between compute nodes of IB cluster.

ANUE/Ixia hardware-based emulator is used to collect MPI measurements over Ethernet connections with 11 Round Trip Times (RTT) in 0–366 ms range. These RTT values are strategically chosen to represent three ranges: (a) smaller values represent cross-country connections, for example, computing facilities distributed across the US, (b) 93–183 ms represent inter-continental connections, and (c) 366 ms represents a connection spanning the globe, which is mainly used as a limiting case. External periodic and random packets losses are introduced using ANUE/Ixia devices at four different loss rates. These emulators delay the packets as per the specified RTT value, and thus closely emulate the physical long distance path. Equally importantly, these emulations closely match TCP dynamics of physical connections with corresponding RTTs, which is a critical factor in assessing MPI performance over long distance connections. In particular, these emulations lead to different TCP dynamics and responses under deterministic (periodic) and random losses of Ethernet segments [14], which result in a wider spread of the execution times at high loss rates under random losses (Sect. 8).

3   Analytical Formulation

We now provide a formal description of the underlying estimation problems to support subsequent analytical treatment of RTT and loss rate estimation methods. Let E be a random variable representing the execution time of MPI Send_Receive primitive; it is distributed according to the joint probability distribution $\mathbb{P}_{E,R,L}$, where R and L are the random variables representing RTT and loss rate, respectively. In general, the distribution $\mathbb{P}_{E,R,L}$ is quite complex since it depends on the properties of the network connection and host systems, and also the software stack consisting of the operating system, networking and

MPI modules. Given an execution time measurement $E = e$, the conditional distribution $\mathbb{P}_{R,L|e} = \mathbb{P}_{R,L|E=e}$, characterizes the distribution of RTT and loss rate at this value e. Then, RTT-regression function f^R is defined as the expected value of RTT at $E = e$ given by

$$f^R(e) = \int R d\mathbb{P}_{R,L|e},$$

which is averaged over both R and L at each e. The loss-regression function f^L is the expected value of the loss rate at $E = e$, which is similarly given by

$$f^L(e) = \int L d\mathbb{P}_{R,L|e}.$$

In general, these regressions cannot be obtained even in theory since the underlying distribution $\mathbb{P}_{E,R,L}$ is unknown. In stead, ML methods are employed to estimate their approximations using a training sample $(E_i, R_i, L_i), i = 1, 2, \ldots, l$, wherein E_i is the execution time measured over a connection with RTT R_i and loss rate L_i. The distributions of the connection parameters R and L are determined by the design of connection configurations, and are fixed while the measurements of E are repeated. Thus, the distribution of E encompasses factors due to the properties of physical connection parameters as well as operating system, TCP and MPI modules.

Then, RTT and loss rate estimation problems can be cast as estimating the regression functions f^R and f^L, respectively, using measurements. We consider that RTT-regression estimate $\hat{f}_A^R$ is obtained by method $A \in \mathscr{A} = \{\text{EOT,GPR,LR,RT,SVM,LR−F,GPR−F}\}$ using the measurement pairs $(E_i, R_i), i = 1, 2, \ldots, l$. Similarly, the loss-regression estimate $\hat{f}_A^L$ is obtained by method $A \in \mathscr{A}$ using the measurement pairs $(E_i, L_i), i = 1, 2, \ldots, l$. At a given execution time $E = e$, $\hat{f}_A^R(e)$ and $\hat{f}_A^L(e)$ are the estimates of RTT and loss rate, respectively, provided by method A.

4   Execution Time Measurements

The execution times of MPI Send_Receive operations collected at the application-level are shown as a function of RTT in Fig. 4 for loss rates, 0.1, 1, 10 and 20%, of externally induced losses under four loss scenarios, one deterministic periodic and three random, namely uniform, Poisson and Gaussian. Their increasing trend as a function of RTT is evident at lower loss rates, 0.1% and 1%, but it becomes less prominent at 10% loss rate, and essentially disappears at 20% loss rate as outliers dominate. Overall, the execution times as well as their variations increase as loss rate is increased, which is an indication of the increased complexity of their estimation at higher loss rates.

In terms of losses, the execution times are shown as function of loss rates 0.1, 1, 10 and 20% in Fig. 5 under the four loss scenarios. The measurements at any loss rate encompass all 11 RTT values, and 10 repeated measurements

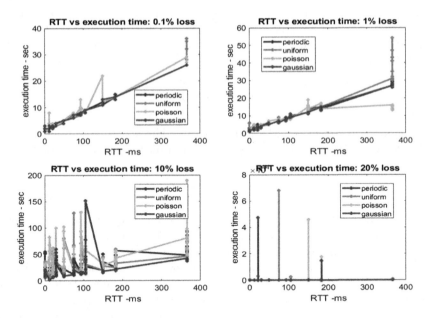

Fig. 4. Execution times of MPI_Sendrecv operations as function of RTT under four external loss scenarios, namely, periodic, uniform, Poisson and Gaussian.

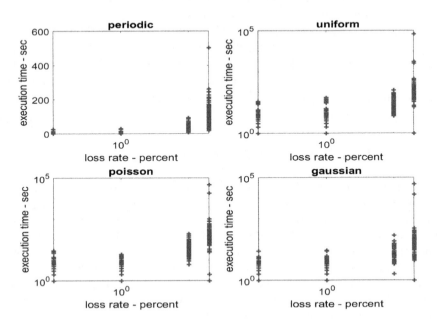

Fig. 5. Execution times of MPI_Sendrecv operations as function of loss rate under four external loss scenarios, namely, periodic, uniform, Poisson and Gaussian.

at each RTT value. Their increasing trend as a function of loss rate is evident overall but is sharper for periodic losses and is more diffused with overlaps across loss rates in all three random loss scenarios. The ranges of execution times are much wider for random losses compared to periodic losses. Also, the execution times as well as their variations increase overall as loss rate is increased for deterministic periodic loss scenario. But, in random loss scenarios the variations are more subtle: their spread is similar at all loss rates except at 20%, wherein a few measurements are very large, which indicate the complexity of loss rate estimation in these scenarios.

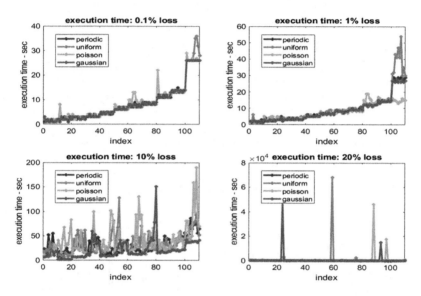

Fig. 6. Traces of execution times in seconds of MPI_Sendrecv operations under four external loss cases.

5   RTT Estimators

We present RTT estimates in the form of traces that are indexed by groups of 440 measurements that correspond to increasing loss rates, and within each group we have 11 sub-groups that correspond to increasing RTT values as shown in Fig. 2; each sub-group corresponds to 10 repeated measurements at a fixed pair of loss rate and RTT. The 440 measurements for each loss scenario shown in Fig. 6 will be used to compare qualitatively with the corresponding RTT estimate traces. We utilize the regression estimation codes from matlab statistics toolbox.

5.1   Five Estimators

The five estimation methods are chosen to reflect different characteristics of the underlying regressions, namely, smooth and non-smooth functions, respectively.

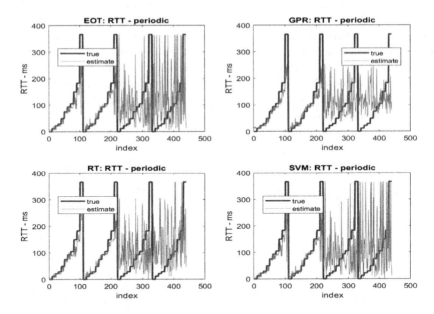

Fig. 7. Periodic losses.

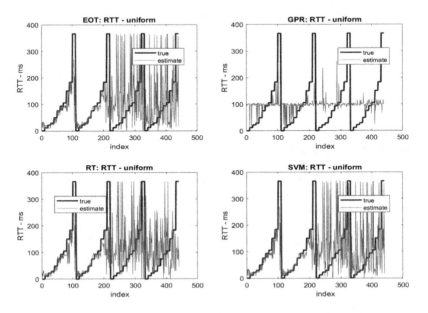

Fig. 8. Uniform losses.

GPR and SVM with Gaussian kernels [16] are based on non-linearly transforming the feature space X into regression space of Y. They both provide smooth regression functions f_{GPR} and f_{SVM}, and their respective function classes $\mathscr{F}_{GPR}$ and $\mathscr{F}_{SVM}$ consist of smooth functions as a result of Gaussian kernels. EOT [2,8] method is based on boosting of a collection of classification trees that are customized to fit the training data using the AdaBoost method. RT [3] methods is also based on trees that are customized to fit the training data. They both lead to a highly non-smooth regression functions f_{EOT} and f_{RT}, and their function classes $\mathscr{F}_{EOT}$ and $\mathscr{F}_{RT}$ consists of a collection of decision tree. LR is a smooth and linear method and leads to f_{LR} from the function classes $\mathscr{F}_{LR}$, which is effective in RTT estimation under no losses [15] but is quite limited under losses as indicated by its RMSE in Tables 1 and 2.

The estimators under periodic, uniform, Poisson and Gaussian loss scenarios are shown in Figs. 7, 8, 9 and 10, respectively. Under periodic and Gaussian losses, all estimates are more accurate at 0.1% and 1% loss rate but are inaccurate at higher loss rates; in particular, they capture the increasing trends in RTT at low loss rate but exhibit high variation at 10% and 20% loss rate. Under unform and Poisson losses, GPR method does not capture the increasing RTT trend at any loss rate, while other non-linear estimates captured it. Interestingly, GPR achieved lowest RMSE among individual estimators which is due to the inclusion of measurements at higher loss rate that resulted in "averaging" across all loss rates. This undesirable artifact of low RMSE but less accurate estimate at low loss rates is illustrated in Figs. 8 and 9. LR and SVM methods have highest and second highest RMSE among the twenty cases in Table 1.

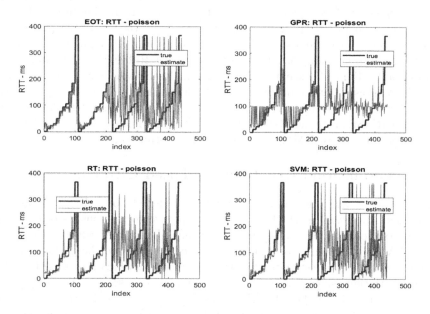

Fig. 9. Poisson losses.

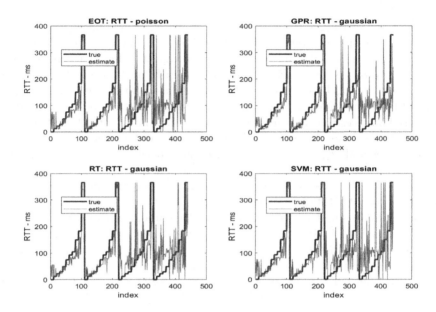

Fig. 10. Gaussian losses.

5.2   Estimator Fusers

The estimators from five individual methods are used as 5-dimensional input to a fuser which produces RTT as its output. The linear regression fuser (LR-F) is a linear combination of the individual estimators, and the GPR fuser (GPR-F) is obtained using GPR method based on the outputs of five estimators corresponding to the training sample. GPR-F achieved lower RMSE than best individual estimator GPR in all except under Gaussian losses, whereas LR-F has lower RMSE for Poisson and uniform losses. As shown in Fig. 1, the fused estimates were able to capture the increasing RTT trend at lower loss rates while achieving lower RMSE error than GPR under uniform and Poisson losses, unlike GPR estimator with lowest RMSE among individual estimators.

6   Generalization Equations for Fused Estimates

We consider five individual estimates, indexed by $A \in \mathscr{A}_I = \{\text{EOT,GPR,LR,RT,SVM}\}$, such that the fuser input vector X consists of five real-valued components, $X^A, A \in \mathscr{A}_I$, and output Y is a real-valued estimate of RTT. RMSE values in Table 1 have been used to compare the performance of fusers and individual estimators in previous sections, which are subject to statistical variations since they are computed based on a sample. We now derive confidence bounds for these RMSE values which provide analytical justification for their use. For simplicity of presentation of analytical results, we use MSE in place of RMSE following the common practice in finite sample theory [17].

6.1   Regression Problem: Finite Sample Generalization

In a generic regression estimation problem the feature vector $X \in \Re^d$ and the output vector $Y \in \Re$ are distributed jointly accordingly to an unknown distribution $\mathbb{P}_{X,Y}$. The *expected error* of a regression function f is

$$I(f) = \int (f(X) - Y)^2 \, d\mathbb{P}_{X,Y}.$$

The *expected best* regression estimator f^* minimizes $I(.)$ over $\mathscr{F}$, i.e., $I(f^*) = \min_{f \in \mathscr{F}} I(f)$. The *empirical error* $\hat{I}(f)$ based on training data $(X_i, Y_i), i = 1, 2, \ldots, l$, is defined as

$$\hat{I}(f) = \frac{1}{l} \sum_{i=1}^{l} (f(X_i) - Y_i)^2$$

It is an approximation of $I(f)$ computed based on the training data. The *empirical best* regression estimator $\tilde{f}$ minimizes $\hat{I}(.)$ over $\mathscr{F}$, i.e., $\hat{I}(\tilde{f}) = \min_{f \in \mathscr{F}} \hat{I}(f)$.

The joint distribution $\mathbb{P}_{X,Y}$ of data is complex, domain specific, and is only partially known. In our context, it depends on the finer details of the underlying software and hardware components, which may manifest as additional random variables. For an individual estimator $A \in \mathscr{A}_I$, X and Y correspond to execution time E and RTT R, respectively, and $\mathbb{P}_{X,Y}$ corresponds to $\mathbb{P}_{E,R,L}$ which involves additional random variable of the loss rate L. For fusers, X and Y correspond to 5-dimensional vector consisting of outputs of estimators and RTT R, respectively. In general, an optimal f^* cannot be computed precisely with probability one even in principle, since $\mathbb{P}_{X,Y}$ is either unknown or not computationally conducive. Under certain conditions, Vapnik's generalization theory [17] establishes that there exists a *confidence function* $\delta(.)$ such that for a "suitable" estimator $\hat{f}$ obtained from training data we have

$$\mathbb{P}_{X,Y}^l \left[I(\hat{f}) - I(f^*) > \varepsilon \right] < \delta(\varepsilon, \hat{\varepsilon}, l) \qquad (1)$$

where $\varepsilon, \hat{\varepsilon} > 0$, $0 < \delta < 1$, and $\hat{I}(\hat{f}) = \min_{f \in \mathscr{F}} \hat{I}(f) + \hat{\varepsilon}$. This condition ensures that "error" of $\hat{f}$ is within ε of optimal error (of f^*) with probability $1 - \delta$, *irrespective* of the underlying measured or computed data distribution $\mathbb{P}_{X,Y}^l$. Furthermore, under these conditions, the confidence parameter $\delta(\varepsilon, \hat{\varepsilon}, l)$ approaches 1 as the sample size l approaches infinity.

Consider the fuser class $\mathscr{F}_F$ used in fusing the estimators $f_A \in \mathscr{F}_A, A \in \mathscr{A}_I$. Let f_F denote the regression function obtained by composing f_A's with the fuser function from $\mathscr{F}_F$. The *error reduction* Δ_F of the fused estimate over the best individual classifier is defined as

$$\Delta_F = \min_{A \in \mathscr{A}_I} I(f_A) - I(f_F).$$

Then, if $\mathscr{F}_F$ has the isolation property [11], then $\Delta_F \geq 0$. The best error reduction is given by

$$\Delta_F^* = \min_{A \in \mathscr{A}_I} I(f_A^*) - I(f_F^*).$$

and its estimate based on a sample is given by

$$\tilde{\Delta}_F = \min_{A \in \mathscr{A}_I} \hat{I}(f_A) - \hat{I}(f_F).$$

The error reduction values Δ_F based on measurements are shown in Table 1 for the two fusers LR-F and GPR-F. GPR-F has positive $\tilde{\Delta}_F$ values in three scenarios indicating that the fused estimate has lower RMSE than the lowest of its constituent estimators (namely, GPR). LR-F has positive $\tilde{\Delta}_F$ values in two scenarios, which might be attributed to the lack of the required statistical independence in estimator outputs. We show in the next section that the estimate $\tilde{\Delta}_F$ reflects the optimal improvement Δ_F^* achievable by the fuser within a formal framework.

6.2   Estimator Fusers: Generalization Equations

The generalization bound $\delta(\varepsilon, \hat{\varepsilon}, l)$ applicable to five individual estimators can be derived using various properties of the corresponding estimator classes [12]. In particular, these bounds for GPR and SVM with Gaussian kernels could be based on fat-shattering index [16], and for EOT and RT they may be based on bounded total variation [1]. In Theorem 1, we assume that these generalization bounds are available from existing works, and their detailed derivations are beyond the scope of this paper.

We now show that the estimate $\tilde{\Delta}_F$ is within ε of the optimal Δ_F^* with a probability that improves with the training data size l independent of the underlying distribution $\mathbb{P}_{Y,X}$.

Theorem 1. *Consider that there exists $\delta_B(\varepsilon, \hat{\varepsilon}_B, l)$ such that based on i.i.d. l-sample, we have*

$$\mathbb{P}_{X,Y}^l \left[I(\hat{f}_B) - I(f_B^*) > \varepsilon \right] < \delta_B(\varepsilon, \hat{\varepsilon}_B, l). \tag{2}$$

for all individual estimators $B \in \mathscr{A}_I$, $N_{\mathscr{A}_I} = |\mathscr{A}_I|$, and both fusers $B = \text{LR-F}, \text{GPR-F}$ such that $\delta_B(\varepsilon, \hat{\varepsilon}_B, l) \to 0$ as $l \to \infty$. Then, the probability that the closeness between $\tilde{\Delta}_F$ and Δ_F^* is within ε is bounded as

$$\mathbb{P}_{X,Y}^l \left[|\tilde{\Delta}_F - \Delta_F^*| < \varepsilon \right]$$
$$> 1 - \delta_D(\varepsilon/2, \hat{\varepsilon}_D, l) - \sum_{A \in \mathscr{A}_I} \delta_A(\varepsilon/(2N_{\mathscr{A}_I}), \hat{\varepsilon}_A, l),$$

for both fusers $D = \text{LR-F}, \text{GPR-F}$.

Proof. We first note that for $D = \text{LR-F}, \text{GPR-F}$

$$|\tilde{\Delta}_F - \Delta_F^*| \le |\hat{I}(\hat{f}_D) - I(f_D^*)| + \left| \min_{A \in \mathscr{A}_I} \hat{I}(\hat{f}_A) - \min_{A \in \mathscr{A}_I} I(f_A^*) \right|,$$

which establishes that the condition $|\tilde{\Delta}_F - \Delta_F^*| > \varepsilon$ implies that at least one term on the right hand side is greater than $\varepsilon/2$. We now have

$$|\hat{I}(\hat{f}_D) - I(f_D^*)| \leq |\hat{I}(\hat{f}_D) - I(\hat{f}_D)| + |I(\hat{f}_D) - I(f_D^*)|,$$

which in turn establishes that the condition $|\hat{I}(\hat{f}_D) - I(f_D^*)| > \varepsilon/2$ implies that at least one term on the right hand side is greater than $\varepsilon/4$. Then, by hypothesis in Eq. (2), both conditions are simultaneously satisfied with probability at most $\delta_D(\varepsilon/4, \hat{\varepsilon}_d, l)$. Similarly, we have

$$\left| \min_{A \in \mathscr{A}_I} \hat{I}(\hat{f}_A) - \min_{A \in \mathscr{A}_I} I(f_A^*) \right| \leq \left| \min_{A \in \mathscr{A}_I} \hat{I}(\hat{f}_A) - \min_{A \in \mathscr{A}_I} I(\hat{f}_A) \right|$$
$$+ \left| \min_{A \in \mathscr{A}_I} I(\hat{f}_A) - \min_{A \in \mathscr{A}} I(f_A^*) \right|,$$

which in turn establishes that the condition $\left| \min_{A \in \mathscr{A}_I} \hat{I}(\hat{f}_A) - \min_{A \in \mathscr{A}_I} I(f_A^*) \right| > \varepsilon/2$ implies that at least one term on the right hand side is greater than $\varepsilon/4$. Then, we consider the two upper bounds

$$\left| \min_{A \in \mathscr{A}_I} \hat{I}(\hat{f}_A) - \min_{A \in \mathscr{A}} I(\hat{f}_A) \right| \leq \sum_{A \in \mathscr{A}_I} \left| \hat{I}(\hat{f}_A) - I(\hat{f}_A) \right|$$

$$\left| \min_{A \in \mathscr{A}_I} I(\hat{f}_A) - \min_{A \in \mathscr{A}_I} I(f_A^*) \right| \leq \sum_{A \in \mathscr{A}_I} \left| I(\hat{f}_A) - I(f_A^*) \right|.$$

In each case, the condition that left hand side is larger than $\varepsilon/2$ implies at least one of the terms under the summation is greater $\varepsilon/(2N_{\mathscr{A}_I})$. Under the hypothesis of this theorem in Eq. (2), both conditions are satisfied with probability at most

$$\sum_{A \in \mathscr{A}_I} \delta_A(\varepsilon/(2N_{\mathscr{A}_I}), \hat{\varepsilon}_a, l).$$

By combining the above terms together, we have

$$\mathbb{P}_{X,Y}^l \left[|\tilde{\Delta}_F - \Delta_F^*| > \varepsilon \right]$$
$$< \delta_D(\varepsilon/2, \hat{\varepsilon}_D, l) + \sum_{A \in \mathscr{A}_I} \delta_A(\varepsilon/(2N_{\mathscr{A}_I}), \hat{\varepsilon}_A, l),$$

which proves the theorem.   □

The confidence bound in this theorem is distribution-free in that it does not depend on $\mathbb{P}_{X,Y}$. It is expressed in terms of the *precision* parameter ε and the *confidence* parameter $\left[1 - \delta_D(\varepsilon/2, \hat{\varepsilon}_D, l) - \sum_{A \in \mathscr{A}_I} \delta_A(\varepsilon/(2N_{\mathscr{A}_I}), \hat{\varepsilon}_A, l) \right]$, which approaches 1 with increasing number of measurements l.

Table 2. Loss rate estimates RMSE of five estimators and linear fuser.

Loss type	EOT	GPR	LR	RT	SVM	LR-F
Gaussian	**5.19**	6.39	5.34	6.82	7.17	6.42
Periodic	7.26	**6.62**	6.82	7.00	6.91	6.81
Poisson	6.99	**6.61**	7.42	6.93	6.72	6.74
Uniform	7.26	**6.62**	6.82	7.00	6.91	6.81

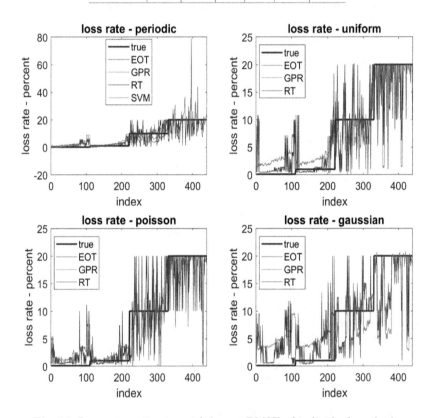

Fig. 11. Loss rate estimates with lowest RMSE of individual methods.

7   Loss Rate Estimators

The loss estimates of four non-linear estimators are shown in top left plot of
Fig. 11 for periodic losses. For random losses, SVM estimates have extreme variations and hence are omitted in the plots. Also, LR estimator is omitted in all
plots due to its extremely large variation under all loss scenarios. Qualitatively,
smooth SVM and non-smooth RT methods both exhibit large variations, which
indicate the underlying properties of the data rather than these methods; indeed
GPR is the only method that did not produce large variations. RMSE of five
estimators and linear fuser LR-F are shown in Table 2. Methods with lowest

RMSE for loss rate estimation are shown in Fig. 12, which are EOT for periodic losses and GPR for all others. Both of them exhibited lower variations at low loss rate and an increasing trend as RTT is increased at a fixed loss rate. As in the case of RTT estimation, the "averaging" by GPR resulted in lower RMSE but less accurate estimates at low loss rates; but, interestingly, this effect is more dominant for Gaussian errors unlike for RTT estimation. In almost all cases, at fixed loss rate, the estimators showed an increasing trend as RTT is increased.

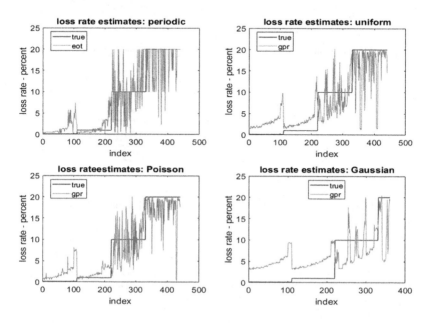

Fig. 12. Loss rate estimates with lowest RMSE of individual methods.

The large variations shown in the scatter plots in Fig. 5 indicate high RMSE by any estimate since its output is a function and data dispersed around it contributes to RMSE. In particular, a smooth estimate will not be able to capture these variations as applicable to GPR and SVM methods. While tree-based methods in principles can capture such variations, they require a large number of leaf nodes, and those with smaller number will result in large RMSE. In summary, the results indicate the challenging nature of the underlying datasets, which in some sense expose the limitations of the conventional ML approaches for loss rate estimation.

8   Execution Time Measurements: Data Regressions

Qualitative insights into the performance of regression estimators can be gained by examining the scatter plots of measurements separately at low and high loss rates. Overall increasing trend of RTT when plotted as a function of execution

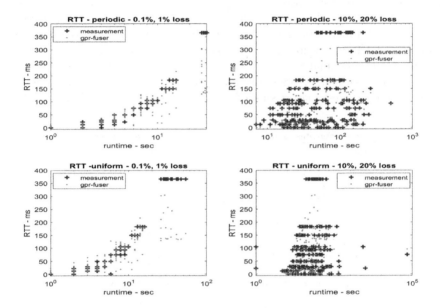

Fig. 13. Data regressions under periodic and uniform losses.

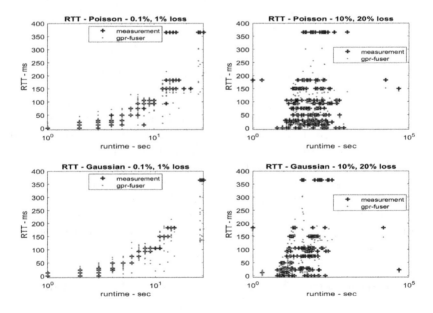

Fig. 14. Data regressions under Poisson and Gaussian losses.

time is evident under 0.1% and 1% loss rates, but not under and 10% and 20% loss rates in all four scenarios, as shown in Fig. 13 for periodic and unform losses, and in Fig. 14 for Poisson and Gaussian losses. Even GPR-F estimate with the lowest overall RMSE captures the increasing trend only in the former case but not in the latter case. Qualitatively, the wide spread of measurements at high loss rates indicates the lack of information needed to estimate RTT by any method that uses regression function, smooth or non-smooth.

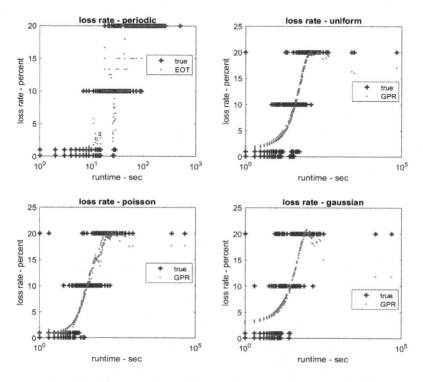

Fig. 15. Data regressions of loss estimators with lowest RMSE.

For loss rates, the scatter plots are shown in Fig. 15, which have significant variations in execution times at fixed values of loss rate. The non-smooth EOT method with lowest RMSE under periodic losses captures several loss rates at 10 and 20% loss rates as shown in top left plot. The smooth GPR estimator is plotted along with data in all three random loss scenarios in which it has lowest RMSE; several estimated points are in between the loss rate values, and the estimator shows a continuous trend in the mapping from measurements to loss rate. Similar to RTT estimates, an overall increasing trend of loss rate when plotted as a function of execution time is evident under 0.1% and 1% loss rate; but, the loss rate is fixed at two values 0.1 and 1% as shown in Fig. 16 for periodic and unform losses, and in Fig. 17 for Poisson and Gaussian losses.

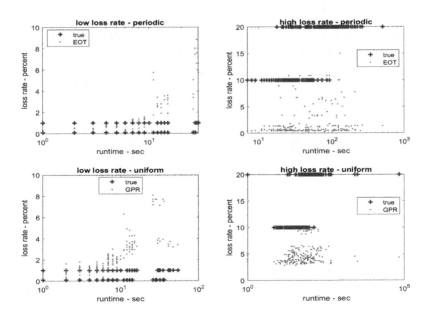

Fig. 16. Loss regression at low and high loss rates for periodic and uniform losses.

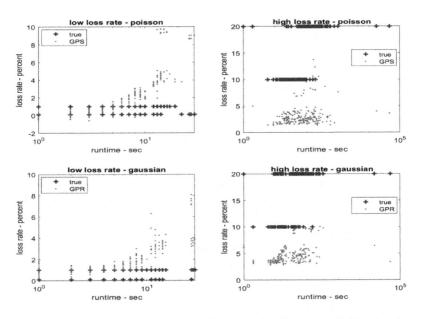

Fig. 17. Loss regression at low and high loss rates for Poisson and Gaussian losses.

This is a "bleed over" artifact due to increasing RTT at both 0.1 and 1%loss rates. Under 10% and 20% loss rates in all four scenarios there is a significant scatter in loss estimates, indicating the underlying complexity of regression estimation.

9   Conclusions

Rich datasets of MPI measurements are becoming increasingly available as more and more computations are distributed over wide-area networks. These measurements exhibit certain characteristics, such as longer executions times and large variations, that are atypical of conventional MPI applications executed on single computing systems with Inifiniband or custom interconnects. Losses are integral to wide-area networks as TCP that supports MPI utilizes self-induced losses to pace its flows. Consequently, these distributed computations need to mitigate the inefficiencies due to network delays and their variations. These computations may be distributed across geographically dispersed nodes that are dynamically identified; consequently, the RTTs and loss rates of the underlying connections may not be a priori known. The MPI measurements collected at the application-level reflect the connection length and losses, and have been shown to be useful in estimating RTT and loss rate using ML methods, albeit accurately only at low loss rates.

Complementing previous works under deterministic periodic loss scenarios, we studied five ML methods to estimate the connection RTT and loss rates under random losses, which are more reflective of practical scenarios. As in previous works [15], the results show that accurate estimates can be generated at low loss rates but they become inaccurate at loss rates 10% and higher. However, this randomness manifests in subtle ways, resulting in different performances of non-linear estimators; in particular, GPR that achieves low RMSE does not provide accurate RTT estimates at low loss levels, unlike others with higher RMSE. These effects are mainly due to the highly non-linear response of the underlying TCP dynamics that "amplify" the randomness of losses. Furthermore, it is equally complex to assess the performance of ML methods due to their non-linear nature, and their fusers are only effective in some scenarios for RTT estimation. In another direction, these results highlight the strengths and limitations of ML methods for network-level estimation problems using application-level measurements.

This work constitutes only initial steps in understanding the complexity of estimating network-level parameters using application-level measurements, and the performance of various ML solutions, including individual and fused estimates. Future work may involve studying the random losses due to external traffic in production networks, which may not follow known random processes. Since there is no universal way to choose among various ML methods from sample performance only, it would be of future interest to investigate into domain specific customizations, hyper-parameter tuning, fusers and other approaches to RTT and loss rate estimation [12,13].

References

1. Anthony, M., Bartlett, P.L.: Neural Network Learning: Theoretical Foundations. Cambridge University Press, Cambridge (1999)
2. Breiman, L.: Random forests. Mach. Learn. **45**(1), 5–32 (2001)
3. Breiman, L., Friedman, J.H., Olshen, R.A., Stone, C.J.: Classification and Regression Trees. Wadsworth and Brooks, Grove (1984)
4. Chhabra, A., Kiran, M.: Classifying elephant and mice flows in high-speed scientific networks. In: IEEE/ACM Workshop on Innovating the Network for Data-Intensive Science (INDIS) (2017)
5. Devroye, L., Gyorfi, L., Lugosi, G.: A Probabilistic Theory of Pattern Recognition. Springer, New York (1996). https://doi.org/10.1007/978-1-4612-0711-5
6. Giannakou, A., Gunter, D., Peisert, S.: Flowzilla: a methodology for detecting data transfer anomalies in research networks. In: IEEE/ACM Workshop on Innovating the Network for Data-Intensive Science (INDIS) (2018)
7. Hassan, M., Jain, R.: High Performance TCP/IP Networking: Concepts, Issues, and Solutions. Prentice Hall, Upper Saddle River (2004)
8. Hastie, T., Tibshirani, R., Friedman, J.: The Elements of Statistical Learning: Data Mining, Inference, and Prediction. Springer, New York (2001). https://doi.org/10.1007/978-0-387-84858-7
9. Lakshman, T.V., Madhow, U., Suter, B.: TCP/IP performance with random loss and bidirectional congestion. IEEE/ACM Trans. Netw. **8**(5), 541–555 (2000)
10. Rao, N.S.V.: On fusers that perform better than best sensor. IEEE Trans. Pattern Anal. Mach. Intell. **23**(8), 904–909 (2001)
11. Rao, N.S.V.: Measurement-based statistical fusion methods for distributed sensor networks. In: Iyengar, S.S., Brooks, R.R. (eds.) Distributed Sensor Networks, 2nd edn. Chapman and Hall/CRC Publishers, Boca Raton (2011)
12. Rao, N.S.V.: Finite-sample generalization theory for machine learning practice for science. In: DOE ASCR Scientific Machine Learning Workshop (2018)
13. Rao, N.S.V., et al.: Classifiers for dissolution events in processing facility using effluents measurements. In: Institute of Nuclear Materials Management Annual Meeting (2019)
14. Rao, N.S.V., Imam, N., Boehm, S.: A case study of MPI over long distance connections. In: 13th Annual IEEE International Systems Conference (2019)
15. Rao, N.S.V., Imam, N., Liu, Z., Kettimuthu, R., Foster, I.: Estimation of RTT and loss rate of wide-area connections using MPI measurements. In: IEEE/ACM Workshop Innovating the Network for Data-Intensive Science (INDIS) (2019)
16. Scholkopf, B., Burges, C.J.C., Smola, A.J. (eds.): Advances in Kernel Methods. MIT Press, Cambridge (1999)
17. Vapnik, V.N.: Statistical Learning Theory. Wiley, New York (1998)

Algorithm Selection and Model Evaluation in Application Design Using Machine Learning

Srikanth Bethu[1,2(✉)], B. Sankara Babu[1,2], K. Madhavi[1,2], and P. Gopala Krishna[1,2]

[1] Department of Computer Science and Engineering, JNTU Hyderabad, Hyderabad 500090, Telangana, India
srikanthbethu@gmail.com, bsankarababu81@gmail.com, bmadhaviranjan@yahoo.com, gopalakrishna.aucsse@gmail.com
[2] Gokaraju Rangaraju Institute of Engineering and Technology, JNTU Hyderabad, Hyderabad 500090, Telangana, India

Abstract. AI has turned into a focal piece of our life – as buyers, clients, and, ideally, as scientists and professionals! Regardless of whether we are applying prescient displaying systems to our examination or business issues, accept we make them thing in like manner: We need to make "great" forecasts! Fitting a model to our preparation information would one say one is a thing, however how would we realize that it sums up well to concealed information? How would we realize that it does not only retain the information we sustained it and neglects to make high forecasts on future examples, tests that it has not seen previously? Additionally, how would we select an appropriate model in any case? Perhaps an alternate learning calculation could be more qualified for the current issue? The right utilization of model assessment, model choice, and calculation choice systems is indispensable in scholarly AI examine just as in numerous mechanical settings. This article audits various systems that can be utilized for every one of these three subtasks and talks about the primary focal points and drawbacks of every method with references to theoretical and observational investigations. Further, suggestions are given to empower best yet plausible practices in research and uses of AI. In this article, we have used applications like Drowsiness detection, Oil price prediction, Election result evaluation as examples to explain algorithm selection and model evaluation.

Keywords: Algorithms · Machine learning · Performance evaluation

1 Introduction

Computer vision [1] is the changing of data from a still or camcorder into either a depiction or another decision. Every such change is associated with achieving a particular target. PC obtains a system of numbers from a camera or the plate, and it is straightforward as that. Ordinarily, there is no worked in model affirmation or modified control of focus and hole, no cross-association with significant lots of inclusion. For the most part, vision structures are still really unsuspecting.

S. Boumerdassi et al. (Eds.): MLN 2019, LNCS 12081, pp. 175–195, 2020.
https://doi.org/10.1007/978-3-030-45778-5_12

PC based insight/Machine Learning [2] is the predictable assessment of calculations and exact models that PC structures use to play out a particular errand without utilizing express appropriately manages, dependent upon models and thinking. It viewed as a subset of false information. Mimicked insight calculations [3] produce a numerical model of test information, known as "arranging information," to pick wants or choices without being unequivocally adjusted to playing out the undertaking. Human-made insight estimations utilized in a comprehensive gathering of employments, for example, email confining and PC vision, where it is infeasible to build up a figuring of express precludes for playing the assignment. Computerized reasoning unflinchingly identifies with computational bits of learning, which spotlights on making wants utilizing PCs. The assessment of numerical improvement passes on techniques, theory, and application domains to the field of AI. Information mining is a field of concentrate inside AI and spotlights on exploratory information assessment through execution learning. In its application, crosswise over business issues, AI is also induced as a quick assessment.

Analytics [4] is the disclosure and correspondence of significant models in data. The immense thing to note presently is that examination is a methodology. It is an interdisciplinary methodology that regularly joins number juggling, bits of statistics [5], programming designing, perceptive techniques, data visualization [6], and various fields of study.

1.1   Estimating the Performance of Machine Learning Model

In any case, we feed the availability information to our learning estimation to get capability with a model. Second, we anticipate the names of our test set. Third, we tally the measure of wrong wants on the test dataset to enroll the model's check accuracy. Subordinate upon our objective, looking over the presentation of a model is not that unimportant, shockingly. Enable us to assemble the primary concerns why we overview the insightful introduction of a model:

- We need to evaluate the hypothesis execution, the intelligent presentation of our model on the future (unpretentious) information.
- We need to produce the farsighted showcase by tweaking the learning calculation and picking the best performing model from a given theory space.
- We need to see the AI check that is most fitting for the present issue; along these lines, we need to consider indisputable figuring's, picking the best-performing one in like way as the best performing model from the tally's hypothesis space.

Dismissing the way where these three sub-assignments recorded the above idea for all plans and reason that we need to audit the presentation of a model, they all require various structures. We will talk about a touch of the various systems for managing these sub-tries in this article. Unquestionably, we need to check the future execution of a model as certainly as could be customary pondering the current condition. Notwithstanding, if there is one key to remove a message from this article, it is that uneven execution evaluations are extraordinarily all right in model choice and check choice if the partiality impacts all models reasonably. If we rank various models or mean something negative for one another to pick the best-performing one, we essentially need to know the "relative" execution. For instance, if all our introduction examinations are suspiciously uneven, and we junk their grandstands by 10%, it would not impact the arranging request. Amazingly more solidly, if we have three models with need exactness measures, for example,

L2: 70% > L1: 65% > L3: 60%,

We would look at present position them a comparable path if we incorporate a 10% pessimistic inclination:

L2: 60% > L1: 55% > L3: half.

Startlingly, on the off chance that we report the future want exactness of the best-arranged model (L2) to be 60%, this would be very mistaken. Looking over the total execution of a model is likely one of the most testing assignments in AI.

Let us understand with the following example, how we design and evaluate a model.

1.2   Driver Drowsiness Detection Application Design and Its Issues

In this application, we watch the Driver Drowsiness [7, 17] ID, which is a vehicle prosperity development which foresees setbacks when the driver is getting worn out. Various assessments have prescribed that around 20% of all road incidents are exhaustion related, up to half on explicit boulevards. Driver fatigue is an essential factor in a large number of vehicle incidents. Driver carelessness might be the eventual outcome of nonattendance of sharpness when driving in light of driver drowsiness and redirection. In perspective on the getting of video from the camera, that is before the driver performs ceaseless planning of a moving toward video stream to deduce the driver's level of shortcoming in case the drowsiness is evaluated, by then it will give the alert by identifying the eyes.

The main explanation behind this concept was to use the retinal reflection as an approach to managing to find the eyes on the face, and starting their forward, using the nonattendance of this reflection as a framework for perceiving when the eyes closed. Applying the computation on the consecutive video housings may help in the figuring of the eye end period. The eye end period for sluggish drivers is more extended than standard flashing. So we will alert the driver when the eye closed distinguished. Nowadays, a consistently expanding number of reasons for a living require a whole deal obsession. Drivers must watch out for the road so that they can react to unforeseen events right away. Driver exhaustion, much of the time, transforms into a quick purpose behind some vehicle crashes. Therefore, there is a need to develop the structures that will perceive and tell a driver of her/him horrendous psycho-physical condition, which could inside and out reduce the amount of exhaustion related minor collision.

Nevertheless, the progression of such structures experiences various difficulties related to brisk and suitable affirmation of a driver's exhaustion symptoms. One of the specific possible results to complete driver laziness area structures is to use the vision-based system. This article presents the at present used driver drowsiness area structures. The particular pieces of using the vision structure to perceive driver lethargy discussed. A couple of assessments have conveyed various evaluations of the level of absence of rest as it relates to road incidents. Moreover, driver preoccupation or carelessness is another fundamental issue for safe driving.

The algorithm selection and performance evaluation was discussed in the coming chapter in related to machine learning.

1.3   Prediction of Election Results Application Design and Its Issues

Sentiment Analysis [9, 24] seen as a grouping of AI and ordinary language getting ready. It is used to evacuate, see, or portray evaluations from different substance structures, including news, surveys, and articles, and sorts them as positive, fair, and harmful. Estimation assessment has overwhelmingly used in data science for the examination of customer reactions on things and reviews. They are used to appreciate customer assessments on different kinds of things, friendliness organizations like travel, motel arrangements. It has, in like manner, ended up being not able to research customer tweets – positive, negative, or fair-minded by crawling twitter through APIs.

In this application, we will separate examples in the Indian General Election 2019 by utilizing idea examination of Twitter data. The accumulated tweets are analyzed using a word reference-based approach to managing to choose the estimations of the all-inclusive community. We choose the furthest point and subjectivity measures for the accumulated tweets that help in understanding the customer supposition for a particular candidate.

The algorithm selection and performance evaluation was discussed in the coming chapter in related to machine learning.

1.4   Crude Oil Price Detection Application Design and Its Issues

Crude oil is the world's driving fuel, and its expenses bigly influence the overall condition, economy similarly as oil examination and abuse works out. Oil Price [10, 26] guesses are especially useful to ventures, governments, and individuals. AI-anyway various procedures have been delivered at predicting oil costs, and it remains one of the most testing assessing issues on account of the high shakiness of oil costs gauging models that foresee future events used in different fields, for instance, budgetary angles and science since they are essential gadgets in essential administration. A perfect guess gives understanding into the implications of an action or inaction and fills in as an estimation to condemn one's ability to affect future events. Appropriately, buyers are in all regards at risk to use more oil and like this increase the carbon spread.

On the other hand, bolstered low oil expenses could incite a drop in overall oil and gas examination, and abuse works out. Fluctuating oil costs furthermore accept a critical activity in the overall economy. The fall in oil expenses would achieve an unassuming lift to worldwide money related development, disregarding the way that the owners of oil parts suffer pay hardships. Progressing assessment from the World Bank exhibits that for each 30% rot of oil costs, the overall GDP (Gross Domestic Product) would extend by 0.5%. At the same time, the drop in oil expenses would lessen the reasonable expense for fundamental things, and in this manner, the development rate would fall.

The algorithm selection and performance evaluation was discussed in the coming chapter in related to machine learning.

2   Literature Survey

Model assessment is a tricky subject. To ensure that we do not wander a lot from within the message, let us make certain suppositions and turn out a touch of the particular terms that we use throughout this article. We shall expect our models are (free and vaguely passed on), which recommends that the total of what tests have been drawn from a relative likelihood of different countries and are autonomous from one another. A situation where tests are not free would work with regular information or time-game-plan information. The issues to be considered for model evaluation are given as Learning and Classification, Prediction accuracy and Loss range 0–1, Variance, Bias., etc.

2.1   Learning and Classification

Here we will concentrate on supervised learning, one of the categories of AI and Machine learning where our objective characteristics known in our open dataset. Even though different considerations in like way apply to fall away from the faith assessment, we will concentrate on depiction, the endeavor of prominent target names to the models.

2.2   Prediction Accuracy and Loss Range 0–1 [12]

In the going with the article, we will concentrate on the longing accuracy, which portrayed as the measure of every single right gauge confined by the measure of tests. We register the longing exactness as to the measure of accurate figures detached by the measure of tests n. Or on the other hand in logically formal terms, we depict the longing exactness ACC as

$$ACC = 1 - ERR$$

Where the check blunders ERR is figured as the run of the mill estimation of the 0–1 hardship over n tests in a dataset S:

$$ERR_S = \frac{1}{N} \sum_{i=1}^{n} L\left(\widehat{Y_1}, Y_i\right)$$

We will most likely get settled with a model h that has a decent speculation execution. Such a model lifts the figure exactness or, the alternate way, likelihood, C (h) of making an off-center want desire

$$C(h) = P_{r(x,y) \sim D}[h(x) \neq y]$$

Where D is the making course our information has drawn from, x is the part vector of a model with class name y.

2.3   Bias

When we utilize the term propensity in this article, we infer the quantifiable tendency (rather than the inclination in an AI structure). Guideline talking terms, the inclination of an estimator $\beta^\wedge$ is the capability between its collective worth $E[\beta^\wedge]E[\beta^\wedge]$ and the veritable estimation of a parameter β surveyed.

$$Bias = E[\beta^\sim] - \beta Bias = E[\beta^\sim] - \beta$$

In like manner, if $E[\beta^\wedge] - \beta = 0 E[\beta^\wedge] - \beta = 0$, by then $\beta^\wedge$ is a reasonable estimator of β. Significantly more solidly, we register the longing inclination as the separation between the commonplace check exactness of our model and the authentic figure accuracy. For instance, on the off chance that we procedure the gauge exactness on the arranging set, this would be an in a perfect world uneven look at of the all accuracy of our model since it would overestimate the absolute precision.

2.4   Variance [14]

The thing that matters is the quantifiable difference in the estimator $\beta^\wedge$ and its ordinary worth $E[\beta^\wedge]$

$$Variance = E([\beta^\sim - E[\beta^\sim])^2$$

The change is a degree of the abnormality of our model's figures on the off chance that we emphasize the learning system on different occasions with little dangers in the course of action set. The touchier the model-building strategy is towards these changes, the higher the capacity. Finally, let us disambiguate the terms model, theory, classifier, learning figuring, and parameters:

- Target work: In sagacious appearing, we are ordinarily amped okay with demonstrating a particular method; we have to learn or surveyed a specific, unknown purpose of imprisonment. Beyond what many would consider possible $f(x) = y$ is quite far $f(\cdot)$ that we have to appear.
- Hypothesis: A theory is a specific work that we perceive (or trust) resembles beyond what many would consider possible; the target farthest arrives at that we have to store neatly. In the setting of spam gathering, it would be a portrayal pick we thought of that enables us to separate spam from non-spam messages.
- Model: In the AI field, the terms theory and model are a critical piece of the time used comparatively. In various sciences, they can have different implications: A hypothesis could be the "educated supposition" by the inspector, and the model would be the closeness of this theory to test this hypothesis.
- Learning figuring [9]: Again, we will probably find or assessed quite far, and the learning estimation is an enormous measure of heading that attempts to show beyond what many would consider possible using our game-plan dataste. A learning count goes with a theory space; the methodology of potential speculations it examines to show the dull target purpose of control by portraying the last theory.

- Classifier [15]: A classifier is an outstanding occasion of a theory (nowadays, usually learned by an AI computation). A classifier is a hypothesis or discrete-regarded limit that used to dole out (prominent) class names to standard particular data centers. In an email portrayal model, this classifier could be a hypothesis for checking messages as spam or non-spam. Be that as it may, a theory must not so much be synonymous with the term classifier. In another application, our hypothesis could be a limit with regards to mapping concentrate time and informational establishments of understudies to their future, relentless regarded, SAT scores – a steady target variable, proper for backsliding examination.

- Hyperparameters [16]: Hyperparameters are the tuning parameters of an AI figuring —for instance, the regularization idea of an L2 discipline in the mean squared blunder cost utmost of straight fall away from the faith, or inspiration for setting the best noteworthiness of a choice tree. Strikingly, model parameters are the param-eters that a learning estimation fits the arranging information – the parameters of the model itself. For instance, the weight coefficients (or tendency) of a straight fall away from the faith line and its tendency (or y-focus get) term are model parameters.

Let us understand with the mentioned applications in the previous chapter, knowing and understanding about how evaluation of a model and algorithm selection was done.

2.5 Related Work on Driver Drowsiness Detection Application

In June 2010, Bin Yang et al. [17] depicted 'Camera-based Drowsiness Reference for Driver State Classification under Real Driving Conditions'. They proposed that degrees of the driver's eyes can see aloofness under test structure or underlying conditions. The display of the most recent eye following composed in-vehicle consumption figure measures assessed. These measures are surveyed indeed and by a get-together method subject to a massive dataset of 90 h of confirmed street drives. The outcomes show that eye-following tiredness unmistakable affirmation limit tolerably for express drivers as long as the squints presentation works fittingly. Purpose of reality, even with some proposed upgrades, regardless, there so far issues with repulsive light conditions and for people wearing glasses. As an arrangement, the camera-based languor evaluations give a significant responsibility to an absence of consideration reference, at any rate, are not satisfactorily prepared to be the standard reference.

In 2013, Kong et al. [28] portrayed 'Visual Analysis of Eye State and Head Pose for Driver Alertness Monitoring'. They demonstrated visual appraisal of eye state and head present (HP) for stable seeing of sharpness of a vehicle driver. Most existing approaches to manage admin visual presentation of non-planned driving models depend either on eye end or head motioning edges to pick the driver's tiredness or redirection level. The proposed course of action utilizes visual highlights, for example, eye list (EI), understudy progress (PA), and HP to confine necessary data on the sharpness of a vehicle driver. A help vector machine (SVM) orders a get-together of video portions into alarm or non-sorted out driving occasions. Exploratory outcomes demonstrate that the virtuoso showed that blueprint offers high demand accuracy with acceptably low astounds and false cautions with individuals of different ethnicity and sexual bearing in generous street driving conditions.

In June 2014, Eyosiyas et al. [19] spread out 'Driver Drowsiness Detection through \HMM based Dynamic Modeling'. They proposed another procedure for confining the outward appearance of the driver through the Hidden Markov Model (HMM) based outstanding appearing to see laziness. They have executed the check utilizing a kept driving course of action. Exploratory outcomes checked the sensibility of the proposed technique.

From the above survey, we have understood that the methods that were proposed by the researchers are not enough to produce accurate results and needs to be proposed more advanced models. The model evaluation and algorithm selection validated based on the accuracy parameters like bias and variance. And also, from the above observations, we found that SVM algorithm is not enough to generate more results in different occasions. So, we have proposed new models to solve this problem and it is discussed in next chapter.

2.6   Related Work on Crude Oil Price Detection Application

Since the oil worth time approach is a nonlinear long-memory strategy, it is a respectable likelihood for utilization of near to estimation structures. Notwithstanding, no assessment in this subject has found in the oil worth choosing the structure. Researchers utilize particular strategies to discover near to neighborhoods in these frameworks, yet k-means and SOM are the most usually utilized packaging techniques around there. In some way, we will pack imitated state space of oil worth time game-plan utilizing the k-induces gathering technique.

In sorting out neural structure plans, there are two or three components, for example, number of layers, number of neurons in each layer, and move limits, which impact shrewd impact the presentation of neural systems. These parts are generally picked utilizing the awkward and troubling method of experimentation with no legitimize. Hereditary estimation (GA) [25] is an astonishing methodology in this setting because of its capacity to look into a large area of plan space and experience promising zones through acquired endeavors. GA was explicitly not utilized in the oil regard measuring with ANN (to the degree we could know).

From the above survey, we have understood that the methods that were proposed by the researchers are K-means and SOM are not enough to produce accurate results and needs to be proposed more advanced models. So, we have proposed new models to solve this problem and it is discussed in next chapter.

3   Algorithm Selection and Model Evaluation

In this chapter, we have discussed the pitfalls of the existing methods and procedures that were used to improve accuracy in the previous chapter are highlighted. From the previous chapter, the model evaluation and algorithm selection validated based on the accuracy parameters like Learning and Classification, Prediction accuracy and Loss range 0–1, Variance, Bias., etc. are not enough to generate accurate results, so here we have separately discussed and proposed new parameters in improvement of algorithm selection and model evaluation. And also, the existing algorithms are not enough and needs to be improved, so here we also highlighted how algorithm selection is chosen for the discussed applications.

Model Selection concerning AI can have different ramifications, contrasting with different degrees of reflection. Hyperparameters are the parameters of the learning procedure, which we have to decide from the prior, i.e., before model fitting. Alternately, model parameters cannot avoid being parameters that rise in light of the fit. In a determined backslide model, for example, the regularization quality (similarly as the regularization type, accepting any) is a hyperparameter that must be resolved before the fitting, while the coefficients of the fitted model cannot avoid being model parameters. Finding the hyper benefit parameters for a model can be necessary for the model execution on the given data. For something different, we ought to pick the best learning procedure (and their corresponding "perfect" hyperparameters) from a great deal of qualified AI systems. In going with, we will insinuate this as a computation decision. With a gathering issue near to, we may contemplate, for instance, paying little respect to whether a determined backslide model or a random forest classifier yields the best course of action execution on the given task.

Model appraisal targets assessing the theory slip-up of the picked model, i.e., how well they picked model performs on unnoticeable data. An incredible AI model is a model that not performs merely well on data seen during getting ready (else an AI model could recall the readiness data), yet furthermore, on unnoticeable data. Accordingly, before conveyance a model to age, we should be genuinely sure that the model's introduction will not degenerate when it looked with new data.

A last articulation of alarm: when overseeing time plan data where the endeavor is to make gauges, train, endorsement, and test sets must pick by separating the data along with the transient turn. That is, the "most prepared" data used for setting up, the later one for endorsement, and the most recent one for testing. Unpredictable examining does not look right for this circumstance.

Starting at now, the holdout system and various sorts of the bootstrap contemplated checking the speculation execution of our watchful models. We split the dataset into two zones: preparing and a test dataset. After the AI calculation fit a model to the organizing set, we investigated it on oneself decision test set that we hold from the AI figuring during model fitting. While we were talking about issues, for example, the inclination change exchange off, we utilized fixed hyperparameter settings in our learning checks, the extent of k in the k-closest neighbors' estimation [26]. We depicted hyperparameters as the parameters of the getting figuring itself, which we need to show up from the earlier – before model fitting. Remarkably, we inferred the parameters of our following model as the model parameters.

Over the long haul, the k-closest neighbors' estimation may not be a perfect decision for advancement dressing the capability between hyperparameters and model parameters, since it is a sleepy understudy and a nonparametric structure. In this stand-separated condition, disengaged learning (or case-based getting) amasses that there is no blueprint or model fitting stage: A k-closest neighbor's model genuinely stores or holds the engineering information and utilizations it precisely at need time. Everything considered every course of action occasion watches out for a parameter in the k-closest neighbors' model. Primarily, nonparametric models cannot avoid being models that cannot be depicted by a fixed number of parameters that changed according to the game plan set. The strategy information doesn't pick the structure of parametric models as opposed to being set from the as of now; non parametric models don't expect that the information looks for after certain likelihood transports not in the humblest degree like parametric techniques (exceptional cases of nonparametric frameworks that make such suppositions are Bayesian nonparametric systems). In like manner, we may express those nonparametric systems to make fewer questions about the information than parametric structures.

Instead of k-closest neighbors, a reasonable occasion of a parametric framework is settled fall away from the certainty, a summed up direct model with a fixed number of model parameters: a weight coefficient for each part factor in the dataset despite an inclination unit. These weight coefficients in decided fall away from the faith, the model parameters, are fortified by extending a log-probability work or confining the critical expense. For fitting a model to the preparation information, a hyperparameter of a decided apostatize calculation could be the measure of cycles or rejects the course of action set (ages) in propensity based streamlining. Another cause of a hyperparameter would be the estimation of a regularization parameter; for example, the lambda-term in L2-regularized decided to lose the faith. Changing the hyperparameter respects when running learning analyze over a game-plan set may appreciate various models. The way toward finding the best-performing model from many models that were made by various hyperparameter settings is called model attestation. The going with a zone changes a progression with the holdout structure that is helpful when completing this attestation method.

The open portal has gotten together to demonstrate the most traditional procedure for model assessment and model attestation in AI practice: k-overlay cross-ensuring. The term cross-support is utilized uninhibitedly recorded as a printed copy, where specialists and examiners a part of the time understanding the train/test holdout structure as a cross-ensuring strategy. Regardless, it may look unbelievable to consider the cross of organizing and support shapes in new rounds. Here, the central thought behind cross-support is that each model in our dataset finds the chance of being endeavored. K-overlay cross-support is an unusual case of research a dataset set k times. In each round, we split the dataset into k parts: one piece utilized for under-structure, and the rest of the $k-1$ piece is joined into a sorting out subset for model examination.

For hyperparameter assurance, we can use K-cover cross-endorsement (CV) [14]. Cross-endorsement fills in as seeks after:

- We split the arrangement set into K humbler sets. Note that the cautions as for imbalanced data in like manner apply here.
- We set aside all of the K overlays one time. We train the equal number of models from there are different mixes of hyper model parameters on the remainder of the $K-1$ cover and figure the endorsement score on the hold-out overlay.
- For every game plan of hyperparameters, we process the mean endorsement score and select the hyperparameter set with the best execution on the holdout endorse-ment sets. Then again, we can apply the "one-standard-screw up guideline" [2],

which suggests that we pick the most miserly model (the model with least multi-faceted nature) whose show is not more than a standard misstep underneath the best performing model.

For count assurance, we need an undeniable, unpredictable method. Here, settled cross-endorsement [15] acts the legend and fills in as seeks after:

- We split the data into K more diminutive sets (outer cover).
- Each of the K folds we set aside one time. For each learning strategy, we by then perform K' - overlay CV (following the framework above) on the K − 1 remaining folds, in which we do we do hyperparameter assurance. For terseness, one denotes settled CV with K outer folds and K' inner overlays as K × K's settled CV. Typical characteristics of K × K' are 5 × 2 or 5 × 3.
- We use the best hyperparameter set for each estimation to evaluate its endorsement score on the holdout cover.
- Then we figure the mean endorsement score (similarly as standard deviation) over the K cover and select the best performing count.

Sub sequent, we pick the best hyperparameter set reliant on CV using the full planning set and check the hypothesis mix-up using the test set. At last, we retrain the model using the united data of getting ready and test set.

4   Implementation of Application Design

In this chapter we have discussed about how the model evaluation is designed for any application using machine learning algorithms, and how algorithm selection has done. Here we also suggested proposed models based on the evaluation factors of any algorithm.

4.1   Driver Drowsiness Detection Application

Face Tracking Searching for Face in each edge in each scale grows the multifaceted computational nature. The persistent presentation of the count can realize if we use the transient information. If the position and size of the Face are known accurately at an edge, then we can pick ROI around that position where we can find the Face in coming about the diagram. The multifaceted computational nature is less since the chase zone lessened. The utilization of SVM computation [19] is used for Face following. Track the Face of the driver. Make a copy of the concentrations used for handling and discovering the ROI of the geometric. Change between the concentrations before and the present housings separately to customer advancement. Get the accompanying Frame in the video gathering. Track the concentrations in the ROI. Check the geometric change between the old concentrations and the new concentrations and shed individual cases using direct translation. (Least Four Frames are required to figure). Show pursued core interests. Reset the concentrations and demonstrate the clarified edge using an android application.

At the point when the Face is restricted, the ensuing stage is to recognize the circumstance of the eye. Henceforth the eye distinguished is orchestrated to open or close. The recognizable proof of eye in the face territory modeled as a thing acknowledgment issue. This classifier set up together eye disclosure as for interesting gained camera plot. One customer described classifiers for the open eye, and shut-eye used in this recognizable proof technique. The classifier for open and close is set up with a database of positive and negative pictures are considered. The ROI decision made, and the area of the eye performed in the restricted region. Picking the territory of interest reduces the computational requirements of the issue. This ROI contains eyes. In the occasion that the eye is distinguished, and no squint occurs, by then the counter is set to 0. If the gleam is perceived, then the counter is expanded, and it demonstrates the prepared driver perceiving languid, and an alarm gets sounded. As opposed to using any computation to perceive yawning, the here essential basis used. At the point when Face distinguishing proof has done, mouth district picture altered from Face perceived picture. After that, one cloak picture prepared, and it covers the mouth zone of altered pictures. A shroud picture is just a white picture containing all of the ones and having the same size of mouth and area cut picture. After that, the farthest point of pixel spots of mouth zone in the cover picture found. By then, apply edge framework (for Male set the edge a motivating force as 250 and Females set the motivator as 10). Finally, count the hard and fast no of the square pixel if the check is more conspicuous than breaking point means yawn perceived.

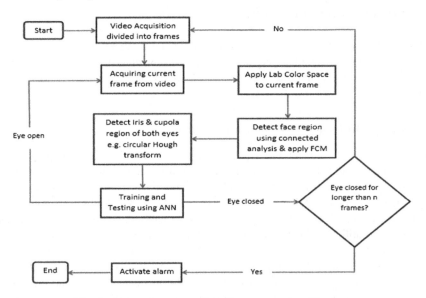

Fig. 1. Driver drowsiness detection proposed architecture.

The above Fig. 1 shows the proposed architecture of driver drowsiness application. Here, the architecture shows the model evaluation and algorithm selection using machine learning concepts. The procedure used to design application is termed as a model evaluation and calculation of results is termed as a algorithm selection.

4.2   Crude Oil Price Prediction Application

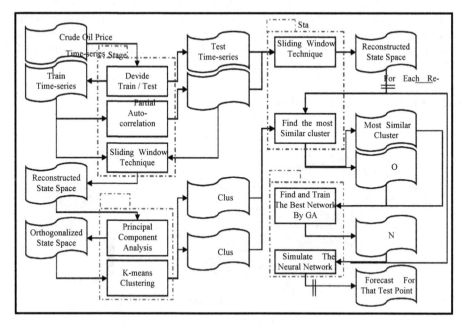

Fig. 2. Crude oil price prediction proposed architecture.

In the above Fig. 2, we have shown how the algorithm selection and model evaluation are done using machine learning. The architecture of the crude oil price prediction also shows the proposed procedure to gain accurate results using machine learning. Here K-means algorithm used as a selection algorithm.

The proposed model fuses four rule stages: information arrangement, crushing, structure engineering, and ANN plan, and checking. The rule stage readies the time plan of the particular oil cost to be utilized in different stages. It wires information division, accreditation of jeans, and state-space redirection of each piece as per saw several pairs of jeans. Each point in the train/test state space is a train/test plan. The ensuing stage sees arranging structures as information, orthogonalizes the space, and after that packs it. Each pack is a social gathering of close to models. The third stage makes each test manual for the closest assembling. At long last in the fourth master-mind, an ANN for every party is made, and the test models made to each pack are endeavored the varying ANNs to figure checks. The going with subparts clarifies these phases in detail.

Information arranging done as in the significant stage, foul oil worth time -approach is considered. A touch of this time strategy is spared something for testing purposes while the rest will be utilizing as a planning set. After the division of the dataset, the Partial Autocorrelation Function (PACF) of the arranging time procedure is desperate down with a 5% criticalness level to locate the most senseless number of jeans (L) to be joined as a guarantee to ANN model. Right, when L depicted as a sliding window of size, L + 1 is utilized to mirror the state space from sorting out (and later simultaneously, testing) time strategy of foul oil cost. Imitated state space of the engineering time methodology focuses taking after clear worth patterns (on an exceptionally essential level, preparing structures) while changed state space of the test time procedure focuses looking like test worth structures (which their first L estimations will be utilized in ANN to predict their targets – future costs).

4.3   Election Results Prediction Application

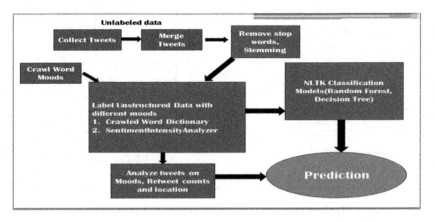

Fig. 3. Election results prediction proposed architecture.

The above Fig. 3 shows the proposed architecture of Election results application using sentimental analysis process. The classification models are suggested to predict the results, and also evaluation procedure is given by connecting with each other.

The precision of an estimation examination system is, on an intermediate level, how well it agrees with human choices. It is by and large evaluated by variety gauges subject to precision and survey over the two target classes of negative and positive compositions. In any case, as demonstrated by human research, raters usually agree about 80% of the time (see Inter-rater unflinching quality). Along these lines, a program which achieves 70% precision in orchestrating supposition is doing practically similarly as individuals, regardless of the way that such accuracy may not sound incredible. If a program were "right" 100% of the time, individuals would regardless not resist repudiating it about 20% of the time, since they vary that much about any answer.

On the other hand, PC structures will make through and through unexpected bungles in comparison to human assessors, and like this, the figures are less indistinguishable. For instance, a computer structure will encounter trouble with refutations, distortions, jokes, or jokes, which customarily is not hard to manage for a human pursuer: a couple of goofs a PC system makes will seem, by all accounts, to be unnecessarily blameless to a human. The utility for practical business tasks of end assessment as it described in academic research has been raised uncertainty about, generally since the essential one-dimensional model of an idea from contrary to positive yields rather negligible noteworthy information for a client worrying over the effect of open chat on for instance brand or corporate reputation.

5   Results and Discussion

In this chapter we have discussed about results generation using the machine learning methods. The data set is taken from koggle.com and applied on our experimental setup using OpenCV and Phython. All the data sets are tested using the proposed architectures of each application to test the machine learning algorithms. The algorithms are proposed based on the pitfalls that were discussed in Literature Survey chapter.

Fig. 4. Extraction of EAR and MAR values using ANN

The above Fig. 4 shows the EAR and MAR values. When the person's eyes are closed, EAR value counted, and when the person yawns, MAR value is counted. The above figure also shows EAR and MARS values when the eyes and mouth of a person are closed. In the above figure, we can also observe the alert message is generated when eyes are closed. Use of tiredness territory with SVM was done, which breakers the going with advances: Successful runtime getting of video with the camera. The gotten video was divided edges, and each edge was inspected. The fruitful revelation of the face looked for after by conspicuous confirmation of eye. On the off chance that completion of an eye for dynamic edges was perceived, by then it is assigned tired condition else it is viewed as a normal glimmer, and the drift of getting a picture and dissecting the condition of the driver is done over and over. In this utilization, during the drowsy express, the eye is not wrapped by a circle, or it is not recognized, and a relating message appears. If the driver is not torpid, by then, the eye is perceived by a circle, and it prints 1 for each convincing zone of the open eye.

This way, we have arranged a model apathy disclosure system using OpenCV programming and arranged classifiers. The structure so made was viably attempted, its hindrances perceived, and a future approach made. Driver Drowsiness Detection was attempted to empower a driver to stay attentive while driving in order to diminish car accidents realized by the languor. This paper was stressed by overtired drivers and their capacity to cause car crashes. The driver exhaustion [23] recognizable proof system registers drowsiness level from the driver using a mix of OpenCV and Camera. OpenCV is an item to figure whether a driver is drowsy. At the same time, it recoups pictures from the camera, which is fast enough to perceive a driver's features logically. The system uses open source programming called an OpenCV picture getting ready libraries; the gets pictures are dealt with in this. Raspberry pi and open cv make the overall system to a simplicity drowsiness disclosure structure.

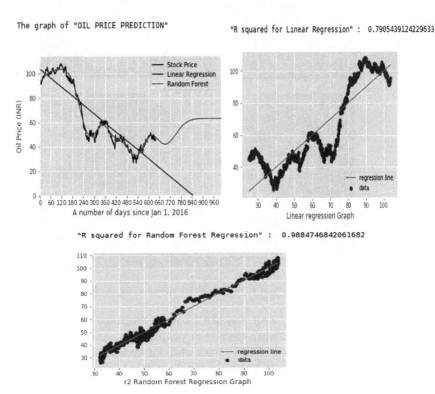

Fig. 5. Oil price prediction graph

From the above Fig. 5. The costs, when foreseen using Linear Regression [24], predicted characteristics are differentiated and actual costs, it is found that 79% exactness results. Right when expenses are foreseen using Random Forest Regression and differentiated and genuine expenses in the dataset, the accuracy was 98.8% results are showed up with the help of r squared worth.

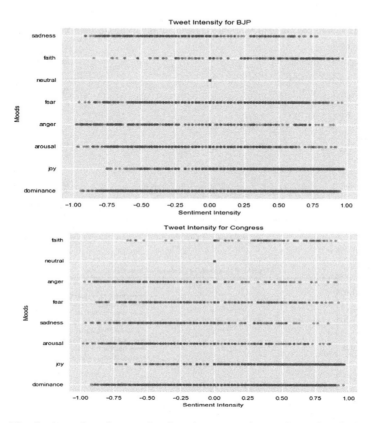

Fig. 6. Arranging of two national parties votes using sentimental analysis

The above Fig. 6 shows the arrangement of two national parties votes using sentimental analysis. The data is taken from koggle.com and implemented using OpenCV.

From the underneath Fig. 6. All tweets [25] vary in power from −1 to +1. As the underneath figures show strong definite suspicions like "Fulfillment" and "Certainty" incline progressively 0 to +1 for both BJP and Congress, while negative emotions like "Shock" and "Issue" incline more between −1 to 0. "Fair-minded" incline is focused on zero. Evaluations like "Fervor" and "Quality" are appropriated correspondingly between −1 to +1, which indicates they can be either tweeted in a constructive or adverse character. They seek after ing above figure outlines tweet that has a spot with both BJP and Congress. "Dominance" is as seen as the overwhelming perspective.

From the below Fig. 7, we have observed that Decision Tree classifiers are used to generate winning prediction graph, and each sentiment is calculated to decide the favorable party at particular location. Based on the every sentiment prediction score the winner is declared. The same theory is applied in recent elections also that are held during in the month of March in India, and tasted good results.

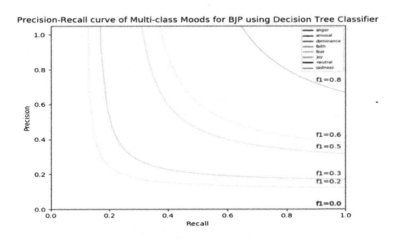

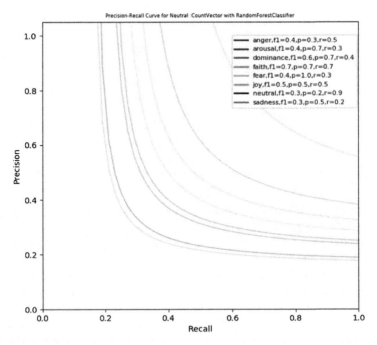

Fig. 7. Winning prediction graph generation using Decision Tree algorithm

From the above Fig. 7. A test tweet including the two get-togethers: "close fight in. The differentiation among BJP and Congress, not many seats," clearly shows close and firm challenge between the two. Decision tree classifier is settled on out of decision centers and leaf center points, where the decision of endorsing part regards and settling on branches are allotted to decision center points, while imprints named to leaf center points. Decision trees [26] create by picking the best decision stumps of most astonishing precision during the request process. Leaf center points with screw up underneath palatable edge are re-put by new decision stumps on a subset of getting ready data that stays away from the route from the establishment of the subtree to the leaf. The best decision stump can be picked by enlisting the information increment or the entropy of the structure. Deductively, entropy is described as the total of the probability [28] of each imprint times the log probability of that identical name. As the perspectives contain imbalanced classes, the model execution is plotted using a precision survey twist. Separating the gauge models [29] of BJP and Congress, it is seen that model precision (exactness, audit, and F-Score) of Congress out-performs BJP by an enormous edge (over 45%).

So, using the application design, we can clearly understand that model evaluation and algorithm selection will be varied based on the proposed methodology. In this article we have observed all the pitfalls from the Literature survey and searched the most accurate algorithms. We have applied machine learning concepts to predict good results. Finally we found Decision Tree Classifier, Random Forest and Regression algorithms are having more accuracy than the existing algorithms like SVM, K-Means and KNN which are discussed in previous chapter. Finally we suggest these machine learning algorithms to predict more results for any type of applications.

6 Conclusion

In every practical sense, regardless, I would in a perfect world prescribe repeating the arrangement test split on different occasions to enlist the confirmation interim on the mean measure (i.e., averaging the individual runs). Regardless, one fascinating clear for the present is that having fewer models in the test set develops the qualification and consequently builds up the conviction between times.

Past assessments have proposed different strategies to recognize drowsiness. In the wake of doing composing study, different frameworks have been found for perceiving driver tiredness, and they use different sorts of data as a commitment for their computation. After the outline of different sorts of strategies, it is found that using a camera is the best system that can be adequately associated and reasonable in all conditions. We researched this procedure for PC vision and proposed an excellent technique to distinguish driver laziness reliant on perceiving eyelid closing and opening using counterfeit neural frameworks as gathering figuring. In this paper, First of all, the video housings are picked up from the camera, which could be fixed with the goal that it should not ruin the road - point of view on the driver.

From the Experiment results, we found that the data requires logically real cleaning and mix (emojis). An inexorably definite classifier is furthermore well inside the space of credibility, to the extent neural frameworks which could amass the attitudes fittingly.

A sensible and fitting desire was beyond the realm of imagination as the data is exceptionally rough, and we required a whole managed dataset with stamped data. As the data is taken mostly from twitter so it could not get the commonplace section of the masses, which is noteworthy. Regardless of the way that evaluations look positive for BJP, we cannot wrap up only reliant on Twitter assessment.

Given up the making examined until this point, various issues can be perceived. In any case, the information utilized in the wants are generally drawn from the WTI cost of Brent cost, and they do not consider different data sources adjusted to the market. The capricious of the foul oil worth market is a result of the dependence of the market on different parts. Ousting these parts in foreseeing the market can restrain the authenticity of a guess instrument, protecting it from being cautious. A model with exceptional figure results shows staggering interconnections among data sources and the yield, which recommends the condition of reliance. Expansion partner examines that weight the insecurity some segment of the market areas of not long ago constrained. The mind-boggling part has concentrated on the value side of the measure as opposed to the sections that caused the upgrades. Among other standard effect segments utilized in the raw petroleum, want models are sales and supply. Despite the way that oil solicitation and supply expect essential employments in the precariousness of the worth, the use of these recognitions obliges the ability of various elements, for instance, input data, achieving a model not being thorough.

By including and partner the key segments included, an inexorably sweeping figure of the market can be cultivated. Third, most of the investigation thought about has utilized the time-course of action data. Data preprocessing and data depiction methodology were absent in most by far of the assessment. These two strategies help to clean and decrease bustles in instructive collections and organize them in the party the arrangement of want, and, later, this assistance to convey cautious outcomes. Without these techniques, the craving instrument will be less reliable. Fourth, breaks down have shown that anticipating the costs' models is more celebrated than imagining the discrete worth itself. The discrete worth will make research logically captivating and wise for specialists despite how the reasonableness of the assessments drove, starting quite recently, is up to this point broken—this framework the models discussed in this part with the data used for the figures.

References

1. Alpaydin, E.: Combined 5x2cv F test for comparing supervised classification learning algorithms. Neural Comput. **11**(8), 1885–1892 (1999)
2. Bengio, Y., Grandvalet, Y.: No unbiased estimator of the variance of k-fold cross-validation. J. Mach. Learn. Res. **5**(Sep), 1089–1105 (2004)
3. Bonferroni, C.: Teoria statistica delle classi e calcolo delle probabilita. Pubblicazioni del R Istituto Superiore di Scienze Economiche e Commericiali di Firenze **8**, 3–62
4. Breiman, L., Friedman, J., Stone, C.J., Olshen, R.A.: Classification and Regression Trees. CRC Press, Boca Raton (2004)
5. Cochran, W.G.: The comparison of percentages in matched samples. Biometrika **37**(3/4), 256–266 (1950)

6. Jung, C.R., Kleber, C.R.: A lane departure warning system based on a linear-parabolic lane model. In: Proceedings of the IEEE Intelligent Vehicles Symposium, pp. 891–895 (2004)
7. Dietterich, T.G.: Approximate statistical tests for comparing supervised classification learning algorithms. Neural Comput. **10**(7), 1895–1923 (2008)
8. Dunn, O.J.: Multiple comparisons among means. J. Am. Stat. Assoc. **56**(293), 52–64 (1961)
9. Efron, B., Tibshirani, R.J.: An Introduction to the Bootstrap. CRC Press, Boca Raton (1994)
10. Edwards, A.L.: Note on the "correction for continuity" in testing the significance of the difference between correlated proportions. Psychometrika **13**(3), 185–187 (1948)
11. Efron, B.: Nonparametric standard errors and confidence intervals. Can. J. Stat. **9**(2), 139–158 (1981)
12. Efron, B.: Estimating the error rate of a prediction rule: improvement on cross-validation. J. Am. Stat. Assoc. **78**(382), 316–331 (1983)
13. Efron, B.: Bootstrap methods: another look at the Jackknife. In: Kotz, S., Johnson, N.L. (eds.) Breakthroughs in Statistics. SSS, pp. 569–593. Springer, New York (1992). https://doi.org/10.1007/978-1-4612-4380-9_41
14. Efron, B., Tibshirani, R.: Improvements on cross-validation: the 632+ bootstrap method. J. Am. Stat. Assoc. **92**(438), 548–560 (1997)
15. Fleiss, J.L., Levin, B., Paik, M.C.: Statistical Methods for Rates and Proportions. Wiley, Hoboken (2013)
16. Hastie, T., Tibshirani, R., Friedman, J.H.: The Elements of Statistical Learning: Data Mining, Inference, and Prediction. Springer, New York (2009). https://doi.org/10.1007/978-0-387-84858-7
17. Hawkins, D.M., Basak, S.C., Mills, D.: Assessing model fit by cross-validation. J. Chem. Inf. Comput. Sci. **43**(2), 579–586 (2003)
18. Batista, J.: A drowsiness and point of attendance monitoring system for driver vigilance. In: Proceedings of the IEEE Intelligent Transportation Systems Conference, pp. 702–708 (2007)
19. Clanton, J.M., Bevly, D.M., Hodel, A.S.: A low-cost solution for an integrated multi-sensor lane departure warning system. IEEE Trans. Intell. Transp. Syst. **10**(1), 47–59 (2009)
20. Kohavi, R.: A study of cross-validation and bootstrap for accuracy estimation and model selection. In: IJCAI, vol. 14, no. 2, pp. 1137–1145 (1995)
21. Bergasa, L.M., Member, A., Nuevo, J., Sotelo, M.A., Barea, R., Lopez, M.E.: Real-time system for monitoring driver vigilance. IEEE Trans. Intell. Transp. Syst. **7**(1), 63–77 (2016)
22. Flores, M.J., Armingol, J.M., de la Escalera, A.: Driver drowsiness detection system under infrared illumination for an intelligent vehicle. IET Intell. Transp. Syst. **5**(4), 241–251 (2011)
23. Nickerson, D., Rogers, T.: Do you have a voting plan? Implementation intentions, voter turnout, and organic plan making. Psychol. Sci. **21**(2), 194–199 (2010)
24. Nickerson, D.W.: Partisan mobilization using volunteer phone banks and door hangers. Ann. Am. Acad. Polit. Soc. Sci. **601**, 10–27 (2015)
25. Nickerson, D.W., Friedrichs, R.F., King, D.C.: Partisan mobilization experiments in the field: results from a statewide turnout experiment in Michigan. Polit. Res. Q. **34**(1), 271–292 (2016)
26. Nickerson, D.W.: Quality is job one: volunteer and professional phone calls. Am. J. Polit. Sci. **51**(2), 269–282 (2017)
27. Kulkarni, S., Haidar, I.: Forecasting model for crude oil price using artificial neural networks and commodity futures prices. Int. J. Comput. Sci. Inf. Secur. **2**(1) (2009)
28. Abdullah, S.N., Zeng, X.: Machine learning approach for crude oil price prediction with Artificial Neural Networks-Quantitative (ANN-Q) model. In: Proceedings of the International Joint Conference on Neural Networks, pp. 1–8 (2010)
29. Shin, H., Hou, T., Park, K., Park, C.K., Choi, S.: Prediction of movement direction in crude oil prices based on semi-supervised learning. Decis. Support Syst. **55**, 348–358 (2013)

GAMPAL: Anomaly Detection for Internet Backbone Traffic by Flow Prediction with LSTM-RNN

Taku Wakui[1]([✉]) [iD], Takao Kondo[1,2] [iD], and Fumio Teraoka[3] [iD]

[1] Graduate School of Science and Technology, Keio University, Minato City, Japan
dona@inl.ics.keio.ac.jp
[2] Headquarters of Information Technology Center,
Keio University, Minato City, Japan
[3] Faculty of Science and Technology, Keio University, Minato City, Japan

Abstract. This paper proposes a general-purpose anomaly detection mechanism for Internet backbone traffic named *GAMPAL (General-purpose Anomaly detection Mechanism using Path Aggregate without Labeled data)*. GAMPAL does not require labeled data to achieve a general-purpose anomaly detection. For scalability to the number of entries in the BGP RIB (Routing Information Base), GAMPAL introduces *path aggregates*. The BGP RIB entries are classified into the path aggregates, each of which is identified with the first three AS numbers in the AS_PATH attribute. GAMPAL establishes a prediction model of traffic throughput based on past traffic throughput. It adopts the LSTM-RNN (Long Short-Term Memory Recurrent Neural Network) model focusing on periodicity in weekly scale of the Internet traffic pattern. The validity of GAMPAL is evaluated using the real traffic information and the BGP RIB exported from the WIDE backbone network (AS2500), a nation-wide backbone network for research and educational organizations in Japan. As a result, GAMPAL successfully detects traffic increases due to events and DDoS attacks targeted to a stub organization.

Keywords: Network Traffic Analysis · General-Purpose Anomaly Detection · Internet Backbone · LSTM-RNN

1 Introduction

The Internet backbone network contains large amount of traffic originated from various kinds of users and services. The traffic pattern is peaky and jaggy, which changes every moment even in ordinary times. On the other hand, the Internet backbone network might encounter anomalies caused by not only failures of network facilities but also disturbances such as flash crowds from social phenomenon and cyber attacks. Because the disturbances are basically observed only in traffic pattern, it is difficult to find each anomaly from the operators'

© IFIP International Federation for Information Processing 2020
Published by Springer Nature Switzerland AG 2020
S. Boumerdassi et al. (Eds.): MLN 2019, LNCS 12081, pp. 196–211, 2020.
https://doi.org/10.1007/978-3-030-45778-5_13

viewpoints. In order to operate the Internet backbone network stably, it is necessary to establish a general-purpose mechanism for finding these anomalies from traffic information.

Anomaly detection mechanism are categorized into two approaches: signature-based approach and behavior-based approach. The signature-based approach can detect known anomalies. It is suitable for real-time detection [1–3]. However, it fails to detect unknown anomalies such as new attacks. The behavior-based approach can detect unknown anomalies. Most of existing mechanisms use labeled data composed of anomaly and non-anomaly traffic information [4]. However, it is difficult to collect such traffic information. In addition, the labeled data causes overfitting to the target network. Therefore, the behavior-based approach is not suitable for general-purposed anomaly detection. Also, Most of existing anomaly detection mechanisms are specialized for a particular environment such as a DC (Data Center) for Internet Services [5] and SDN (Software-Defined Networking) [4] or they focus on a particular anomaly such as DDoS (Distributed Denial of Service) [6]. This paper proposes a general-purpose anomaly detection mechanism for Internet backbone traffic named *GAMPAL (General-purpose Anomaly detection Mechanism using Path Aggregate without Labeled data)*. GAMPAL establishes a prediction model of traffic throughput based on the past traffic throughput and utilizes the LSTM-RNN (Long Short-Term Memory Recurrent Neural Network) model focusing on periodicity in daily or weekly scale of the Internet traffic pattern. For scalability to the number of entries in the BGP RIB (Routing Information Base), GAMPAL introduces *path aggregates*. The BGP RIB entries are classified into the path aggregates, each of which is identified with the first three AS numbers in the AS_PATH attribute. GAMPAL generates predicted throughput for each path aggregate. In GAMPAL, an indicator named *NSD (Normalized Summation of Differences)* is introduced, which reflects the difference between the predicted throughput and the observed throughput. Anomaly is detected if the NSD value is larger than the threshold.

This paper implements a parser of traffic information produced by NetFlow version 9 and the BGP RIB in the MRT format [7] and a learning mechanism for a prediction model of traffic throughput based on LSTM-RNN model. The learning mechanism utilizes the cuDNN (CUDA Deep Neural Network) [8] library and Chainer library [9] in order to support a GPU computing environment. The evaluation utilizes the real traffic and the BGP RIBs exported from the WIDE backbone network (AS2500) [10], a nation-wide backbone network for research and educational organizations in Japan.

2   Related Work

Anomaly detection mechanisms are categorized into two approaches: signature-based approach and behavior-based approach. The signature-based approach [1] defines some rules to detect anomalies and applies these rules to logging outputs of servers and network facilities. The behavior-based approach monitors activities of end hosts or communication sessions in a networked system and detects some changes compared with the past ones. Because it is almost impossible to define rules to detect any kinds of anomalies in the Internet traffic [2,3], this paper discusses the existing work based on the latter approach.

For enterprise/DC (Data Center) scale network, [5] proposes a performance anomaly detection mechanism for cloud and Internet services. This mechanism is based on statistical behavior analysis which includes two techniques: a behavior-based technique with adaptive learning and a prediction-based technique with statistically robust control charts. [11] proposes a general-purpose anomaly detection mechanism for an enterprise network. This mechanism is based on CNN-based classification of visualization of traffic information. The traffic information is categorized with the MCODT (Micro-Cluster Outlier Detection in Time series) cluster algorithm and visualized by the SOM (Self Organization Map) dimentionality reduction algorithm. [4] is an intrusion detection mechanism for SDN (Software-Defined Networking). This mechanism utilizes GRU (Gated Recurrent Unit) RNN based classification which is learned by the NSL-KDD[12] labeled data set.

For Internet scale network, [6] proposes a botnet traffic detection mechanism based on traffic information in P2P networks. This mechanism includes CNN-based classification and a decision tree method for enhancing anomaly detection rate. [13] proposes a framework for real-time anomaly detection of cyber-attacks focusing on the Internet traffic. This framework combines unsupervised and supervised classification mechanisms. The former is based on an auto-encoder neural network while the latter is based on a nearest neighbor classifier model in which the manual operation is required.

Table 1 shows the comparison between GAMPAL and the existing mechanisms [4–6,11,13]. There are four metrics as follows: (i) scalability to the Internet, (ii) versatility to any kinds of anomalies, (iii) consideration on periodicity of the traffic pattern especially for Internet-scale network, and (iv) necessity of labeled learning data. In terms of scalability, [4] proposes an anomaly detection for small scale network. The SOM used in [11] does not have an aggregation mechanism because it focuses only on an enterprise network, not an Internet-scale network, and does not consider scaling. In terms of versatility, [4–6] are not versatile to anomaly types. [4] proposes an intrusion detection for SDN. [5] focuses on anomalies in cloud and Internet services. [6] is a mechanism specialized for botnet detection. [11] proposes a general-propose anomaly detection mechanism for an enterprise network. [13] proposes a general-purpose anomaly detection mechanism. In terms of consideration on periodicity, [4,11] focus on periodicity of traffic. [4] uses GRU RNN which can learn data for a longer period than simple RNN. [11] uses MCODT, a clustering algorithm for time-series data. [6,13] do not focus on periodicity of traffic. In terms of necessity of labeled data, most of existing mechanisms use labeled data. [5] uses real-world datasets of Web services and evaluates the validity of anomaly detection by comparing with that of an open source package. [11] does not use labeled data. The detection validity is evaluated by comparing the time when the proposed method detects behavior changes and the time when an event occurs in the real-world. [13] uses

Table 1. Comparison of related work.

Related work	Enterprise/DC Scale			Internet Scale		
	[4]	[5]	[11]	[6]	[13]	GAMPAL
Scalability	No	-	No	-	Yes	Yes
Versatile to the types of anomaly	No	No	Yes	No	Yes	Yes
Consideration on periodicity of traffic	Yes	-	Yes	No	No	Yes
Necessity of labeled data	Yes	No	No	Yes	Middle	No

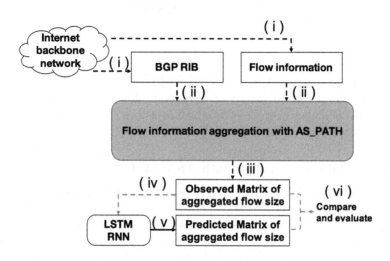

Fig. 1. Overview of GAMPAL methodology.

labeled data in supervised classification and un-labeled data in unsupervised classification. In contrast to existing mechanisms, GAMPAL satisfies the four metrics.

3   Methodology

3.1   Overview of GAMPAL Methodology

Figure 1 shows the overview of the GAMPAL methodology. GAMPAL is an anomaly detection mechanism using a prediction model based on the LSTM-RNN model. First, the flow information and the BGP RIB used in flow information aggregation are exported from an Internet backbone network (Fig. 1-(i)). *The observed matrix of aggregated flow size* is generated from the flow information and the AS_PATH attribute of the BGP RIB (Fig. 1-(ii), (iii)). Next,

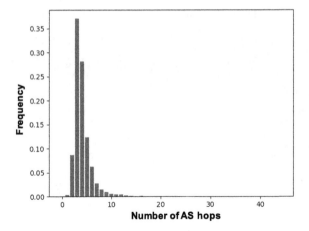

Fig. 2. Histogram of AS_PATH length.

the matrix of aggregated flow size is inputted to the LSTM-RNN (Fig. 1-(iv)). As a result, the predicted matrix of aggregated flow size is outputted. GAMPAL detects anomalies with a metric which measures the difference between the predicted flow size and the observed flow size (Fig. 1-(vi)).

3.2   Flow Data Aggregation with AS_PATH

GAMPAL adopts throughput of each flow as a general-purpose metric of traffic pattern in the Internet backbone network. A flow can be identified with the five tuples, i.e., source/destination IP addresses, source/destination ports, and protocol number. In a backbone network in which the BGP full routes are maintained, the order of the number of flows will be the square of the number of the BGP full routes. To make GAMPAL scalable to the Internet, the observed flows are mapped into groups named *the path aggregates*.

GAMPAL utilizes the AS_PATH attribute of the BGP RIB to define the path aggregates. At a traffic measurement node in a backbone network, a large number of destination addresses close to the IP address of the measurement node will be observed while a small number of destination addresses distant from the IP address of the measurement node will be observed. Therefore, the observed flows that have destination addresses close to the IP address of the measurement node should be classified in more detail to effectively detect anomalies. In contrast, it is sufficient to roughly classify the observed flows that have destination addresses distant from the IP address of the measurement node to detect anomalies. Figure 2 shows the distribution of the AS_PATH length of the IPv4 BGP full routes observed in AS2500 on June 17, 2018. The minimum value, the maximum value, the mode value, and the median value are 0 (iGP routes), 44, 3, and 4, respectively. Since the distribution of the AS_PATH length is heavily biased to small values and has a long and thin tail, it is sufficient to define path aggregates with a short AS_PATH length.

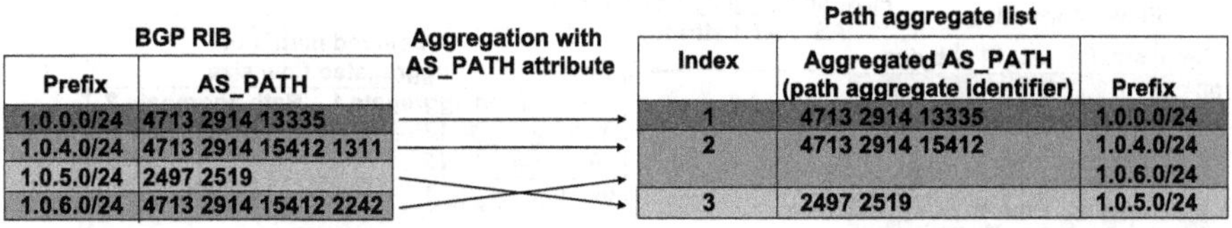

BGP RIB		Aggregation with AS_PATH attribute	Path aggregate list		
Prefix	AS_PATH		Index	Aggregated AS_PATH (path aggregate identifier)	Prefix
1.0.0.0/24	4713 2914 13335	→	1	4713 2914 13335	1.0.0.0/24
1.0.4.0/24	4713 2914 15412 1311	→	2	4713 2914 15412	1.0.4.0/24
1.0.5.0/24	2497 2519				1.0.6.0/24
1.0.6.0/24	4713 2914 15412 2242	→	3	2497 2519	1.0.5.0/24

Fig. 3. Example of AS_PATH aggregation.

GAMPAL adopts the mode value of the AS_PATH length, i.e., 3, to define the path aggregates. That is, the first three AS numbers of the AS_PATH attribute defines a single path aggregate and they are used as *the path aggregate identifier*. Consequently, 727,261 IPv4 BGP full routes (as of in January 2019) can be classified into 31,258 path aggregates.

Each observed flow is mapped to a single path aggregate to which the BGP route for the destination address prefix of the observed flow is classified. Thus, a path aggregate is composed of the path aggregate identifier and IP address prefixes that are mapped to the path aggregate. As a result, the number of observed flows can be aggregated to the number of the path aggregates at the most.

3.3 Training Approach: The Day of the Week

An Internet backbone network, such as a nation-wid backbone network usually consists of several branch NOCs (Network Operation Centers). As the Internet traffic pattern per NOC typically has periodicity in a daily or weekly scale, there are two approaches for training the prediction model: *the weekly training model* and *the day of the week training model*. The former uses continuous data of a week, e.g., from Sunday to Saturday, as the training data and predicts the traffic of the next week. The latter uses past data on the same day of the week, e.g., every Monday of the past two months, as training data. In a preliminary measurement, we made prediction models based on both approaches and compared them. As a result, the latter approach showed more valid prediction than the former one. Furthermore, the traffic pattern of the commodity Internet in Japan shows a weekly periodicity [14]. Therefore, GAMPAL adopts the latter approach, i.e., the day of the week training approach.

3.4 Overview of Prediction Procedures

Figure 3 shows an example of AS_PATH aggregation. First, GAMPAL creates the path aggregate list with the flow aggregation method described in Sect. 3.2. As shown in Fig. 3, the entries in the BGP RIB are classified into the path aggregates with the first three AS numbers of the AS_PATH attribute. For example, the two entries of the prefix 1.0.4.0/24 and the prefix 1.0.6.0/24 in the BGP RIB are classified to a single path aggregate (the *Path aggregate 2* in the table

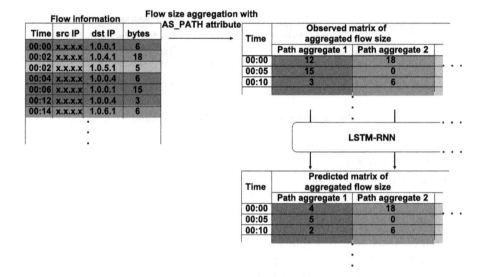

Fig. 4. Example of flow data aggregation by AS_PATH.

of the path aggregate list), because the first three AS numbers of the AS_PATH attribute are the same.

After creating the path aggregate list, the observed matrix of aggregated flow size are created with the path aggregate list. As shown in Fig. 4, the observed matrix of aggregated flow size has time-series entries, each of which contains the sum of the flow size during the time period. The data size of an observed flow is aggregated into an entry of the observed matrix of aggregated flow size. For example, as shown in Fig. 4, the entries whose destination address matches the prefix 1.0.4.0/24 and the prefix 1.0.6.0/24 in the Flow information table are mapped to the Path aggregate 2 in the observed matrix of aggregated flow size. Each entry of the observed matrix of aggregated flow size contains the sum of the bytes for 5 min.

Finally, GAMPAL generates the predicted matrix of aggregate flow size per path aggregate with the LSTM-RNN model.

4   Implementation

Figure 5 shows overall procedures of GAMPAL. This section describes the implementation of GAMPAL.

4.1   Implementation Environment

GAMPAL is implemented in Python 3.7.0 on a server running Ubuntu Server 18.04.1. Chainer 5.1.0 is used to implement LSTM for training and prediction. nfdump version 1.6.17 [15] is used to convert the flow information.

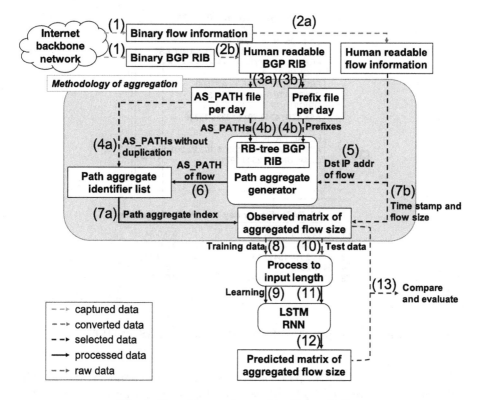

Fig. 5. Overall procedures of traffic prediction

bgpdump version 1.4.99.13 [16] is used to convert the BGP RIBs. GPU (Graphics Processing Unit) is used for calculations of LSTM-RNN. The GPU platform is CUDA 9.0.

4.2   Data Pre-processing

First, *binary flow information* and *binary BGP RIB* exported from the Internet backbone network are converted to *human readable flow information* and *human readable BGP RIB* (Fig. 5-(1),(2a),(2b)).

Processing of NetFlow. The NetFlow, which is used as the flow information format in this paper, is recorded in a binary file format. The binary flow information contains time stamp, five tuples, and data size of the flow. It is converted to a text file, the human readable flow information, using nfdump (Fig. 5-(2a)). Because the binary file is recorded per hour, the text file also contains flow information for an hour.

Processing of BGP RIB. The BGP RIB is recorded in the MRT format. This binary BGP RIB is converted to the human readable BGP RIB using bgpdump

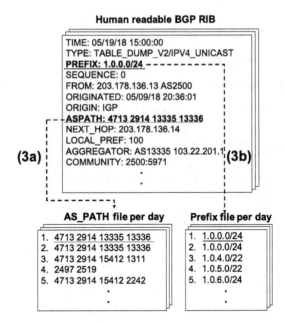

Fig. 6. Examples of BGP RIB, Prefix file, and AS_PATH file.

(Fig. 5-(2b)). Next, the AS_PATHs are extracted from the human readable BGP RIB and saved in *the AS_PATH file per day* (Fig. 5-(3a)). Prefixes are extracted from the human readable BGP RIB and saved in *the Prefix file per day* (Fig. 5-(3b)). Figure 6 shows a part of the human readable BGP RIB, a part of the AS_PATH file per day, and a part of the Prefix file per day. The procedure numbers in Fig. 6 correspond to those in Fig. 5. From each BGP RIB entry, the AS_PATH is extracted and saved in the AS_PATH file per day while the prefix is extracted and saved in the Prefix file per day. Thus, an entry in the AS_PATH file per day corresponds to the entry in the Prefix file per day at the same line number. For example, as shown in Fig. 6, the first line of the AS_PATH file per day (4713 2914 13335 13336) corresponds to the first line of the Prefix file per day (1.0.0.0/24).

4.3   Generating Path Aggregate Identifier List and Matrix of Aggregate Flow Size

The blue area in Fig. 5 shows the procedure after the pre-processing of the flow information. This section describes the definition and generation of a *path aggregate identifier list*, generation of a matrix of aggregate flow size (Fig. 5-(4)–(7)).

Generating Path Aggregate Identifier List. The AS_PATH file per day created from the human readable BGP RIB of the latest date in the training data is used to define the path aggregate identifier and create the path aggregate identifier list. The path aggregate identifier list includes all of the aggregated

AS_PATH in the BGP RIB without duplication (Fig. 5-(4a)). As described in Sect. 3.2, the combination of the first three AS numbers is defined as the path aggregate identifier. Figure 7 shows a part of the path aggregate identifier list created from the AS_PATH file on May 19, 2018. For example, the line 1 of the Path aggregate identifier list in Fig. 7 shows a path aggregate identifier defined with AS4713, AS2914, and AS13335.

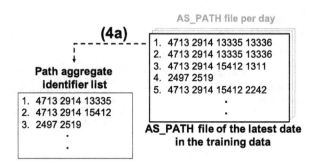

Fig. 7. Example of the path aggregate identifier list.

Generating Observed Matrix of Aggregated Flow Size. Figure 8 shows the structure of the observed matrix of aggregated flow size. It has a two dimensional structure. Each row of the matrix corresponds to a specific time period (e.g., 5 min). Each column of the matrix corresponds to a path aggregate. Each element of the matrix contains the sum of bytes of the corresponding flow for the time period.

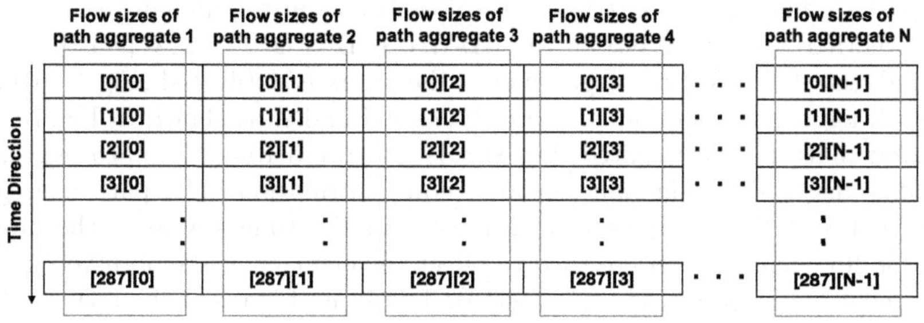

Flow sizes of path aggregate 1	Flow sizes of path aggregate 2	Flow sizes of path aggregate 3	Flow sizes of path aggregate 4		Flow sizes of path aggregate N
[0][0]	[0][1]	[0][2]	[0][3]	・ ・ ・	[0][N-1]
[1][0]	[1][1]	[1][2]	[1][3]	・ ・ ・	[1][N-1]
[2][0]	[2][1]	[2][2]	[2][3]	・ ・ ・	[2][N-1]
[3][0]	[3][1]	[3][2]	[3][3]	・ ・ ・	[3][N-1]
・	・	・	・		↕
[287][0]	[287][1]	[287][2]	[287][3]	・ ・ ・	[287][N-1]

Time Direction

Fig. 8. The structure of observed matrix of aggregated flow size.

Figure 8 shows that the number of the path aggregates in the observed matrix of aggregated flow size is N. GAMPAL adopts 5 min as the time period of each row. In case that the observed matrix of aggregated flow size are divided per day, the number of rows is 288 as shown in Fig. 8.

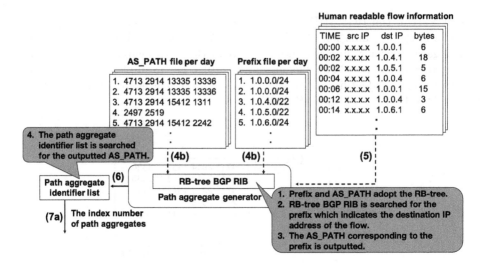

Fig. 9. Overview of path aggregate index generation.

Figure 9 shows a detailed diagram for generating *the path aggregate index*, which is the index in the AS_PATH file per day and the Prefix file per day. The procedure numbers in Fig. 9 correspond to those in Fig. 5. The RB-tree RIB file is converted from the corresponding Prefix file and the AS_PATH file (Fig. 9-(4a), (4b)). The RB-Tree RIB file adopts a self-balancing binary search tree (Red-Black-Tree [17]) in which the prefixes are the main values. Since the number of prefixes in the BGP RIB will be in the order of the number of the BGP full routes, it is necessary to reduce the search time for the destination IP addresses in the human readable flow information. The observed matrix of aggregated flow size is generated from the human readable flow file and the RB-tree RIB file of the same date. The destination IP address of each flow in the human readable flow file is queried with the prefix in the RB-tree RIB (Fig. 9-(5)). When the prefix is found, the AS_PATH corresponding to the prefix is outputted (Fig. 9-(6)) and the path aggregate identifier list (Fig. 9-(7a)). Finally, as shown in Fig. 10, the observed matrix of aggregated flow size is generated from the path aggregate identifier list and the human readable flow information. The path aggregate index in the path aggregate identifier list and the time stamp in the human readable flow information are used to select the element in the observed matrix of aggregated flow size (Fig. 5-(7a), (7b)). The sum of bytes of the flow is added to the corresponding element of the observed matrix of aggregated flow size.

4.4   Training of Traffic Prediction Model

The LSTM-RNN model for traffic prediction is implemented with Chainer [9], an open source deep learning framework and the NstepLSTM class, a class for supporting LSTM-based learning in Chainer. The implementation is optimized

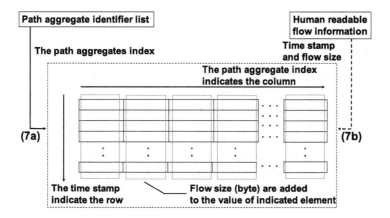

Fig. 10. The matrix of aggregated flow size generation.

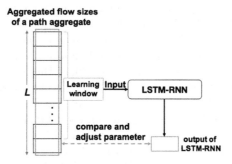

Fig. 11. Input data to LSTM-RNN and training.

to use cuDNN (CUDA Deep Neural Network) [8] library for a GPU computing environment.

In the LSTM-RNN model, the time period of the learning data must be longer than that of expected periodicity. As described in Sect. 3.3, since the traffic pattern of the commodity Internet in Japan shows weekly periodicity, it is sufficient to focus on daily periodicity in GAMPAL. Because Sect. 4.3 describes that each element in the observed matrix of aggregated flow size is the sum of the bytes per path aggregate within 5 min, the number of rows of the observed matrix of aggregated flow size is 288. Therefore, the time period of expected periodicity is 288 in GAMPAL.

Figure 11 shows the way to input the elements of a path aggregate in the observed matrix of aggregated flow size. Suppose that the value of L is larger than the expected periodicity (i.e., 288 elements in the matrix of aggregated flow size) of the traffic pattern. The learning window specifies $L-1$ out of L elements. The specified elements can be inputted and the remaining element is compared with the output. The parameters for LSTM-RNN are adjusted according to the result of this comparison. The learning window slides forward one by one.

5 Evaluation

5.1 Datasets

In the evaluation, the flow data (NetFlow) and the BGP RIB exported from WIDE backbone Network (AS2500) [10] are used. The backbone network is a nation-wide Layer-2 and Layer-3 network and includes branch NOCs, some of which provide connectivity to stub organizations such as universities. The backbone network is not only used as an external connection network for each organization, but also frequently used as a testbed for experimentation of new technologies. NetFlow is observed at a branch NOC accommodated in a university and the BGP RIB is observed at a route server in the backbone network.

5.2 Evaluation Indicator

GAMPAL predicts throughput, i.e., the number of bytes per unit time, for each of approximately 30,000 path aggregates. The number of bytes per unit time varies for each path aggregate. Some path aggregates have zero to several bytes while some path aggregates record hundred thousands or millions bytes. It is necessary to define an indicator that can evaluate these path aggregates in the same scale. Therefore, indicators with different scales depending on the data such as MSE (*Mean Square Error*) are not suitable. In addition, the measured and predicted values may include zero, which means there was no flow for 5 min. Therefore, indicators that cannot be calculated with data containing zero such as $RMSPE$ (*Root Mean Square Percentage Error*) are not suitable. Thus, this paper defines an indicator named NSD (*Normalized Summation of Differences*) where m_i denotes the i th observed value, p_i denotes the i th predicted value, and T denotes the number of input values.

$$NSD = \frac{\sum_{i=1}^{T} |m_i - p_i|}{\sum_{i=1}^{T} \max(m_i, p_i)} \tag{1}$$

NSD is the ratio of the sum of the differences between the observed and predicted values to the sum of the larger value of the observed and predicted values. NSD takes a value between 0 and 1 regardless of the scale of value. Also, NSD is the indicator that can be calculated even if the observed or predicted value is zero. NSD shows how much the predicted value is different from the observed value, that is, it shows the validity of prediction. If the difference between the observed value and the predicted value is small, the NSD value is small.

5.3 Validity of General-Purpose Anomaly Detection

In the evaluation, the NSD value is calculated for normal and abnormal days. On normal days, there seems to be no incident affecting the network. On abnormal days, an incident may have occurred. In the evaluation, June 24–25, 2018, and June 22–24, 2019 are selected as normal days, while October 17, 2018, November

Table 2. Dates of event traffic and normal traffic.

Attribute	Target date of evaluation	Training data
Normal	Jun. 24, 2018	May 6, 13, 20, 27, Jun. 3, 10, 17, 2018
Normal	Jun. 25, 2018	May 5, 14, 21, 28, Jun. 4, 11, 18, 2018
Event	Oct. 17, 2018	Sep. 5, 12, 19, 26, Oct. 3, 10, 2018
Event	Nov. 22, 2018	Oct. 11, 18, 25, Nov. 1, 8, 15, 2018

Table 3. Dates of DDoS traffic and normal traffic.

Attribute	Target date of evaluation	Training data
Normal	Jun. 22, 2019	Jun. 1, 8, 15, 2019
Normal	Jun. 23, 2019	Jun. 2, 9, 16, 2019
Normal	Jun. 24, 2019	Jun. 3, 10, 17, 2019
DDoS	Jul. 6, 2019	Jun. 8, 15, 22, 2019
DDoS	Jul. 7, 2019	Jun. 2, 9, 16, 23, 2019
DDoS	Jul. 8, 2019	Jun. 3, 10, 17, 24, 2019

22, 2018, and July 6–8, 2019 are selected as abnormal days. Using the data on those days, this paper tries to detect event traffic and DDoS attacks. On October 17, 2018, connection failure to YouTube [18] occurred. On November 22, 2018, there was a campus festival of the university that accommodates the measurement NOC. At the end of June 2019, a UDP reflection/amplification attack using ARMS (Apple Remote Management Service) was observed around the world [19]. This attack was also observed at the university. The university blocked communications for ARMS on July 9, 2019. Therefore, it is assumed that an abnormal state due to the attack was observed just before July 9, 2019. Tables 2 and 3 show the normal and abnormal dates and their training data. If the prediction model created with the data of the normal days is used to predict the data of the abnormal days, the difference between the measured data and the predicted data should be large.

Figure 12 shows the result of the evaluation. The value on top of a bar is the average NSD value of all "path aggregates" on each day. The NSD values on the days marked as "Event" (October 17 and November 22, 2018) are larger than those of the normal days. The NSD values on the days marked as DDoS attack are larger than those of the normal days. The NSD values on June 22–25 are all below 0.40, but those on July 6–8 are all above 0.43. Furthermore, the maximum NSD value for the six days is observed on July 8 (0.443), the day before the university settled the DDoS attacks. This indicates that the flows on the abnormal days cannot accurately be predicted. In other words, the behavior on the abnormal days was different from that of the normal days. This result shows that GAMPAL can detect anomalies caused by the event traffic and the DDoS attack.

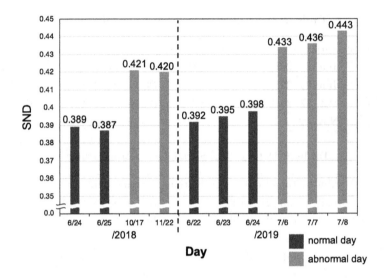

Fig. 12. Result of evaluation.

6   Conclusion

This paper proposed a general-purpose anomaly detection mechanism for Internet backbone traffic based on a LSTM-RNN-based prediction model. To make GAMPAL scalable to the number of the Internet full routes, each flow is mapped to a single path aggregates identified with the first three AS numbers of the AS_PATH attribute of the BGP RIB. This paper evaluated the validity of GAMPAL using the observed flow data and the BGP RIBs exported from the WIDE backbone network (AS2500), a nation-wide backbone network for research and educational organizations in Japan. The evaluation showed that when a stub organization of the backbone network suffers from DDoS attacks, the difference between the predicted and observed values is significantly different. Therefore, GAMPAL properly reflected the state of the Internet backbone with only the traffic throughput.

References

1. Liao, H., Lin, C.R., Lin, Y., Tung, K.: Intrusion detection system: a comprehensive review. J. Netw. Comput. Appl. **36**(1), 16–24 (2016)
2. Kumar, R., Sharma, D.: HyINT: signature-anomaly intrusion detection system. In: Proceedings of ICCCNT 2018, pp. 1–7 (2018)
3. Kwon, J., Leea, J., Lee, H., Perrig, A.: PsyBoG: a scalable botnet detection method for large-scale DNS traffic. Comput. Netw. **97**, 48–73 (2016)
4. Tang, T.A., Mhamdi, L., McLernon, D., Zaidi, S., Ghogho, M.: Deep recurrent neural network for intrusion detection in SDN-based networks. In: Proceedings of IEEE NetSoft 2018, pp. 202–206 (2018)

5. Ibidunmoye, O., Rezaie, A., Elmroth, E.: Adaptive anomaly detection in performance metric streams. IEEE Trans. Netw. Serv. Manag. **15**(1), 217–231 (2018)
6. Chen, S., Chen, Y., Tzeng, W.: Effective botnet detection through neural networks on convolutional features. In: Proceedings of IEEE TrustCom/BigDataSE 2018, pp. 372–378 (2018)
7. Petrie, C., King, T.: Multi-Threaded Routing Toolkit (MRT) routing information export format with BGP additional path extensions. RFC 8050, sl IETF (2017)
8. NAVIDIA cuDNN. https://developer.nvidia.com/cudnn. Accessed 20 Aug 2019
9. Chainer: A flexible framework for neural networks. https://chainer.org/
10. WIDE backbone. http://two.wide.ad.jp/
11. Flanagan, K., Fallon, E., Jacob, P., Awad, A., Connolly, P.: 2D2N: a dynamic degenerative neural network for classification of images of live network data. In: Proceeding of IEEE CCNC 2019, pp. 1–7 (2019)
12. NSL-KDD dataset. https://www.unb.ca/cic/datasets/nsl.html. Accessed 20 Aug 2019
13. Kathareios, G., Anghel, A., Mate, A., Clauberg, R., Gusat, M.: Catch it if you can: real-time network anomaly detection with low false alarm rates. In: Proceedings of IEEE (ICMLA 2017), pp. 924–929 (2017)
14. Cho, K., Fukuda, K., Esaki, H., Kato, A.: The impact and implications of the growth in residential user-to-user traffic. In: Proceedings of ACM SIGCOMM 2006, pp. 207–218 (2006)
15. nfdump. http://nfdump.sourceforge.net. Accessed 20 Aug 2019
16. bgpdump. https://bitbucket.org/ripencc/bgpdump/wiki/Home. Accessed 20 Aug 2019
17. Red-Black-Tree. https://developer.nvidia.com/cudnn. Accessed 20 Aug 2019
18. TeamYoutube. https://twitter.com/TeamYouTube/status/1052393799815589889?ref_src=twsrc
19. NETSCOUT. https://www.netscout.com/blog/asert/call-arms-apple-remote-management-service-udp

Revealing User Behavior by Analyzing DNS Traffic

Martín Panza[(✉)], Diego Madariaga, and Javier Bustos-Jiménez

NIC Chile Research Labs, University of Chile, Santiago, Chile
{martin,diego,jbustos}@niclabs.cl

Abstract. The Domain Name System (DNS) is today a fundamental part of Internet's working. Considering that Internet has grown in the last decades as part of human's culture, user patterns regarding their behavior are present in the network data. As a consequence, some of these human behavior patterns are present as well in DNS data. With real data from the '.cl' ccTLD, this work seeks to detect those human patterns by using Machine Learning techniques. As DNS traffic is described by a time series, particular and complex techniques have to be used in order to process the data and extract this information. The procedure that we apply in order to achieve this goal is divided in two stages. The first one consists of using clustering to group DNS domains basing on the similarity between their users' activity. The second stage establishes a comparison between the obtained groups by using Association Rules. Finding human patterns in the data could be of high interest to researchers that analyze the human behavior regarding Internet's usage. The procedure was able to detect some trends and patterns in the data that are discussed along with proper evaluation measures for further comparison.

Keywords: DNS · Clustering · Association rules · Human behavior

1 Introduction

As a critical component in Internet's infrastructure, the Domain Name System (DNS) plays a vital role in Internet's working. As the system that translates the domain names to IP addresses, every web service relies on it to operate. For its part, with the continuous growth of users, Internet is nowadays an important element that affects humans' life and culture in an undeniable way. Taking this into consideration, human behavior patterns can be recognized in the Internet's data flow; and as a consequence, in the DNS traffic. These patterns make this source of data highly valuable for the analysis and understanding of the human conduct over the usage activity on Internet.

As an example of this statement, one can identify a strong periodic behavior when simply visualizing the amount of queries in DNS traffic. The periodicity showed in Fig. 1 is caused by the high traffic that people generate during the

© IFIP International Federation for Information Processing 2020
Published by Springer Nature Switzerland AG 2020
S. Boumerdassi et al. (Eds.): MLN 2019, LNCS 12081, pp. 212–226, 2020.
https://doi.org/10.1007/978-3-030-45778-5_14

day, and the low traffic at night when the majority of people rest. Likewise, human activity is higher during weekdays rather than during the weekend, as can also be seen in Fig. 1, where Saturday and Sunday correspond to the two lower peaks of the time series. This data corresponds to the Chilean country code top-level domain (ccTLD): '.cl', and it is the data that we will use later for experimentation in this work, with a further description in Sect. 3.

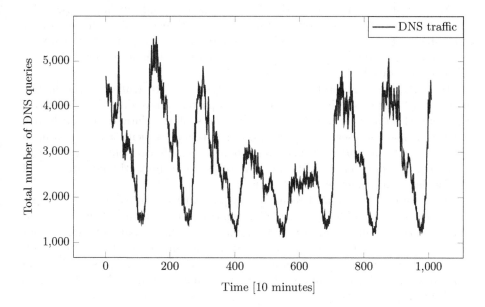

Fig. 1. DNS traffic time series

Moreover, given the purpose of DNS as a way of using IP addresses in a human-understandable way, it contains even more information regarding these behavior patterns. Many times one can easily speculate on the content of a webpage by looking at its domain name.

Recognizing and studying these patterns could be of high relevance for researchers interested in analyzing the human behavior on Internet usage. As well as for resources managing, that DNS operators might be interested in using to improve the service provided by their systems.

This work seeks to use Machine Learning techniques on real DNS traffic from authoritative servers in order to discover and analyze human patterns, showing a useful process for this purpose based on methods and evaluation measures from relevant related work.

Considering that DNS traffic could be described by time series, we apply methods and distance measures specifically designed for this purpose; as time series analysis is a topic of research itself that has acquired huge relevance in the literature in recent years. Mainly due to its applicability on several and diverse topics, for example, financial markets, brain activity, and astronomy.

2   Related Work

The study of human behavior has always been of great interest to researchers, mainly in the field of sociology [2]. However, understanding human behavior in computer science is an emergent research field that has significantly benefited from the rapid proliferation of wireless devices that frequently report status, and location updates [3].

Most of state of the art works that address the study of human behavior through the analysis of networking-related data exploit the high periodicity present in network data. This high periodicity and, consequently, low entropy, is mostly attributed to the impact of the regularity of human patterns [7,13] on the network state [16].

Recently, the temporal and spatial analysis of data traffic on the mobile network [16] has shown how different human patterns have different effects over the network state, generating distinct patterns of the data traffic in diverse locations. Also, as researches have shown how this periodicity is also present in DNS data [12], important conclusions about user behavior can be deduced at analyzing this portion of the network, in order to optimize the performance of this critical component of the Internet.

Time series analysis has become a very popular topic of research lately. Specially because of its usage on popular topics, such as financial markets; and because concepts like similarity and summarization have many different visions depending on the problem [4]. On top of it, data mining on time series studies have developed various adaptations of the common techniques [6] since, in general, each problem is addressed with an original procedure depending on its conditions.

3   Data-Set Overview

The data-set used in this work consists of a week of normal operation traffic of one of the authoritative DNS servers of the '.cl'. It starts on 7 November, 2018, until 14 November of the same year. '.cl' is the country code top-level domain (ccTLD) of Chile, administrated by NIC Chile. Every DNS packet from queries to the server and responses to users is present in the data-set. The server studied belongs to an anycast configuration along with other servers.

A time series of DNS traffic was built by aggregating all the successful server responses into 10-min intervals. Therefore, each point of the time series corresponds to the number of DNS packets from server responses with record types 1, 2, 15, or 28 (A, NS, AAAA, MX) obtained in ten minutes of data. For the purpose of this work, only the most important domains on '.cl' were considered; in view of the vast number of domains that this ccTLD is responsible for, most of which contain low activity. We based on Amazon Alexa's top sites [1] to determine the most relevant domains for our study. We made a further selection based on the number of queries received for those domains, resulting in 82 high activity domains of the Amazon Alexa's top sites. All the time series together manifest a total of 2,854,260 DNS packets.

The Fig. 1 in Sect. 1 shows an aggregation of the whole time series used in this work, showing the total number of DNS queries every 10 min during the studied week.

Since the data comes from a normal working of the system, it takes on great importance in the analysis of this work and gives relevance to the results obtained as users patterns are captured in the traffic.

4   Methodology

Considering the domains that were taken into account basing on the criteria described in Sect. 3, an experimental procedure was made consisting of two stages that are further described in the following sections.

The first stage corresponds to a clustering analysis on all the time series, in order to find groups of domains according to their traffic activity from the number of queries received from the users. Each domain's time series was pre-processed by applying a Simple Moving Average (SMA) method and a Z-Score normalization to them, with the purpose of reducing noise and capturing the regular shape of the time series, as well as reducing the scale, which was convenient for the distance measures that were used. In this way, giving the clustering algorithm a smoother and consistent input. The time series clustering algorithm used in the experiments was the Partitioning Around Medoids (PAM) [10] for multiples values of k. The selected value of k used for further analysis was determined by the internal clustering validation measure: Davies-Bouldin Index [5]. With regard to the time series distance metric used by the algorithm, the Shape-Based Distance proposed by Paparrizos and Gravano [14] was established in a sliding window of 12 h, i.e. half-a-day. Before the execution of the experiment, different tags were assigned to each domain as a way of both give a description about the domain's content type, and to evaluate the results using an external clustering validation measure: Rand Index. Lastly, after obtaining the results and selecting k, we display the groups given by the algorithm and discuss the nature of their domain members.

The objective in the second stage was to establish a comparison between the groups obtained in the clustering analysis. To achieve this goal, an association rules analysis was made on a representative of each of the groups, corresponding to the centroid from each cluster obtained in the previous stage. The algorithm used in this phase was the Apriori algorithm [9]. However, to properly feed this algorithm with the time series, a previous procedure to transform time series to a set transactions was done. The most relevant rules were showed in the Results section, and later discussed in the Discussion section.

Some important aspects of this process were implemented using the R packages *dtwclust* [15], containing time series clustering tools, and *apriori* [8] for association rules analysis.

Finally, some conclusions and future work are proposed in the final section.

5   Clustering Analysis

5.1   Algorithm and Configuration

The clustering algorithm used for the experiments is the Partitioning Around Medoids, using the distance measure Shape-Based Distance.

Partitioning Around Medoids. Partitioning Around Medoids (PAM) is different from k-means algorithm since it uses elements from the data-set as centroids. The advantage is that it is less sensitive to outliers as it minimizes dissimilarities between the clustering members, and not squared euclidean distances as k-means does. It does require a similarity measure.

The algorithm proceeds as follows:

1. Select k domains as medoid-domain.
2. Link all the other domains to their closest medoid-domain.
3. Calculate the total cost (sum of dissimilarities).
4. While (total cost decreases) do:
   - For each medoid-domain do:
     - For each non-medoid-domain do:
       * Use the non-medoid-domain as medoid-domain instead of the current medoid-domain.
       * Link all the other domains to their closest medoid-domain.
       * Recalculate the total cost.
       * If the total cost increased, then undo the substitution between the medoid-domain and the non-medoid-domains.

A specific advantage of this algorithm to the benefit of this work is that, due to its nature, the final centroids are members from one of each cluster. In Sect. 6 we use this aspect to directly choose candidates for the Association Rules analysis.

Shape-Based Distance. The Shape-Based Distance (SBD) is a similarity measure for time series. It is less costly than the popular Dynamic Time Warping (DTW). It is described by the following equation:

$$SBD(\boldsymbol{x}, \boldsymbol{y}) = 1 - \max_{w}\left(\frac{CC_w(\boldsymbol{x}, \boldsymbol{y})}{\sqrt{\|x\| \cdot \|y\|}}\right) \tag{1}$$

where $CC(x, y)$ is the cross-correlation and w is a value that maximizes $CC_w(x, y)$ based on the convenient shift of the time series with regard to the other one.

This measure reaches values between 0 to 2, and it is highly sensitive to scale. That is why a normalization is required. We used Z-Score normalization as suggested by the distance's authors. In addition, we used a half-a-day window size for the calculations of the similarity.

5.2   Evaluation

Clustering validation measures are divided in two types regarding the information that they require: internal and external. Both have the objective of determining how good the clusters obtained by a clustering algorithm are.

While internal validation measures only require spatial information of the clusters themselves, external validation measures use information that instructs how the result is expected to be, such as what cluster members should or should not be together.

Since we are not interested in adjusting the algorithm to obtain a particular result, an internal validation measure was used for the evaluation of the clustering algorithm: Davies-Bouldin measure. More specifically, it was used to compare the quality of the clusters obtained for different values of k (number of clusters).

Nonetheless, tags were still given to each domain as a way of providing a description of what the domains are related to, allowing further discussion, and also allowing an additional external evaluation.

The tags assigned to each domain are showed in Table 1.

Table 1. Descriptive tags assigned to the domains

Tag	Description
BA	Banking
BS	Big Stores
EC	E-Commerce
ED	Educational
GO	Governmental
JS	Job Sites
OS	Online shopping
NP	Newspaper
PD	Postal Delivery
RS	Radio Station
SE	Search Engine
SU	Supermarket
TC	Telecommunication
TO	Tourism
TV	Television

Davies-Bouldin Index. Davies-Bouldin Index (D-B) is given by the following equation:

$$DB = \frac{1}{N} \sum_{i=1}^{N} D_i \tag{2}$$

where N is the number of clusters, and:

$$D_i = \max_{i \neq j}(R_{i,j}) \tag{3}$$

$$R_{i,j} = \frac{S_i + S_j}{M_{i,j}} \tag{4}$$

$$S_i = \frac{1}{T_i} \sum_{j=1}^{T_i} d(x, c_i) \tag{5}$$

$$M_{i,j} = d(c_i, c_j) \tag{6}$$

where c_i is the centroid of the cluster i, T_i is the size of the cluster i, and $d(c_i, c_j)$ is the distance between the two clusters.

This index measures the average distance between each cluster and its most similar one. Thus, a lower score means that the quality of the clusters is better.

Rand Index. The Rand Index (RI) is a similarity measure between two clustering solutions. It is given by the following equation:

$$RI = \frac{TP + TN}{TP + TN + FP + FN} \tag{7}$$

where TP corresponds to the True Positives, i.e. the number of elements that are grouped together in both clustering results. TN are the True Negatives, elements that are separated in different clusters in both clustering results. FP and FN are False Positives and False Negatives. They represent the elements that belong to the same cluster only in one of the two clustering solutions, but don't belong to the same cluster in the other clustering solution. In which one of the clustering solutions this happens determines what would be a FP or a FN.

In this case, our tags compose a clustering solution that will be compared to the corresponding clustering solution after selecting the k value, in order to obtain the Rand Index.

5.3   Data Pre-processing

With the purpose of reducing the noise in the time series and capturing the essence of their shape to facilitate the establishment of comparisons between the clustering algorithm, a smoothing and normalization process was made on every time series. First, a Simple Moving Average (SMA) was performed with five as the number of periods, in order to reduce noise. Secondly, a Z-Score normalization was applied to modify the scale of the data, as the distance measure to be used is sensitive to scale.

5.4 Experimental Results

The clustering algorithm was performed for different values of k (number of clusters) in the range from 2 to 10. Davies-Bouldin Index was obtained for each execution. Table 2 shows the score for each value of k. As denoted on it, the minimum score was obtained by k = 6, which corresponds to the best number of clusters according to the evaluation measure. Therefore, the clustering results for k = 6 were considered for the following experiments in this work.

Table 2. Davies-Bouldin Index for number of clusters k.

k	D-B Index
2	0.483
3	0.448
4	0.348
5	0.333
6	**0.292**
7	0.462
8	0.493
9	0.403
10	0.397

Table 3 displays the groups obtained by the clustering algorithm for six different groups, listing all their members by their domain names. It also presents the Rand Index described in Sect. 5.2.

As way of visualizing what is contained inside the clusters, Fig. 2 shows plots for each cluster with all the time series of the domains that belong to that particular cluster together.

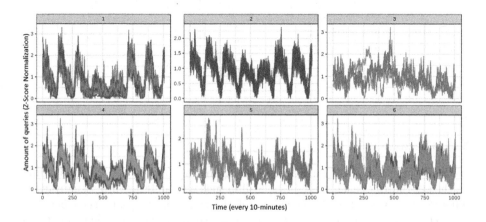

Fig. 2. Clusters members plot

Table 3. Domains and tags by clusters

Cluster	Domain	Tag	Cluster	Domain	Tag	Cluster	Domain	Tag
	aiep	ED		bancochile	BA		abcdin	BS
	bancoedwards	BA		bancoestado	BA		bsale	OS
	bancosantiago	BA		bancofalabella	BA		buscalibre	OS
	bci	BA		bancoripley	BA		chileautos	EC
	bluex	PD		claveunica.gob	GO		chiletrabajos	JS
	chileatiende.gob	GO		cmr	BS		chilevision	TV
	chilexpress	PD		dafiti	OS		comunidadescolar	ED
	correos	PD		despegar	TO		conicyt	ED
1	dt.gob	GO		emisora	RS		cooperativa	RS
	entel	TC	4	lider	SU		curriculumnacional	ED
	mercadopublico	GO		mercadolibre	EC		duoc	ED
	officebanking	BA		pjud	GO	6	easy	BS
	scotiabankazul	BA		publimetro	NP		extranjeria.gob	GO
	scotiabankchile	BA		registrocivil	GO		inacap	ED
	scotiabank	BA		ripley	BS		laborum	JS
	sii	GO		santander	BA		mercadopago	BA
	sistemadeadmision	GO		sodimac	BS		mineduc	GO
	13	TV		trabajando	JS		mitarjetacencosud	BA
	24horas	TV		transbank	BA		movistar	TC
	adnradio	RS		yapo	EC		santotomas	ED
	biobiochile	RS		airbnb	TO		uc	ED
	elmostrador	NP	3	google	SE		uchile	ED
	mega	TV		redgol	NP		udec	ED
2	paris	BS		tripadvisor	TO		webescuela	ED
	pcfactory	BS		clarochile	TC			
	soychile	NP		df	NP			
	t13	TV	5	groupon	TO			
	tvn	TV		itau	BA			
	wom	TC		linio	OS			

Rand Index
0.772

5.5   Discussion

As observed in Fig. 2, the process successfully made groups of domains depending on the attributes of each time series. Still in such a straightforward visualization, differences between the arrangements of the time series can be seen between distinct clusters. One clear aspect is on the weekend, that can be easily identified as the lower peaks in the middle zone of the time series in Cluster 1. These peaks indeed correspond to Saturday and Sunday in the data. Meaning that users of those domains reduce their activity on weekends. On the other hand, members from others clusters, such as Cluster 6 do not clearly demonstrate these distinctions between weekdays and weekends, as users of those domains maintain a uniform usage throughout the whole week. Moreover, members from Cluster 3 show a completely opposite behavior, with peaks on weekends. Nevertheless, all the domains seem to share in common a decrease of activity during nighttime.

The clusters listed in Table 3 also demonstrate a valuable outcome as patterns can be observed when taking into account the content type of the domains, specially when considering our initial descriptive tags. For instance, every domain originally tagged as Educational [ED] was grouped together in Cluster 6, just for *aiep* who was assigned to Cluster 1. This tag considers many of the most important universities and institutes in Chile. Such as Universidad de Chile (*uchile*), Universidad Católica de Chile (*uc*), Universidad de Concepción (*udec*), and Departamento Universitario Obrero y Campesino (*duoc*). As well as some government educational-related domains, such as *conycit* (National Commission for Scientific and Technological Research), and also *mineduc* (Ministry of Education) and *curriculumnacional* (National Curriculum) that were originally tagged as Governmental [GO]. Logically, this kind of domains should present similar traffic, and this is successfully recognized by the algorithm. However, some other not-related domains are also included in the cluster, such as *chilevision* [TV] or *chileautos* [EC].

Another estimable result is the group formed on Cluster 2. As it contains all the domains tagged as Television [TV], except for one. It also incorporates two Radio-Station [RS], and two Newspaper [NP] tagged domains. If we consider that all these tags fit as part of mass media, then we distinguish an interesting pattern captured by our procedure.

We can also observe that the three domains that we manually tagged as Postal-Delivery [PD] were grouped together in the Cluster 1. In this cluster there are also five Governmental [GO] domains and seven Banking [BA] domains.

Additionally, two of the four domains previously tagged as Tourism [TO] were grouped in the smallest cluster along two other domains. One of them is *google* that has a unique tag Search-Engine [SE], expected by us to be distinguished from the rest, assumption that was partially fulfilled.

Given all the above, it is possible to assure that human behavior patterns influence the DNS traffic of the domains, establishing important differences between them, that can be detected by the used time series distance measure. Moreover, these patterns can be detected by the clustering algorithm to successfully create groups whose members show similar behavior and are very likely to share content meaningful to humans. Thus, detecting human patterns in DNS is feasible by employing clustering techniques.

6   Association Rules

In order to establish comparisons between the clusters obtained from the procedure of Sect. 5, association rules are expected to highlight the trends and patterns within the time series. The resulting centroid from every cluster, which corresponds to a domain's time series, was considered as the representative for the experiments and analysis performed in this section. In this way, the association rules procedure was applied on six time series representing the members of each cluster.

The association rules algorithm used is the popular Apriori algorithm. In order to feed it with our data, some transformations were required as a pre-processing stage. That is why an SAX was used to convert the time series to symbols, in addition to a rule for feature extraction.

6.1   Apriori Algorithm

Apriori algorithm was designed to generate association rules that indicate patterns and trends inside a data-set composed by multiple collections of items, commonly associated with transactions. It focuses on the frequency with which the items appear in the transactions, and with what other items they are usually present.

The algorithm receives a minimum support as input, as well as the transactions, and generates candidate itemsets whose appearances in the transactions are filtered by the minimum support given. Finally, it outputs all the association rules that remain. Selecting the relevant rules after this process falls completely to the user criteria, depending on some common evaluation indicators for these rules:

1. Support:

$$Supp(X) = \frac{|\{t \, \epsilon \, T; X \subseteq t\}|}{T} \tag{8}$$

2. Confidence:

$$Conf(X \to Y) = \frac{Supp(X \cup Y)}{X} \tag{9}$$

3. Lift:

$$Lift(X \to Y) = \frac{Supp(X \cup Y)}{Supp(X) \times Supp(Y)} \tag{10}$$

where T is the total number of transactions and t is a single transaction.

6.2   Data Pre-processing

Given that Apriori algorithm receives a list of transactions as input, a transformation is needed to be previously made to the time series. A direct solution is transforming the time series to symbols and pass collections of symbols to the algorithm. This is taken care of by the Symbolic Aggregate approXimation (SAX) [11].

However, SAX uses Piecewise Aggregate Approximation (PAA) to obtain the symbolic values. This procedure reduced the time series length from 1008 points to 168. Five symbols were used in the transformation, resulting in the following time series:

$$S = \{ s_t : t \, \epsilon \, T, \, s \, \epsilon \, \{a, b, c, d, e\} \} \tag{11}$$

where e corresponds to the highest values of the previous time series, and a to lowest ones. Also, $|T| = 168$.

Additionally, one last feature was added to the time series to maintain some relevant information. Using the remaining time series 12, each symbol was assigned an integer in the following way:

a = 0	
b = 1	
c = 2	
d = 3	
e = 4	

This with the purpose of obtaining the difference every two points in the time series as a way to establish a measure of flow change in the traffic to not only know its position at a given time, but also its direction.

For example, if a time series has the symbol b at a given point, and in the next point it changes to d, we will note this change as $d - b = 4 - 2 = 2$, and we will say that it increased by 2.

Adding this feature and grouping by every two points leaves our final DNS traffic time series as:

$$S = \{ (s, n)_t : t \, \epsilon \, T; \, s \, \epsilon \, \{a, b, c, d, e\} \, ; \, n \, \epsilon \, \mathbb{N}; \, n \, \epsilon \, [-4, 4] \} \tag{12}$$

With $|T| = 84$.

This is the final form of the time series that the Apriori algorithm received as a transactions array.

6.3 Results

Table 4 shows what we consider as the most relevant rules after mining the association rules resulting from the Apriori algorithm. The table is subdivided by rules that contain only numeric values, only alphabetic values, and both of them.

6.4 Discussion

The rules showed in Table 4 indicate some patterns in the comparison between the members of each cluster obtained in Sect. 5.

For example, rule number 3 tells us that every time there was a big increase (magnitude 2) experienced in the Clusters 2 and 4, there was also the same

Table 4. Relevant Association Rules obtained from the Apriori algorithm

Number	Body	Head	Support	Confidence	Lift
1	C3 = 0, C4 = −1	C5 = −1	0.11	0.818	2.864
2	C4 = −1, C5 = −1, C6 = −1	C2 = −1	0.06	1	5.600
3	C2 = 2, C4 = 2	C1 = 2	0.04	1	28
4	C2 = 1	C3 = 0	0.13	0.917	1.510
5	C3 = 0, C5 = 0, C6 = 0	C4 = 0	0.18	1	1.615
6	C2 = 0, C3 = −1	C5 = 0	0.13	0.917	1.878
7	C1 = 0, C2 = 0, C3 = 0, C5 = 0, C6 = 0	C4 = 0	0.10	1	1.615
8	C5 = −1, C6 = −1	C2 = −1	0.07	0.857	4.800
9	C2 = −1, C3 = 0, C5 = −1	C6 = −1	0.06	0.833	4.118
10	C1 = 0, C2 = −1	C4 = −1	0.07	1	4
11	C2 = −1, C5 = −1	C4 = −1	0.10	0.889	3.556
12	C2 = −1, C4 = −1	C5 = −1	0.010	0.800	2.800
13	C2 = 0, C3 = −1, C6 = 0	C5 = 0	0.11	1	2.049
14	C1 = 1, C2 = 0	C6 = 0	0.08	1	1.474
15	C1 = 0, C2 = 0, C4 = 1	C6 = 1	0.036	1	12
16	C1 = a, C2 = a	C6 = a	0.18	1	5.600
17	C1 = a, C4 = a, C5 = a, C6 = a	C2 = a	0.12	1	5.250
18	C2 = c, C3 = e	C6 = c	0.08	1	4.941
19	C2 = b, C3 = a, C6 = b	C4 = b	0.07	1	4.667
20	C1 = e, C4 = e	C6 = e	0.14	0.800	4.200
21	C5 = e, C6 = e	C1 = e	0.13	0.917	4.053
22	C6 = e	C4 = e	0.18	0.938	3.938
23	C1 = e, C4 = e, C6 = e	C5 = e	0.12	0.833	3.684
24	C4 = b, C6 = b	C2 = b	0.13	0.917	3.667
25	C1 = c, C6 = b	C2 = b	0.07	0.857	3.429
26	C1 = b, C6 = c	C5 = b	0.07	0.857	3.130
27	C1 = c, C6 = b	C2 = b	0.07	0.857	3.429
28	C3 = c, C6 = c	C4 = b	0.06	0.833	3.889
29	C1 = d, C4 = c	C1 = −1	0.06	1	3.652
30	C2 = c, C5 = b	C5 = 1	0.06	0.833	5
31	C2 = 0, C5 = 0, C1 = e, C5 = e	C3 = −1	0.07	0.857	4.500
32	C4 = 0, C3 = a	C3 = 0	0.08	1	1.647

increase in Cluster 1 with a tremendously high value of lift. However, not with a big value of support.

Number 4 tells with high support that an increase in Cluster 2 will be likely to be accompanied with no change in Cluster number 3. This sets up some differences between the clusters' behavior that would not be easy to see otherwise.

Rule number 2 states, with a high lift value, that if Clusters 4, 5, and 6 experience a decrease, you can safely expect that Cluster 2 will decrease too. With higher support but lower lift, rule number 8 states that if only Clusters

5 and 6 decrease, Cluster 2 will decrease likewise. Rules number 11 and 12 indicate that this behavior will also occur in the other way. That is, if Cluster 2 experiences a decrease along with 4 or 5, the remaining one will be very likely to decrease as well.

Rule number 3 says that if Cluster 3, 5 and 6 maintain their value, Cluster 4 will maintain its value too. However, rule number 13 says that Cluster 5 will maintain its value when both Cluster 2 and 6 do not change, and Cluster 3 is experiencing a decrease.

As for the rules containing symbols, some rules like 16, 17, 21, and 22 tell us what clusters tend to stay in their peaks or valleys when other clusters experience the same. However, other rules such as number 18 tell us that when some clusters are currently in their top or bottom values, others can be found in their middle values; in this case Cluster 6 always obtained c value when Cluster 2 was in c, but Cluster 3 was in his peak e.

Rule number 28 tells us that when Clusters 3 and 6 stay in their middle values, Cluster 4 is very likely to be lower on activity than them.

Finally, some more complex rules regarding both symbols and numeric changes were obtained in the last rows. For example, they tell us that when Cluster 2 has value c and Cluster 5 has value b, Cluster 5 tends to increase with very high lift index. (Rule number 30).

Another case is in rule number 32, saying that when Cluster 4 is not changing its activity and Cluster 3 is at its lowest activity, Cluster 3 tends to maintain its behavior as well. This corresponds to information that is tremendously hard to obtain by other means.

7   Conclusions and Future Work

The procedure proposed in this work was able to identify some patterns in the used time series data. The first stage of our experimentation was able to group domains that have similar content meaningful for humans, obtaining an acceptable external evaluation index as a way for further comparison, but most importantly demonstrating semantic coherence in the domains that were grouped together. As for the second stage, association rules showed interesting trends when comparing the centroids from each cluster that could be useful for performing further analysis and pattern mining.

Taking these results into account, we conclude that human patterns are present in the DNS data, and that these techniques were able to find some of them. This demonstrates that they could be mined and recognized using the appropriate methods and data processing.

Every step from our procedure was associated with an evaluation index as a way of comparison. We suggest as future work the use of other methods that could both find different patterns in the data, and improve the quality of their extraction. Moreover, we claim an achievement of our goal of finding human patterns present in DNS data, however we encourage a more in-depth analysis of the patterns singularly, with the purpose of recognizing more detailed information about them. We strongly believe that these patterns could be of interest for researchers that analyze the human behavior, in this case over activity on the Internet.

References

1. Amazon: Amazon Alexa Topsites (2019). https://www.alexa.com/topsites
2. Berelson, B., Steiner, G.A.: Human behavior: an inventory of scientific findings (1964)
3. Bui, N., Cesana, M., Hosseini, S.A., Liao, Q., Malanchini, I., Widmer, J.: A survey of anticipatory mobile networking: context-based classification, prediction methodologies, and optimization techniques. IEEE Commun. Surv. Tutor. **19**(3), 1790–1821 (2017)
4. Cassisi, C., Montalto, P., Aliotta, M., Cannata, A., Pulvirenti, A., et al.: Similarity measures and dimensionality reduction techniques for time series data mining. In: Advances in Data Mining Knowledge Discovery and Applications, pp. 71–96 (2012)
5. Davies, D.L., Bouldin, D.W.: A cluster separation measure. IEEE Trans. Pattern Anal. Mach. Intell. **2**, 224–227 (1979)
6. Fu, T.C.: A review on time series data mining. Eng. Appl. Artif. Intell. **24**(1), 164–181 (2011)
7. Gonzalez, M.C., Hidalgo, C.A., Barabasi, A.L.: Understanding individual human mobility patterns. Nature **453**(7196), 779 (2008)
8. Hahsler, M., Chelluboina, S., Hornik, K., Buchta, C.: The arules R-package ecosystem: analyzing interesting patterns from large transaction datasets. J. Mach. Learn. Res. **12**, 1977–1981 (2011). http://jmlr.csail.mit.edu/papers/v12/hahsler11a.html
9. Jiawei, H., Kamber, M., Kaufmann, M.: Data mining: concepts and techniques. University of Simon Fraser (2001)
10. Kaufman, L., Rousseeuw, P.J.: Partitioning around medoids (program PAM). In: Finding Groups in Data: An Introduction to Cluster Analysis, vol. 344, pp. 68–125 (1990)
11. Lin, J., Keogh, E., Lonardi, S., Chiu, B.: A symbolic representation of time series, with implications for streaming algorithms. In: Proceedings of the 8th ACM SIGMOD Workshop on Research Issues in Data Mining and Knowledge Discovery, pp. 2–11. ACM (2003)
12. Madariaga, D., Panza, M., Bustos-Jiménez, J.: DNS traffic forecasting using deep neural networks. In: Renault, É., Mühlethaler, P., Boumerdassi, S. (eds.) MLN 2018. LNCS, vol. 11407, pp. 181–192. Springer, Cham (2019). https://doi.org/10.1007/978-3-030-19945-6_12
13. Oliveira, E.M.R., Viana, A.C., Sarraute, C., Brea, J., Alvarez-Hamelin, I.: On the regularity of human mobility. Pervasive Mob. Comput. **33**, 73–90 (2016)
14. Paparrizos, J., Gravano, L.: k-shape: efficient and accurate clustering of time series. In: Proceedings of the 2015 ACM SIGMOD International Conference on Management of Data, pp. 1855–1870. ACM (2015)
15. Sarda-Espinosa, A.: dtwclust: Time Series Clustering Along with Optimizations for the Dynamic Time Warping Distance (2019). https://CRAN.R-project.org/package=dtwclust, r package version 5.5.4
16. Wang, H., Xu, F., Li, Y., Zhang, P., Jin, D.: Understanding mobile traffic patterns of large scale cellular towers in urban environment. In: Proceedings of the 2015 Internet Measurement Conference, pp. 225–238. ACM (2015)

A New Approach to Determine the Optimal Number of Clusters Based on the Gap Statistic

Jaekyung Yang[1]([⊠]) [ID], Jong-Yeong Lee[1] [ID], Myoungjin Choi[2] [ID], and Yeongin Joo[1] [ID]

[1] Department of Industrial and Information Systems Engineering, Jeonbuk National University, Jeonju, Jeonbuk 54896, Republic of Korea
jkyang@jbnu.ac.kr
[2] Howon University, 64 Howondae 3gil, Impi, Gunsan City, Jeonbuk 54058, Republic of Korea

Abstract. Data clustering is one of the most important unsupervised classification method. It aims at organizing objects into groups (or clusters), in such a way that members in the same cluster are similar in some way and members belonging to different cluster are distinctive. Among other general clustering method, k-means is arguably the most popular one. However, it still has some inherent weaknesses. One of the biggest challenges when using k-means is to determine the optimal number of clusters, k. Although many approaches have been suggested in the literature, this is still considered as an unsolved problem. In this study, we propose a new technique to improve the gap statistic approach for selecting k. It has been tested on different datasets, on which it yields superior results compared to the original gap statistic. We expect our new method to also work well on other clustering algorithms where the number k is required. This is because our new approach, like the gap statistic, can work with any clustering method.

Keywords: Clustering · Number of clusters · Data mining

1 Introduction

There are still many open challenges in the clustering task. Those challenges are getting even worse in the current big data era, where data is collected from many sources at high speed. This paper focuses on answering the question: how to decide on the number of clusters k? Being one of the oldest question in the clustering literature, the question has been tackled by hundreds of researchers with many solutions that have been proposed. Among these solutions, the gap statistic is one of the most modern approaches. It is backed by the rigorous theoretical foundation and has been shown to outperform many other heuristic-based approaches such as elbow or silhouette. However, there are still several drawbacks to the original design of the gap statistic, which limits its applicability in real applications. This paper introduces a new technique to mitigate those limitations. The technique can improve the effectiveness of the gap statistic in multiple dimensions. The gap statistic that uses the newly proposed

S. Boumerdassi et al. (Eds.): MLN 2019, LNCS 12081, pp. 227–239, 2020.
https://doi.org/10.1007/978-3-030-45778-5_15

technique is called the "new gap" for short. The following few subsections describe literature reviews.

The Elbow Approach

The oldest method called 'elbow' has been proposed to determine the number of clusters for k-mean clustering algorithm [6]. This is a visual method. The idea of the elbow method is to run clustering method on the dataset for a range of values of k (for example from 1 to 10), and for each value of k calculate clusters and internal index (it could be the sum of squared error (SSE), the percentage of variance, etc.). Then plot a line chart of the internal index for each value of k. At some value of k the value of internal index drops dramatically, and after that, it reaches a plateau when k is increased further. This is the best k value we can expect. Figure 1 illustrates how the elbow method work. In Fig. 1, the line chart goes down rapidly with k increasing from 1 to 2, and from 2 to 3, and reaches an elbow at $k = 3$. After that, it decreases very slowly. Looking at the chart, it looks like maybe the right number of cluster is three because that is the elbow of this curve.

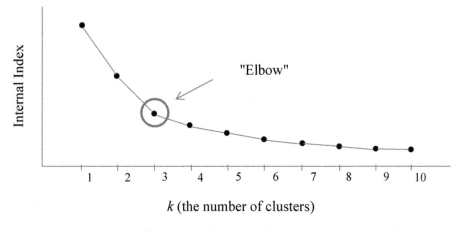

Fig. 1. Identification of Elbow point

However, the elbow method does not always work well. Sometimes, there are more than one elbow, or no elbow at all.

Average Silhouette Approach

Average silhouette method computes the average silhouette of observations for different values of k [2, 3]. The optimal number of clusters k is the one that maximizes the average silhouette over a range of possible values for k [7]. Given a clustering result with k clusters ($k > 1$), we can estimate how well an observation i is clustered by calculating its silhouette statistic $s^k(i)$. Let $a(i)$ be the average distance from observation i to other points in its cluster, and $b(i)$ be the average distance from observation i to points in its the nearest cluster, then the silhouette statistic $s^k(i)$ is calculated by:

$$s^k(i) = \frac{b(i) - a(i)}{\max\{a(i), b(i)\}}$$

A point is well clustered if $s^k(i)$ is large. The average silhouette score $avgS(k)$ gives an estimation of the overall clustering quality when clustering the dataset into k clusters:

$$avgS(k) = \frac{1}{n} \sum_{i=1}^{n} s^k(i),$$

where n is the number of data points.

Therefore, we select k so that it maximizes the average silhouette score. However, this average silhouette is only a heuristic metric, which can be shown to perform poorly in many cases. Note that avgS(k) is not defined at $\underline{k} = 1$.

Hartigan Statistic

Hartigan proposed the statistic [1]:

$$H(k) = \frac{\frac{W_k}{W_{k+1}} - 1}{n - k - 1},$$

where W_k is the average within-cluster sum of squares around the cluster means. The formula to calculate W_k is given in the next section about the gap statistic.

The idea is to start with $k = 1$ and keep adding a cluster until $H(k)$ is sufficiently large. Hartigan suggested the "sufficiently large" cut-off is 10. Hence the estimated number of clusters is the smallest $k \geq 1$ such that $H(k) \leq 10$.

Gap Statistic

Gap statistic was introduced in 2001 by Tibshirani et al. [4] and is still a state-of-the-art method for estimating k. It has been shown to outperform the elbow, average silhouette, and Hartigan methods in both synthesized and real datasets [4, 5]. The method works by assuming a null reference distribution. It then compares the change in within-cluster dispersion with the expected change if the null distribution is true. If when $k = K$ and the within-cluster dispersion starts decreasing slower than the expected rate of the reference distribution, the gap statistic returns k as the expected number of clusters. The formal definition of the gap statistic is given as follows:

Let $d_{ij} = \|x_i - x_j\|^2$ denotes the Euclidean distance between observation i and j, D_r is the sum of the pairwise distance for all points in a given cluster C_r containing n_r points.

$$D_r = \sum_{i \in C_r} \sum_{j \in C_r} d_{ij}$$

Then measure of compactness of clusters W_k is the average within – cluster sum of squares around the cluster means:

$$W_k = \sum_{r=1}^{k} \frac{1}{2n_r} D_r$$

The purpose of clustering is with a given K finding the optimal W_k, when k increases, W_k decreases. But the speed reduction of W_k also decreases. The idea of elbow method is to choose the k corresponding to the "elbow" (finding k that point has the most significant increase in goodness-of-fit). The problems when using elbow method is no reference clustering to compare, and the differences $W_k - W_{k-1}$'s are not normalized for comparison.

The main idea of the gap statistic is to standardize the graph of $\log(W_k)$ by comparing it with its expectation under an appropriate null reference distribution of the data. Estimate of the optimal number of clusters is then the value of k for which $\log\left(W_k^{data}\right)$ falls the farthest below this reference curve $\log\left(W_k^{null}\right)$:

$$Gap_n(k) = E_n^*\left\{\log\left(W_k^{null}\right) - \log\left(W_k^{data}\right)\right\}$$

With E_n^* is the expectation under a sample size of n from reference distribution, we estimate $E_n^*\left\{\log\left(W_k^{null}\right)\right\}$ by an average of B copies $\log\left(W_k^{null}\right)$, each of which is computed from a Monte Carlo sample from reference distribution. Cluster the Monte Carlo samples into k groups and compute $logW_{kb}$, $b = 1, 2 \ldots, B$, $k = 1, 2 \ldots,$ K. Compute the (estimated) gap statistic:

$$Gap(k) = \frac{1}{B}\sum_{b=1}^{B} logW_{kb}^{null} - \log\left(W_k^{data}\right)$$

Those $logW_{kb}^{null}$ from the B Monte Carlo replicates exhibit a standard error $sd(k)$ which, accounting for the simulation error, is turned into the quantity

$$s_k = \sqrt{1 + \frac{1}{B}}.sd(k)$$

Finally, the optimal number of cluster K is the smallest k such that

$$Gap(k) \geq Gap(k+1) - s_{k+1}$$

The above rule to select k is presented in the original gap statistic paper and called the "Tibs2001SEmax" rule in the R clustering implementation of the gap statistic. Since 2001, several other alternatives to this rule have been proposed, such as the "firstSEmax" rule [8] or the "globalSEmax" rule [9]. In this study, the Tibs2001SEmax rule in all experiments was used as the baseline approach. In this paper, the term "gap statistic" refers to the function Gap(k) with the Tibs2001SEmax is used as the k-selecting rule.

Figure 2 provides an example of how the gap statistic works. Figure 2a plots the example dataset with two well-separated clusters. Figure 2b shows the line

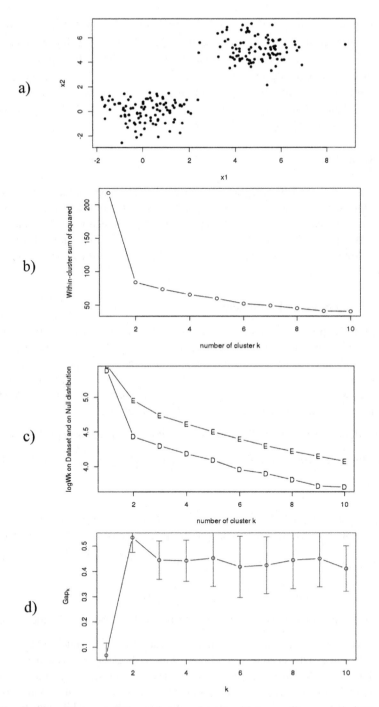

Fig. 2. How the gap statistic works on a dataset with two well-separated clusters

representing the within sum of squares W_k^{data}, which is a downward trend in number of cluster k. Figure 2c shows the log of the expected rate $\log(W_{kb}^{null})$ using an assumed null distribution (uniform distribution in this case). Figure 2d shows the gap statistic, which is calculated by subtracting the log expected rate $\log(W_k^{null})$ for the $\log(W_{kb}^{data})$. The optimal number of k is the smallest k such that there is a significant chance that Gap (k) is higher than $Gap(k + 1)$, which is $k = 2$ in this case. Tibshirani used one standard deviation s_{k+1} to determine when the chance is significant.

2 Methodology

Although being backed by a rigorous theoretical foundation (unlike other heuristic-based methods like elbow or silhouette), the Gap statistic still has several drawbacks that limit its applicability to practical applications. In this section, we conduct several experiments with synthesized datasets to demonstrate those limitations. Based on the insights learned from those experiments, we then introduced a new technique to improve the gap statistic.

2.1 The Gap Statistic Limitations

By design, the gap statistics can only work well when all the clusters in the dataset are well-separated from each other. However, this is rarely the case in practice, where clusters usually overlap up to a certain degree. This "non-overlapping" assumption is one of the main reason that limits the gap statistics effectiveness in real applications. Figure 3 shows how the gap statistics fail to identify the correct K in simple synthesized datasets, that the clusters only barely overlap each other.

(a) the ovl2Gauss dataset: 400 data points in 2 dimensions that sampled equally from the two 2D Gaussian distributions: $\mathcal{N}\left(\begin{bmatrix} 0 \\ 0 \end{bmatrix}, \begin{bmatrix} 1 & 0.7 \\ 0.7 & 1 \end{bmatrix}\right)$ and $\mathcal{N}\left(\begin{bmatrix} 4 \\ 0 \end{bmatrix}, \begin{bmatrix} 1 & -0.7 \\ -0.7 & 1 \end{bmatrix}\right)$.

(b) gap statistic with Tibs2001SE rule suggests $k = 3$ instead of 2 for the ovl2Gauss.

(c) the ovl3Gauss dataset: 600 data points in 2 dimensions that sampled equally from the three Gaussian distributions: $\mathcal{N}\left(\begin{bmatrix} 0 \\ 0 \end{bmatrix}, \begin{bmatrix} 1 & 0.7 \\ 0.7 & 1 \end{bmatrix}\right)$, $\mathcal{N}\left(\begin{bmatrix} 0 \\ 8 \end{bmatrix}, \begin{bmatrix} 1 & 0.7 \\ 0.7 & 1 \end{bmatrix}\right)$, and $\mathcal{N}\left(\begin{bmatrix} 0 \\ 4 \end{bmatrix}, \begin{bmatrix} 1 & -0.7 \\ -0.7 & 1 \end{bmatrix}\right)$.

(d) gap statistic with Tibs2001Se rule suggests $k = 4$ instead of 3 for the ovl3Gauss.

However, clusters should not overlap with each other too much. Otherwise, the notion of "cluster" will become very fuzzy. This is because the data density in the overlapping area is the sum of the data density of the two clusters in that area. This can potentially make the overlapping area become another cluster. In some applications, we indeed want to recognize that overlapping space as a cluster, while that behavior is unexpected in other applications. Figure 4 illustrates this confusion in the case of two strongly overlapping clusters.

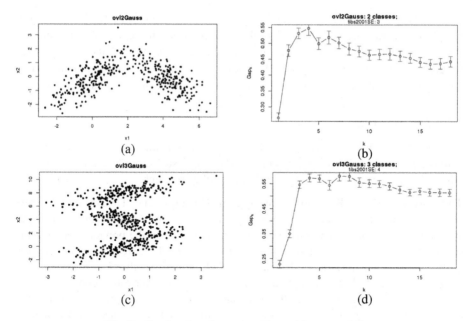

Fig. 3. Overlapping clusters problem with gap statistic

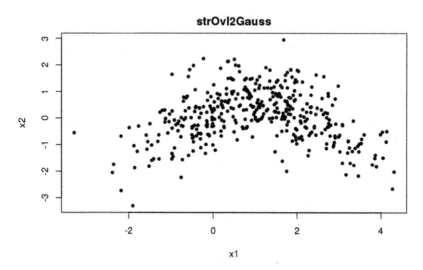

Fig. 4. Two strongly overlapping clusters can be correctly seen as one, two, or 3 clusters.

Besides the non-overlapping assumption, the gap statistic also assumes that there is no hierarchical clustering structure in the dataset. This means in the dataset; there is no cluster that consists of many smaller clusters. In addition, the gap statistics require a lot of computing power to compute the expected W_k under the null reference distribution

$E_n^*\{\log(W_k)\}$. It has to sample the null reference distribution B times $(B \geq 50)$, for each sample b, we run the clustering algorithm. In this case, the clustering algorithm is PAM, which takes $O(n^2)$ with n is the number of data points. In total, the complexity of the algorithm to estimate the $E_n^*\{\log(W_k)\}$ is $O(Bn^2)$. This would make it impossible to apply gap statistic on dataset with more than several thousands of data points.

2.2   The New Gap

As described in the previous section, the gap statistic method has largely three limitations. However, we only focus on the overlapping issue to produce a new gap. The other limitation issues will be covered in the further research.

The 1stDaccSEmax Rule for Overlapping Clusters. The Tibs2001SEmax rule returns the smallest k such that the gap at that point has a significant chance (one standard error) to be higher than the next gap. As shown in the previous section, this rule is very sensitive to overlapped clusters. In fact, when there are overlapping clusters in the dataset, the gap does not decrease but slightly increase after $k = K$ (where K is the real number of clusters in the dataset). This results in over-estimation of K.

Therefore, instead of using the gap statistic directly, we propose to use the deceleration of the gap statistic (Dacc statistic for short). The Dacc is calculated as follows:

$$Dacc(k) = [Gap(k) - Gap(k-1)] - [Gap(k+1) - Gap(k)]$$
$$= 2Gap(k) - Gap(k-1) - Gap(k+1)$$

Figure 5 shows how the $Dacc(k)$ statistic can be computed from the $Gap(k)$ statistic.

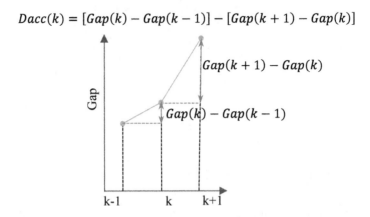

Fig. 5. How to compute Dacc(k) from Gap(k)

We designed this statistic based on the insight that when k is going from 1 to K, the $Gap(k)$ increases with constant or accelerated speed, up to the point where $k = K$. At that point, the $Gap(k)$ will suddenly slow down its speed of increasing or start to decreasing (negative speed). Figure 6 illustrates how the $Dacc(k)$ looks like in different scenarios.

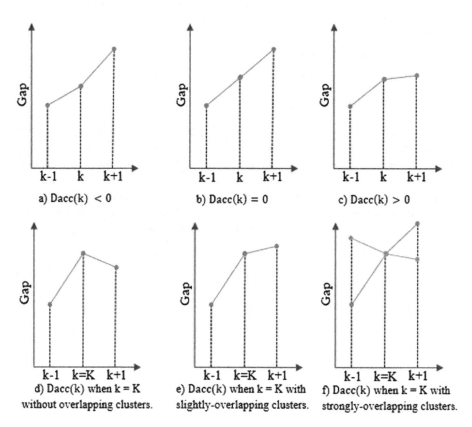

Fig. 6. The Dacc(k) value in different scenarios;

Figure 6(a–c) Different cases where $Dacc(k) < 0$; $Dacc(k) = 0$ and $Dacc(k) > 0$.

Figure 6(d) In dataset with K non-overlapping clusters: $Gap(k)$ increases when $k < K$, reaches its first local maxima at $k = K$, and starts decreasing when $k = K + 1$. Therefore, $k = K$ is also the first local maxima of $Dacc(k)$.

Figure 6(c) In dataset with K clusters where some clusters slightly overlap each other: $Gap(k)$ still increases from $k = K$ to $k = K + 1$, making $Gap(k = K)$ no longer the first local maxima. However, since the overlapping area is small (slightly-overlapping assumption), the increasing speed from $Gap(K)$ to $Gap(K + 1)$ is significantly smaller than the increasing speed from $Gap(K - 1)$ to $Gap(K)$, making the Dacc statistic still maximize at $k = K$. Therefore, the $Dacc(k)$ is more robust than the $Gap(k)$ in a dataset with slightly-overlapping clusters.

Figure 6(f) In dataset with K clusters where some clusters strongly overlap each other: the definition between clusters becomes very fuzzy. Two strongly overlapping clusters can be correctly considered as one, two, or three clusters. Therefore, both Dacc and Gap statistic behave unpredictably in this case.

To take into account the sampling error occurring when estimating the expected W_k under the null distribution, I incorporate the standard error s_k to the $Dacc(k)$ to get the $DaccSE(k)$ as follows:

$$DaccSE(k) = [(Gap(k) - 0.5s_k) - (Gap(k-1) + 0.5s_{k-1})]$$
$$- [(Gap(k+1) + 0.5s_{k+1}) - (Gap(k) - 0.5s_k)]$$

$$DaccSE(k) = 2Gap(k) - Gap(k-1) - Gap(k+1) - 0.5s_{k-1} - 0.5s_{k+1} - s_k$$

As we can see, the higher the sampling errors at k - 1, k, or k + 1, the more DaccSE penalizes the Dacc estimation. Note that I used half standard error in the $DaccSE$ (k) formula. We can choose to use different factor for the standard error based on how "aggressive" or "conservative" you want the DaccSE to behave. Figure 7 illustrates how the DaccSE(k) is calculated. While the Dacc is calculated based on the green line, the DaccSE is calculated based on the dashed orange line. The DaccSE penalizes the $Gap(k - 1)$, $Gap(k)$ and $Gap(k + 1)$ estimation according to how big the s_{k-1}, s_k, and s_{k+1} are.

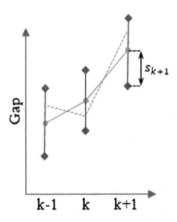

Fig. 7. How the DaccSE(k) is derived from $Gap(k)$ and s_k.

The $Gap(k)$ chart can have multiple peaks, especially when the dataset has a hierarchical clustering structure. Therefore, instead of selecting k where the $DaccSE$ (k) reaches its global maxima, we select the k where $DaccSE(k)$ reaches its first local maxima. This is similar to the idea of searching for the first local maxima of the Tibs2001SEmax rule introduced in the original gap paper. This new rule is called the 1stDaccSEmax rule. Generally, the 1stDaccSEmax rule keeps looking for k with the highest positive DaccSE, with k sequentially running from $k = 2$ to $k = kmax$ and stop at the point where $Gap(k)$ higher than $Gap(k) - s_{k+1}$. Figure 8 shows how the 1stDaccSEmax rule works in different situations.

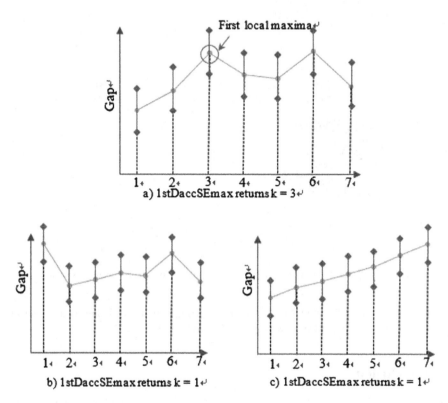

Fig. 8. How the 1stDaccSEmax works in different kinds of Gap charts

Note that although the $DaccSE(k)$ statistic does not define when $k = 1$, the 1stDaccSEmax rule can still detect if there is no cluster in the dataset. This can happen in two situations, which are illustrated in Fig. 8. In Fig. 8b, Gap(1) > Gap(2) by a margin bigger than s_2. Therefore, we stop looking for k right from the beginning and return $k = 1$ right away. In Fig. 8c, all the DaccSE is negative (there is no k at which the gap decreases). Therefore, we also return $k = 1$ in this case.

Figure 9 shows the effectiveness of the 1stDaccSEmax rule on synthesized datasets with overlapping clusters.

Figure 9(a) The ovl2Gauss dataset.

Figure 9(b) Tibs2001SEmax suggests $k = 3$ because $Gap(k)$ still increases from Gap(2) to Gap(3) due to the overlapping. The 1stDaccSEmax predicts correctly that $k = 2$, because the decrease at $k = 3$ is smaller than the decrease at $k = 2$.

Figure 9(c) The ovl3Gauss dataset.

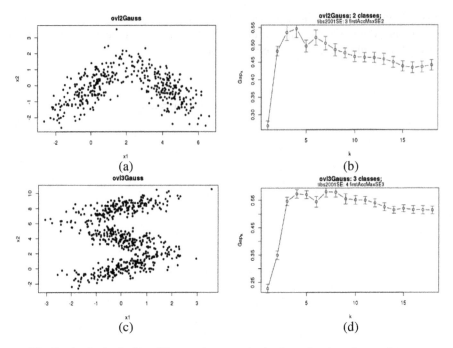

Fig. 9. Apply the 1stDaccSEmax rule on synthesized overlapping clusters datasets.

Figure 9(d) The Tibs2001SEmax predicts wrongly that $k = 4$ due to the overlapping issue. The 1stDaccSEmax rule correctly predicts that $k = 3$.

3   Conclusion

This study focuses on improving the gap statistic for the task of predicting the number of clusters k of a dataset. It identifies and demonstrates three main limitations of the gap statistic, including the overlapping clusters problem, the hierarchical clustering structure problem, and the big dataset problem. Based on these insights, we proposed the new technique to tackle the overlapping problem: the 1stDaccSEmax rule. The performance of the new method is evaluated with several synthetic datasets. It is believed that the performance of the new gap method would be shown to be better than all other traditional approaches. The further numerical experiments will be done on several real datasets with some other new techniques to overcome the other gap limitations.

Acknowledgment. This work was supported by the Industry-Academic Cooperation R&D grant (Dec. 24 2018–Dec. 23 2019) funded by the LX Spatial Information Research Institute (SIRI).

This work was supported by the National Research Foundation of Korea (NRF) grant funded by the Korea government (MSIT) (2017R1D1A1B03034475).

References

1. Hartigan, J.A.: Clustering Algorithms. Wiley Series in Probability and Mathematical Statistics xiii, 351 p. Wiley, New York (1975)
2. Wang, F., Franco-Penya, H.-H., Kelleher, J., Pugh, J., Ross, R.: An analysis of the application of simplified silhouette to the evaluation of k-means clustering validity (2017). https://doi.org/10.1007/978-3-319-62416-7_21
3. Rousseeuw, J.P.: Silhouettes: a graphical aid to the interpretation and validation of cluster analysis. J. Comput. Appl. Math. **20**(1987), 53–65 (1987)
4. Tibshirani, R., Walther, G., Hastie, T.: Estimating the number of data clusters via the Gap statistic. J. Roy. Stat. Soc. B **63**, 411–423 (2001)
5. Chiang, M., Mirkin, B.: Intelligent choice of the number of clusters in k-means clustering: an experimental study with different cluster spreads. J. Classif. **27**, 3–40 (2010). https://doi.org/10.1007/s00357-010-9049-5
6. Kodinariya, T.M., Makwana, P.R.: Review on determining number of cluster in k-means clustering. Int. J. Adv. Res. Comput. Sci. Manage. Stud. **1**(6), 90–95 (2013)
7. Kaufman, L., Rousseeuw, P.J.: Finding Groups in Data: An Introduction to Cluster Analysis. Wiley, New York (1990)
8. Maechler, M.: firstSEMax rule in R document (2012). https://www.rdocumentation.org/collaborators/name/Martin%20Maechler
9. Dudoit, S., Fridlyand, J.: A prediction-based resampling method for estimating the number of clusters in a dataset. Genome Biol. **3**, Article number: research0036.1 (2002)

MLP4NIDS: An Efficient MLP-Based Network Intrusion Detection for CICIDS2017 Dataset

Arnaud Rosay[1(✉)], Florent Carlier[2], and Pascal Leroux[2]

[1] STMicroelectronics, 11 rue Pierre-Félix Delarue, 72100 Le Mans, France
arnaud.rosay@st.com
[2] CREN, Le Mans Université, Avenue Olivier Messiaen, 72085 Le Mans, France
{florent.carlier,pascal.leroux}@univ-lemans.fr
http://www.st.com, http://cren.univ-nantes.fr/

Abstract. More and more embedded devices are connected to the internet and therefore are potential victims of intrusion. While machine learning algorithms have proven to be robust techniques, it is mainly achieved with traditional processing, neural network giving worse results. In this paper, we propose usage of a multi-layer perceptron neural network for intrusion detection and provide a detailed description of our methodology. We detail all steps to achieve better performances than traditional machine learning techniques with a detection of intrusion accuracy above 99% and a low false positive rate kept below 0.7%. Results of previous works are analyzed and compared with the performances of the proposed solution.

Keywords: Machine Learning · Multi-Layer Perceptron · Network intrusion detection · CICIDS2017 dataset

1 Introduction

In recent years, IoT is growing in all areas. This is typically the case for smart agents (IoT-a) that proliferates to provide complex functionalities. Such smart agents rely on communication to cooperate [1]. This is also true for cars that shall embed a cellular connection for emergency call. This is mandatory in Europe from April 1st, 2018 for all passengers cars and light commercial vehicles [12] and also in Russia from January 2017. The presence of a modem enables the emergence of new services requiring data transfer through this cellular connection. In such a context, attacks against IoT-a may lead to loss of functionality or even worse open access to other elements of the network potentially compromising privacy. Cars become potential victims of hackers able to run remote attacks on entire fleet of vehicles. In response to these risks, a first approach is to use network intrusion detection system (IDS) tools such as Snort (open source IDS tool) to analyze network traffic and extract signatures that can be compared

© IFIP International Federation for Information Processing 2020
Published by Springer Nature Switzerland AG 2020
S. Boumerdassi et al. (Eds.): MLN 2019, LNCS 12081, pp. 240–254, 2020.
https://doi.org/10.1007/978-3-030-45778-5_16

to known attack signatures. This type of approach has two limitations: on the one hand, the signature of an attack must already be known and on the other hand, it generates a high false positive rate [4]. An alternative approach is to use Machine Learning (ML) techniques. These techniques include supervised learning to classify network flows into different categories and unsupervised learning to detect anomalies.

Supervised learning is only possible by having a dataset available, in this case the recording of network frames containing normal traffic and attacks. Cellular or WLAN connectivity are quite common in IoT and connected cars and we can consider that the intrusion is similar to the attack that would be carried out on a conventional computer network. Common network intrusion detection datasets like KDD-Cup99 [9], NSL-KDD [17] or CICIDS2017 [14] containing different types of attacks against computers or servers can be used for IoT and connected cars.

Traditional machine learning algorithms and neural network-based techniques (also called deep learning) can be used as supervised learning methods. The former group contains algorithms like Decision Tree, Random Forest, or SVM. Multi-Layer Perceptron (MLP), Convolutional Neural Network (CNN) or Recurrent Neural Network (RNN) are examples of the latter group. In IDS researches, many papers cover deeply the traditional approaches while a smaller amount of papers concentrate on neural network techniques.

Our main contributions consist in two parts. Firstly, we propose an approach based on multi-layer perceptron exercised on recent dataset. CICIDS2017 contains intrusion attacks and traffics that are representative of current network usage. All steps are detailed from dataset analysis to tuning of neural network. Secondly, we analyse previous works on the same dataset and compare their performances with our experimental results.

This paper presents in Sect. 2 the related work. Our proposed methodology is described in Sect. 3. Section 4 provides the results of our approach using several metrics and compares it to previous works. A link to the source code is provided so that it can be used to reproduce results and for further improvements. Finally, Sect. 5 concludes this paper and identifies ideas for future work.

2   Related Work

Several studies have been conducted on multiple network intrusion detection datasets using various methods. Dhanabal and Shantharajah [3] analyzed the NSL-KDD dataset content and studied several classifiers from the traditional machine learning techniques. They obtained pretty good intrusion detection with accuracy around 99% with SVM and J48 (C4.5) decision tree. Tang et al. [16] used a deep learning approach on the same dataset. They proposed a small neural network with a very limited number of features in input and reached an accuracy of almost 76%.

With the release of CICIDS2017, Sharafaldin et al. [14] provided performances of intrusion detection on their dataset. A significant gap is observed on precision, recall and f_1-measure between all traditional ML algorithms and multi-layer perceptron. K-Nearest Neighbors and Quadratic Discriminant Analysis outperformed MLP by almost 20%. Feature selection, data pre-processing and details about the classifiers are not described and therefore results cannot be reproduced. In their paper, Jiang et al. [6] focused only on denial of service, 4 classes among the 14 of CICIDS2017 dataset. They proposed new features and compared the results of a neural network with features provided in original CICIDS2017 CSV files. A two-level model was proposed by Ullah et Mahmoud [18]. The first level classes traffic either as normal or attack with a decision tree and the second one identifies the attack type with a random forest after data augmentation based on synthetic minority oversampling technique [2] and edited nearest neighbors. The method has been tested on CICIDS2017 and UNSW-NB15 datasets. Ustebay et al. [19] proposed a two-step approach for CICIDS2017 dataset. Recursive Feature Elimination (RFE) based on Random Forest is used to identify the most useful features that are then injected in a neural network.

Those works either obtain lower performances with neural network or when achieving good results do not consider all types of attacks. We propose a detailed methodology to achieve high performance with neural network without removing any type of attacks.

3   Our Methodology and MLP Solution

After selecting a dataset, we propose an approach consisting in training a multilayer perceptron neural network in order to quantify the benefits and potential limitations of using deep learning in IDS. Figure 1 give the overall framework.

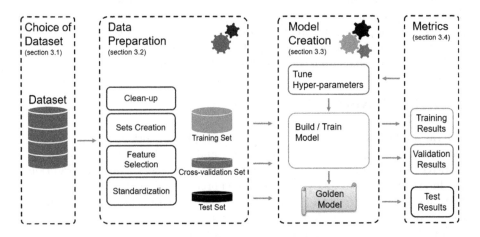

Fig. 1. Overview of the proposed approach.

3.1   Choice of Dataset

KDD-Cup99 is a dataset for IDS publicly released in 1999. It is derived from DARPA-98 dataset that contains raw data corresponding to captured network frames with TCPDUMP. The training set is composed of 24 types of attacks. Raw data have been processed to produce 41 features in KDD-cup99. A first critique of this dataset has been carried out by McHugh [10]. Tavallaee et al. [17] provided a detailed analysis and proposed a derived dataset referred to as NSL-KDD with the goal to solve some of the shortcomings of KDD-Cup99 described McHugh. NSL-KDD dataset has been widely used for IDS since 2009. Despite that attacks has evolved over the time, KDD-Cup99 and NSL-KDD remains a subject of study [13,15,20]. We can consider that most of current attacks are not present in a dataset relying on traffic recorded 20 years ago.

In 2015, Moustafa et Slay proposed a new dataset called UNSW-NB15 [11] that has been generated by simulation and representing 31 h of traffic. As a major difference with NSL-KDD, it contains low footprint and modern attacks retrieved from CVE site[1] that are grouped in nine different families of attacks. UNSW-NB15 is composed of 49 features. A part has been extracted from packet headers while some others have been generated specifically.

Gharib et al. [5] reviewed different datasets and evaluated them with respects to 11 criteria: complete network configuration, complete traffic, labeled dataset, complete interaction, complete capture, available protocols, attack diversity, anonymity, heterogeneity, feature set, metadata. On top of elements already identified in [17], KDD-Cup99 and therefore its NSL-KDD derivative suffer from the lack of some important protocols like HTTPS.

The University of New Brunswick's Canadian Institute for Cybersecurity released the CICIDS2017 dataset based on the framework defined in [14] intending to solve shortcomings of previous datasets. Raw data in the form of PCAP files are provided together with a set of 84 features in CSV files. Network traffic has been recorded over 5 days. The first one only contains normal traffic while 14 types of attack appear for the other days.

Due to the shortcomings of KDD-cup99 and NSL-KDD, we restricted the choice to either UNSW-NB15 or CICIDS2017. Table 1 presents 7 parameters. The first one is year of creation of each dataset. Then the number of features gives us the quantity of input data and labels characterizing each record. The number of records and the distribution between normal traffic and data are important information for dataset selection.

We observe that CICIDS2017 is more recent, include more features and contains more instances than UNSW-NB15. Date of creation and quantity of data are key elements. Firstly, old dataset are no longer representative of current network traffic. As example NSL-KDD does not contain HTTPS exchanges while it represents more than 70% of traffic today. Secondly, more data leads to better learning. For these reasons, CICIDS2017 dataset is selected for the evaluation of our intrusion detection solution.

---

[1] http://cve.mitre.org/.

Table 1. Comparison of UNSW-NB15 and CICIDS2017.

Parameters	UNSW-NB15	CICIDS2017
Year of creation	2015	2017
Features	49 (+2 labels)	84 (+1 label)
Attack families	9	14
Duration	31 h	5 days
# of instances	2,540,044	2,830,743
# of normal instances	2,218,764	2,273,097
# of attack instances	321,283 (12.65%)	557,646 (19.70%)

3.2 Data Preparation

Original Dataset Clean-Up. A series of operations has been carried out to detect the presence of empty lines, redundant features and non numeric values in numeric fields. This systematic verification allowed to drop more than 280,000 instances whose all features were empty. One column corresponding to forward header length appearing twice, one instance has been removed. Most of features are numeric but some values are not a number. Such cases appear in 6 different traffic types: BENIGN, FTP-PATATOR, DoS Hulk, Bot, PortScan and DDoS. As the number of instances containing 'NaN' or 'Infinity' is negligible in each traffic type, these instances have simply been removed. Table 2 provides the number of records for each traffic type and the number of 'NaN'/'Infinity'.

Table 2. CICIDS2017 traffic instances.

Traffic	Original instances	NaN/Infinity	After clean-up
BENIGN	2,273,097	1,777	2,271,320
Bot	1,966	10	1,956
DDoS	128,027	2	128,025
DoS GoldenEye	10,293	0	10,293
DoS Hulk	231,073	949	230,124
DoS Slowhttptest	5,499	0	5,499
DoS Slowloris	5,796	0	5,796
FTP-PATATOR	7,938	3	7,935
Heartbleed	11	0	11
Infiltration	36	0	36
PortScan	158,930	126	158,804
SSH-PATATOR	2,897	0	2,897
WebAttack BruteForce	1,507	0	1,507
WebAttack SQL Injection	21	0	21
WebAttack XSS	652	0	652

Training Set, Cross-Validation Set and Test Set Creation. As shown in Table 2, the dataset is highly imbalanced at two different levels. First, the normal traffic (BENIGN) accounts for 80% and second, some attacks (Heartbleed, Infiltration, WebAttack SQL injection) are represented by a very limited number of instances. This is known to be a challenge for supervised learning. The proposal to create dataset for MLP training and testing consists in choosing randomly 50% of each attacks for the training set, 25% for the cross-validation set and 25% for the test set, ensuring that each instance is used only once. Then each set is completed by adding randomly selected instances of normal traffic. This results in a dataset that is balanced in term of normal traffic versus attacks but still imbalanced in term of attack types. The exact composition of the training, cross-validation and test set is provided in Table 3.

Table 3. Dataset split.

Traffic	Training set	Cross-validation set	Test set
BENIGN	278,274	139,135	139,135
Bot	978	489	489
DDoS	64,012	32,006	32,006
DoS GoldenEye	5,146	2,573	2,573
DoS Hulk	115,062	57,531	57,531
DoS Slowhttptest	2,749	1,374	1,374
DoS Slowloris	2,898	1,449	1,449
FTP-PATATOR	3,967	1,983	1,983
Heartbleed	5	2	2
Infiltration	18	9	9
PortScan	79,402	39,701	39,701
SSH-PATATOR	2,948	1,474	1,474
WebAttack BruteForce	753	376	376
WebAttack SQL Injection	10	5	5
WebAttack XSS	326	163	163
Total	**556,548**	**278,270**	**278,270**

Feature Selection. Feature selection is one of the fundamental concepts of machine learning that greatly influences the model performance. Irrelevant or partially relevant features can have a negative impact and lead to a decrease in accuracy. An analysis of the dataset revealed 8 features that are not informative as their value is constant whatever the traffic types. Consequently, features 'Bwd PSH Flags', 'Bwd URG Flags', 'Fwd Avg Bytes/Bulk', 'Fwd Avg Packets/Bulk', 'Fwd Avg Bulk Rate', 'Bwd Avg Bytes/Bulk', 'Bwd Avg Packets/Bulk', 'Bwd Avg Bulk Rate' have been dropped. As we don't want the model to learn when attack occurs, the 'Timestamp' feature cannot be considered as informative.

Each instance is characterized by its source and destination port and IP address, its protocol and a flow identifier containing the same information. As 'FlowID' feature is redundant with other features, it has been removed from the dataset.

Our model will be trained with 2 feature sets in order to ease comparison with previous works on the same dataset. Variant-1 contains 73 characteristics: all features except those listed above. In variant-2, source/destination IP address and source port have be dropped, resulting in 70 features.

Standardization. Data pre-processing is essential to prepare the dataset for an efficient training. In particular, the neural network can better learn when all features are scaled within the same range. This is especially useful when the inputs are on very different scales. Several normalization techniques exist. In our implementation, Z-score normalization (also called standardization) has been selected as it presents better results than other techniques on the selected dataset. For each feature $\mathcal{F}_j$, the transformation of each value x_i is given by Eq. 1 where $\mu^{(\mathcal{F}_j)}$ and $\sigma^{(\mathcal{F}_j)}$ are respectively the mean and standard deviation values of feature $\mathcal{F}_j$.

$$x_i^{(\mathcal{F}_j)} = \frac{x_i^{(\mathcal{F}_j)} - \mu^{(\mathcal{F}_j)}}{\sigma^{(\mathcal{F}_j)}} \tag{1}$$

3.3   Model Creation

Model Description. Multi-layer perceptron is a fully connected, feed-forward neural network classifier. Figure 2 shows the architectural design of our model. Inputs correspond to the normalized values of the selected features of the original dataset. Both hidden layers contain 256 nodes. The classifier has 15 outputs for the 14 types of attacks and the benign traffic. Dropout is used as regularization technique for hidden layers to prevent over-adjustment on training data by dropping units randomly in the MLP with a probability *keep_prob*.

Neural network output $h^{(3)}$ is calculated by chaining outputs of the different layers according to Eqs. 2, 3 and 4 in which x, $W^{(i)}$ and $b^{(i)}$ are respectively the input vector containing selected features, the matrix of weights and the vector of biases for layer i.

$$h^{(1)} = g_1\left(W^{(1)T}.x + b^{(1)}\right) \tag{2}$$

$$h^{(2)} = g_1\left(W^{(2)T}.h^{(1)} + b^{(2)}\right) \tag{3}$$

$$h^{(3)} = g_2\left(W^{(3)T}.h^{(2)} + b^{(3)}\right)' \tag{4}$$

Activation functions g_1 and g_2 bring non linearity to neural network. As MLP does not provide an intrinsic normalization of its outputs, scaled exponential linear unit given in Eq. 5 is used as activation function for hidden layers with values $\lambda = 1.0507$ and $\alpha = 1.6733$ as defined in [8] to take advantage of self

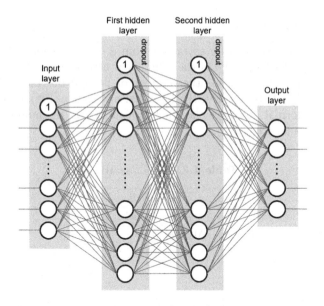

Fig. 2. MLP architectural design.

normalization. Output layer is a softmax activation function g_2 as defined in Eq. 6 so that each output can be interpreted as the probability of predicting a given class. The predicted label $\hat{y}$ is given by $\hat{y} = argmax\, \boldsymbol{h}^{(3)}$.

$$g_1(z) = \lambda. \begin{cases} \alpha.(e^z - 1) & \text{for } z < 0 \\ z & \text{for } z \geq 0 \end{cases} \tag{5}$$

$$g_2(z)_j = \frac{e^{z_j}}{\sum_{k=1}^{N} e^{z_k}} \text{ for } j \in [1; 15] \tag{6}$$

Training. For model implementation and training, we have used python and Tensorflow as deep learning framework. Training set is divided in mini-batch of 32 instances. MLP learns classification by tuning the weights w between neural network nodes in order to reduce the cross-entropy loss function $\mathcal{L}(w)$ as defined in Eq. 7 where y the ground truth label and $\hat{y}$ the predicted class. $\mathcal{L}(w)$ is optimized with Adam algorithm. Three parameters α, β_1 and β_2 described in [7] allow to configure this optimizer.

$$\mathcal{L}(w) = -\left[y.\log(\hat{y}) + (1 - y).\log(1 - \hat{y})\right] \tag{7}$$

Two different models corresponding to the two variants described in Sect. 3.2 have been trained.

3.4   Evaluation Metrics

Several key information can be extracted from a confusion matrix as depicted in Table 4: True Positive (TP) is the number of attacks correctly predicted as attacks, True Negative (TN) is the number of normal instances classified as normal traffic while False Positive (FP) and False Negative (FN) are respectively the number of normal instances classified as attacks and the number of attacks predicted as normal traffic.

Table 4. Confusion matrix.

		Predicted labels	
		Attacks	Normal
Actual labels	Attacks	TP	FN
	Normal	FP	TN

Different metrics can be derived from the information contained in the confusion matrix. As some existing papers use only a subset of metrics, we propose to cover more metrics in order to enable comparison of our work with future studies on the same dataset.

TN_{rate} given in Eq. 8 is the percentage of traffic classified as benign over the actual number of benign instances and FP_{rate} in Eq. 9 is the percentage of benign traffic classified as attacks.

$$TN_{rate} = \frac{TN}{TN + FP} \tag{8}$$

$$FP_{rate} = \frac{FP}{TN + FP} \tag{9}$$

Precision and *Recall* given in Eqs. 10 and 11 correspond respectively to the percentage of attacks correctly detected over the number of instances predicted as attacks and the percentage of attacks correctly detected over the total number of actual attacks.

$$Precision = \frac{TP}{TP + FP} \tag{10}$$

$$Recall = \frac{TP}{TP + FN} \tag{11}$$

Accuracy measures the proportion of total number of correct classifications.

$$Accuracy = \frac{TP + TN}{TP + TN + FP + FN} \tag{12}$$

$f1_{score}$ is the harmonic mean of *Precision* and *Recall*

$$f1_{score} = \frac{2 \times precision \times recall}{precision + recall} \tag{13}$$

Matthews Correlation Coefficient (MCC) is an interesting measure taking into account all elements of the confusion matrix in a correct way for imbalanced dataset as opposed to the accuracy which may report high value even when the whole minority class is wrongly classified. As intrusion detection datasets are intrinsically imbalanced, this is an key metric for such application. MCC provides a value between -1 and $+1$. A perfect prediction corresponds to $MCC = 1$. At the opposite, $MCC = -1$ denotes a full disagreement between predictions and actual classes. A random prediction would result to $MCC = 0$.

$$MCC = \frac{TP \times TN + FP \times FN}{\sqrt{(TP + FP) \times (TP + FN) \times (TN + FP) \times (TN + FN)}} \quad (14)$$

4   Experimental Results

4.1   Performance Evaluation

After training a golden model, the test set has been used to measure the MLP performance as a 15-class classifier. The resulting confusion matrices for the 2 variants are simplified in Tables 5 and 6 to show the estimated class in 3 columns: benign, correct attack and other attacks. In multi-class classification, an attack can be predicted as another attack. This denotes an issue if the correct class is required. In intrusion detection, such a misclassification is acceptable as it is still considered as an attack. Consequently, predictions of all attack types have been merged in one single class to get a binary classifier to calculate metrics in Table 7.

Table 5 shows that MLP performs well on all classes except 'Infiltration'. This specific issue is most likely due to the extremely low number of instances. The neural network didn't succeed in learning from the only 18 examples available in the training set. A significant difference can be observed in Table 6. On the one hand, 1244 normal traffic instances are classified as error, generating false positive detection. On the other hand, more attack instances are classified as normal traffic for almost all classes, meaning that the classifier failed to detect some attacks. In particular, 'infiltration' and 'WebAttack' are not successfully detected. Again, it corresponds to classes with low number of instances. It confirms the well-know fact that the amount of data is a key point for success in machine learning.

The best results are obtained with variant-1 but we can guess the MLP learns the IP address of the machine conducting the attack. By removing IP addresses and source ports in variant-2, the model requires a longer training to reach good performances without being as good as the previous one. This clearly show the importance of these 3 features. Even if the accuracy is above 99%, the recall falls from 99.99% to 99.34% and FP_{rate} goes up from 0.08% to 0.62%. Our classifier provides an MCC value close to 1 and therefore indicate a pretty good performance.

All results can be fully reproduced by using the source code available on Github (https://github.com/ArnaudRosay/mlp4nids) to generate the training, cross-validation and test sets, build the golden model and obtain the values of all the metrics described in this document.

Table 5. Simplified confusion matrix for variant-1.

Actual classes	Predictions		
	BENIGN	Correct attack	Other attack
BENIGN	139,017	–	118
Bot	5	472	12
DDoS	1	31,994	11
DoS GoldenEye	0	2,544	29
DoS Hulk	0	57,529	2
DoS Slowhttptest	0	1,345	29
DoS Slowloris	0	1,434	15
FTP-PATATOR	0	1,969	14
Heartbleed	0	2	0
Infiltration	4	3	2
PortScan	0	39,675	26
SSH-PATATOR	0	1,459	15
WebAttack BruteForce	0	349	27
WebAttack SQL Injection	0	0	5
WebAttack XSS	0	0	163

Table 6. Simplified confusion matrix for variant-2.

Actual classes	Predictions		
	BENIGN	Correct attack	Other attack
BENIGN	138,277	–	858
Bot	193	296	0
DDoS	30	31,970	6
DoS GoldenEye	33	2,537	3
DoS Hulk	45	57,486	0
DoS Slowhttptest	18	1,348	8
DoS Slowloris	7	1,420	22
FTP-PATATOR	12	1,967	4
Heartbleed	0	2	0
Infiltration	7	0	2
PortScan	43	39,639	19
SSH-PATATOR	18	1,455	1
WebAttack BruteForce	340	0	36
WebAttack SQL Injection	2	0	3
WebAttack XSS	162	0	1

Table 7. Performance results.

Metrics	Variant-1 (73 feat.)		Variant-2 (70 feat.)	
	Training	Test	Training	Test
TP	278,256	139,125	276,444	138,225
FP	257	118	1,630	858
FN	18	10	1,830	910
TN	278,017	139,017	276,644	138,277
TN_{rate} (%)	99.91	99.92	99.41	99.34
FP_{rate} (%)	0.09	0.08	0.59	0.62
Recall (%)	99.99	99.99	99.34	99.35
Precision (%)	99.91	99.92	99.42	99.38
Accuracy (%)	99.95	99.95	99.38	99.36
F1-score (%)	99.95	99.95	99.38	99.36
MCC	0.9990	0.9991	0.9876	0.9873

4.2 Comparison with Prior Results

Table 8 compares results with reference papers. As different metrics are used, it is not possible to fill in all cells of the table. In addition to neural network solutions, traditional machine learning algorithms are also considered for this comparison.

Our model outperforms all the results reported by Sharafaldin et al, both for neural network and traditional machine learning techniques. Jiang et al. achieves a slightly worse performances than our solution. It should be noted that the type of attacks has been limited to application layer DoS (slowloris, slowhttptest, hulk, DDos GoldenEye) and with a very limited amount of instances (4171). Therefore it is difficult to make an exact comparison.

A two-stage approach using decision trees proposed by Ullah and Mahmoud [18] reports metrics with an average of 100%, a closer look at the details reveals that 4 attack types are not perfectly detected by the decision trees. These are the exact same classes for which our classifier encounter some difficulties. The lack of significant digits and averaging method in [18] do not allow a fine-grain analysis but the overall comparison shows that our MLP-based solution reaches the same performance level as traditional machine learning algorithms.

The neural network based solution of Jiang et al. [6] achieved results similar to our proposed method but only cover DoS attacks. As shown in Tables 5 and 6, those attacks are not the most difficult to detect. We can expect a drop of their overall performance when all other attacks are taken into account.

Ustebay et al. [19] also used multi-layer perceptron but did not achieved high performance. Nevertheless, their results cannot be directly compared with [6] and [18] as the latter use IP addresses of the machines conducting attacks. In a real life scenario, addresses of hackers are not known and cannot be used for

Table 8. Performance comparison.

Paper	Algorithm	Accuracy (%)	Precision (%)	Recall (%)	FP_{rate} (%)	MCC
Our work - variant-1	MLP (45 epochs)	99.95	99.92	99.99	0.08	0.9991
Our work - variant-2	MLP (715 epochs)	99.36	99.38	99.35	0.62	0.9873
Sharafaldin et al. [14]	MLP	–	77	83	–	–
	Quadratic Discriminant Analysis	–	97	88	–	–
	K-Nearest Neighbors	–	96	96	–	–
Jiang et al. [6]	MLP	99.23	99.87	99.60	0.77	–
Ullah and Mahmoud [18]	Decision Tree + Random Forest	–	100	100	–	–
Ustebay et al. [19]	MLP	91	–	–	–	–

intrusion detection. It should be noted that once source IP address is removed, RFE reports the source port as the most important feature. More generally, this proves that usage of source/destination ports and addresses features may improve intrusion detection on CICIDS2017 but may not be realistic for application to a real network.

5   Conclusion

This paper proposed an approach based on multi-layer perceptron for network intrusion detection system covering analysis of the dataset, definition of the MLP and its training. As in any machine learning project, cleaning the dataset and selecting features is an important step before optimization of a neural network. The experiment has shown that MLP is a viable solution reaching top performance. Our approach provides better results than previous implementations with neural networks. It should be noted that IP addresses and destination port are important features, helping to detect intrusion detection. Nevertheless this is not suitable for real world implementation. Without these features, our approach reaches high performance at the cost of a longer training phase. Deep learning techniques are computationally expensive. We only focused on performance but in a constrained system it may be a drawback pushing to use traditional machine learning algorithms. This may evolve with the increase of hardware accelerators for deep learning in many electronic devices.

In future work, neural network in supervised learning may be improved by data augmentation techniques. Number of instances in classes that are not well detected is clearly a point to address to obtain better performances. Beyond this, supervised learning does not allow detection of attacks unseen during the training phase. Unsupervised learning methods may be considered to overcome this limitation.

References

1. Carlier, F., Renault, V.: IoT-a, embedded agents for smart internet of things. application on a display wall. In: 2016 IEEE/WIC/ACM International Conference on Web Intelligence Workshops (WIW), pp. 80–83, October 2016. https://doi.org/10.1109/WIW.2016.034
2. Chawla, N.V., Bowyer, K.W., Hall, L.O., Kegelmeyer, W.P.: SMOTE: synthetic minority over-sampling technique. J. Artif. Intell. Res. **16**, 321–357 (2002)
3. Dhanabal, L., Shantharajah, D.S.P.: A study on NSL-KDD dataset for intrusion detection system based on classification algorithms. Int. J. Adv. Res. Comput. Commun. Eng. **4**, 446–452 (2015)
4. Garg, A., Maheshwari, P.: Performance analysis of Snort-based intrusion detection system. In: 2016 3rd International Conference on Advanced Computing and Communication Systems (ICACCS), vol. 01, pp. 1–5, January 2016. https://doi.org/10.1109/ICACCS.2016.7586351
5. Gharib, A., Sharafaldin, I., Lashkari, A.H., Ghorbani, A.A.: An evaluation framework for intrusion detection dataset. In: 2016 International Conference on Information Science and Security (ICISS), pp. 1–6, December 2016. https://doi.org/10.1109/ICISSEC.2016.7885840
6. Jiang, J., et al.: ALDD: a hybrid traffic-user behavior detection method for application layer DDoS. In: 2018 17th IEEE International Conference on Trust, Security and Privacy in Computing and Communications/12th IEEE International Conference on Big Data Science and Engineering (TrustCom/BigDataSE), pp. 1565–1569, August 2018. https://doi.org/10.1109/TrustCom/BigDataSE.2018.00225
7. Kingma, D.P., Ba, J.: Adam: a method for stochastic optimization. In: 2015 3rd International Conference for Learning Representations (2014)
8. Klambauer, G., Unterthiner, T., Mayr, A., Hochreiter, S.: Self-normalizing neural networks. In: 2017 Advances in Neural Information Processing Systems, pp. 971–980 (2017)
9. Lee, W., Stolfo, S.J., Mok, K.W.: Mining in a data-flow environment: experience in network intrusion detection. In: Proceedings of the Fifth ACM SIGKDD International Conference on Knowledge Discovery and Data Mining, KDD 1999, pp. 114–124. ACM, New York (1999). https://doi.org/10.1145/312129.312212
10. McHugh, J.: Testing intrusion detection systems: a critique of the 1998 and 1999 DARPA intrusion detection system evaluations as performed by Lincoln laboratory. ACM Trans. Inf. Syst. Secur. **3**(4), 262–294 (2000). https://doi.org/10.1145/382912.382923
11. Moustafa, N., Slay, J.: UNSW-NB15: a comprehensive data set for network intrusion detection systems (UNSW-NB15 network data set). In: 2015 Military Communications and Information Systems Conference (MilCIS), pp. 1–6, November 2015. https://doi.org/10.1109/MilCIS.2015.7348942
12. European Parliament: Regulation (EU) 2015/758 of the European Parliament and of the Council of 29 April 2015 concerning type-approval requirements for the deployment of the eCall in-vehicle system based on the 112 service and amending Directive 2007/46/EC. Official Journal of the European Union, May 2015
13. Riyaz, B., Ganapathy, S.: An intelligent fuzzy rule based feature selection for effective intrusion detection. In: 2018 International Conference on Recent Trends in Advance Computing (ICRTAC), pp. 206–211, September 2018

14. Sharafaldin, I., Lashkari, A.H., Ghorbani, A.A.: Toward generating a new intrusion detection dataset and intrusion traffic characterization. In: Proceedings of the 4th International Conference on Information Systems Security and Privacy - Volume 1: ICISSP, pp. 108–116. SciTePress, January 2018. https://doi.org/10.5220/0006639801080116

15. Shone, N., Ngoc, T.N., Phai, V.D., Shi, Q.: A deep learning approach to network intrusion detection. IEEE Trans. Emerg. Top. Comput. Intell. **2**(1), 41–50 (2018). https://doi.org/10.1109/TETCI.2017.2772792

16. Tang, T.A., Mhamdi, L., McLernon, D., Zaidi, S.A.R., Ghogho, M.: Deep learning approach for Network Intrusion Detection in Software Defined Networking. In: 2016 International Conference on Wireless Networks and Mobile Communications (WINCOM), pp. 258–263, October 2016. https://doi.org/10.1109/WINCOM.2016.7777224

17. Tavallaee, M., Bagheri, E., Lu, W., Ghorbani, A.A.: A detailed analysis of the KDD CUP 99 data set. In: 2009 IEEE Symposium on Computational Intelligence for Security and Defense Applications, pp. 1–6, July 2009. https://doi.org/10.1109/CISDA.2009.5356528

18. Ullah, I., Mahmoud, Q.H.: A two-level hybrid model for anomalous activity detection in IoT networks. In: 2019 16th IEEE Annual Consumer Communications Networking Conference (CCNC), pp. 1–6, January 2019. https://doi.org/10.1109/CCNC.2019.8651782

19. Ustebay, S., Turgut, Z., Aydin, M.A.: Intrusion detection system with recursive feature elimination by using random forest and deep learning classifier. In: 2018 International Congress on Big Data, Deep Learning and Fighting Cyber Terrorism (IBIGDELFT), pp. 71–76, December 2018. https://doi.org/10.1109/IBIGDELFT.2018.8625318

20. Zyad, E., Taha, A., Mohammed, B.: Improve R2L attack detection using trimmed PCA. In: 2019 International Conference on Advanced Communication Technologies and Networking (CommNet), pp. 1–5, April 2019. https://doi.org/10.1109/COMMNET.2019.8742361

Random Forests with a Steepend Gini-Index Split Function and Feature Coherence Injection

Mandlenkosi Victor Gwetu[✉], Jules-Raymond Tapamo, and Serestina Viriri

University of KwaZulu-Natal, Private Bag X54001, Durban 4000, South Africa
gwetum@ukzn.ac.za

Abstract. Although Random Forests (RFs) are an effective and scalable ensemble machine learning approach, they are highly dependent on the discriminative ability of the available individual features. Since most data mining problems occur in the context of pre-existing data, there is little room to choose the original input features. Individual RF decision trees follow a greedy algorithm that iteratively selects the feature with the highest potential for achieving subsample purity. Common heuristics for ranking this potential include the gini-index and information gain metrics. This study seeks to improve the effectiveness of RFs through an adapted gini-index splitting function and a feature engineering technique. Using a structured framework for comparative evaluation of RFs, the study demonstrates that the effectiveness of the proposed methods is comparable with conventional gini-index based RFs. Improvements in the minimum accuracy recorded over some UCI data sets, demonstrate the potential for a hybrid set of splitting functions.

Keywords: Random Forest · Gini-index · Feature engineering · Feature coherence · Circularity

1 Introduction

Legacy supervised machine learning algorithms that are commonly used in pattern recognition tasks include Bayesian Networks, Neural Networks, Support Vector Machines, k-Nearest Neighbours and Decision Trees (DTs) [16]. Although sustained research into each of these individual classification and regression algorithms has led to improved effectiveness, it is widely accepted that ensemble methods generally out perform them [26]. Ensemble methods such as bagging, boosting and Random Forests (RFs) combine the individual outputs of several base learners into a more reliable aggregate committee decision [26]. RFs are ensembles of DTs generated from a bootstrapped training data set [6]; they have gained cross-disciplinary popularity due to their high accuracy rates and simple interpretation [32].

© IFIP International Federation for Information Processing 2020
Published by Springer Nature Switzerland AG 2020
S. Boumerdassi et al. (Eds.): MLN 2019, LNCS 12081, pp. 255–272, 2020.
https://doi.org/10.1007/978-3-030-45778-5_17

The history of RFs can be traced back to the first breed of DTs: CART [7], ID3 and C4.5 [21], which use a common recursive divide-and-conquer approach to partition the training data set until class homogeneity is achieved. These DT types differ in terms of splitting criteria, types of attributes allowed, type of output provided (regression and/or classification), support for missing values, tree pruning strategy and ability to detect outliers [25]. Tree pruning is normally required to reduce the problem of overfitting the classification model on the given training instances. Although the concept of bagging was one of the first techniques to combine outputs from several random DTs, the resultant trees were found to be highly correlated [26]. RFs seek to improve the effectiveness of bagging by reducing the resemblance between trees and thus, simultaneously reduce variance and bias errors. Bias error reflects how inaccurate, learned models are at capturing the significant trends in the training data while variance error captures how monotonous the models are at labeling training instances [18, p. 311]. Simultaneously low, bias and variance errors, are desirable in ensemble classifiers as they indicate that individual base classifiers are highly accurate but diverse[1].

An alternative DT-based ensemble technique is boosting, which aims to improve weak learning algorithms through a committee method that gives more focus to incorrectly classified instances [14,26]. This is done by specifically assigning weights to: (1) individual classifiers in the committee based on training set error and (2) misclassified instances in the training set in order to increase their influence in the next iteration. Boosting has however been found to be less popular than RFs and bagging due to its lack of consistency and low convergence likelihood [26, p. 150]. Furthermore, RFs offer greater opportunity for parallel execution than boosting which has sequential iterations that are dependent on their predecessors.

In DT induction, the criteria used to decide which attribute is the most suitable for partitioning the data set portion at each node, is crucial for achieving high classification accuracy. A study by Raileanu et al. [22] sought to theoretically analyze the two most commonly used splitting criteria/metrics: the gini-index and information gain functions in order to solve the general problem of selecting the most suitable criteria for a given data set. Their findings revealed that the frequency of disagreement between the two criteria is only 2%, thus confirming previously published empirical results which assert that it is impossible to decide on which of the two tests is preferable [22]. This however does not mean, the two criteria can not be optimized individually, neither should it preclude the search for other criteria that may be more effective. Although RFs are an elegant and effective classification technique, there is room for achieving locally optimal attributes [23] and a need for capturing other attribute relations besides conditional interactions [32].

Feature engineering/construction refers to the common practice of creating new features from an existing feature set in a bid to assist classification algorithms to better distinguish between similar previously encountered instances [15]. It is a part of the broader topic of data representation, which seeks to

---

[1] The few misclassifications that individual classifiers make, are in different contexts.

unravel more informative perspectives of a data set by mainly using dimensionality reduction methods such as Linear Discriminant Analysis (LDA) and Principal Component analysis (PCA) [2]. In practise, the new augmented feature set from feature engineering may be used as is or subsequently complemented by a feature selection stage which identifies the most suitable subset of features. Typical feature engineering strategies include computations such as rule based conjunctions, rational differences and polynomial relations based on existing features [15].

This study proposes an optimized gini-index metric and a shape based feature engineering technique as a means towards improving the effectiveness of RFs. A steepend gini-index function is used to replace the conventional gini-index function in order to induce a preferential bias towards probabilities that suggest purity. A novel feature coherence model, based on the shape of a synthetic radial feature contour, is proposed for injecting new attributes to reflect general feature correlation within an instance. The specific question that the study seeks to answer is whether this steepened gini-index splitting function and the proposed shape feature injection can improve the effectiveness of RFs.

The remainder of this paper is structured as follows. Section 2 outlines the RF algorithm in detail and reports on previous work centred on its optimization. Section 3 explains the proposed new methods while Sect. 4 describes the framework used for experimentation. The results of the study are presented in Sect. 5 then Sect. 6 concludes the study and looks at proposed future work.

2   Random Forests

Although RFs also use sampling of training instances with replacement (bootstrapped sampling) like bagging, they introduce additional stochastic behaviour by choosing a random set of predictors (attributes or features) without replacement at each DT node. The conventional RF algorithm (commonly referred to as Forest-RI) can be formally described by Algorithm 1; alternative descriptions can be found in [6, 14, 26].

The maximum permissible purity of DT nodes, along with the parameters ns and d, can be used as constraints for limiting the size and sensitivity of each DT in the RF. The purity of a node is the highest proportion of any class present in its sample. A choice can be made between the constraints imposed by parameters ns and d, since either constraint can effectively limit tree depth, albeit though different criteria. The most commonly used values in literature for the parameter m are 1, $sqrt(M)$ and $log_2(M) + 1$ [3]. The sample size, n is normally set to N, the size of the training set [6], to yield a sampling ratio of 1. In each of the $Ntree$ iterations, a DT is induced based on the given inputs; thus creating a forest of DTs, $\{T_t,\ t = 1, \ldots, Ntree\}$.

Each DT, T_t can be considered as a classifier, $\{h(x, \theta_t)\}$ where θ_t is a set of independent but identically distributed stochastic vectors and x is a new instance to be classified [6]. θ_t is determined during induction by the set of random samples and features chosen with and without replacement respectively.

Algorithm 1. The Forest-RI Algorithm

Input: $S_T = \left\{ (x_i, y_i) \middle| \begin{array}{l} i = 1, \ldots, N, \\ x_i = \{x_{i_j} \mid j = 1, \ldots, M\}, \\ y_i \in \{1, \ldots, C\} \end{array} \right\}$, where S_T is the training set.

$m \leq M$, where m is the number of features to select.

$n <= N$, where n is the sample size.

$n_s <= n$, where n_s is the minimum node size.

$d \in \mathbb{N}$, where d is the maximum tree depth.

$Ntree \in \mathbb{N}$, where $Ntree$ is the number of trees in the forest.

Output: A forest $\{T_t,\ t = 1, \ldots, Ntree\}$ of DT classifiers

    for $t = 1, \ldots, Ntree$ **do**

        create a bootstrapped sample, S_t by randomly selecting (with replacement) n elements from S_T such that $|S_t| = n$.

        create a root node to a DT, T_t based on S_t.

        add the root node to the set, L_{nt} of non-terminal leaf nodes of T_t.

        while L_{nt} is not empty **do**

            let $l_{nt}.depth$ and $l_{nt}.sample$ represent the depth of l_{nt} in T_t and the sample associated with l_{nt} respectively, where l_{nt} is a non-terminal leaf node in L_{nt}. Calculate $l_{nt}.depth$, $|l_{nt}.sample|$ and $purity(l_{nt}.sample)$.

            if $|l_{nt}.sample| < n_s$ or $purity(l_{nt}.sample) == 1$ or $l_{nt}.depth > d$ **then**

                calculate $l_{nt}.class$ as the majority class of $l_{nt}.sample$.

                remove l_{nt} from L_{nt}.

            else

                choose a random feature set $\{f_j, 1 \leq j \leq M\}$ of size m, without replacement.

                find the best feature, $f_b \in \{f_j\}$ for splitting $l_{nt}.sample$.

                split l_{nt} into l_{nta} and l_{ntb} using f_b.

                remove l_{nt} from L_{nt}.

                add l_{nta} and l_{ntb} to L_{nt}.

            end if

        end while

    end for

An individual DT classifier is considered to have perfectly mastered its training set if $h(x_i, \theta_t) = y_i\ \forall\ (x_i, y_i) \in S_T$. The algorithm terminates when L_{nt} is empty; at this point all leaf nodes in a given DT are labelled with one of the C possible classifications based on the majority class within their node's sample. For a given test instance, a RF solicits predictions from each of its $Ntree$ DTs and the ensemble prediction is usually determined by a majority vote.

Although the gini-index was used in the Forest-RI algorithm to determine the attribute value yielding the best split, it has since been found to be weak at identifying strong conditional associations among features [17]. A study by Rob-nik and Sikonja [23] sought to improve the effectiveness of RFs by using several attribute evaluation measures instead of just one, then aggregating DT votes using the margin achieved on similar out-of-bag instances as a weight. The evaluation heuristics used were: gini-index, gain ratio, Minimum Description Length

(MDL), ReliefF or Myopic ReliefF; with each heuristic being applied to its fifth of the trees in a RF. A slight increase in effectiveness is observed when comparing the use of five heuristics against the Gini index alone. This improvement is especially visible on data sets with strong feature dependencies and is attributed to the use of the ReliefF algorithm which manages to decrease DT correlation while retaining prediction strength. A more significant improvement in RF classification accuracy is achieved across several data sets when adopting the new voting strategy.

The pioneering study on RFs by Breiman [6] also proposed a variation in the induction of RFs by using random linear combinations of inputs, a procedure known as Forest-RC. At any given node, L existing numerical features are randomly chosen and added together using random weights in the range $[-1,1]$ to form a new feature. F such features are generated and the best splitting condition is chosen from the range of all possible feature values. Experiments on 19 data sets using $L = 3$ and $F = 2$ or 8 show that although Forest-RC is generally more comparable to Adaboost than Forest-RI, it is not necessarily superior.

Due to the multiple steps in the Forest-RI algorithm, there are several options for improving performance and effectiveness. Improvement in performance can be facilitated through an implementation that exploits parallelization opportunities presented by modern multi-core processors and GPUs like the FastRF, LibRF and the CUDARF algorithms [12]. Improvement in RF effectiveness is generally facilitated by creating a committee of diverse but highly accurate tree based classifiers; a comprehensive survey of such RF variants can be found in [11,17].

3   Proposed Methods

This study aims to improve the effectiveness of RFs by using an alternative splitting function and a feature construction technique that captures inter-feature relationships. The foundations for these two concepts are elaborated below.

3.1   Split Functions

In the original Forest-RI algorithm, the gini-index [7] is used to obtain the best splitting criteria from the random subset of features that represent the instances internal to each DT node. Since the gini-index is a measure of impurity, it can be used to estimate how well a given splitting condition separates the instances within a node into their different classes. The gini-index (also known as gini impurity[2]) was proposed by Breiman et al. [8] as a splitting criteria for DTs known as Classification and Regression Trees (CART). Although 4 other splitting criteria (symmetric Gini, twoing, ordered twoing and class probability) were also proposed, the Gini index generally performs best. Other impurity measures that can be used as alternatives to the gini-index include entropy and classification

---

[2] Higher scores are achieved on impure data sets, so it can be seen as measuring impurity.

error [30]. Because the CART algorithm is generally used to build binary DTs, it is more applicable to binary, numerical and ordinal attributes as their values can be partitioned into 2 groups. For numerical attributes, all possible values of each feature are evaluated as possible thresholds for splitting a node and the value that yields the highest gini-index is chosen.

The Gini index of a DT node can be formulated as follows [11]:

$$Gini(S) = 1 - \sum_{i=1}^{C} p(i)^2, \tag{1}$$

which is equivalent to:

$$Gini(S) = \sum_{i=1}^{C} p(i) * (1 - p(i)), \tag{2}$$

where S is the sample of instances in a node and C is the number of class labels in the data set. If all the classes in the data set are enumerated from 1 to C, then $p(i)$ is the probability of the ith class in a particular node. Likewise the entropy and classification error metrics can be formulated as in Eqs. 3 and 4 respectively [24,30].

$$E(S) = - \sum_{i=1}^{N} p(i) * log(pi). \tag{3}$$

$$CE(S) = 1 - \max_{1 \leq i \leq C} p(i). \tag{4}$$

When considering the viability of an attribute splitting condition or value, the chosen impurity metric is calculated for each potential child node resulting from such a split and normalized using the probability of the child node. These normalized total impurity values are then summed up to represent the combined impurity for the splitting condition in question. The splitting condition yielding the lowest normalized total impurity is potentially the best condition as it corresponds to the highest purity in the given context. The general approach in literature is to use the impurity gained at a particular child node relative to its parent as opposed to the absolute impurity of the child node [23,31]. The generic formulation of this approach for all impurity based measures in the context of binary DTs is as follows [23]:

$$\Delta impurity(a_v) = impurity(S_0) - \sum_{j=1}^{2} p(S_j|a_v) * impurity(S_j|a_v), \tag{5}$$

where a is the attribute to split on using a_v as the specific condition or value, S_0 is the parent node and S_j is one of the two child nodes. $\Delta impurity$ is normally referred to as impurity gain; when applied to the gini-index it is referred to as gini gain. Despite the availability of other gain splitting functions such as Information Gain and Gain Ratio, the Gini Gain was chosen by Breiman for use in RFs because of its simplicity and effectiveness [23]. Information Gain and Gain Ratio are both based on entropy, with the latter being a normalized derivation of the former. The complexity of Information Gain and Gain Ratio arises mainly from their logarithmic computations.

The approach adopted towards formulating alternative impurity measures was to visually analyse the behaviour of existing metrics (gini-index, entropy and class error) in the context of different probabilities. The assumption made was that the gini-index and entropy were superior to class error [8,30], with the gini-index being preferred primarily because of its computational simplicity [22]. The task was then to identify or formulate alternative functions that had a similar shape to the gini-index and entropy functions for input in the range [0,1]. Two proposals were made: the Gaussian and Steepened Gini Index (SGI) functions, with the latter proving more effective than the former.

Gaussian Impurity Function. The Gaussian function was proposed due to its graph which is a symmetric bell curve shape that is similar to the graph of the gini-index. It is widely used in statistics to describe normal distributions and can be formulated as follows [13]:

$$G(x) = a * e^{-\left(\frac{(x-b)^2}{2c^2}\right)}, \tag{6}$$

where a is the amplitude, b is the position of the peak within the bell shape and c is the standard deviation which controls the spread of the bell shape. Since the input of an impurity measure is a probability, this Gaussian distribution is over input values between 0 and 1. In this study, the parameter b was set to 0.5 since probabilities at the tails of the distribution are generally indicative of lower impurity while those in the middle are more likely to correspond with diversity. In a statistical normal distribution, about 99.7% of the data values lie within three standard deviation [9], hence the standard deviation is one third of the distance from the mean to either of the tail ends. We thus set parameter c to 0.1667 (which is $\frac{0.5}{3}$). After considering a few options (0.125, 0.25, 0.5, and 1), the parameter a was set to 0.5 as this value seemed to yield higher accuracy rates on the sonar UCI data set[3] [19]. The resulting Gaussian impurity function, shown in Eq. 7, simply replaces class probabilities with corresponding Gaussian distribution outputs.

$$Gaussian(S) = \sum_{i=1}^{C} G(i). \tag{7}$$

Figure 1 shows the behaviour of the Gini index, entropy, classification error, SGI and the proposed Gaussian metric for probabilities encountered in two-class nodes of a dichotomous DT[4]. It can be observed that all metrics are:

---

[3] Since this data set was used for parameter tuning, final evaluation is mainly based on other data sets to ensure an unbiased experimental context.

[4] Each node has at most two child nodes and each node has at most two classes. The permutations of nodes with more than two classes were not explored due to the computational overhead of computing them.

1. symmetric, showing that a $p(i)$ vs $1 - p(i)$ node class split has the same impurity as its inverse, $1 - p(i)$ vs $p(i)$;
2. maximized at probability 0.5 when classes are equally represented within a node and,
3. minimized for pure nodes which are represented at the tail ends where all metrics give an impurity value of 0 except the Gaussian metric which slightly deviates from the trend with a value of 0.011.

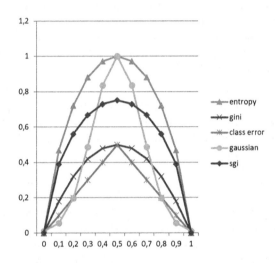

Fig. 1. Impurity functions for two-class node distributions

The Gaussian metric was anticipated to emphasize the difference between the impurity of probabilities in the middle of the distribution and that of those towards the tails; this was expected to favour splits where one of the classes is clearly dominant. Although preliminary experiments on the sonar data set showed the Gaussian metric to be comparable with the gini-index, the former was consistently lower than the latter. Hence, the Gaussian metric was not explored in further experiments and an alternative was sort.

Steepend Gini Index. Our interpretation of the inferiority of the Gaussian metric to the gini-index is based on the fact that split scores are determined using the relative as opposed to absolute impurity of a node. This highlights the importance of comparing the metrics based on gradient in order to model the concept of relative impurity. Since nodes are expected to gain in purity as we move down a DT, it is proposed that a desirable metric would be one that has a steeper negative impurity gradient towards the leaves of the DT. At this stage, a split that leads to greater node purity should be favoured since this node is unlikely to undergo further purification.

This reasoning led us to explore the possibility of modifying the gini-index function such that it yields a steeper gradient towards the two ends of its symmetric shape. After a few attempts at adapting the gini-index function in order to achieve this behaviour, the following function was adopted:

$$SGI(S) = \sum_{i=1}^{C} \frac{p(i) * (1 - p(i)) + \sqrt{p(i) * (1 - p(i))}}{2}. \tag{8}$$

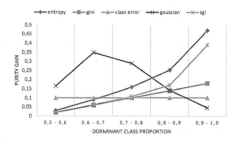

Fig. 2. Gradual change in purity for two-class node splits

Figure 1 shows that the SGI metric is indeed a steepend version of the gini-index. Figure 2 simulates the situation where the dorminant class within a node consistently achieves an increase of 10% in purity as we move down a DT. The root node is assumed to have an equal proportion of two classes while the leaf nodes are pure. From this simulation it is evident that although the Gaussian metric has a similar shape to other metrics, the same can not be said for its gradient function. The gini-index is observed to maintain a consistant gradient while the SGI tries to initially mimick this consistancy but then steepens towards the end. Although entropy appears to have similar behaviour to SGI, it has an initially less consistant gradiant change than SGI and is less steeper at the final node transformation.

3.2   Shape Based Feature Engineering

It was initially envisaged that the adoption of the SGI metric would be enough to yield a significant improvement in the effectiveness of RFs; previous literature however seems to suggest that alternative impurity measures alone may only provide minor improvement [22]. Robnik [23] observes that although the gini-index offers good performance, it evaluates attributes separately and does not take attribute inter-relationships into account. The ReliefF measure was then proposed as a solution for alleviating misclassifications due to high feature interactions. Our work explores feature engineering as an alternative approach for capturing important information from multiple features simultaneously.

Feature engineering/construction entails transforming a given input feature set in order to give a more adequate representation of instances during the training and testing of machine learning models [27]. In some cases, the generated set of new features is deliberately smaller than the original feature set for improved computational efficiency and in order to remove irrelevant features. Examples of such approaches include clustering, LDA and PCA; which are also used for data compression [27]. In other cases, the size of the set of generated features is not constrained and may even be bigger than that of the original feature set. Examples of such approaches in the context of DTs include the FRINGE [20] and Forest-RC [6] algorithms which used Boolean operators and linear combinations respectively to generate a new set of composite features.

Instead of constructing new features that are each based on a selected subset of the original input feature set, this study proposes the formulation of new features that capture the level of coherence between all the feature values in a given instance. The underlying assumption is that if such properties can be empirically quantified, they could be used as extra features for improved class discrimination. This proposed formulation is inspired by the statistical radar/spider chart, which graphically displays multivariate data using a two-dimensional chart of multiple numerical variables represented on a radial axes with a common starting line [29]. Normalized input attributes are allocated fixed orientations from the centre using the order provided by the data set and the contour produced by a given instance is used to characterize its level of feature coherence. These contours can then be analyzed using existing shape descriptors; at this stage only the circularity property has been used as the viability of this concept is still being explored. A circularity score is normally deduced by measuring the perimeter of a closed contour as well as the area of the region it encloses, then computing [5]

$$\frac{perimeter^2}{4 * \pi * area}. \tag{9}$$

The fact that the axes of radar charts are numerical, restricts the application of this method to data sets with only numerical feature values. A hypothetical radar chart is shown in Fig. 3 as an illustration of how normalized feature values can be used to plot radial graphs. The main conclusion that can be drawn from the plots is that spherical contours should be more indicative of greater feature coherence than jagged shapes; it is also expected that instances from the same class should have similarly shaped radial graphs. In cases where instance feature values are in the range [0,1), the generated shapes were observed to be generally the same. This effectively meant that there was no improvement in instance differentiation after adding circularity as an extra feature, especially for data sets with few attributes. This problem was alleviated by scaling the normalized values to the range [0,10). It was envisage that there would be a need for the normalization of feature values so as to reduce the bias of circularity towards any one feature. Indeed, experiments on the sonar data set showed an improvement in classification effectiveness when the injected circularity score was calculated from normalized feature values. The original feature values were however, left un-normalized to avoid distorting the data sets.

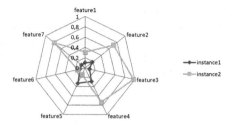

Fig. 3. Hypothetical feature coherence example. Instance1 has greater circularity than Instance2

4   Experimental Protocol

The main of objective of this study is to improve the effectiveness of RFs through the use of a SGI feature evaluation heuristic and a shape-based feature set characterization method. In order to ascertain the effectiveness of these two techniques on RFs, four experiments are conducted in line with the experimental design shown in Fig. 4. The remainder of this section ellaborates on the subcomponents of this design.

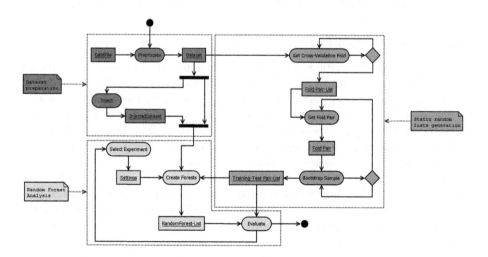

Fig. 4. Experimental design

4.1   Dataset Preparation

The effectiveness of the methods proposed in this study is tested using 10^5 data sets from the UCI repository [19]. These data sets are drawn from Robnik [23] and Breiman's [6] studies on RFs; we exclude data sets with nominal attributes and missing values as these properties are beyond the scope of the present study. An additional constraint is enforced to exclude data sets with more than 3000 instances, for computational reasons. The characteristics of the chosen data sets are summarized in Table 1, which reveals the diversity of the problems represented, in terms of data set size (N), number of features (M) and number of classes (C).

Table 1. UCI datasets

Dataset	N	M	C	Dataset	N	M	C
bupa	345	6	2	iris	150	4	3
ecoli	336	7	8	segmentation	2310	19	7
german-numeric	1000	24	2	sonar	208	60	2
glass	214	9	7	vehicle	846	18	4
ionosphere	351	34	2	yeast	1484	8	10

Each of the chosen data sets is preprocessed by converting all feature values to floating points and class labels to integers, then saved in a consistant format. The output of the data preparation phase is two data set files: one with the original feature set only and another with the circularity feature included.

4.2   Static Random List Generation

Given the highly stochastic nature of RFs, it is imperative to ensure a significant level of contextual consistancy in their induction and evaluation in order to minimize the effect of uncertainty on the outcome of comparative analyses. A deliberate effort is made to subject each RF variant taking part in a comparative experiment, to the same training and testing instances. This is accomplished by generating static/fixed lists of cross validation folds and training samples, before any tree induction or experimentation takes place.

The lists that are generated only contain indexes of instances in the data set, for the sake of efficiency. One random number generator seed is used to produce an entire collection of static random lists. This collection is subsequently used to ensure RF training and testing consistancy in four evaluation experiments. The random number generator is reset at the start of each experiment to enable flexible reenactment of all 4 experiments. The same results can be reproduced repeatedly regardless of the order in which the individual experiments are conducted.

---

[5] The sonar data set is used for parameter tuning.

Although the adopted experimental design (including the RF algorithm), offers several opportunities for parallelization on multi-core processors, this generally comes at the expense of a deterministic execution cycle since parallel loop iterations are typically non-deterministic and may differ from run to run [4, p. 670]. In our case, we require loop iterations to follow a specific order of execution in order to ensure that experimental results can be reproduced verbatim. As a result, no parallelization is implemented in this study. This unfortunately forces us to cap the size of data sets to be investigated, in order to avoid computational overhead.

4.3   Random Forest Analysis

In line with the methodology adopted by Robnik [23] we execute all experiments under the following settings: (1) The recommended RF parameter values are used: the number of trees, $Ntrees = 100$; the number of attributes randomly chosen at each DT node, $m = sqrt(M)^6$; and the cut-off node size, $n_s = 5$. (2) All data sets are evaluated using 10-fold cross-validation. The following additional default parameter values are used: a sampling ratio of 1 and a maximum tree depth, d equal to n.

The four experiments conducted, test the effectiveness of the standard Forest-RI algorithm using either the gini or SCI impurity measure, with and without shape feature injection. We refer to each of these four modifications to the Forest-RI algorithm as RF variants. The use of 10-fold cross validation means that each experiment trains and tests 10 RFs. The accuracy of a RF is calculated as the percentage of test set instances that it correctly classifies. To capture the overall effectiveness of a particular RF approach, we record statistics such as the minimum, maximum, mean and median from the 10 fold accuracies. The code to facilitate all these experiments was implemented in C++, with the opencv library being used for shape feature calculation.

The output of the four experiments is used to compare the effectiveness of the Forest-RI algorithm in the following five contexts: (1) using gini impurity with and without shape feature injection, (2) using SGI impurity with and without shape feature injection, (3) using gini impurity vs SGI, both without shape feature injection, (4) using gini impurity vs SGI, both with shape feature injection and (5) using gini impurity without shape feature injection vs SGI with shape feature injection. We consider this comparison to be objective, since all Forest-RI variants are trained and tested on precisely the same instances.

Since preliminary experiments revealed that it was possible for results to differ on different runs of cross validation, it seemed appropriate to adopt repeated cross validation [33]. For each data set, the cross validations of four experiments are repeated 30 times [28,34], with a different seed in each case. To avoid ambiguity, we propose some terminology for referring to the statistics recorded in this study. For each complete run of 10-fold cross validation using one of the four experiments, we record the following fold accuracies (FAs): minimum-FA,

---

[6] M is the number of attributes used to represent each instance in the data set.

maximum-FA, mean-FA and median-FA. After the 30 cross validation repeats, we calculate the cross validation accuracies (CVAs) for each of the four experiments. Specifically, minimum-CVA, maximum-CVA and mean-CVA are derived using the minimum minimum-FA, maximum maximum-FA and mean mean-FA respectively.

We ultimately seek to establish whether any comparative difference in effectiveness can be attributed to the proposed techniques under investigation or the stochastic properties of RFs and training vs. testing set splits. The Wilcoxon signed-ranks test [10, 23] is used to evaluate the statistical significance of such a difference in RF effectiveness. For each of the five effectiveness comparisons conducted, the null hypothesis assumes there is no difference in the performance of the two RF approaches in question.

5   Results

Table 2 shows the p-values associated with the Wilcoxon signed-ranks test comparing pairs of median-FAs from 30 repeated experiments. Each pair either compares RFs based on splitting function or inclusion of shape feature injection; the training and testing sets used in both cases are however identical. We thus have two groups of 30 corresponding median-FAs and the p-values indicate the overall probability that the difference in corresponding median-FAs is due to chance. A lower p-value is indicative of a more significant difference in median-FAs of the two groups of RFs. We used this test to demonstrate whether the RF variants proposed in this study yield significant differences in effectiveness. Out of the 45 tests done, 5 and 13 tests were significant at an α level of 0.1 and 0.05 respectively. Although this means that in the majority of tests, the RF variants under comparison showed insignificant differences in effectiveness; we note that the +gini vs +sgi and -gini vs +sgi tests each recorded significant differences in 5 of the 9 data sets used. For the bupa data set all tests yielded insignificant differences while the yeast data set attained significant differences in all but the -sgi vs +sgi test.

Table 2. Wilcoxon signed-ranks test p-values from comparing 30 repeated experiment pairs. p-values are classified as either significant at 0.05 level$^\checkmark$, significant at 0.1 level$^\checkmark$, or non-significant^x. +gini and -gini represent RFs using the gini-index with and without shape feature injection respectively. Likewise with steepend gini-index (sgi).

RF pair	Dataset								
	bupa	ecoli	german-numeric	glass	ionosphere	iris	segmentation	vehicle	yeast
-gini vs +gini	0.866^x	0.178^x	0.137^x	0.118^x	0.091$^\checkmark$	0.208^x	0.05$^\checkmark$	0.232^x	<0.01$^\checkmark$
-sgi vs +sgi	0.15^x	0.125^x	<0.01$^\checkmark$	0.021$^\checkmark$	0.851^x	0.08$^\checkmark$	0.268^x	0.551^x	0.838^x
-gini vs -sgi	0.461^x	0.144^x	0.732^x	0.187^x	0.035$^\checkmark$	0.779^x	0.315^x	0.316^x	<0.01$^\checkmark$
+gini vs +sgi	0.757^x	<0.01$^\checkmark$	0.018$^\checkmark$	0.407^x	0.155^x	0.333^x	0.025$^\checkmark$	0.078$^\checkmark$	<0.01$^\checkmark$
-gini vs +sgi	0.232^x	0.011$^\checkmark$	<0.01$^\checkmark$	0.275^x	0.028$^\checkmark$	0.067$^\checkmark$	0.834^x	0.202^x	<0.01$^\checkmark$

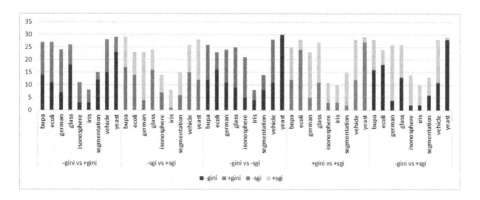

Fig. 5. Frequencies of median-FA superiority after 30 experiment repeats

Table 3. CVA statistics (%) over 30 repeats

CVAs		Dataset								
		bupa	ecoli	german-numeric	glass	ionosphere	iris	segmentation	sonar[a]	vehicle	yeast
minimum-CVA	-gini	52.94	66.67	64	42.86	80	73.33	52.38	52.38	60.71	49.32
	+gini	55.88	66.67	65	33.33	82.86	80	66.67	55	61.18	50.68
	-sgi	47.06	66.67	64	47.62	80	73.33	66.67	55	60	45.27
	+sgi	44.12	63.64	65	42.86	80	80	66.67	57.14	63.53	43.24
mean-CVA	-gini	71.63	85.21	76.47	76.38	92.94	94.78	87.67	83.18	74.98	61.52
	+gini	71.36	85.43	76.97	74.9	93.17	95.02	87.38	83.25	74.73	60.96
	-sgi	72.45	84.63	76.69	76.8	93.55	94.84	87.75	84.08	75.01	59.49
	+sgi	70.95	84.57	77.26	75.48	93.53	95.2	87.73	84.36	75.07	59.28
	robnik [23]	71.90	86.60	75.80	78.10	94	96	98.10	84.10	74.60	61.40
	bader [1]	–	70.51	–		77.10	97.24	95.08	–	87.97	75.31	–
maximum-CVA	-gini	91.43	100	90	100	100	100	100	100	87.06	71.81
	+gini	88.57	100	91	100	100	100	100	100	87.06	70.95
	-sgi	91.43	97.06	91	100	100	100	100	100	88.24	72.48
	+sgi	91.43	100	89	95.45	100	100	100	100	89.29	70.47

[a] Results after parameter tuning.

Since the Wilcoxon signed-ranks test merely highlights the significance of differences in performance, we rely on median-FA comparisons to infer on the superiority of one RF variant over another. Figure 5 shows the number of times the median-FA of a RF variant was higher than that of its competitor over 30 cross validation repeats. Although no clear trend of superiority is demonstrated in the 9 data sets used, we note that the RF variants considered have strengths in different contexts. For example, the german-numeric data set seems to favour +gini over -gini, +sgi over -sgi, +sgi over +gini and +sgi over -gini; which is the exact opposite of the yeast data set context. For some data sets (for example isonosphere, iris and segmentation), most of the 30 experiment repeats yielded exactly the same median-FA while the remaining repeats favoured one RF variant.

Table 3 shows the range of FAs over the 300 (30 cross validation repeats, each with 10 folds) times RFs are trained and tested for each data set. In previous literature such as [1, 23], classification effectiveness is reported using the mean-FA of just one cross validation cycle. Although our mean-CVA results are shown to be comparable with accuracies in previous literature, this statistic alone does not give a comprehensive and reliable picture of RF performance. By reporting the minimum-CVA, maximum-CVA and mean-CVA, we give an indication of the worst, best and average performance of a given RF variant. The distribution of our mean-CVAs confirm the finding of a disagreement of only 2% between splitting criteria, made in previous literature [22]. However, the minimum-CVAs show a greater level of variance. For example, a difference of −14% is shown between the mean-CVAs of -gini and other RF variants.

6   Conclusion

This study sought to improve the effectiveness of RFs through the use of a steepend gini-index and shape feature injection. Although such improvements are indeed recorded over some data sets, the general trend is that of an insignificant difference in effectiveness. When considering the mean-CVA and minimum-CVA results of -gini vs +sgi, we note that the latter outperforms the former over more datasets; we therefore conclude that the steepened gini-index splitting function and the proposed shape feature injection can improve the effectiveness of RFs.

In addition to the proposed RF variants, a major contribution of this study is an experimental framework which allows for a high level of contextual consistancy and repeatability in the induction and evaluation of RFs. Previous studies such as [1, 23] have used the outcome of single runs of cross validation on multiple data sets as evidence of apparent algorithm optimization. We have argued that any claimed superiority should be demonstrated under highly controlled conditions that limit unnecessary stochastic variation, and sustained over multiple repetitions.

Over the course of this study, some opportunities for further work have been identified, we conclude by outlining some of these areas. The large differences in minimum-CVA over several data sets, highlight the potential of creating a RF that uses a hybrid of the RF variants considered in this study. In such a case, the hybrid RF would be equiped to deal with the varying level of complexity in different data sets. Additionally, the extreme weaknesses of one RF variant in some contexts could be compensated for by the better performance of another. Future work will focus on exploring this idea of a hybrid set of RF variants in conjunction with weighted voting. Since some of the minimum-CVAs may have been caused by unfavourable random cross validation splits, the use of stratified cross validation in future work may provide a slightly more controlled training and testing environment. A simple shape descriptor has been adopted in this study; extensions to this work may consider other more advanced shape characterization methods such as moments.

References

1. Bader-El-Den, M.: Self-adaptive heterogeneous random forest. In: 2014 IEEE/ACS 11th International Conference on Computer Systems and Applications (AICCSA), pp. 640–646. IEEE (2014)
2. Bengio, Y., Courville, A., Vincent, P.: Representation learning: a review and new perspectives. IEEE Trans. Pattern Anal. Mach. Intell. **35**(8), 1798–1828 (2013)
3. Bernard, S., Heutte, L., Adam, S.: Forest-RK: a new random forest induction method. In: Huang, D.-S., Wunsch, D.C., Levine, D.S., Jo, K.-H. (eds.) ICIC 2008. LNCS (LNAI), vol. 5227, pp. 430–437. Springer, Heidelberg (2008). https://doi.org/10.1007/978-3-540-85984-0_52
4. Bischof, C.: Parallel Computing: Architectures, Algorithms, and Applications, vol. 15. IOS Press, Amsterdam (2008)
5. Bottema, M.J.: Circularity of objects in images. In: 2000 Proceedings of the IEEE International Conference on Acoustics, Speech, and Signal Processing. ICASSP 2000, vol. 4, pp. 2247–2250. IEEE (2000)
6. Breiman, L.: Random forests. Mach. Learn. **45**(1), 5–32 (2001)
7. Breiman, L., Friedman, J., Stone, C.J., Olshen, R.A.: Classification and Regression Trees. Wadsworth, Belmont (1984)
8. Breiman, L., Friedman, J., Stone, C.J., Olshen, R.A.: Classification and Regression Trees. CRC Press, Boca Raton (1984)
9. Carroll, T.A., Pinnick, H.A., Carroll, W.E.: Probability and the westgard rules. Ann. Clin. Lab. Sci. **33**(1), 113–114 (2003)
10. Demšar, J.: Statistical comparisons of classifiers over multiple data sets. J. Mach. Learn. Res. **7**(Jan), 1–30 (2006)
11. Fawagreh, K., Gaber, M.M., Elyan, E.: Random forests: from early developments to recent advancements. Syst. Sci. Control Eng.: Open Access J. **2**(1), 602–609 (2014)
12. Grahn, H., Lavesson, N., Lapajne, M.H., Slat, D.: CudaRF: a CUDA-based implementation of random forests. In: 2011 9th IEEE/ACS International Conference on Computer Systems and Applications (AICCSA), pp. 95–101. IEEE (2011)
13. Guo, H.: A simple algorithm for fitting a gaussian function. IEEE Signal Process. Mag. **28**(5), 134–137 (2011)
14. Hastie, T., Tibshirani, R., Friedman, J.: The Elements of Statistical Learning. Springer, New York (2001). https://doi.org/10.1007/978-0-387-21606-5
15. Heaton, J.: An empirical analysis of feature engineering for predictive modeling. In: SoutheastCon 2016, pp. 1–6. IEEE (2016)
16. Kotsiantis, S.B., Zaharakis, I., Pintelas, P.: Supervised machine learning: a review of classification techniques (2007)
17. Kulkarni, V.Y., Sinha, P.K.: Random forest classifiers: a survey and future research directions. Int. J. Adv. Comput. **36**(1), 1144–1153 (2013)
18. Manning, C.D., Raghavan, P., Schütze, H., et al.: Introduction to Information Retrieval, vol. 1. Cambridge University Press, Cambridge (2008)
19. Newman, C.B.D., Merz, C.: UCI repository of machine learning databases (1998). http://www.ics.uci.edu/~mlearn/MLRepository.html
20. Pagallo, G.: Learning DNF by decision trees. In: IJCAI, vol. 89, pp. 639–644 (1989)
21. Quinlan, J.: Building classification models: Id3 i c4. 5. Dane udostepnione pod adresem: http://yoda.cis.temple.edu (1993). 8080
22. Raileanu, L.E., Stoffel, K.: Theoretical comparison between the gini index and information gain criteria. Ann. Math. Artif. Intell. **41**(1), 77–93 (2004). https://doi.org/10.1023/B:AMAI.0000018580.96245.c6

23. Robnik-Šikonja, M.: Improving random forests. In: Boulicaut, J.-F., Esposito, F., Giannotti, F., Pedreschi, D. (eds.) ECML 2004. LNCS (LNAI), vol. 3201, pp. 359–370. Springer, Heidelberg (2004). https://doi.org/10.1007/978-3-540-30115-8_34
24. Rokach, L., Maimon, O.: Decision trees. In: Maimon, O., Rokach, L. (eds.) Data Mining and Knowledge Discovery Handbook, pp. 165–192. Springer, Boston (2005). https://doi.org/10.1007/0-387-25465-X_9
25. Singh, S., Gupta, P.: Comparative study ID3, cart and C4. 5 decision tree algorithm: a survey. Int. J. Adv. Inf. Sci. Technol. (IJAIST) **27**, 97–103 (2014)
26. Siroky, D.S., et al.: Navigating random forests and related advances in algorithmic modeling. Stat. Surv. **3**, 147–163 (2009)
27. Sondhi, P.: Feature construction methods: a survey. sifaka. cs. uiuc. edu **69**, 70–71 (2009)
28. de Sousa, J.M., Pereira, E.T., Veloso, L.R.: A robust music genre classification approach for global and regional music datasets evaluation. In: 2016 IEEE International Conference on Digital Signal Processing (DSP), pp. 109–113 (2016). https://doi.org/10.1109/ICDSP.2016.7868526
29. Tague, N.R.: The Quality Toolbox, vol. 600. ASQ Quality Press, Milwaukee (2005)
30. Teknomo, K.: Tutorial on decision tree (2009). http://people.revoledu.com/kardi/tutorial/decisiontree
31. Timofeev, R.: Classification and regression trees (cart) theory and applications. Ph.D. thesis, Humboldt University, Berlin (2004)
32. Touw, W.G., et al.: Data mining in the life sciences with random forest: a walk in the park or lost in the jungle? Brief. Bioinform. **14**, 315–326 (2012). https://doi.org/10.1093/bib/bbs034
33. Vanwinckelen, G., Blockeel, H.: On estimating model accuracy with repeated cross-validation. In: BeneLearn 2012: Proceedings of the 21st Belgian-Dutch Conference on Machine Learning, pp. 39–44 (2012)
34. Zhou, S., Chen, Q., Wang, X.: Active deep networks for semi-supervised sentiment classification. In: Proceedings of the 23rd International Conference on Computational Linguistics: Posters, pp. 1515–1523. Association for Computational Linguistics (2010)

Emotion-Based Adaptive Learning Systems

Sai Prithvisingh Taurah[1], Jeshta Bhoyedhur[2],
and Roopesh Kevin Sungkur[3(✉)]

[1] Tylers Ltd., Quatre Bornes, Mauritius
kta@tylers.mu
[2] LSL Digital, Tombeau Bay, Mauritius
[3] Department of Software and Information Systems, University of Mauritius,
Reduit, Mauritius
r.sungkur@uom.ac.mu

Abstract. Right from our primary school to professional academic level, the classical education system modus operandi, forces us to follow a series of predefined steps to climb the stairs of academic levels. Traditionally those predefined steps forces students to go through the beginner level to advanced level and then specialized in a specific level. The main problem was that the teaching styles and content delivery was not tailored to every learning styles and student personalities. The traditional education system is moving towards adaptive learning system where students are not bound only to one predefined set of contents. Therefore the traditional "one size fits all" approach is no longer valid as it were before. Each student has their curriculum based on their unique needs and personality. Adaptive learning may be referred as the process of creating unique learning experience for each and every learner based upon the learner's personality, interests and performance. This research presents a novel approach of adaptive learning by presenting an emotion-based adaptive learning system where the emotion and psychological traits of the learner is considered to provide learning materials that would be most appropriate at that particular instance of time. It shall demonstrate an intelligent agent based expert system using artificial intelligence and emotion detections capabilities to measure the user learning rate and find an optimum learning scheme for the latter.

Keywords: Adaptive learning · Personalisation · Emotion · Neural networks · Machine learning

1 Introduction

Given the uproar of distance learning through Massive Open Online Courses (MOOC), top universities in the world such as Harvard and MIT joined the craze to propagate knowledge. Despite the millions of subscription for the Harvard MOC, only 10% of students were completing the courses. Feedback from student's show that the content of the courses did not suit their current knowledge level and that the way the course content was presented decelerated their learning rate. Hence the concept of "one size fits all" was questioned by researchers who brought forward the concept of learning styles and prior knowledge relationship to learning process. The idea of adaptive

S. Boumerdassi et al. (Eds.): MLN 2019, LNCS 12081, pp. 273–286, 2020.
https://doi.org/10.1007/978-3-030-45778-5_18

learning system was the answer to mitigate the dropout rate from MOOC. The idea of adaptation is described as the concept of making changes in the educational environment to match variety in the learner needs and abilities to sustain suitable context for interaction. Adaptive hypermedia systems build a model of the goals, preferences and knowledge of each individual user; this model is used throughout the interaction with the user in order to adapt to the needs of that particular user (Brusilovsky et al. 2000). One aspect of adaptive system is adaptive learning, which is the use of technology to derive correct learning pattern for the different learner's intrinsic and extrinsic factors and delivering the contents in a personalized way. Components of an adaptive learning system include a content model, a learner model and an instructional model.

1.1 Content Model

This refers to the way the specific topic, or content domain, is structured, with thoroughly detailed learning outcomes. It is responsible for adaptation of the content as per the user requirements for better interaction. The structure of the domain knowledge relies on symbolic methods. This is often represented as a semantic network of domain concepts, or generally elementary pieces of knowledge for the given domain related with different kinds of links (Oxman and Wong 2014).

1.2 Learner Model

This is also known as the Student Model. The aim is to guide the tutor in taking the pedagogical decisions better adapted to a learner. It models the statistical implications of the knowledge, complications and misapprehensions of the person. It reflects the learner's understanding in a particular field and is prone to changes. The learner information can also be stored such as name, personality style, learning style, age etc.

1.3 Instructional Model

It is the interface provided by the system depending on the individual differences such that the learning process is facilitated. Information from both learner and content model is used to deliver responsive response.

2 Literature Review

2.1 Types of Adaptation

The different types of adaptation are briefly discussed in the section below.

Adaptive Interaction: Adaptation occurs at the graphical user interface and are planned to simplify the user's interaction with the system, without, however changing in any way the learning content itself. Examples: alternative color scheme, font sizes to accommodate user preferences (Paramythis and Loidi-Reisinger 2003).

Adaptive Course Delivery: Adaptations are envisioned to customize a course to the individual learner. The intention is to adjust the gap between course contents and the user requirement so that the ideal learning result is attained (Paramythis and Loidi-Reisinger 2003). Examples of adaptation in this category are dynamic course (re-) structuring; adaptive navigation; and, adaptive selection of alternatives course material (Brusilovsky 2000). Adaptive navigation tends to show the content of an on-line course in enhanced order, where the enhancement criteria considers the learner's background and performance.

Dynamic Courseware Generation: It generates a customized course by considering explicit learning goals, as well as, the basic level of the student's knowledge. The system with dynamic generation studies and adapts to the students' advancement during his interaction with the generated course in real-time.

Content Discovery and Assembly: Application of adaptive methods in the discovery and assembly of learning material from the content model (Paramythis and Loidi-Reisinger 2003). Information collected on the user learning style and prior knowledge on the corpus are the parameters that allow rules defined to be triggered.

Adaptive Collaboration Support: It involves apprehending adaptive support in learning processes that includes communication between multiple persons (Paramythis and Loidi-Reisinger 2003).

2.2 Emotions

"Emotions are basic psychological systems regulating an individual's adaptation to personal and environmental demands. Emotions are closely related to cognitive, behavioral, motivational and physiological processes; therefore they are generally important for learning and achievement" (Seel 2012; Khalfallah and Slama 2018). An influential research on human behaviour put forward that the learning process is more effective when it is associated with positive relations than it is with building negative one. Several researches have indicated that one way for effective learning to take place is by having positive emotions while learning takes place (Corradino and Fogarty 2017; Fatahi 2019; Lane and D'Mello 2019). Furthermore brain imaging has showed that these positive emotions are very important to efficient learning; instructional styles that backs up positive emotions have been correlated with more efficient cognitive processing (Hinton et al. 2008). Researchers claimed that positive mood assists difficult cognitive functions that require elasticity, integrations and use of cognitive material such as memory, classification, creative problem solving, decision-making and learning (Febrilia et al. 2011). However from the results of a recent research, it was stated that good mood does not really guarantee that the student is able to focus. On the other hand, it does show that a bad mood do affect learning process subsequently.

2.3 Studies in the Field

Several adaptive learning tools have been developed till now. These tools focus on different aspects that contribute to learning. Thus they use one of these approaches: adaptive content, adaptive assessments, and adaptive sequences or they use a combination of two of these approaches stated (EdSurge 2016). Knewton is a web learning platform which focuses on adaptive sequences. It records feedback and responds to changes on a real time basis. According to Knewton (n.d.), the learning materials are built on thousands of observations consisting of theories, structure, and difficulty level. Knewton analyses these learning materials and uses sophisticated algorithms to render the most appropriate content to the user. Knewton also added that data are collected from a network of students and these data are recorded, analysed and applied to optimize the next output to each student.

2.4   Machine Learning Techniques

A number of different learning styles classification algorithms have been used since the last decades in adaptive learning system. As per (Truong 2016) which has reviewed 51 studies, the following were the most used in the last ten years: Bayesian network, Rules (Association rules), Neural Network, and Naïve Bayes Network.

Bayesian Network
"A Bayesian network is a graphical model that encodes probabilistic relationships among variables of interest" (Heckerman et al. 1995). Bayesian networks can be used to establish learners profile according to activities that they are selected and realized.

Rule Based Algorithm
Rule based system also known as expert system uses rules knowledge representation for knowledge coded into the system. A rule based system is a way of encoding a human expert's knowledge in a fairly narrow area into an automated system.

Neural Network
Artificial neural network models have specific properties such as capability to adapt, learn or to cluster data. ANN has been modeled from the human cortex but in a less complex way. It contains several nodes arranged in layers (input, hidden and output layer). Activation of a layer is done by the activation function.

Naïve Bayes
"Naive Bayes classifiers, a family of classifiers that are based on the popular Bayes' probability theorem, are known for creating simple yet well performing models" (Raschka et al. 2014). A Naïve-Bayes classifier is built by using the training data to approximate the probability of each group according the examples.

3   Proposed Solution

3.1   Overview of System

The system developed is an online adaptive learning platform which takes into consideration the human psychological factor and human emotional behavior. To increase the efficiency of the system and provide contribution to the field, a multimodal approach for adaptivity is chosen. Upon registration, the user will be provided with a prior knowledge test which is performed to situate the user knowledge level in the domain model space and after a learning style questionnaire is used to identify the user learning style preference, and then, At this point the personalized learning path is generated for optimum learning rate which is achieved by a neural network. During the learning phase, the user emotion is tracked to identify the user current mood that is bored, neutral, surprise among others. This data is used to predict a time table showing exactly when it is optimum for him to study. This is done by a second neural network. After each section completed the user will take a test whereby if his performance is low a reinforcement rule will apply where he will be given additional personalized content to master this section. Those methods will not require resource intensive hardware except a camera and access to a good internet connection. These allow for a wider audience to be acquired and educational academies can use the system at minimal cost.

3.2 Architecture of Proposed System

Three tier architecture is privileged where the system is broken down into Presentation, Application, and Data tiers. This helps in the maintainability of the system and the agility to cope with changing requirements. On the Presentation layer a web interface is provided based on the bootstrap library for adaptivity on different screen sizes. Data tier consist of a content and learner model. The content model contains the learning objects in different version and the learner model contains the student static data such as personal information and dynamic data like student performance and learning paths. Springboot has been used as the web development framework it acts as a middleware between the presentation tier and the data tier. At the application layer, two neural network form the core logic of the system. The content prediction neural network take input such a student performance, learning style and prior knowledge to predict best learning object to learn. The time table neural network uses recorded emotion of user during hourly interval and performance to predict favorable hours for student to learn. Hence a personalized time table is created for the user which suggest hours of days best to learn (Fig. 1).

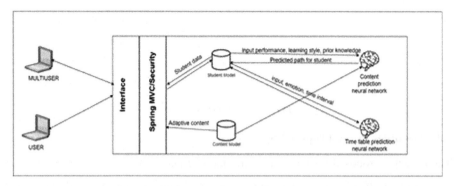

Fig. 1. System architecture

3.3   Choice of Final Tools

IntelliJ Ultimate was the IDE of choice in conjunction with tomcat server, XAMPP and MySQL database. SpringBoot Framework was used as the backbone for the web application. This bundle provide ease of development for web application using Spring technologies such as Spring security, Spring MVC, Hibernate and view resolvers such as Thymeleaf. Libraries used were bootstrap and JQuery for frontend and Neuroph for the neural network on the server side. Microsoft cognitive emotion API was used for emotion detection (Fig. 2).

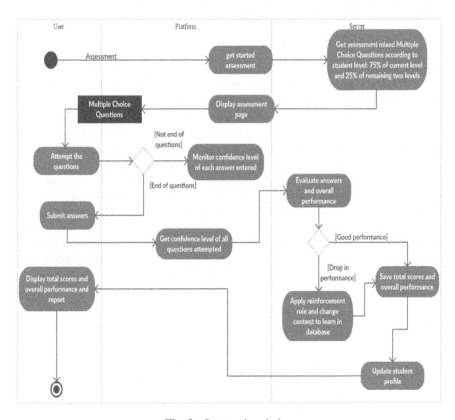

Fig. 2. System description

3.4   Modules in the Proposed System

The modules proposed for the system are outlined below.

Module 1: Predict user's learning content (Training of Data Set and testing the neural with dataset to get the predicting of learning content)
Module 2: Psychometric model
Module 3: Predict user's time table
Module 4: User prior knowledge
Module 5: Monitoring emotions of user while learning takes place
Module 6: Displaying time table (html)

4 Results and Interpretation

4.1 Training of Data Set

The personalised content page for the learner is shown below (Fig. 3).

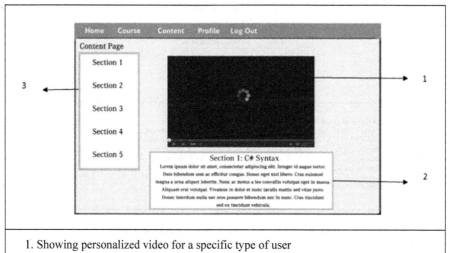

1. Showing personalized video for a specific type of user
2. Transcript of video used as a summary and to provide additional resources
3. Showing navigation to all the section, each link point to a personalized page for the user

Fig. 3. Personalised content page

In this training of dataset, nine inputs, ten neurons and four outputs are used to build the neural network. The neural is run for each interval learned to get the classification of the interval. The average emotion and performance for the specific time interval is used as input (Table 1).

Each emotion is represented as:

Table 1. Emotions

Emotion	Binary representation
Anger	00000001
Contempt	00000010
Disgust	00001000
Fear	00010000
Happiness	00010000
Neutral	00100000
Sadness	01000000
Surprise	10000000

The performance is represented as double integer. An example of input to the neural network would be 0, 0, 0, 1, 0, 0, 0, 0, 0.80 (Table 2 and Fig. 4).

Table 2. Output

Output	Meaning
001	Most favorable
010	Favorable
100	Not favorable

```
Testing trained perceptron
Input: [0.0, 0.0, 0.0, 0.0, 0.0, 0.0, 0.0, 1.0, 0.55] Output: [0.0, 1.0, 0.0]
Input: [0.0, 0.0, 0.0, 0.0, 0.0, 0.0, 1.0, 0.0, 0.68] Output: [0.0, 1.0, 0.0]
Input: [0.0, 0.0, 0.0, 0.0, 0.0, 0.0, 0.0, 1.0, 0.8] Output: [0.0, 0.0, 1.0]
Input: [0.0, 0.0, 0.0, 0.0, 0.0, 0.0, 1.0, 0.0, 0.9] Output: [0.0, 0.0, 1.0]
Input: [0.0, 0.0, 0.0, 0.0, 0.0, 0.0, 1.0, 0.0, 0.35] Output: [1.0, 0.0, 0.0]
Input: [0.0, 0.0, 0.0, 0.0, 1.0, 0.0, 0.0, 0.0, 0.18] Output: [1.0, 0.0, 0.0]
Input: [0.0, 0.0, 0.0, 0.0, 0.0, 0.0, 0.0, 1.0, 0.28] Output: [1.0, 0.0, 0.0]
```

Fig. 4. Input and output using neural network

4.2   Results

A number of experiments have been performed to see if the emotion of the learner is correctly detected. Since the monitoring works in background, this module has been tested where the face of the person is visible on the page. The image in left is the live streaming and the one in the right is the image shot. An alert is displayed when the emotion is recognized (Fig. 5).

Fig. 5. Testing the emotion monitoring system

Figure 6 below shows a personalized time table where learning would be most conducive for a specific learner (Fig. 7).

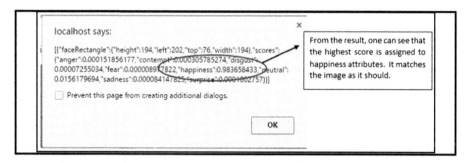

Fig. 6. Testing the emotion monitoring system - 2

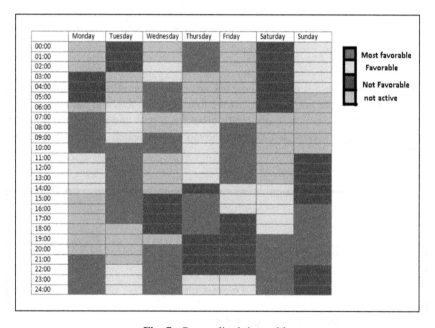

Fig. 7. Personalised time table

5   Discussions

5.1   Testing the Accuracy

For the accuracy of training data set, the total network error for each iteration of the neural network is observed. As the number of iteration performed increased, the total network error decreases. The figure below shows the relations between iteration performed a total network error (Fig. 8).

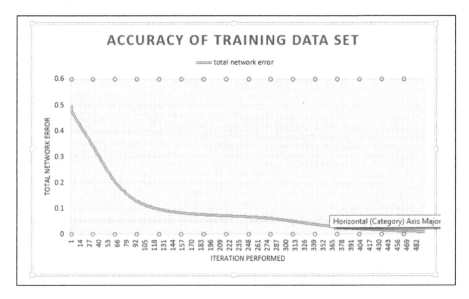

Fig. 8. Accuracy

The table below shows the total network error for the iteration performed. It shows the same concept as the graph above. At the first iteration, the total network error was 0.5009843578823818, which is not really good for a start however at the 493th iteration the total network error is 0.009950455956534508 which is really good (Fig. 9 and Table 3).

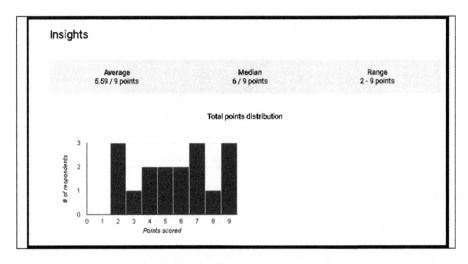

Fig. 9. Results for non-adaptive scenario

Table 3. Total network error at each iteration

Iteration number	Total network error
1	0.5009843578823818
2	0.47491693836473853
3	0.4696483547660897
4	0.46478039443432384
5	0.45985672557279855
..	..
20	0.3932929797443801
..	..
50	0.2555228021830213
..	..
100	0.11904842671535672
..	..
150	0.08475942876046237
..	..
200	0.07397413704542873
..	..
250	0.06740893346060287
..	..
300	0.054403972658013805
..	
350	0.03493140578900644
..	
400	0.021648853428465915
..	
450	0.013942979512677475
..	
493	0.009950455956534508

5.2   Critical Analysis of Proposed System

The proposed system uses a form of supervised learning because it is based on storing data from the user in a database and then generating data. A questionnaire has been designed to get the learning profile of the user. Based on the answers provided, the learning profile was determined and saved in the database for later for generating learning content. Whenever the user is learning, his/her emotions are determined and an algorithm is used to determine the emotion expressed for the longest time period in the interval. The average emotion for the time interval, together with the performance is used to predict if the time interval learned is *favourable*, m*ost favourable* or *not favourable* using Neural Network. For each test and for each question, the probability that a particular learner gets the right answer is calculated using the Item Response Theory (IRT) model, with the knowledge level and difficulty of question as parameters.

The expected performance is calculated using those probabilities (all questions have the same marks). Actual performance is then compared with expected one. Any increase or decrease in performance is reported and the user profile is updated. Additionally the learning profile of each user is updated each time the system is used. For example, when a test is taken, the learning profile is updated by using reinforcement rule to lower the content level if performance is low.

Comparison Against Knewton Adaptive Learning by Knewton

This section compares the proposed system with Knewton which had been introduced in the previous sections of this report. Knewton uses the unsupervised learning method to get its data points whereas the proposed system uses supervised learning method. While Knewton delivers courses based on the most recent student profile, the proposed system uses learning style to determine which learning content is most appropriate for the user. However, one major advantage is that Knewton depends on a large volume of data to do precise clustering and prediction whereas the proposed system accommodates for smaller user population and does not require a large sample size to determine learning profiles. Most importantly the proposed system does something that Knewton does not. The proposed system predicts the customised timetable of the learner by adding a new dimension to adaptivity which is that of emotion. The system recognizes the learner's emotion while learning takes place and this emotion is evaluated to get a favourable time interval for effective learning to take place.

5.3   User Testing

Testing of Non-adaptive System Scenario

Student was provided a PowerPoint presentation on the topic to learn and upon completion was given a multiple choice test to ascertain his newly acquired knowledge. Student were allow fifteen minute to learn the content and five minute to do the test. Google form was used to create PowerPoint and questionnaire. Since topic to be learned was C# programming, users with no prior background in programming was chosen. A sample of 17 students was used (Fig. 9 and Table 4).

Table 4. Statistical analysis of non-adaptive scenario

Statistical operation	Explanation	Result
Mean	Average score of student	5.78
Mode	Maximum occurrence of a score	2,7,9
Median	Middle value separating the distribution	6
Variance	How far a dataset is spread out	6.47
Standard deviation	How spread out numbers are	2.54

Testing of -Adaptive System Scenario

Student performance was tested using the adaptive system. A sample of 17 students with no prior knowledge in programming was used. Student was given fifteen minute to learn the content and five minute to complete the test. The topic to be learned was C# programming. Test Question was the same as in the scenario of the non-adaptive system scenario below (Table 5).

Table 5. Statistical analysis of adaptive scenario

Statistical operation	Explanation	Result
Mean	Average score of student	6.18
Mode	Maximum occurrence of a score	7
Median	Middle value separating the distribution	5
Variance	How far a dataset is spread out	2.10
Standard deviation	How spread out numbers are	1.45

Discussion of Result

Student given personalized learning content had an average score above passing mark and are nearly clustered. The result was satisfactory given users were put under a time limit to learn a whole new concept. User feedback was used to improve details in each section and more examples were suggested.

6 Conclusion

Nowadays e-learning application is being very responsive but there is a problem as each individual's needs is different. The one-size-fits-all is not the solution to build learning platform as each learner is different. These were what has been confirmed at the end of testing phase. It is found to be true that adaptivity in software is the key for future application building success. This research brings forward a novel aspect of adaptivity by considering another intrinsic factor that is that of the learner's emotion. Current research have so far been concentrating on intrinsic factors such as learner's prior knowledge, learning pace and learning style. Machine learning techniques such as neural network to make prediction of personalized content and classification of time interval proved to be very useful. The accuracy of the proposed emotion-based adaptive learning system is also a conclusive factor. Future works may include the use of body language recognition and human gesture recognition to get the mood of the user. Additionally identifying hand gesture to navigate through content and speech-based assessment of user would be interesting features to implement in the future to make the content rendering mechanism more adaptive and responsive.

References

Brusilovsky, P., Stock, O., Strapparava, C. (eds.): AH 2000. LNCS, vol. 1892. Springer, Heidelberg (2000). https://doi.org/10.1007/3-540-44595-1

Corradino, C., Fogarty, K.: Positive emotions and academic achievement - applied psychology OPUS - NYU Steinhardt. Steinhardt.nyu.edu (2017). http://steinhardt.nyu.edu/appsych/opus/issues/2016/spring/corradino_fogarty. Accessed 9 Aug 2019

EdSurge: Software | Adaptive Learning | EdSurge (2016). https://www.edsurge.com/research/special-reports/adaptive-learning/software. Accessed 29 July 2019

Fatahi, S.: Educ. Inf. Technol. 24, 2225 (2019). https://doi.org/10.1007/s10639-019-09868-5

Febrilia, I., Warokka, A., Abdullah, H.H.: University students' emotion state and academic performance: new insights of managing complex cognitive. J. e-Learn. High. Educ. (2011). https://doi.org/10.5171/2011.879553

Heckerman, D., Geiger, D., Chickering, D.: Learning Bayesian networks: the combination of knowledge and statistical data. Mach. Learn. 20(3), 197–243 (1995). https://www.microsoft.com/en-us/research/wp-content/uploads/2016/02/tr-95-06.pdf. Accessed 7 Aug 2019

Hinton, C., Miyamoto, K., Della-Chiesa, B.: Brain research, learning and emotions: implications for education research, policy and practice. Eur. J. Educ. 43(1), 87–103 (2008)

Khalfallah, J., Slama, J.B.H.: The effect of emotional analysis on the improvement of experimental e-learning systems. Comput. Appl. Eng. Educ. (2018). https://doi.org/10.1002/cae.22075

Knewton Adaptive Learning, 1st edn. [ebook] Knewton (n.d.). https://www.knewton.com. Accessed 30 July 2019

Lane, H.C., D'Mello, S.K.: Uses of physiological monitoring in intelligent learning environments: a review of research, evidence, and technologies. In: Parsons, T.D., Lin, L., Cockerham, D. (eds.) Mind, Brain and Technology. ECTII, pp. 67–86. Springer, Cham (2019). https://doi.org/10.1007/978-3-030-02631-8_5

Oxman, S., Wong, W.: White paper: adaptive learning systems. DV X Innovations DeVry Education Group (2014)

Paramythis, A., Loidi-Reisinger, S.: Adaptive learning environment and e-learning standards. In: Williams, R. (ed.), Proceedings of the 2m1 European Conference on e-Learning (ECEL2003), Glasgow, Scotland, 6–7 November, pp. 369–379. Academic Conferences International Reading (2003). ISBN O-9544577-4-9

Raschka, S.: Naive Bayes and text classification Sebastian Raschka's Website (2014). http://sebastianraschka.com/Articles/2014_naive_bayes_1.html. Accessed 7 Aug 2019

Seel, N.: Encyclopedia of the Sciences of Learning, 1st edn. [S.l.], pp. 166–167. Springer (2012)

Truong, H.M.: Integrating learning styles and adaptive e-learning system: current developments, problems and opportunities. Comput. Hum. Behav. 55(Part B), 1185–1193 (2016)

Machine Learning Methods for Anomaly Detection in IoT Networks, with Illustrations

Vassia Bonandrini[1], Jean-François Bercher[2(✉)], and Nawel Zangar[2]

[1] ESIEE Paris, Marne-la-Vallée, France
vassia.bonandrini@edu.esiee.fr
[2] LIGM, UMR 8049, École Des Ponts, UPEM, ESIEE Paris,
CNRS, UPE, Marne-la-Vallée, France
{jf.bercher,nawel.zangar}@esiee.fr

Abstract. IoT devices have been the target of 100 million attacks in the first half of 2019 [1]. According to [2], there will be more than 64 billion Internet of Things (IoT) devices by 2025. It is thus crucial to secure IoT networks and devices, which include significant devices like medical kit or autonomous car. The problem is complicated by the wide range of possible attacks and their evolution, by the limited computing resources and storage resources available on devices. We begin by introducing the context and a survey of Intrusion Detection System (IDS) for IoT networks with a state of the art. So as to test and compare solutions, we consider available public datasets and select the CIDDS-001 Dataset. We implement and test several machine learning algorithms and show that it is relatively easy to obtain reproducible results [20] at the state-of-the-art. Finally, we discuss embedding such algorithms in the IoT context and point-out the possible interest of very simple rules.

Keywords: Internet of Things · IoT · IDS · NIDS · Intrusion detection system · Rules · CIDDS-001

1 Introduction

The problem of securing electronic devices is as old as computers exist, but with time computers have gained more and more resources, so IDS in these devices became more efficient. Now, a lot of different small devices without the power of modern computers are connected to a network and are the target of many attacks. Moreover, every IoT system is different and has specific worries depending on the type of the attack (DDoS, Blackhole, Sybil Attack...) they want to be protected from. Wireless Sensor Networks (WSN) for instance has unique characteristics such as limited power supply, low transmission bandwidth, small memory size, and data storage [3]. It is thus crucial to develop and deploy new IDS.

Section 2 presents a brief review of the literature and Sect. 3 presents the problem of selecting or simulating a dataset to test IDS. Section 4 presents the implementation, tests, and results of several machine learning algorithms for outlier detection for CIDDS-001 dataset. Section 5 presents some simple decision rules that can be

© IFIP International Federation for Information Processing 2020
Published by Springer Nature Switzerland AG 2020
S. Boumerdassi et al. (Eds.): MLN 2019, LNCS 12081, pp. 287–295, 2020.
https://doi.org/10.1007/978-3-030-45778-5_19

introduced in an IoT network to work like an IDS. Section 6 presents a summary of the conclusions and future work.

2   Related Work

Doshi et al. [4] simulate IoT networks with raspberry Pi and virtual machines. They collect network data from their system and test this data with 5 algorithms: K-nearest neighbors, Support Vector Machine (SVM) with linear kernel, decision tree, random forest, and neural network. The specificity of their work is to use stateful features and they get up to 30% better performance compared to without these features.

Hussain et al. [5] list for each problem many surveys that use machine learning techniques (Table 4 on the document). For anomaly and intrusion detection:

– K-means clustering and Decision Tree [6]
– Artificial Neural Network ANN [7]
– Novelty and Outlier Detection [8]
– Decision Tree [9]
– Naive Bayes [9, 10]

Butun et al. [3] classify the IDS methodology of IDS in 3 categories:

1 Anomaly based detection:
   We create an activity profile for each member of the network and a certain amount of deviation is reported as an anomaly. This method is adequate to detect never known attacks but we need to update the profiles periodically because the network behavior can change rapidly.
2 Misuse based detection
   A signature (profile) of the previously known attack is used and is used as a reference to flag the next attacks. The disadvantage of this method is that it cannot detect new type of attacks, but the false positive rate is very low.
3 Specification based detection
   That's a mix of the previous ones, "a set of specifications and constraints that describe the correct operation of a program or protocol is defined." [3] But it takes a lot of time to develop special rules to get a low false-positive rate.

Some surveys have unclear results, sometimes there is no result. There are also very few simulations and implementations in real systems.

3   Dataset

Selecting a dataset to design and evaluate NIDS ML-based algorithms is not immediate and may be a full part challenge.

One of the most used datasets is the KDD cup99 set, but it still presents defaults, as emphasized by Tavallaee et al. [11]:

- a lot of redundant measures
- some parts of the train set were used as test sets in some studies
- set is too long forcing to take only part of the set.

Ring et al. [12] did an exhaustive list of network-based detection data set and compared them. One of the recent and not too heavy dataset is the CIDDS-001 (Coburg Intrusion Detection Data Set) [13] which was described as follows:

"The CIDDS-001 data set was captured within an emulated small business environment in 2017, contains four weeks of unidirectional flow-based network traffic, and comes along with a detailed technical report with additional information. As special feature, the data set encompasses an external server which was attacked in the internet. In contrast to honeypots, this server was also regularly used by the clients from the emulated environment. The CIDDS-001 data set is publicly available and contains SSH brute force, DoS and port scan attacks as well as several attacks captured from the wild" [12].

The dataset contains 14 features as follow (Table 1).

Table 1. Features within the CIDDS-001 data set, from [13]

Id	Attribute name	Attribute description
1	Src IP	Source IP Address
2	Src Port	Source Port
3	Dest IP	Destination IP Address
4	Dest Port	Destination Port
5	Proto	Transport Protocol (e.g. ICMP, TCP, or UDP)
6	Date first seen	Start time flow first seen
7	Duration	Duration of the flow
8	Bytes	Number of transmitted bytes
9	Packets	Number of transmitted packets
10	Flags	OR concatenation of all TCP Flags
11	Class	Class label (Normal, Attacker, Victim)
12	AttackType	Type of Attack (PortScan, DoS, Bruteforce, PingScan)
13	AttackID	Unique Attack id
14	AttackDescription	Additional information about the set attack parameters

In our experiment, we used the "Class" attribute as the target for classification, removed the AttackType, AttackID and AttackDescription features which are obviously correlated with the "attacker" class. Furthermore, since IPs were anonymized, they do not convey information so we also removed them. We also use "Date first seen" as the x-axis. Finally, we transformed Flags, Class and Proto, which are categorical features, into "dummy variables" by one-hot encoding.

In the CIDDS-01 dataset, we used the internal-week1 subset of observations, as it contains 42 of the 92 attacks on the entire dataset.

Anomalies are labeled as victim or attacker. However, this file has more than 8 million rows and less than 20% are anomalies. Hence we face a case of imbalanced classes. In such a case, the more represented class can have a "masking effect" on the others; this has been studied in [14] for this dataset. Given the high number of instances available, rebalancing the classes can be simply done by subsampling the majority class (otherwise, one can also oversample the minority classes by creating new, synthetic, instances). In our case, we decided to (a) shuffle the data, (b) keep half of the data for a final evaluation (c) subsample the other half to keep about 180000 instances per class.

4   Experiments and Results

The experiments were carried out using Google Colaboratory with 32 GB of ram and the Tensor Processing Unit acceleration material.

The metrics that were used to evaluate the performance of the algorithm include the classification accuracy, precision, recall and F1-score. These metrics are expressed by the equations below,

$$\text{Accuracy} = \frac{TP + TN}{TP + TN + FP + FN}$$

$$\text{Precision} = \frac{TP}{TP + FP}$$

$$\text{Recall} = \frac{TP}{TP + FN}$$

$$\text{F1-score} = \frac{2.Precision.Recall}{Precision + Recall}$$

where, TP, TN, FP and FN stand for true positives, true negatives, false positives and false negatives, respectively.

We shuffle the set and we take 33% of the set as the test set and 66% as the train set. Then, we classify the traffic with 4 algorithms: K-Nearest Neighbors (KNN), Decision Tree (DT), Random Forest (RF) and Neural Network (NN). We used the python sklearn package to do our test.

For the KNN, we use only 1 neighbour with a uniform weight function. We get a global accuracy of 99.27%; other metrics are reported Table 2.

Table 2. Results with the K Nearest Neighbour algorithm

KNN	Precision	Recall	F1-score
Attacker	0.9633	0.9972	0.9799
Normal	0.9996	0.9916	0.9956
Victim	0.9573	0.9988	0.9976

For the Decision Tree, we used the default parameters, which is the Gini criterion for measuring the quality of a split and maximum expansion of nodes. We already obtained a global accuracy of 99.89%; other performance metrics are given Table 3.

Table 3. Results with the Decision Tree algorithm

DT	Precision	Recall	F1-score
Attacker	0.9956	0.9982	0.9969
Normal	0.9998	0.9990	0.9994
Victim	0.9946	0.9996	0.9970

For instance, we get something as shown in Fig. 1 for the beginning of the tree:

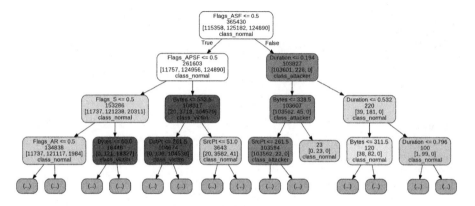

Fig. 1. First nodes of the decision tree

The decision tree has a depth of 31, a total of 563 nodes and 282 leaves, thus 281 tests. From this tree, it is possible to deduce, and even generate automatically, a classification script (see the source code [20] for an example). Running this code on an instance will predict the classification with 99.88% accuracy.

For the RF, we selected the best parameters using a grid search strategy, which consists of computing the performance, by cross validation, on a grid of possible parameters values, and then selecting the best estimator. We used global accuracy as the performance metric. In particular, we used 800 trees with a max depth of 20. We get a global accuracy of 99.95%, with the performances reported in Table 4.

Table 4. Results with the Random Forest algorithm

RF	Precision	Recall	F1-score
Attacker	0.9982	0.9981	0.9982
Normal	0.9997	0.9996	0.9997
Victim	0.9981	0.9993	0.9997

An interesting outcome of the random forest is that we can extract which are the most important features in computing the classification. The features are shown by importance on Fig. 2.

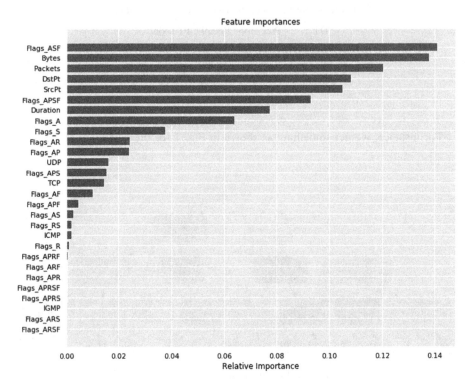

Fig. 2. Relative importance of features in the CIDDS-01 dataset.

For the Neural Network, we used the multi-layer perceptron classifier. This model optimizes the log-loss function using LBFGS or stochastic gradient descent. We used 100 neurons in the hidden layer with the rectified linear unit function for the hidden layer activation function. The maximum iteration was set to 200. We finally obtained a global accuracy of 99.25%; performances metrics for the different classes are shown Table 5.

Table 5. Results with the Neural Network algorithm

NN	Precision	Recall	F1-score
Attacker	0.9929	0.9938	0.9933
Normal	0.9906	0.9962	0.9934
Victim	0.9914	0.9883	0.9899

For the CIDDS-01 dataset and the ML-based algorithms, we obtained very high accuracies. As mentioned in [12], for this particular dataset, balancing the classes or not has a very tight influence on the accuracy, which is already so high. Anyway, by a careful selection of the hyperparameters we recover results similar to the best RF-WHICD in [14] (where only two classes normal/attacker were considered) (Table 6).

Table 6. Comparison with related work

Survey	Approach	Accuracy (%)
Verma and Ranga [15]	2NN	99.60
Verma and Ranga [16]	DT	99.90
Tama and Rhee [17]	DNN-10-FCV	99.90
Idhammad et al. [18]	Entropy + RF	99.54
Abdulhammed et al. [14]	RF-WHICD	99.99
Proposed KNN	KNN-1	99.27
Proposed DT	DT	99.89
Proposed RF	RF	99.95
Proposed NN	NN	99.25

5 Rules

The most accurate algorithm is the RF with 99.95% accuracy. However, it may be difficult to embed on an IoT and it lacks interpretability. The problem of extracting simple rules from a forest of decision trees has been considered in the machine learning community. The goal was to find a trade-off between the modelization power of random forests and some simple rules interpretable as in a (small) decision tree. The Skope rules Python library [19] enables us to extract such rules from a random forest. In our experiments, we took all the instances to train the model and extract the rules. For the victim class, the skope rules are:

- Bytes > 100
- Duration <= 0.03749999962747097
- Flags == ASF

And for the attacker class, the identified rules are:

- Dst Pt <= 261
- Duration <= 0.032500000670552254
- Flags == APSF

With these very simple rules, we already get a global accuracy of 86.88%. Furthermore, by a simple inspection of data, we discovered that adding the instances flagged with AR (class victim) or S (class attacker) TCP flags enables us to improve the accuracy to 98.45%, with only 0.35% are miss classifications.

6 Conclusion

We have considered the problem of detecting anomalies or intrusions in IoT networks. We have first presented the context and reviewed approaches in the literature. We have focused on machine learning-based methods that can learn directly from data and find what are the important features, without resorting to specialized models of the network or specialized signatures. We have then selected a dataset of network activity with several attacks, which is regularly used to develop NIDS and as a benchmark of proposals. Using standard open-source libraries, we have implemented and evaluated several ML-based algorithms, with performances that are at the state-of-the-art. The sources are available and results easily reproducible [20].

Using and implementing such solutions in an IoT network requires to consider the possible computational overhead. For a router or network supervisor, we need a network traffic module to capture the incoming network, and the classifiers, once trained by a decision tree or a random forest, can probably be implemented. For the IoT devices themselves, where consumption and computational costs can be more severely constrained, it is possible to use a decision tree classifier (at most 20 comparison tests and a 840 lines program in Python) or even to use the very simple rules (3 tests) derived from a random forest, with a 98.5% accuracy and a low false-positive rate. Testing these ideas on real devices and real data is the objective of future efforts.

References

1. IoT under fire: Kaspersky detects more than 100 million attacks on smart devices in H1 (2019). https://www.kaspersky.com/about/press-releases/2019_iot-under-fire-kaspersky-detects-more-than-100-million-attacks-on-smart-devices-in-h1-2019. Accessed 29 Oct 2019
2. Newman, P.: IoT Report: How Internet of Things technology growth is reaching mainstream companies and consumers. Business Insider, 28 January 2019
3. Butun, I., Morgera, S.D., Sankar, R.: A survey of intrusion detection systems in wireless sensor networks. IEEE Commun. Surv. Tutor. 1(16), 266–282 (2014)
4. Doshi, R., Apthorpe, N., Feamster, N.: Machine learning DDoS detection for consumer Internet of Things devices. In: 2018 IEEE Security and Privacy Workshops (SPW), pp. 29–35 (2018)
5. Hussain, F., Hussain, R., Hassan, S.A., Hossain, E.: Machine Learning in IoT Security: Current Solutions and Future Challenges. arXiv [cs.CR], 14 March 2019
6. Shukla, P.: ML-IDS: a machine learning approach to detect wormhole attacks in Internet of Things. In: 2017 Intelligent Systems Conference (IntelliSys), pp. 234–240 (2017)
7. Cañedo, J., Skjellum, A.: Using machine learning to secure IoT systems. In: 2016 14th Annual Conference on Privacy, Security and Trust (PST), pp. 219–222 (2016)
8. Nesa, N., Ghosh, T., Banerjee, I.: Non-parametric sequence-based learning approach for outlier detection in IoT. Fut. Gener. Comput. Syst. 82, 412–421 (2018)
9. Viegas, E., Santin, A., Oliveira, L., França, A., Jasinski, R., Pedroni, V.: A reliable and energy-efficient classifier combination scheme for intrusion detection in embedded systems. Comput. Secur. 78, 16–32 (2018)

10. Pajouh, H.H., Javidan, R., Khayami, R., Ali, D., Choo, K.-K.R.: A two-layer dimension reduction and two-tier classification model for anomaly-based intrusion detection in IoT backbone networks. IEEE Trans. Emerg. Top. Comput. **2**(7), 314–323 (2019)
11. Tavallaee, M., Bagheri, E., Lu, W., Ghorbani, A.A.: A detailed analysis of the KDD CUP 99 data set. In: 2009 IEEE Symposium on Computational Intelligence for Security and Defense Applications, pp. 1–6 (2009)
12. Ring, M., Wunderlich, S., Scheuring, D., Landes, D., Hotho, A.: A survey of network-based intrusion detection data sets. Comput. Secur. **86**, 147–167 (2019)
13. Ring, M., Wunderlich, S., Gruedl, D., Landes, D., Hotho, A.: Technical Report CIDDS-001 data set (2017)
14. Abdulhammed, R., Faezipour, M., Abuzneid, A., AbuMallouh, A.: Deep and machine learning approaches for anomaly-based intrusion detection of imbalanced network traffic. IEEE Sensors Lett. **1**(3), 1–4 (2019)
15. Verma, A., Ranga, V.: On evaluation of network intrusion detection systems: Statistical analysis of CIDDS-001 dataset using machine learning techniques. Pertanika J. Sci. Technol. **3**(26), 1307–1322 (2018)
16. Verma, A., Ranga, V.: Statistical analysis of CIDDS-001 dataset for network intrusion detection systems using distance-based machine learning. Proc. Comput. Sci. **125**, 709–716 (2018)
17. Tama, B.A., Rhee, K.-H.: Attack classification analysis of IoT network via deep learning approach. Res. Briefs Inf. Commun. Technol. Evol. (ReBICTE) **3**, 1–9 (2017)
18. Idhammad, M., Afdel, K., Belouch, M.: Detection system of HTTP DDoS attacks in a cloud environment based on information theoretic entropy and random forest. Secur. Commun. Netw. **2018** (2018). Article ID 1263123
19. https://github.com/scikit-learn-contrib/skope-rules
20. Bonandrini, V., et al.: https://github.com/VasgoTheTotoroo/IDS_IoT

DeepRoute: Herding Elephant and Mice Flows with Reinforcement Learning

Mariam Kiran[1], Bashir Mohammed[2(✉)], and Nandini Krishnaswamy[2(✉)]

[1] Energy Sciences Network (ESnet), Lawrence Berkeley National Laboratory,
Berkeley, CA, USA
mkiran@lbl.gov
[2] Scientific Data Management, Lawrence Berkeley National Laboratory,
Berkeley, CA, USA
{bmohammed,nkrishnaswamy}@lbl.gov

Abstract. Wide area networks are built to have enough resilience and flexibility, such as offering many paths between multiple pairs of end-hosts. To prevent congestion, current practices involve numerous tweaking of routing tables to optimize path computation, such as flow diversion to alternate paths or load balancing. However, this process is slow, costly and require difficult online decision-making to learn appropriate settings, such as flow arrival rate, workload, and current network environment. Inspired by recent advances in AI to manage resources, we present DeepRoute, a model-less reinforcement learning approach that translates the path computation problem to a learning problem. Learning from the network environment, DeepRoute learns strategies to manage arriving elephant and mice flows to improve the average path utilization in the network. Comparing to other strategies such as prioritizing certain flows and random decisions, DeepRoute is shown to improve average network path utilization to 30% and potentially reduce possible congestion across the whole network. This paper presents results in simulation and also how DeepRoute can be demonstrated by a Mininet implementation.

Keywords: Reinforcement learning · Route optimization · Path computation

1 Introduction

The rise of data hungry services such as mobile, video streaming and Cloud/Internet applications are bringing unprecedented demands to underlying network backbones [9]. Wide area networks (WANs) are investigating intelligent and efficient network management techniques to do load balancing, improve used bandwidth and overall optimize network performance. Traffic congestion can

This work is funded under DOE ASCR Early Career Grant Deep Learning: FP00006145.

S. Boumerdassi et al. (Eds.): MLN 2019, LNCS 12081, pp. 296–314, 2020.
https://doi.org/10.1007/978-3-030-45778-5_20

directly cause performance deterioration, such as when links are oversubscribed causing bottlenecks [10]. Many services rely on having high-throughput transfers and need high capacity links such as 100s Gbps. However, even for the busiest link, the current average utilization is only between 40–60%, to account for unanticipated peaks [1]. Traffic engineering and path computation techniques such as MPLS-TE (Multiprotocol Label Switching Traffic Engineering) [5], Google's B4 [18] and Microsoft's SWAN (Software Driven WAN) [1] have proposed manners in which routers can greedily select routing patterns for arriving flows, both locally and globally, to increase path utilization. However, these techniques require meticulously designed heuristics to calculate optimal routes and also do not distinguish between arriving flow characteristics.

Path computation has a number of real-world networking implications. Examples such as load balancing, minimizing congestion and utilizing maximum bandwidth as some cases that can be explored as a reward. WAN networks al-low a number of pathways to exist between pairs of end-hosts.

Internet and WAN traffic usually contains a mixture of flow characteristics, such as long and short flows, which if on the same path, can have detrimental effects on each other. Known as bulk long-living file transfers (elephant flows) are bandwidth-sensitive and short transfers (mice flows) are latency-sensitive, significantly impacting user experience and require innovative ways to manage them [34]. WAN traffic usually contains a 80%:20% flow distribution (mice to elephant ratio) and if mixed on same paths, can cause queuing delays impacted by high latency and low throughput [2]. Isolating flows to dedicated routes [38] is challenging in real-time [32,34] and has led to under-utilized paths. Researchers have attempted to recognize arriving flows to efficiently manage them [22]. In data center networks, these are often prioritized, such as mice latency [6,7] or elephant throughput [4] to improve data center network performance. In WAN, path computation uses optimization to calculate paths taken between end-hosts.

Leveraging research in software defined networking (SDN) and reinforcement learning, we explore this problem of path computation and finding optimal routes in general, as a path resource allocation problem. *We use machine learning algorithms to learn and provide viable solutions for dynamic flow management for both elephant and mice flows.*

Network routing and machine learning is not new. Broadly speaking, there are two main approaches used here [33]:

1. optimize routing configurations by predicting future traffic conditions depending on past traffic patterns or
2. optimize routing configurations based on number of feasible traffic scenarios with aim to improve performance parameters.

While SDN's centralized control offers great promise, these calculations cause overhead for high-performance networks and need global management to work, which is difficult in a large network [12]. Recent success of machine learning in complex decision making problems such as Alpha-Go [29], cooling datacenters and self-driving cars [37] suggest feasible applications to our problem. Particularly reinforcement learning (RL), actively being applied in robotics [31], allows agents to learn how to make better decisions by interacting directly with the environment. Using concepts of rewards and penalties it can learn through experience to optimize its objective function.

Revisiting the path computation challenge, in this paper we use RL to find optimal paths (or routes) as a resource management problem. First, the system learns optimal paths by repeatedly selecting available paths between source and destination, given the arriving flow distribution. RL directly interacts with the network environment and learns best paths to minimize the flow completion time given the current network conditions. Our RL approach (DeepRoute) is developed using Q-learning as a value-based gradient reinforcement learning [39].

Our experiments are focused on WAN scenarios, with two implementations,

1. We simulate an environment with synthetic data set of 100 flows, with 80:20 (elephant:mice) distributions, being allocated on 4 possible paths. We compare DeepRoute to either randomly selecting paths or prioritizing one flow type over the other. Here we use the analogy of 100 flows waiting to be allocated on paths and aim to quickly empty the wait queue.
2. We translate the experiment on Mininet network emulator environment. Doing so, we remove some assumptions we drew in the simulation and also understand how DeepRoute will work in a real network environment.

In both cases, we train DeepRoute via experience generating abundance of flow interaction data with the network. Evaluation of the agent's performance is done on test data set, which is previously unseen flow data, to see how well DeepRoute fares against other techniques.

2   Background

2.1   Path Computation and Flow Management in WAN

We explain why path computation is a challenging problem:

- Network traffic demands are continuously changing and are often difficult to predict. Routing tables are continuously changing with new devices joining/leaving the network. However, paths between two endpoints, usually follows one optimized route. If there are too many flows, this leads to potential congestion on the path [13].
- Underlying network systems are complex and distributed, with multiple links between source-destination pairs. Understanding link properties such as bandwidth or latency is difficult to model accurately [14].
- Most flow management techniques are developed for data center networks [3]. WANs, on the other hand, prioritize performance metrics such as minimizing packet loss and bandwidth utilization for traffic diversion [36]. Google's B4 and Microsoft's SWAN, both, make decisions on application characteristics and heuristics.

Compared to approaches designed by B4 and SWAN [1,18], we leverage reinforcement learning to provide alternative to heuristic-based path computation problem. We aim to allow *networks to learn best routes such to minimize flow completion times of both elephant and mice flows.*

Path computation has a number of real-world networking implications. Examples such as load balancing, minimizing congestion and utilizing maximum bandwidth as some cases that can be explored as a reward. WAN networks allow a number of pathways to exist between pairs of end-hosts. These paths can have equal or different cost distributions such as settings for bandwidth, latency, throughput and more. These settings can determine how quickly the arriving flow will reach its destination, and can be allocated using different egress ports to choose path to take. ECMP routing is an example of this where it uniformly picks an egress port to reduce congestion on one path, however has seen to suffer from hash collisions [17] and unbalance on different cost distribution paths.

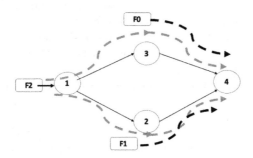

Fig. 1. Example of routing a new flow F1 from 1 → 4. There are two possible paths to take through Node-3 or Node-2.

Another problem is the changing traffic conditions in the network. Figure 1 shows a new flow F2 being allocated to one of the paths. However, there might be other previous flows running F0 and F1, already allocated on part of the paths. This means that while the links costs can be set in advance, the available bandwidth on the links is continuously changing and difficult to anticipate when selecting paths.

Traffic engineering such as MPLS-TE, B4 and SWAN use heuristics to design optimal allocations for paths using local and global optimization functions. For DeepRoute, we focus on WAN network path computation, particularly where we want flow-based routing decisions to improve flow completions times for mice and elephants that can allow all available paths to be chosen and utilized.

2.2 Reinforcement Learning

A reinforcement learning problem is formulated with an agent, situated in a partially observable environment, learning from past interaction data to make current decisions. The agent receives data in form of environment snapshots, processed in some manner, with specific relevant features. After receiving information and computing the value for future actions given the current state, an agent then acts to change its environment, subsequently receiving feedback on its action in form of rewards, until terminal state is reached. The objective is to maximize the cumulative reward over all actions in the time agent is active. Reinforcement learning research has investigated multiple techniques such as in multi-armed bandit problems, resource allocation or finding routes through a maze [30]. Deep reinforcement learning builds upon classical models, replacing the learning with a neural network to approximate policy and value functions. Here, the function approximates the environment state space with actions and rewards. Particularly when the state space is too large to store, this approach has proved feasible in learning approximate conditions. In future, we plan to expand DeepRoute to adapt from Q-learning to neural network learning (deep Q-networks), but is currently out of scope of the work presented here.

Reinforcement learning can be expressed as a Markov Decision Process (MDP) involving sequential decisions. This is given as a tuple (S, A, R, P), where $s \in S$ set of states, $a \in A$ set of actions, $R(s, a, s')$ represents reward given for executing action a at s and moving to new state s'. There is probability P for executing action in state s.

In our problem, the network is modeled as a MDP. We model 4 paths and their current allocations as a state. The agent selects a path and receives a reward after a flow has completed. In the simulation model, this is delayed reward as flows take longer to finish. In the mininet model, we get the reward at every iteration. The reward act as signals to adjust the forwarding-link priorities to enhance or diminish the probability a specific next-hop is selected for the flow. Our agent learns to adjust path selection policies based on experience through continuous modification and rewards.

2.3 Q-Learning Formulation

A Q-value represents state-action combinations. Better Q-values shows better chances of getting higher rewards which are earned at the end of a complete episode. The Q-value is calculated using a Q-function. It approximates the Q-value using prior Q-values, a short-term and a discounted future reward. This way the find optimal control policies across all environment states. Q-learning is an off-policy reinforcement learning algorithm that uses a table to store all Q-values with possible states and action pairs. This table is updated using the Bellman equation, allowing the action to be chosen using a greedy policy, given as with γ is discounting factor.

$$Q(s, a) = R(s, a) + \gamma \max_{a'} Q(s', a') \tag{1}$$

Temporal Difference (TD) Learning. In our model, we enable the agent to learn in every action taken, despite it being end of the episode or not. We define an episode to end after 100 flows have been allocated. The TD learning factor updates current Q-value where α is the learning rate,

$$Q_t(s,a) = Q_{t-1}(s,a) + \alpha TD_t(a,s) \tag{2}$$

Therefore, our final equation becomes,

$$Q_t(s,a) = Q_{t-1}(s,a) + \alpha(R(s,a) + \gamma \max Q(s',a') - Q_{t-1}(s,a)) \tag{3}$$

Where $\gamma \in [0,1]$ represents discounting factor that scales importance of the immediate reward obtained for the action and rewards R obtainable for actions at the new state s'. The learning rate $\alpha \in [0,1]$ models the rate at which of Q-values are updated.

3 Related Work

Efficient selection of paths for short and long flows has also shown to reduce congestion [26]. Additionally, adaptive traffic management using congestion-based routing has proven to improve overall network utilization [8,18]. However, adaptive algorithms in heavy traffic load, could cause oscillatory behavior and cause performance degradation [35]. OWAN was developed to optimize bulk transfers on WAN by re-configuring the optical layer showing a 4-times faster flow completion rate [20].

Flow completion time has been used as a vital metric to improve network congestion [27]. Additionally, separating elephant and mice flows can have a direct impact on network quality [15]. Other approaches have used calendaring to improve bandwidth utilization and reduce congestion [21].

In comparison to above approaches, machine learning has been used for network routing. The authors [33] show how learning can help improve the average congestion rate through softmin learning.

Reinforcement learning, in particular, has given interesting results in compute resource allocations such as CAPES [23] learning to optimize the Luster file system and DeepRM [24] learning to allocate jobs on processors. Within networks, implementations of reinforcement learning in internet congestion control [19] and developing intelligent TCP congestion algorithms [36] have shown game-changing results in managing bandwidth and bottleneck across individual links. But nearly all of these demonstrations have only been shown on small networks and simulation only. There is also little evidence on how these can be expanded to real WAN environments.

In this paper, we prove the usefulness of reinforcement learning for path computation, but this is presented as early work on how Q-learning approaches can provide benefit and be translated in WAN environments, which we will expand in future work.

4   Design of DeepRoute

We define a general network topology with unidirectional links denoted as paths $p \in P$ with varying bandwidth capacity and latency. Flows arrive at time step $t = 1$ and are assigned a fixed path, for example either $1 \rightarrow 2 \rightarrow 4$ or $1 \rightarrow 3 \rightarrow 4$ (in Fig. 1). The arriving flows are generated with a specified size and duration allowing the flow to exist on the path for a time period (longer for elephant flows). Given every flow f_i from source s_i to destination d_i, is defined with flow size v_i and duration r_i. The flow is assigned to a path p_i which would have latency lat_{pi} and available capacity ac_{pi}. The latency influences the actual flow completion time $c_i = r_i + lat_{pi}$. At a given time, flows are allocated based on available bandwidth ac_{pi}, which changes depending on traffic patterns.

Similar to SDN [1], we assume a centrally existing routing control. The controller maintains information on current allocations across paths and makes routing decisions based on completion times of previous flows.

Table 1. Elephant and mice flow distributions used in simulation case.

Flow type	Size of flow (bandwidth units occupied)	Duration of flow (time units occupied)	Distribution %
Elephant	Range (3, 5)	Range (5, 8)	20%
Mice	Range (1, 2)	Range (1, 3)	80%

4.1   DeepRoute in Simulation Model

Figure 2 shows the network topology used in the simulation. The goal is to move all arriving flows from source to destination as quickly as possible. There are 4 possible paths with different bandwidth and latency settings. In the model, we assume 100 flows are arriving together in one timestep, and the controller allocates each flow to the next available path. The 100 flows contain a distribution of 80% mice flows and 20% elephant flows (Table 1). Once allocated, the time step progresses to $t+1$, where the controller tries to allocate the remaining flows.

Objective. When a flow duration r_i (in time units) finishes, it computes its completion time c_i by adding its duration with path latency. We then inverse this, to give the flow's slowness rate by $l_i = c_i/r_i$. Similar to [24], we normalize this, to prevent skewing results for longer flows. The objective of DeepRoute is to get as many flows completed soon as possible.

Paths as Resources. We assume 100 flows arrive at Node-1 going to Node-5. There are 4 possible paths $1 \rightarrow 2 \rightarrow 5$, $1 \rightarrow 5$, $1 \rightarrow 3 \rightarrow 5$ or $1 \rightarrow 4 \rightarrow 5$, with each path having different bandwidths and latency attached. **bu** means total **bandwidth units** available to be allocated. This changes as flows are allocated

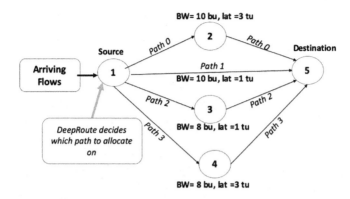

Fig. 2. Topology used in the Simulation Model (bu = bandwidth units occupied and tu = time units occupied).

on them consuming part of the bu equal to flow size. The allocations last for the flow's duration, and the completion time is computed summing path latency given in **tu (time units)**.

State Space. The state of the environment is what DeepRoute learns against. We define this as current available bandwidth across all paths, size of the flow being allocated and current allocations on the paths. For example, after an allocation on path 0 ($1 \rightarrow 2 \rightarrow 5$), the bandwidth availability is now $ac_{p0} = 10 - v_1$ and this part of state becomes $(ac_{p0}, ac_{p1}, ac_{p2}, ac_{p3})$.

Action Space. There are 4 paths so 4 possible actions. We assume one action is taken per one flow, within one time step. If there are no available paths, the controller skips a time step with no allocations. Once all 100 flows are allocated, it finishes a complete episode. The simulation is run for number of iterations containing many episodes. The total reward is calculated per episode when all 100 flows are allocated.

Reward Calculation. At the end of each episode, the RL agent calculates if any flows have finished and total completion time is recorded.

4.2   DeepRoute in Mininet

Figure 3 shows mininet topology forming multiple paths to chose from one end-host to the other. Here, we configure three metrics for the links - capacity, packet loss and latency. Mininet allows us to add delays on links. The capacities (bandwidth) are setup as 2×10 Gbps links for path0 and path1 and 2×8 Gbps for path2 and path3.

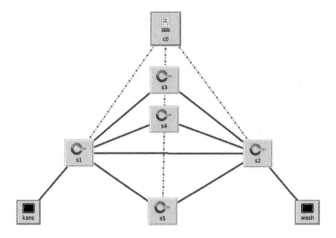

Fig. 3. Mininet topology.

Sending Flows. We ping from one end-host to the other and record the time in which the ping reaches the other host. The controller learns which egress port to use, deciding which path to take. This time is recorded as the reward against the path taken.

SDN Control. We emulate an SDN network using an Ryu OpenFlow switch. To have a global view, we configure a network controller, Openflow switches, linux hosts and network links. The hosts run a standard Linux kernel and network stack to emulate running real network applications. Traceroute measurements are recorded on the route taken and the specific gateway at each hop. We calculate the total time across each hop as a reward. In this case, every episode is represented by one flow (one ping).

State and Action Space. For dynamic multipath selection, we send a packet from source to destination via one of the four network paths, dynamically allocating flows on links. The flows go through paths: $1 \rightarrow 2 \rightarrow 5$, $1 \rightarrow 5$, $1 \rightarrow 3 \rightarrow 5$ and $1 \rightarrow 4 \rightarrow 5$.

For initial learning, we use Link Layer Discovery Protocol (LLDP) [11] to obtain link and switch states in the topology, it uses to advertise device identity and abilities, and other devices connected within. LLDP helps maintain a global view of network topology and also retains a multipath environment.

5 Training DeepRoute

The DeepRoute agent runs in an episodic fashion (with 100 flows in simulation and 1 flow in mininet). The episode terminates when all flows have been allocated.

A Q-value is added with each state and action taken and saved into a Q-table. We design our algorithm based on [30]. As the reinforcement learning algorithm uses Bellman's equation, there is a possibility of overfitting to ideal conditions. To prevent this, during the testing phase, we use ϵ for allowing DeepRoute to select random action rather than Q-table values.

Algorithm 1. Q-learning for Training DeepRoute

Initialize Q-table
for each Iteration: **do**
   for each Episode: **do**
      Generate 100 flows
      for each flow i=1,..., 100: **do**
         Get available bandwidth $(ac_{p1}, ac_{p2}, ac_{p3}, ac_{p4})$
         Get flow to allocate v_i
         Get   current   flow   allocations   across   the   4   paths
         $m_{p1}, e_{p1}, m_{p2}, e_{p2}, m_{p3}, e_{p3}, m_{p4}, e_{p4}$
         State $s_i = ((ac_{p1}, ac_{p2}, ac_{p3}, ac_{p4}), v_i, m_{p1}, e_{p1}, m_{p2},$
         $e_{p2}, m_{p3}, e_{p3}, m_{p4}, e_{p4})$
         if randomnumber $< \epsilon$
         Select any action $a_i \in (a_1, a_2, a_3, a_4)$
         Else
         Check if Q-table has this state and select best action with highest Q-Value
         Update Q-value
         Check expired flows and add reward
         If (s_i, a_i) not found in Q-table, add new entry to Q-table.
      end for
   end for
end for
For each episode:
Print Reward

5.1 Training in Simulation

DeepRoute is trained for multiple iterations as shown in Algorithm 1. Each iteration generates new 100 flows and DeepRoute learns by allocating these flows on the paths. We record the network state (available bandwidth on all 4 paths, current allocations, size of the flow to allocate), action taken (which path is chosen) and reward collected in the episode.

Training Iterations. Training for more iterations, allows the size of the Q-table to grow (from 3532 for 50 iterations, 6238 for 100 and 21640 for 500 iterations). However, Fig. 4 shows that the maximum score is achieved by 400 iterations. Therefore this is chosen as the ideal training iterations.

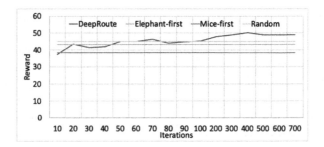

Fig. 4. Changing number of iterations during training gives different rewards.

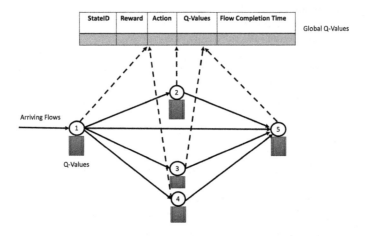

Fig. 5. Q-values for Local and Global levels.

5.2   Training in Mininet

The implementation in Mininet removes some assumptions drawn in the simulation model. Here, along with using an OpenFlow switch, we use packet loss and latency, to calculate the reward for every ping received at other host. The Q-table is recorded by monitoring Wireshark logs across all interfaces on the controller.

Initially, the OpenFlow switch performs active and passive measurements across all egress links, building the Q-table. The utilization data across the interfaces is collected. Here, we also have a local Q-table at each network node, as well as a global aggregation table managed by the network controller, shown in Fig. 5. The wireshark logs allow to collect data on latency, throughput and packet loss. Rewards or the completion time is added afterwards to populate the Q-values in the table constructed. Table 2 shows an example of the Q-table constructed in Mininet. Showing only 4 entries, 1 in each path. We transfer a mixture of flow distributions (8 and 16 GB). The arrival time (or completion time) is considered as the reward, added to the Q-table for that state and action.

6   Evaluation

We evaluate DeepRoute for the following objectives:

- With a distribution of network path parameters, how does DeepRoute compare with other allocation techniques in simulation model?
- How can DeepRoute be translated into real network environments as in a mininet model?
- How much of utilized capacity improvements are observed for the overall network?

Table 2. Sample Q-table showing capacity, latency and flow arrival time.

Capacity (Gbps)	Latency (ms)	Path	Transfer (GB)	Arrival time (ms)
10	300	0	16	60290
10	100	1	8	42086
8	100	2	8	50819
8	300	3	16	59732

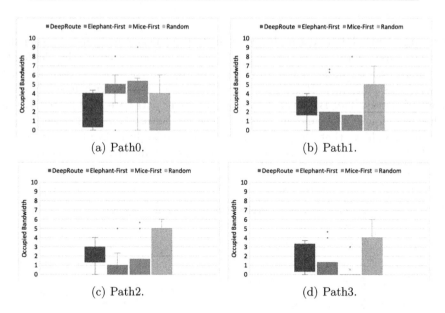

(a) Path0.

(b) Path1.

(c) Path2.

(d) Path3.

Fig. 6. Occupied bandwidth across the paths with arriving flows. Occupied bandwidth refers to the percentage bandwidth occupied on the path.

6.1   Simulation Model Results

Testing Data. We generate a new set of 100 flows of 80:20 (mice:elephant) distribution. This data is unseen by the DeepRoute during training phase. The test data is consistent for all other comparison schemes to validate the results.

Comparing with Other Techniques. We compare DeepRoute with other algorithms - Random, to randomly assign flows on available paths, Prioritize-Mice, to allocate mice flows before allocating elephant flows, and Prioritize-Elephant, allocate elephant flows before mice flows. These have been published in path computation problems [3,32].

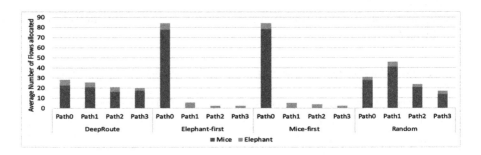

Fig. 7. Average flows allocated to all paths.

Average Path Utilization. Figure 6 shows the path capacity used during the testing phase. Here we see that while Path0 is the most used by the elephant- and mice-first techniques, DeepRoute is able to show a more spread of using other paths efficiently. It learns to use Path1 and Path2, with lower latency more efficiently. This is also shown in Fig. 7, where paths are used more than in other techniques. The random technique just spread use across all paths.

Figure 8 shows that DeepRoute is able to completely utilize the network at stable 30% as compared to the other techniques as number of flows increase.

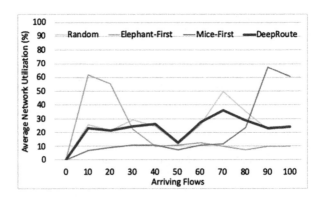

Fig. 8. Average network utilization.

Mice and Elephant Allocations. Figure 7 shows that DeepRoute is able to spread elephant and mice flows more uniformly across all the paths. We also see that by spreading the load, there is less congestion on one path, which was the case with other techniques.

Comparing with Shortest Path Route. Based on the path configurations, the network would use Path 1 the most due to higher bandwidth and lower latency. However, in the simulation DeepRoute learns to allocate on alternate paths, based on the current allocations already active on the paths.

6.2   Mininet Model Results

After training, we configure the Ryu SDN controller to select egress based on Q-table. This adjusts forwarding rules dynamically as new flows arrive.

Testing Data. We generate 100 pings from one end-host to the other and record the utilization across all the paths.

Throughputs Recorded. Figure 9 shows the throughput measurements along the 4 paths. Here, because the mininet model is extremely simple, the states learned are the available paths, flow size sent and the time recorded. This allows the controller to learn the shortest, less cost path, Path 1, as the most optimal path to use. As a result, all future traffic goes through this path and less evenly distributed at other paths. This result shows, if the controller was able to learn more features of the current state, we could enhance the controller decisions. This will form basis for future investigations of implementing DeepRoute in real networks.

7   Discussion

7.1   Modelling Realistic WAN Flows

In this section we explained how we model realistic WAN flows, using different size parameters in mininet simulation. We use NETEM [16] which provides network emulation functionality for testing protocols by emulating properties of wide area networks. To emulate the real-world network scenario, with control on parameters that affect network performance, we change the path delay on all four paths to (300 ms, 100 ms, 100 ms, 300 ms). In running our emulation, we consider four key metrics - link capacity (bandwidth), latency, transfer size and flow arrival time (presented in Table 2). To calculate latency, link delays were assigned paths.

For link capacities (bandwidth), the DeepRoute topology has a total link capacity of 36 Gbps linking the path 0 to path 3. This includes $(2 \times 10 \, \text{Gbps})$ on path 0 and 1 as well as $(2 \times 8 \, \text{Gbps})$ on path 2 and 3 respectively. Iperf3 [25] is

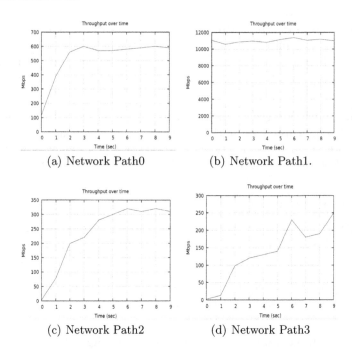

(a) Network Path0   (b) Network Path1.

(c) Network Path2   (d) Network Path3

Fig. 9. Throughput measurement from Source-Destination.

used for measuring performance characteristics, specifically, the TCP version of Iperf3. Each TCP Iperf throughput test is initiated for 60 s per path. All end hosts loop through this process for at least 100 flow rounds, thereby measuring throughput on all the network paths (path 0–3) with different flow distributions. We use wireshark [28], a free and open-source packet analyzer to capture packets on all interfaces.

7.2   Performance Measurement with Load and Delay

As flow are sent, the network controller starts the learning process by performing active and passive measurements between all switches (S1–S5). The active measurements are used to measure latency between the switches, while passive measurements are used to obtain load and residual capacity in each link.

Having successfully deployed emulated WAN using NETEM and Mininet, we conducted some performance measurement. With a set of flows, we specify a normal distribution for delay, but since delays are not always uniform, we specify a Pareto distribution (non-uniform delay distribution). As a result, all packets leaving source to destination via path 1 and 2 will experience a delay time of 100 ms, while those leaving via path 0 and 3 will experience 300 ms. The final results show minimum, average, maximum and standard deviation of the Round-Trip-Time (RTT) and if packet loss is recorded.

7.3   Comparing to Optimization Approaches

The current implementation of Q-table shows that similar results could have been achieved by optimization techniques. However, this implementation is limited by modelling the system as a simple MDP approach. In a real network setting, conditions are much more dynamic with unseen environment conditions. This means the Q-table approach will have to updated the build approximate value functions, such as deep Q-networks, to make decisions in unseen conditions.

8   Conclusions and Future Work

Recent breakthroughs in deep learning research, made possible by accelerated hardware and big data, in many fields. However, there is still a lack of understanding on how this can be used in network routing research.

By utilizing Q-learning we allow the controller to learn from the environment about the paths and best hops between source and destination. With network environments being very dynamic, with possible packet loss and traffic congestion across some of the best paths, we explore how a DeepRoute controller can learn best possible combinations depending on the traffic arriving and the current network conditions to optimally utilize the network.

Commercial systems that promise improved network performance tend to focus on average typical flows and through exploring edges (when compared to average) press exploration of new traffic engineering models. Our work highlights the need to change approaches to path computation and flow management for new applications like hybrid cloud computing and other use cases cited. While networks are challenged to strike the balance between capacity, throughput, latency and cost, AI applications can have an impact on future deployments. Our results show promise on how DeepRoute can allow efficient use of path capacity and the mininet implementation shows how it can be adapted in a real network environment.

References

1. Achieving high utilization with software-driven WAN. Technical report MSR-TR-2013-54, August 2013. https://www.microsoft.com/en-us/research/publication/achieving-high-utilization-with-software-driven-wan/
2. Akella, A., Seshan, S., Shaikh, A.: An empirical evaluation of wide-area internet bottlenecks. In: SIGMETRICS, pp. 316–317 (2003). https://doi.org/10.1145/781027.781075, http://doi.acm.org/10.1145/781027.781075
3. Al-Fares, M., Radhakrishnan, S., Raghavan, B., Huang, N., Vahdat, A.: Hedera: dynamic flow scheduling for data center networks. In: USENIX Conference Networked Systems Design and Implementation, p. 19 (2010). http://dl.acm.org/citation.cfm?id=1855711.1855730
4. Alizadeh, M., et al.: Data center TCP (DCTCP). In: SIGCOMM, pp. 63–74 (2010). https://doi.org/10.1145/1851182.1851192, http://doi.acm.org/10.1145/1851182.1851192

5. Awduche, D., Malcolm, J., Agogbua, J., O'Dell, M., McManus, J.: Requirements for traffic engineering over MPLS (1999)
6. Benson, T., Anand, A., Akella, A., Zhang, M.: MicroTE: fine grained traffic engineering for data centers. In: Conference on Emerging Networking Experiments and Technologies, pp. 8:1–8:12 (2011). https://doi.org/10.1145/2079296.2079304, http://doi.acm.org/10.1145/2079296.2079304
7. Carpio, F., Engelmann, A., Jukan, A.: DiffFlow: differentiating short and long flows for load balancing in data center networks. In: GLOBECOM, pp. 1–6, December 2016. https://doi.org/10.1109/GLOCOM.2016.7841733
8. Chakravorty, R., Banerjee, S., Rodriguez, P., Chesterfield, J., Pratt, I.: Performance optimizations for wireless wide-area networks: comparative study and experimental evaluation. In: International Conference on Mobile Computing and Networking, pp. 159–173 (2004). https://doi.org/10.1145/1023720.1023737, http://doi.acm.org/10.1145/1023720.1023737
9. Cisco: Trending analysis (2019). https://www.cisco.com/c/en/us/solutions/collateral/service-provider/visual-networking-index-vni/vni-hyperconnectivity-wp.html
10. Clark, D., et al.: Measurement and analysis of Internet interconnection and congestion. In: Telecommunication Policy Research Conference, September 2014
11. Congdon, P.: Link layer discovery protocol and MIB. V1. 0, 20 May 2002, pp. 1–20 (2002)
12. Curtis, A.R., Mogul, J.C., Tourrilhes, J., Yalagandula, P., Sharma, P., Banerjee, S.: DevoFlow: scaling flow management for high-performance networks. SIGCOMM Comput. Commun. Rev. **41**
13. Domzal, J., Jajszczyk, A.: New congestion control mechanisms for flow-aware networks. In: International Conference Communications, May 2008. https://doi.org/10.1109/ICC.2008.11
14. Floyd, S., Jacobson, V.: Link-sharing and resource management models for packet networks. IEEE/ACM Trans. Netw. **3**(4), 365–386 (1995). https://doi.org/10.1109/90.413212
15. Habibi Gharakheili, H., Sivaraman, V., Moors, T., Vishwanath, A., Matthews, J., Russell, C.: Enabling fast and slow lanes for content providers using software defined networking. IEEE/ACM Trans. Netw. **25**(3), 1373–1385 (2017). https://doi.org/10.1109/TNET.2016.2627005
16. Hemminger, S., et al.: Network emulation with NetEm. In: Linux Conference Au, pp. 18–23 (2005)
17. Iselt, A., Kirstadter, A., Pardigon, A., Schwabe, T.: Resilient routing using MPLS and ECMP. In: 2004 Workshop on High Performance Switching and Routing 2004. HPSR, pp. 345–349, April 2004. https://doi.org/10.1109/HPSR.2004.1303507
18. Jain, S., et al.: B4: experience with a globally-deployed software defined WAN. SIGCOMM Comput. Commun. Rev. **43**(4), 3–14 (2013). https://doi.org/10.1145/2534169.2486019, http://doi.acm.org/10.1145/2534169.2486019
19. Jay, N., Rotman, N., Godfrey, B., Schapira, M., Tamar, A.: A deep reinforcement learning perspective on internet congestion control. In: International Conference Machine Learning, vol. 97, pp. 3050–3059, 09–15 June 2019. http://proceedings.mlr.press/v97/jay19a.html
20. Jin, X., et al.: Optimizing bulk transfers with software-defined optical WAN. In: SIGCOMM, pp. 87–100 (2016). https://doi.org/10.1145/2934872.2934904, http://doi.acm.org/10.1145/2934872.2934904

21. Kandula, S., Menache, I., Schwartz, R., Babbula, S.R.: Calendaring for wide area networks. In: SIGCOMM, pp. 515–526 (2014). https://doi.org/10.1145/2619239. 2626336, http://doi.acm.org/10.1145/2619239.2626336
22. Kiran, M., Chhabra, A.: Understanding flows in high-speed scientific networks: a NetFlow data study. Future Gener. Comput. Syst. **94**, 72–79 (2019). https://doi. org/10.1016/j.future.2018.11.006, http://www.sciencedirect.com/science/article/ pii/S0167739X18302322
23. Li, Y., Chang, K., Bel, O., Miller, E.L., Long, D.D.E.: CAPES: unsupervised storage performance tuning using neural network-based deep reinforcement learning. In: International Conference for High Performance Computing, Networking, Storage and Analysis, pp. 42:1–42:14 (2017). https://doi.org/10.1145/3126908. 3126951, http://doi.acm.org/10.1145/3126908.3126951
24. Mao, H., Alizadeh, M., Menache, I., Kandula, S.: Resource management with deep reinforcement learning. In: ACM Workshop on Hot Topics in Networks, pp. 50–56 (2016). https://doi.org/10.1145/3005745.3005750, http://doi.acm.org/10.1145/3005745.3005750
25. Mortimer, M.: iPerf3 documentation (2018)
26. Munir, A., et al.: Minimizing flow completion times in data centers. In: IEEE INFOCOM, pp. 2157–2165, April 2013. https://doi.org/10.1109/INFCOM.2013. 6567018
27. Noormohammadpour, M., Srivastava, A., Raghavendra, C.S.: On minimizing the completion times of long flows over inter-datacenter WAN. IEEE Commun. Lett. **22**(12), 2475–2478 (2018). https://doi.org/10.1109/LCOMM.2018.2872980
28. Orebaugh, A., Ramirez, G., Beale, J.: Wireshark & Ethereal Network Protocol Analyzer Toolkit. Elsevier, Amsterdam (2006)
29. Silver, D., et al.: Mastering the game of go with deep neural networks and tree search. Nature **529**, 484 (2016). https://doi.org/10.1038/nature16961
30. Sutton, R.S., Barto, A.G.: Reinforcement Learning: An Introduction. A Bradford Book, USA (2018)
31. Tomar, M., Sathuluri, A., Ravindran, B.: MaMiC: macro and micro curriculum for robotic reinforcement learning. In: International Conference on Autonomous Agents and MultiAgent Systems, pp. 2226–2228 (2019). http://dl.acm.org/citation.cfm?id=3306127.3332066
32. Trestian, R., Muntean, G., Katrinis, K.: MiceTrap: scalable traffic engineering of datacenter mice flows using OpenFlow, pp. 904–907, May 2013
33. Valadarsky, A., Schapira, M., Shahaf, D., Tamar, A.: Learning to route. In: Hot Topics in Networks, pp. 185–191 (2017). https://doi.org/10.1145/3152434.3152441, http://doi.acm.org/10.1145/3152434.3152441
34. Wang, W., Sun, Y., Salamatian, K., Li, Z.: Adaptive path isolation for elephant and mice flows by exploiting path diversity in datacenters. IEEE Trans. Netw. Serv. Manag. **13**(1), 5–18 (2016). https://doi.org/10.1109/TNSM.2016.2517087
35. Wang, Z., Crowcroft, J.: Analysis of shortest-path routing algorithms in a dynamic network environment. SIGCOMM Comput. Commun. Rev. **22**(2), 63–71 (1992). https://doi.org/10.1145/141800.141805, http://doi.acm.org/10.1145/141800.141805
36. Winstein, K., Balakrishnan, H.: TCP ex machina: computer-generated congestion control. In: SIGCOMM, Hong Kong (2013)
37. Xiao, J.: Learning affordance for autonomous driving. In: Workshop on Smart, Autonomous, and Connected Vehicular Systems and Services (2017). https://doi. org/10.1145/3131944.3133941, http://doi.acm.org/10.1145/3131944.3133941

38. Yan, Z., Tracy, C., Veeraraghavan, M., Jin, T., Liu, Z.: A network management system for handling scientific data flows. J. Netw. Syst. Manag. **24**(1), 1–33 (2016). https://doi.org/10.1007/s10922-014-9336-2, http://dx.doi.org/10.1007/s10922-014-9336-2
39. Zhu, C., Leung, H., Hu, S., Cai, Y.: A Q-values sharing framework for multiple independent Q-learners. In: International Conference on Autonomous Agents and MultiAgent Systems, pp. 2324–2326 (2019). http://dl.acm.org/citation.cfm?id=3306127.3332099

Arguments Against Using the 1998 DARPA Dataset for Cloud IDS Design and Evaluation and Some Alternative

Onyekachi Nwamuo[1], Paulo Magella de Faria Quinan[1],
Issa Traore[1(✉)], Isaac Woungang[2], and Abdulaziz Aldribi[3]

[1] ECE Department, University of Victoria, Victoria, BC, Canada
{onyekachien, quinan}@uvic.ca, itraore@ece.uvic.ca
[2] Department of Computer Science, Ryerson University, Toronto, ON, Canada
iwoungan@ryerson.ca
[3] Qassim University, Buraydah, Saudi Arabia

Abstract. Due to the lack of adequate public datasets, the proponents of many existing cloud intrusion detection systems (IDS) have relied on the DARPA dataset to design and evaluate their models. In the current paper, we show empirically that the DARPA dataset by failing to meet important statistical characteristics of real world cloud traffic data center is inadequate for evaluating cloud IDS. We present, as alternative, a new public dataset collected through a cooperation between our lab and a non-profit cloud service provider, which contains benign data and a wide variety of attack data. We present a new hypervisor-based cloud IDS using instance-oriented feature model and super-vised machine learning techniques. We investigate 3 different classifiers: Logistic Regression (LR), Random Forest (RF), and Support Vector Machine (SVM) algorithms. Experimental evaluation on a diversified dataset yields a detection rate of 92.08% and a false positive rate of 1.49% for random forest, the best performing of the three classifiers.

Keywords: Cloud IDS · Cloud security · Machine learning · IDS evaluation · Hypervisor-based IDS

1 Introduction

In today's IT and business world, there has been a significant increase in the public adoption of cloud computing for the production systems and services support, and there seems to be no end in sight [33]. However, the growth in the public adoption of the cloud paradigm has increased organizations exposure to a wide variety of cyber attacks and vulnerabilities. Intrusion detection system (IDS) is one of the key tools being used or explored in combatting cloud attacks.

Until now, the availability of a cloud dataset has been one of the major challenges hampering the progress of the research on cloud IDS. The majority of the works done so far on cloud IDS was done using conventional datasets like the DARPA 1998 or the KDD'99 datasets [1, 3, 5, 10]. More so, the datasets used in the works done on a cloud environment are not made available for public use, in some cases for privacy concerns.

© IFIP International Federation for Information Processing 2020
Published by Springer Nature Switzerland AG 2020
S. Boumerdassi et al. (Eds.): MLN 2019, LNCS 12081, pp. 315–332, 2020.
https://doi.org/10.1007/978-3-030-45778-5_21

These factors have denied the cloud researchers of an all-encompassing real-world cloud intrusion dataset to carry out their work on. As shown in previous studies [4, 7], there are strong differences between cloud network data and conventional network data in terms of their characteristics such as flow inter-arrival time, packet-level communication, load ratios of the internal/external traffic flow, and so forth. On the other hand, the design of anomaly detection models involves constructing normal activity baselines from previously collected sample activity data. Hence, constructing cloud anomaly detection models using conventional network data would fail to capture adequately cloud network behavior considering the aforementioned differences between cloud network and conventional network data.

The objective of the current work is to provide an empirical justification for the need for a dataset collected specifically in a real cloud environment compared with using a conventional network dataset in developing cloud IDS. Furthermore, we explore the design of cloud anomaly detection using supervised machine learning techniques. Specifically, three machine learning algorithms are studied: logistic regression (LR), random forest (RF) and support vector machine (SVM).

The rest of the paper is structured as follows. Section 2 discusses related work. Section 3 highlights informally the deficiencies of the DARPA dataset and introduces as alternative a real Cloud IDS dataset. Section 4 provides empirical evidence supporting the claim that the DARPA dataset does not meet key characteristics of cloud data. Section 5 presents a new hypervisor-based cloud IDS model using supervised machine learning. Finally, Sect. 6 makes concluding remarks.

2   Related Works

To address the lack of public cloud-specific datasets, some researchers have focused on generating new datasets, such as [9, 13, 14]; however, to the best of our knowledge none of these datasets are openly available. As a result, many existing cloud IDS proposals have relied on conventional IDS datasets for development and evaluation, using primarily the DARPA IDS dataset or the KDD CUP dataset. We discuss some of these proposals in the following.

Bhat et al. [3] proposed an approach for detecting intrusions in virtual machine environment on cloud using traditional and multiclass (hybrid) machine learning algorithms. The following machine learning algorithms were considered: Naïve Bayes Tree (NB Tree) classifier, hybrid of NB Tree and Random Forest. The NSL-KDD'99 dataset was used for evaluation and it was observed that hybrid machine learning models perform better than the traditional or individual algorithms. In using single classifiers, their evaluation generated accuracies of 95%, 91% and 98% for each of Random Forest, K-NN and SVM, respectively, while the combination of NB Tree and hybrid of NB Tree and Random Forest resulted in a high accuracy of 99% and low false positive rate of 2%. More so, the hybrid of Random Forest and weighted K-Means amounted to 94.7% accuracy and 12% false positive rates.

Modi and Patel [12] presented an approach that integrates hybrid Network Intrusion Detection Systems to cloud computing environment. The experimental set up involves using the Eucalyptus infrastructure for the simulation of a cloud computing environment, while the KDD IDS dataset was used for the evaluation of their work. Their research framework involved the integration of signature-based detection and anomaly detection. They utilized Snort, for the signature-based intrusion detection and three machine learning classifiers viz the Bayesian, Associative (a machine learning model using association rule) and Decision Tree classifiers singularly and collectively for the network anomaly detection. The experimental result of their proposal for the three classifiers and their collective ability yielded a true positive rate (TPR) of 97.14%, and a false positive rate (FPR) of 1.17%.

Muthurajkumar et al. [15], used the combination of fuzzy SVM and random feature selection algorithms (RSFSA) to propose a cloud intrusion detection model. In their experiment, they built two sets of intrusion detection model, one with the whole data features and the other after introducing feature selection. A dataset consisting of 10% of KDDCUP was used for the experiment and analysis of their approach. The average detection rate from their experimental results before and after applying the RSFSA to the Fuzzy SVM classifier are 86.88% and 94.15%, respectively. Their work confirmed that feature selection plays an important role in the classifiers' detection accuracy. It would have ben interesting to evaluate this proposal using a real cloud intrusion detection dataset.

Chou et al. [5] proposed an adaptive network-based intrusion detection system for the cloud environment using the DARPA 2000 and the KDD Cup 1999 datasets. Their approach used spectral clustering, an unsupervised learning algorithm to build a decision tree-based detection model for detecting an anomaly in an unlabeled network connection data. They used Bro-IDS to generate connections records from the raw packets. Their experimental result on the DARPA dataset yielded a detection rate of 95% and a false positive rate of 4.5% while the KDD Cup 1999 dataset yielded a detection rate of 90% and a false positive rate of 5%. Their approach is not enough robust as it could not detect DOS and some probing attacks which create a great amount of connections.

Ahmad et al. [1] presented an intrusion detection model that uses Dendritic Cell Algorithm for detecting intrusions in cloud computing environment. The experimental evaluation was conducted using the DARPA 1999 dataset. The network-based attributes were used as signals in their experiments. They carried out their experiment on a total of 187 threat events of Week 4 and Week 5 of the DARPA 1999 dataset and the algorithm achieved a detection rate of 79.43% and a false positive rate of 13.43%. In their work, they demonstrated that using Dendritic Cell algorithm could provide a solution in detecting attacks in the cloud environment.

Kannan et al. [8] proposed a host-based cloud intrusion detection system which uses a genetic algorithm based feature selection and a Fuzzy SVM based classifier for deciding if an event is intrusion or not. The cloud environment was simulated with Proxmox VE 1.8 which is an open source virtualization environment while the evaluation was done using the KDD'99 cup dataset. In the experimental results, a detection rate of 98.51% and a false positive rate of 3.13% were obtained.

Zhao et al. [17] put forward an anomaly detection system based on an unsupervised learning algorithm, namely the K-means clustering algorithm. The dataset chosen for their experiment was the KDD Cup 99 dataset. For the comparative analysis of the performance of their proposed approach, they used the Particle Swarm Optimization

(PSO) and Backpropagation (BP) Neural Network algorithms to test the performance of their proposed algorithm. The K-means algorithm performed better than the other two algorithms, yielding a false positive rate (FPR) of 3.56% as against FPR of 6.78% and 5.75% for PSO and BP neural network algorithms, respectively. And in terms of false negative rate (FNR), 7.65% was achieved in contrast to 10.46% and 13.75% obtained using PSO and BP neural network algorithm, respectively. Their study was not carried out with a real cloud IDS dataset but it however worth its salt as it highlighted the possibility of predicting several types of attacks in the cloud.

Xiong et al. [16] proposed an anomaly detection method for cloud computing systems based on two approaches, viz the Synergetic Neural Network (SNN) algorithm and the Catastrophe theory (CT) algorithm. They used the DARPA dataset for their experiment and focused their work on the network traffic information. Their experiment yielded an overall average detection rate of 83% on the SNN algorithm and 86.62% overall average detection rate on the CT algorithm. The experiment also yielded an overall average of 8.3% false positive rate on the SNN algorithm and an overall false positive rate of 9.06% on the CT algorithm.

Li et al. [11] proposed an artificial neural network (ANN) based cloud IDS. The experimental part of their proposal involved simulating a cloud environment using Ubuntu Enterprise Cloud (UEC), a Eucalyptus-powered cloud platform and evaluating the result on 10% of the KDD'99 dataset. The experiment yielded an average detection rate of 99% and an average detection time of 37.1 s. One of the drawbacks here is that the ANN takes huge training time for large databases, therefore, the anomaly detection algorithm may incur an increased cost if retraining is required due to change in traffic behaviour as in the case of the cloud computing environment. More so, the simulated dataset can not stand in as a real cloud dataset.

3 Datasets for Cloud IDS Evaluation

In this section, we compare the characteristics of conventional network data with cloud network data, and give an overview of the ISOT cloud IDS evaluation dataset.

3.1 Cloud Network vs. Conventional Network Data Characteristics

Great disparity exists when considering the proximity of both the cloud and conventional network data centers. While the cloud data centers are distributed globally, the non-cloud data centers are always situated in a close proximity to their users or on the premises of the serving organizations. The global placement of the cloud datacenters satisfies the requirements for geo-diversity, geo-redundancy and regulatory constraints [4]. Studies [4, 7] have shown that the characteristics of cloud network traffic are different from the conventional network traffic in so many ways as explained in the following.

Network Flow: Empirical studies in [4, 7] have shown that the inter-arrival time for 80% of the traffics in the cloud network is usually under 1 ms while with the conventional network, this can be between 4 ms and 40 ms as the traffic does not change

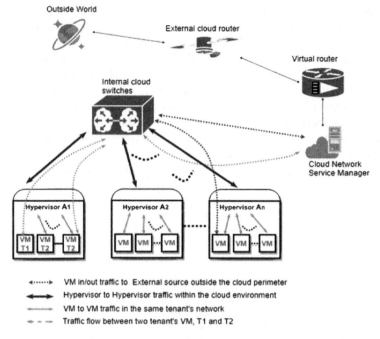

Fig. 1. Network flow in ISOT-CID

that quickly. In their work, it was also noted that the number of active flows for any given second at a switch is at most 10,000 flows and that new flows can also be highly instantaneous in arrival. These studies also went on to explain how the flow inter-arrival time affects the kind of processing that can be done for each new flow and the usefulness of logically centralized controllers for the flow placement. The cloud network traffic is usually bursty in nature with the ON/OFF intervals being characterized by heavy-tailed distributions. Their analysis also shows that in a cloud computing environment, the load ratios of the internal/external traffic flow between the instance to instance or instance and other sources are usually high. In [7], it was also discovered that in a conventional network data, 80% of flows are usually smaller than 10 Kb in size as compared with a cloud network data. On the one hand, the flow of communication patterns in a cloud network is usually high due to the numerous applications being hosted and high link utilization across the cloud's multiple layers. On the other hand, with the traditional network, the communication flow pattern and the link utilization are usually small in size. Figures 1 and 2 show the network flows for a typical cloud environment based on the ISOT-CID and the DARPA 1999 IDS evaluation dataset which was collected by simulating a conventional network environment [2]. In the ISOT-CID environment shown in Fig. 1, we can see significant variability in the network flows including the hypervisor to hypervisor network flows, the in/out traffic flows from VM to external source not within the cloud environment, the VM to VM network flows, and traffic flow between tenants VMs [2]. The conventional network comprises of limited network flows as can be seen in Fig. 1 which has only two network flows viz external and internal traffic flows [2].

Topology: The physical topology of a cloud data center follows a canonical 3-Tier architecture which consists of the core layer or the uppermost layer, aggregation layer or the middle layer and the edge layer or the lower link layer. In contrast, the traditional data centers follow a 2-Tiered topology in which the core layer and the aggregation layers are collapsed to form one layer [4]. In a typical cloud network, data is either centralized or outsourced and provided to the users on-demand irrespective of their geographic location. This relieves the data owner of the full control of their data as the cloud service providers now manage and maintain the data. The cloud data also has the flexibility of being scaled up or down by automated means. Some giant cloud service providers such as Amazon, Google and Microsoft do have cloud data centres dispersed geographically for the provision of universal data access to the various users [4].

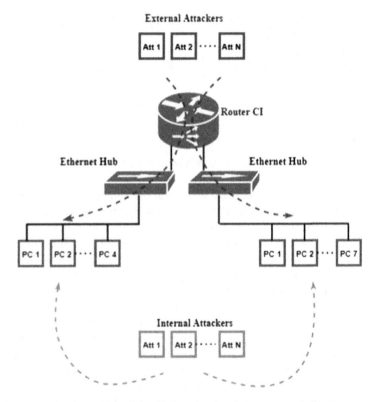

Fig. 2. 1999 DARPA IDS evaluation dataset network flow

3.2   Overview of the ISOT Cloud IDS Dataset

The ISOT-CID is a publicly available dataset that was collected in a real world environment using the infrastructure of Compute Canada, a nonprofit cloud service provider that extends its services in the areas of providing the computational needs of researchers [2]. There were two phases involved in the ISOT-CID data collection procedure, namely Phase 1 in 2016 and Phase 2 in 2018. The data in the two phases were collected on the same production environment based on OpenStack from various cloud layers such as hypervisors, guest hosts layers and the network layer. The dataset size is more than 8 terabytes and it contains data of different formats such as the memory dumps, CPU and disk utilizations, system call traces, system logs and network traffic [6]. Another advantage of the ISOT-CID is that it is labelled and includes both normal and attack activities. The current work is based on the network traffic attributes of the ISOT-CID.

The ISOT-CID collection environment contains three hypervisor nodes viz, node A, node B, and node C. The collecting environment is also composed of 10 virtual machines or instances (VM1 to VM10) launched in three different cloud zones A, B, and C [2].

The benign data in the ISOT-CID came from web applications/traffic and administrative activities [6]. Some of the web application activities include account registration, blog activities, and web browsing. The web traffic statistics revealed that more than 160 legitimate users were involved in the generation of the normal data which comprises of 60 human users and 100 robots [2]. While the administrative activities cut across instance routine maintenance, system rebooting, application updates, file creation, machine access via SSH and remote server access.

Table 1 presents all the attacks covered in ISOT-CID, such as probing, DoS, information disclosure, R2L, input validation, backdoors and authentication breach, etc. These attacks were grouped into insider or outsider attacks depending on its source [2]. On the one hand, the inside malicious activities were perpetrated by either an insider within the cloud environment who had a root access on the hypervisor nodes or by a compromised VM within the cloud environment used as a stepping stone. Some of the inside attacks were backdoor and Trojan horse, network scanning, password cracking, DoS attacks, and so forth. On the other hand, the outside malicious activities emanated from outside the cloud environment with the ISOT-cloud environment being the primary target. Some of the outside attacks are made up of the application layer (layer 7) and network layer (layer 3) DoS attacks, input validation attacks, SQL injection, path/directory traversal and cryptojacking (unauthorized cryptomining). For instance, Fig. 3 shows a timeline for the attack scenario in Phase 1 Day 2 (2016-12-15).

Composition of ISOT-CID Network Traffic Data

The ISOT-CID is composed of three levels of network communications namely: external, internal and local traffic [6]. In the ISOT-CID context, the external traffic is the traffic between the instance and an outside machine. The internal traffic or the hypervisor traffic is between the hypervisor nodes. And finally, the local traffic is the traffic between two VMs on the same hypervisor node. The ISOT-CID network data also comes in kinds, one being without payload on both hypervisors and VMs and the other involving the full network traffic only on the hypervisors [2]. The ISOT-CID network traffic/packet statistics are shown in Table 2.

Table 1. Attacks covered in the ISOT-CID dataset [2]

Attack target layer	Insider attack types	Outsider attack types
Application layer		• SQL Injection • Web Vulnerabilities Scanning • Cross-site Scripting (XSS) • Dictionary/Brute Force login attack • Fuzzers • HTTP Flood DOS • Directory/Path Traversal
Network layer	• Trojan Horse • Backdoor (reverse shell) • Unauthorized Cryptomining (download/install/run cryptominer) • UDP Flood DOS • Stepping Stone Attack • Ports and Network scanning • Synflood DOS • Revealing Users Credentials and Confidential Data by Insider • Dictionary/Brute Force login attack	• Synflood Dos • Unclassified (unsolicited traffic) • DNS amplification DOS • Ports and Network scanning • Dictionary/Brute Force login attack

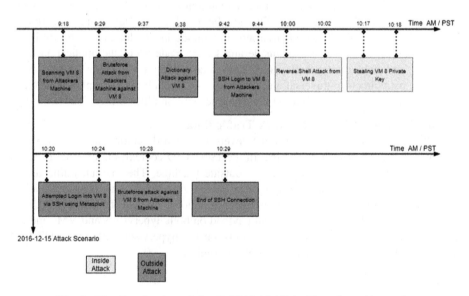

Fig. 3. Timeline for phase 1 day 2 (2016-12-15) inside and outside attacks

Table 2. ISOT-CID Network Traffic Distribution [2]

Phase	Total normal traffic	Total malicious traffic	Total packets
1	22,356,769	15,649	22,372,418
2	9502872	2,006,382	11,509,254

4   Traffic Characterization

We analyzed the network traffic data of the ISOT-CID and the DARPA 1998 datasets by looking at the network communication patterns at the flow-level. The idea is to show how similar or different they are in terms of data transmission behaviour at the flow level. This is also in line with the work done in [4] where the authors used traffic engineering techniques to distinguish between cloud data center networks and conventional or traditional networks. We considered three metrics in our empirical data analysis, viz Number of Flows, Flow Inter-arrival Times, and Flow of Traffic characteristics of the two datasets.

4.1   Number of Active Flows Characteristics

Figure 4 represents the empirical cumulative distribution function (CDF) of the number of active flows at different switches within 120 s time window for both ISOT-CID Phase 2 Day 1 (2018-02-16) dataset and DARPA 1998 Tuesday week 4 Training dataset. Our findings based on the distribution reveals that, the number of active flows for the ISOT-CID is between 2,000 to 6000 about 90% of the time. In the case of the DARPA 1998 dataset, the number of active flows is between 20 and 1000 in 90% of the time interval. This empirical observation supports the results of a prior work on data center traffic [4]. It is also considerable to note that the latency assigned by a controller to a new flow is determinant on the lengths of the flows [4].

4.2   Flow Inter-arrival Time Characteristics

Additionally, we examined the empirical CDF of the flow inter-arrival times under a 120 s time window on ISOT-CID Phase 2 Day 1 (2018-02-16) dataset and DARPA 1998 Tuesday week 4 Training dataset as represented in Fig. 5. We discovered that the flow inter-arrival time for 80% of the new flows arriving at the monitored switch is 1 ms for the ISOT-CID dataset and 4 ms for the DARPA dataset. These results suggest that DARPA is characterized by a smaller number of flows than the ISOT_CID dataset. This empirical observation also supports the results of a prior work [4].

The flow inter-arrival times affects the scalability of the controller because a significant number of new flows arrive at a given switch within an interval of few microseconds [4]. Therefore, it is recommended to use multiple CPU's per controller and multiple controllers to compute the routes in order to scale the throughput of a centralized control framework.

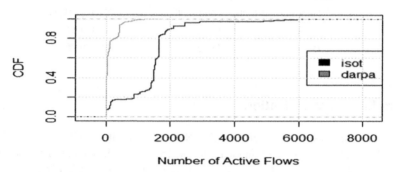

Fig. 4. The CDF of the distribution of the number of flows at the edge switch in ISOT and DARPA

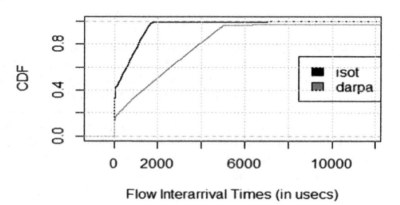

Fig. 5. The CDF of the distribution of the flow inter-arrival time in ISOT and DARPA

4.3 Flow-Level Communication Characteristics

The aggregate network transmission behaviour of the two datasets were examined using a day's traffic from each respectively. We based the analysis on two network flow metrics, viz the *extra-rack* traffic and the *intra-rack* traffic. The *extra-rack* traffic signifies the traffic leaving the switch rack or internal hosts for other internal hosts or external destinations, this is easily measured while the *intra-rack* traffic represents the amount of traffic that stays within the rack or node [4]. On the one hand, the *intra-rack* traffic for ISOT-CID, was computed by taking the difference between the volume of traffic generated by the instances attached to the hypervisor nodes and the traffic exiting the nodes.

On the other hand, the *intra-rack* traffic for DARPA 1998 dataset was computed by taking the difference between the volume of traffic generated by the servers or host attached to the switches and the traffic exiting the switches. Table 3 shows the *extra-rack* and *intra-rack* traffic compositions, while Table 4 provides the percentage representation of these two metrics. Figure 6 depicts a bar graph showing the ratio of *extra-rack* to *intra-rack* traffic in the selected day traffic of both the ISOT-CID and the DARPA 1998 datasets.

Table 3. Extra-Rack and Intra-Rack traffic composition for ISOT-CID and DARPA 1998, showing the number of packets.

Flow metric	ISOT-CID phase 2 day 1 (2018-02-16) dataset	DARPA Tuesday week 4 training dataset
Extra-rack traffic	693426	932843
Intra-rack traffic	2177148	864810
Total traffic	2870574	1797653

Table 4. Percentage composition of Extra-Rack and Intra-Rack traffic for ISOT-CID and DARPA 1998

Flow Metric (%)	ISOT-CID phase 2 day 16 dataset	DARPA Tuesday week 4 training dataset
Extra-rack traffic (%)	24.16	51.89
Intra-rack traffic (%)	75.84	48.11
Total traffic (%)	100	100

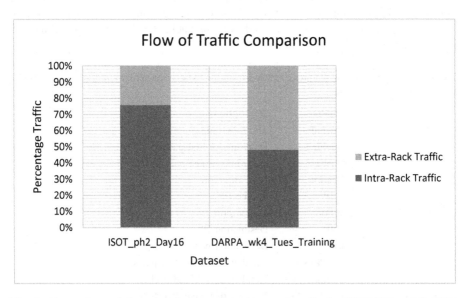

Fig. 6. Comparison of the ratio of extra-rack to intra-rack traffic for ISOT-CID and DARPA 1998 datasets

The result of the analysis shows that for the ISOT-CID dataset, 75.84% of the traffic is confined to within the hypervisor node in which it was generated while 24.16% of the traffic leaves the nodes. This result is in contrast with the DARPA 1998 dataset in which only 48.11% of the traffic stays within the communication nodes and 51.89% of the traffic leaves the nodes. The result of this network traffic analysis supports the observations made in prior studies [4, 7] of network traffic characterization of data centers.

5 Hypervisor-Based Cloud IDS Using Supervised Machine Learning

In this section, we explore the effectiveness of hypervisor-based cloud anomaly intrusion detection using supervised machine learning. Specifically, three machine learning algorithms are studied: logistic regression (LR), random forest (RF) and support vector machine (SVM).

5.1 Feature Model

Because the VM instances in the cloud environment share the same hypervisor, to improve the cloud computing intrusion detection process, the feature extraction should be such that it takes into account the correlated behavior of the instances. A network flow on the other hand can be seen as a bidirectional packet streams between two hosts or the movement of network traffic across different network points, usually from a source to a destination and vice versa. We first grouped the captured hypervisor packets in the pcap file formats into a stream of packet flows based on a time window δt using a flow based forensic and network troubleshooting traffic analyzing tool called Tranalyzer. Eighty raw features were extracted from the packet headers and some of the raw features are represented in Table 5.

Table 5. Some of the raw features extracted from the hypervisor network traffic using the traffic analyzer tool (Tranalyzer)

Feature	Description
flowInd	The flow index
timeFirst	Date/time of first packet
timeLast	Date/time of last packet
duration	Flow duration
srcIP	Source IP
srcPort	Source port
dstIP	Destination IP
dstPort	Destination port
numPksSnt	Number of transmitted packets
numPksRcvd	Number of received packets
numBytesSnt	Number of transmitted bytes
numBytesRcvd	Number of received bytes

We used a three-dimensional features space in this thesis work namely frequency-based features, entropy-based features and load-based features based on the work of Aldribi et al. [2].

Two categories of frequency-based features were adopted for each VM instance, namely, the 'in-frequency' and the 'out-frequency' features. The in-frequency represents the frequency of the packets incoming to a specific instance from any source or endpoint while the out-frequency represents the outgoing packets from a specific instance back to the respective sources.

The load-based features were extracted by taking the ratio of the matching *in* an *out* frequency features as proposed in [2].

The entropy associated with the probability distribution of network traffic occurrences at ingress and egress points during the observation time window was computed for the specific instances. Given a distribution of probabilities $P = \{p_1, p_2, \ldots, p_N\}$ having N variables, entropy is defined as

$$H_s = -\sum_{i=1}^{N} p_i \, log_2 p_i \tag{1}$$

Where $0 \leq p_i \leq 1$ and $\sum_{i=1}^{N} p_i = 1$. And in our case, p_i represents the probability of the distinct frequency features of the traffic during the observation time window. For instance, the entropy of the source IP is calculated by first computing the appearing probability associated with the source IP which is gotten by taking the ratio of the number of packets with the specified source IP address and the total number of packets observed in the flow, after that the entropy equation of (1) is adopted to get the value. The entropy is minimum ($H_s = 0$) at maximum flow concentration or when the features exhibit a deterministic behaviour. On the other hand, the entropy H_s is maximum ($H_s = log_2 N$) at maximum flow dispersion or when the feature is fully at random.

5.2 Data Preparation

The data preparation steps undertaken in this research work in order to get the best features in the dataset are shown in Fig. 7.

We leveraged the computational power of pandas, an open source software library for *Python* programming language to developed a python script to extract the aforementioned three feature categories of our feature model. The extracted features were transformed to the same scale between 0 and 1 using the *min-max* normalization approach. To reduce the dimensionality and complexity of the feature space, we used the *CARET* R-library, a tree-based feature selection technique. The *CARET* R-package gives a percentage score to all the features with the noisy features having a percentage score of zero (0) in accordance to their statistical significance, that is information gain. The most important features are then used to train the machine learning model. Table 6 show the features and their overall significance to the model prediction as indicated by the *CARET* R-package.

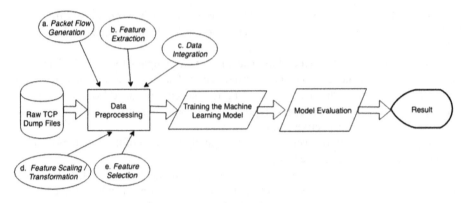

Fig. 7. Data preparation subsystem

5.3   Model Evaluation Using ISOT-CID

In this research work, the performance of the three machine learning algorithms was measured using the detection rate (DR) (also known as true positive rate (TPR)) and the false positive rate (FPR) which are two metrics commonly used for IDS performance computation.

Table 6. Features and their overall importance

Feature	Definition	Overall significance (%)
$f_i^{in}(t)$	The total number of packets flowing to e_i during $[t, t + \delta t]$ divided by δt	100
$L_i(t)$	Load feature matching the ratio of the total number of packets flowing to and from endpoint e_i during $[t, t + \delta t]$	97.382
$f_{i,iP,d,dp}^{out}(t)$	The number of packets flowing from the endpoint e_i to e_d during $[t, t + \delta t]$ divided by δt	87.251
$f_i^{out}(t)$	The total number of packets flowing from e_i during $[t, t + \delta t]$ divided by δt	86.431
$f_{s,sp,i,iP}^{in}(t)$	The number of packets flowing from the endpoint e_s to e_i during $[t, t + \delta t]$ divided by δt	85.590
$max_{ip}\left\{f_{i,ip}^{out}(t)\right\}$	The maximum number of packets over ip flowing out of e_i during $[t, t + \delta t]$ divided by δt	8.298
entropy_dp	Entropy of the destination port	8.023
$f_{i,ip}^{out}(t)$	The number of packets flowing from specific ip in e_i to all dp in all endpoints e_d during $[t, t + \delta t]$ divided by δt	7.877
numPksRcvd	Number of received packets	7.877
numPktsSnt	Number of transmitted packets	6.992

(continued)

Table 6. (*continued*)

Feature	Definition	Overall significance (%)
$L_{max(ip)}(t)$	Load feature matching the ratio of the maximum number of packets over the instant port (ip) flowing to and from endpoint e_i during $[t, t + \delta t]$	3.098
$f_{i,ip}^{in}(t)$	The number of packets flowing from all sp in all endpoints e_s to specific ip in e_i during $[t, t + \delta t]$ divided by δt	2.915
$L_{i,ip}(t)$	Load feature matching the ratio of the number of packets flowing from all source ports in all endpoints e_s to and from specific instance port (ip) in e_i during $[t, t + \delta t]$	2.750
$f_{sp,i}^{in}(t)$	The number of packets flowing from specific sp in all endpoints e_s to all ip e_i during $[t, t + \delta t]$ divided by δt	0.00
$L_{s,sp,i,ip}(t)$	Load feature matching the ratio of numbers of packets flowing to and from the endpoint e_s to e_i during $[t, t + \delta t]$	0.00
$L_{sp,i}(t)$	Load feature matching the numbers of packets flowing from specific source port in all endpoints e_s to and from all instance ports in endpoint e_i during $[t, t + \delta t]$	0.00
entropy_srcPort	Entropy of the source port	0.00
entropy_srcIP	Entropy of the source IP	0.00
$max_{ip}\left\{f_{i,ip}^{in}(t)\right\}$	The maximum number of packets over ip flowing to e_i during $[t, t + \delta t]$ divided by δt	0.00
$f_i^{out}(t)$	The total number of packets flowing from e_i during $[t, t + \delta t]$ divided by δt	0.00

Table 7. Comparison of overall performance for ISOT-CID

Algorithm	Overall	
	FPR (%)	DR (%)
Logistic regression	2.61	90.52
Random forest	1.49	92.08
SVM	1.84	92.06

Table 8. Confusion matrix for random classifier based on ISOT-CID

		Prediction	
		Attack	Normal
Reference	Attack	TP = 114291	FN = 9831
	Normal	FP = 4428	TN = 292775

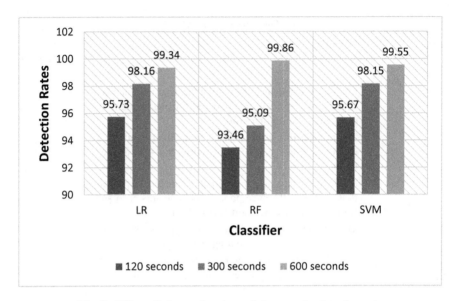

Fig. 8. Effect of observation time window on the detection rate

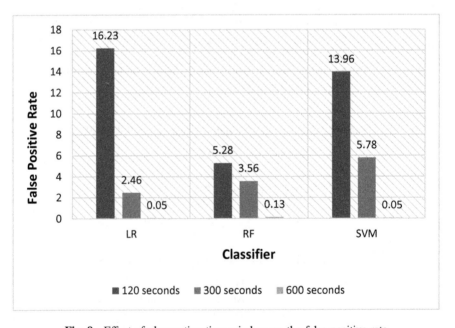

Fig. 9. Effect of observation time window on the false positive rate

The observation time window δt was set at 120 s for the network flow aggregation and feature extraction. The data was processed following the steps described in Fig. 7. The DR and the FPR obtained from the three machine learning algorithms for each VM instances over different attack days and their respective overall results were computed. The overall performance of the machine learning classification algorithms is summarised in Table 7 for ISOT-CID (covering both phases 1 and 2).

The random forest algorithm was the best of the three-machine learning algorithms in terms of performance with a detection rate of 92.08% and a false positive rate of 1.49%.

Table 8 shows the confusion matrix for the random forest classifier. The matrix shows the number of flows being classified; the cells contain the numbers of true positives (TP), false positives (FP), false negatives (FN), and true negatives (TN).

We examined also the effect of varying the size of the observation time window δt on the DR and FPR of the three machine learning models. Day 5 of phase 2 hypervisor B data was used for this analysis. Figures 8 and 9 show the effect on the detection rate and the false positive rate, respectively. On the one hand, as the observation time window increases, the amount of data available for decision increase thereby presenting the model with a more balanced dataset which will in turn aid in better decision making. On the other hand, as the time window increases, the decision time is delayed which represents an increased window of vulnerability.

6   Conclusion

Security and privacy remain one of the main issues faced by cloud computing adopters and consequently, there is an urgent need for the IT professionals and subject matter experts to come up with a system that can both detect and protect the cloud infrastructure from malicious activities. In this paper, we carried out an empirical analysis on the DARPA intrusion evaluation dataset and showed its deficiencies when compared to the ISOT-CID which is a real cloud computing dataset. The results support previous work done on network traffic characterization of data centres. It is the claim of this work that due the deficiencies, the DARPA dataset should not be used as a genuine dataset in the design and evaluation of cloud IDS.

Also, we investigated cloud intrusion detection using different supervised machine learning models. The performance results obtained using the machine learning algorithms are encouraging meaning that if more effort and study is channeled into it, academia and researchers can come up with a better way to protect the cloud computing environment against intrusions.

In this paper, the empirical study for the characterization of network traffic to substantiate the difference between a cloud dataset and a conventional dataset was only limited to three flow-level metrics viz, the number of active flows, flow inter-arrival time, and flow-level communication patterns. Our future work will consist of extending the presented work by exploring other empirical means to further understand the nature of network traffic of the cloud and conventional dataset/datacenters like in the areas of packet-level communication, link utilizations and many more.

References

1. Ahmad, A., Kama, M.N.: CloudIDS: cloud intrusion detection model inspired by dendritic cell mechanism (2017)
2. Aldribi, A., Traore, I., Moa, B.: Hypervisor-based cloud intrusion detection through online multivariate statistical change tracking. Comput. Secur. **88**, 101646 (2020)
3. Bhat, A.H., Patra, S., Jena, D.: Machine learning approach for intrusion detection on cloud virtua machines. IJAIEM **2**(6), 56–66 (2013)
4. Benson, T., Akella, A., Maltz, D.A.: Network traffic characteristics of data centers in the wild. In: Proceedings of the 10th ACM SIGCOMM Conference on Internet Measurement, IMC 2010, pp. 267–280 (2010)
5. Chou, H.-H., Wang, S.-D.: An adaptive network intrusion detection approach for the cloud environment. In: 2015 International Carnahan Conference on Security Technology (ICCST), pp. 1–6 (2015)
6. Equinix: "To maximize the cloud, focus on the network" Equinix whitepaper on Cloud. https://equinix.app.box.com/embed/s/fakirceasoamu4ofxnt9rv6h4ujjytjm. Accessed 28 June 2019
7. Kandula, S., Sengupta, S., Greenberg, A., Patel, P., Chaiken, R.: The nature of datacenter traffic: measurements and analysis. In: Proceedings of the 9th ACM SIGCOMM Conference on Internet Measurement, IMC 2009, pp. 202–208 (2009)
8. Kannan, A., Maguire, G.Q., Sharma, A., Schoo, P.: Genetic algorithm based feature selection algorithm for effective intrusion detection in cloud networks. In: Proceedings of 12th IEEE International Conference Data Mining Workshop, ICDMW 2012, pp. 416–423 (2012)
9. Kholidy, H.A., Baiardi, F.: CIDD: a cloud intrusion detection dataset for cloud computing and masquerade attacks. In: 2012 Ninth International Conference on Information Technology: New Generations (ITNG), pp. 397–402. IEEE (2012)
10. Kwon, H., Kim, T., Yu, S.J., Kim, H.K.: Self-similarity based lightweight intrusion detection method for cloud computing. In: Nguyen, N.T., Kim, C.-G., Janiak, A. (eds.) ACIIDS 2011. LNCS (LNAI), vol. 6592, pp. 353–362. Springer, Heidelberg (2011). https://doi.org/10.1007/978-3-642-20042-7_36
11. Li, Z., Sun, W., Wang, L.: A neural network based distributed intrusion detection system on cloud platform. In: Proceedings of 2012 IEEE 2nd International Conference Cloud Computing and Intelligence Systems, CCIS 2012, vol. 1, pp. 75–79. IEEE (2013)
12. Modi, C.N., Patel, D.: A novel hybrid-network intrusion detection system (H-NIDS) in cloud computing. In: Proceedings of 2013 IEEE Symposium on Computational. Intelligence Cyber Security CICS 2013, 2013 IEEE Symposium Series on Computational Intelligence, SSCI 2013, pp. 23–30 (2013)
13. Moorthy, M., Rajeswari, M.: Virtual host based intrusion detection system for cloud. Int. J. Eng. Technol. (IJET) **5**, 5023–5029 (2013)
14. Mukkavilli, S., Shetty, S., Hong, L.: Generation of labelled datasets to quantify the impact of security threats to cloud data centers. J. Inf. Secur. **7**, 172–184 (2016)
15. Muthurajkumar, S., Kulothungan, K., Vijayalakshmi, M., Jaisankar, N., Kannan, A.: A rough set based feature selection algorithm for effective intrusion detection in cloud model (2013)
16. Xiong, W., et al.: Anomaly secure detection methods by analyzing dynamic characteristics of the network traffic in cloud communications. Inf. Sci. **258**, 403–415 (2014)
17. Zhao, X., Zhang, W.: An anomaly intrusion detection method based on improved K-means of cloud computing. In: Proceedings of 2016 6th International Conference on Instrumentation & Measurement, Computer, Communication and Control (IMCCC) 2016, no. 61272172, pp. 284–288 (2016)

Estimation of the Hidden Message Length in Steganography: A Deep Learning Approach

François Kasséné Gomis[1](✉), Thierry Bouwmans[2], Mamadou Samba Camara[1], and Idy Diop[1]

[1] Cheikh Anta Diop University, Dakar, Senegal
gomisfk@gmail.com
[2] La Rochelle Univ., La Rochelle, France
thierry.bouwmans@univ-lr.fr

Abstract. Steganography is a science which helps to hide secret data inside multimedia supports like image, audio and video files to ensure secure communication between two parts of a channel. Steganalysis is the discipline which detects the presence of data hidden by a steganographic algorithm. There are two types of steganalysis: targeted steganalysis and universal steganalysis. In targeted steganalysis, the steganographic algorithm used to hide data is known. In the case of universal steganalysis, the detection of hidden data doesn't depend on any specific algorithm used in the process of steganography. In this paper, we focus on universal steganalysis of images in a database with an eventual cover-source mismatch problem. It is shown that combining both unsupervised and supervised machine learning algorithms helps to improve the performance of classifiers in the case of universal steganalysis by reducing the cover-source mismatch problem. In the unsupervised step, the k-means algorithm is generally used to group similar images. When the number of features extracted from the image is very large it becomes difficult to compute the k-means algorithm properly. We propose, in that case, to use Deep Learning with Convolutional Neural Network (CNN) to group similar images at first and implement a Multilayer Perceptron (MLP) neural network to estimate the hidden message length in all the different groups of images. The first step of this approach prevents the cover-source mismatch problem. Reducing this issue boost the performance of classifiers in the second step which consists of estimating the hidden message length.

Keywords: Steganography · Steganalysis · Machine learning · Deep learning · Convolutional Neural Networks · Multilayer Perceptron

1 Introduction

Research in universal steganalysis domain become very interesting since researchers discover that deep learning with Convolutional Neural Networks

© IFIP International Federation for Information Processing 2020
Published by Springer Nature Switzerland AG 2020
S. Boumerdassi et al. (Eds.): MLN 2019, LNCS 12081, pp. 333–341, 2020.
https://doi.org/10.1007/978-3-030-45778-5_22

(CNNs) helps to obtain better results in the classification between cover and stego images. Till now CNNs have been never used for regression to estimate the hidden message length. Various types of materials used to capture images and various steganographic algorithms available for hiding data cause a problem called cover-source mismatch in universal steganalysis. In previous studies, it is demonstrated that clustering can be used, as a prior step in the process of steganalysis, to improve the performance of the classifiers in a database with cover-source mismatch [9], before implementing a classification or regression algorithm for universal steganalysis. Generally, authors used clustering with the k-means algorithm to group images into clusters. However, if the number of features extracted is big, it becomes computationally difficult to compute them with the k-means algorithm. In this context, we propose to employ a deep learning-based approach for estimation of the hidden message length in steganography. We called the proposed method DeepStego. To estimate the hidden message length, we use in the first step a CNN for grouping similar images into different categories. Then, in the second step, we implement an MLP neural network to estimate the hidden message length.

The rest of the paper is organized as follows: In Sect. 2, we give the theory in universal steganalysis. In Sect. 3, we present our original method of universal steganalysis. Then, we illustrate our scheme in Sect. 4. In Sect. 5, experiments are conducted on a database and we discuss the results. Concluding remarks and future directions are provided in Sect. 6.

2   Related Works

Research in the universal steganalysis domain focuses either on the extraction of relevant features which are sensitive to any steganographic algorithm or in the machine learning algorithms used to build models for classification or regression. The goal in both cases is to help to boost the performance of classifiers. About relevant features for universal steganalysis of JPEG images, authors use First-order statistics, Inter-block, and Intra-block features. Table 1 is a summary of different categories of features and some authors who proposed them for universal steganalysis of images.

Table 1. Features for universal steganalysis: an overview.

Authors	Categories	Feature names
Ashu and Chhikara [1]	First-order statistics	Global histogram
		AC histograms
		Dual histograms
Chen and Shi [4]	Inter-block features	Co-occurrence matrix
		Variation
		Blockiness
Chen and Shi [4]	Intra-block features	Average Markov matrix

All these features are sensitive to steganographic algorithms embedding impact while at the same time insensitive to the image content. According to the steganographic algorithm used to embed messages, a category of features can be more useful than others. Thus, some authors proposed methods that combine different categories of features [4]. Many different blind steganalysis methods have been proposed in the literature [8]. After choosing a set of features, we need to find a strong algorithm for binary classification or regression (to separate stego and cover images or to estimate the hidden message length). Support Vector Machine [6] and classical Neural Networks are very used for classification between stego and cover images. Recently, some authors start to use Convolutional Neural Network for the same task [5]. To estimate the relative payload, Multiple Linear Regression is also used but the cover-source mismatch problem and the huge number of features extracted make its implementation difficult. It is shown that applying clustering is a good solution before using it [9]. About clustering when the number of features extracted is huge and when the database is big (more than 10,000 images), the computation becomes difficult. Some papers related to deep learning for universal steganalysis have been published. Chaumont et al. made a recapitulation of those methods in their paper [3]. Deep learning with CNN has better performance than usual machine learning algorithms. However, to estimate the hidden message length in the case of universal steganalysis, there are still some difficult challenges to overcome to boost the performance of blind steganalyzers. Some methods which deal with estimation of hidden message length have been proposed in the literature [11]. CNN is a classification algorithm that has never been used in the perspective of estimating the hidden message length.

3   DeepStego: A Deep Learning Methodology

In this section, we detail the proposed approach named DeepStego for universal steganalysis in a database with a cover-source mismatch problem. Figure 1 shows

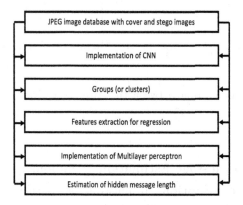

Fig. 1. DeepStego (proposed method): data pipeline.

an overview of the data pipeline. The different steps are described and illustrated in the following.

The proposed approach consists of combining deep learning with CNNs [10] and MLP neural network [6] to build strong and robust models that estimate the hidden message length in universal steganalysis. This method consists of three main steps:

- **Step 1:** Implementation of CNN in a JPEG database containing cover and stego images. The objective here is to group similar images into different groups (or clusters). Grouping images is a strategy to prevent an eventual cover-source mismatch problem in the database. That problem can occur when there is a variety of materials used to get images and a lot of different steganographic algorithms used to hide data into images.
- **Step 2:** Implementation of MLP in all the groups to build models for estimating the hidden message length.
- **Step 3:** Utilization of the models for prediction.

4   Experimental Illustrations

4.1   Cover and Stego Images

We use a steganography Python module called Stegano [2] to generate stego images with different payloads. As shown in Fig. 2, after the embedding process, changes between stego and cover images are not visually detectable. Histograms of the cover image and its stego image are generated by the Stegano module algorithm.

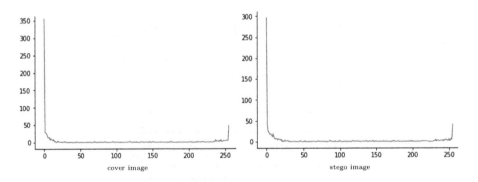

Fig. 2. Histograms comparison between the cover image (left) and the stego image (right).

4.2   The Hidden Message Estimation Technique

Image Database Description. To illustrate our purpose, we use the MNIST Database [7]. This database is very practical in our case. It contains $70,000$ 28×28 grayscale JPEG images divided into 10 categories.

Stego Images Generation. To generate stego images for simulation, we use the module Stegano of Python to embed messages with different lengths inside images. So, we obtain 35, 000 stego images. After that process, we create a vector of labels which contains the lengths of the hidden messages of all the images of the database. That vector will be used for the regression part.

Outcome vector L for CNN   Outcome vector Y for MLP

$$\begin{pmatrix} L_1 \\ L_2 \\ \cdot \\ \cdot \\ L_{n-1} \\ L_n \end{pmatrix} \qquad\qquad \begin{pmatrix} Y_1 \\ Y_2 \\ \cdot \\ \cdot \\ Y_{n-1} \\ Y_n \end{pmatrix}$$

The vector L will be used for classification with CNN and the vector Y for regression with MLP.

Convolutional Neural Networks in the Database. To group images of the MNIST database into different categories, we implement a Convolutional Neural Network classifier with four hidden layers. This is a practical and very convenient database to highlight the proposed method. We perform CNN on the data before using Multilayer Perceptron for regression in the different groups. This architecture of CNN gives a good classification of the images into 10 groups (Fig. 3)

Fig. 3. CNN architecture on MNIST dataset.

This architecture can be changed. It depends on the database we use to perform universal steganalysis of images. The goal in this step is to reduce an eventual cover-source mismatch issue.

Features for Regression with Multilayer Perceptron (MLP) in the Different Categories of Images. To implement an MLP neural network, we use both intra-block and inter-block correlations [4]. It consists of 486 features extracted from a JPEG image. At this step, we need the labels (vector Y) containing the lengths of the embedded messages of all images in the database.

Steganalysis on a Category of Images. To perform universal steganalysis on clusters, we use regression with an MLP neural network architecture with the most relevant features from both intra-block and inter-block correlations.

5   Experimental Results

Estimating the hidden message length is not an easy task. In the case of universal steganalysis combining CNN and MLP neural networks is a good approach to perform that task. In our experiments on the MNIST database, we got interesting results. This database is very convenient to illustrate our method of doing universal steganalysis in a database with a cover-source mismatch problem.

5.1   Deep Learning with CNN for Classification

By applying CNN with a standard architecture, we obtain easily 10 groups of images. Here an illustration of the model performance.

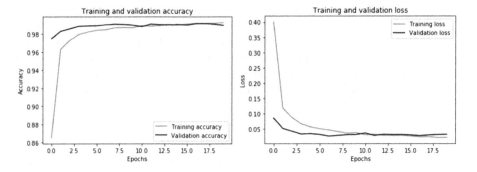

Fig. 4. Accuracy and loss in training and validation datasets.

In Fig. 4, we can observe the evolution of the accuracy score in the training and validation data. We can note that they are very close.

5.2   MLP Neural Network for Regression

In this step, we implement in all the clusters an MLP neural network for estimating the lengths of the hidden messages. Here the architecture of our neural network which consists of an input layer of 12 nodes (12 features selected from the 486 extracted features), three hidden layers of 13 nodes and an output layer of 1 node (estimation of the hidden message length) (Fig. 5).

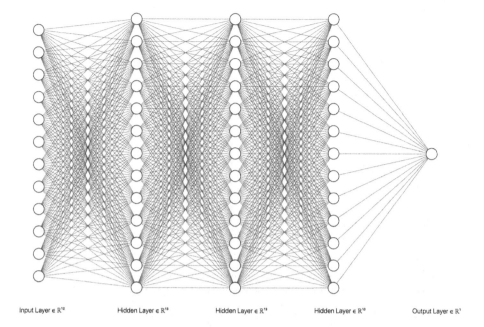

Input Layer ∈ R¹²   Hidden Layer ∈ R¹³   Hidden Layer ∈ R¹³   Hidden Layer ∈ R¹³   Output Layer ∈ R¹

Fig. 5. MLP architecture employed in DeepStego

Table 2 shows the MLP models accuracy scores in the groups generated by CNN.

Table 2. Accuracy score value in the 10 groups for DeepStego.

Groups	Accuracy score
Group 1	0.99
Group 2	0.99
Group 3	0.98
Group 4	0.99
Group 5	0.98
Group 6	0.98
Group 7	0.99
Group 8	0.98
Group 9	0.99
Group 10	0.99

Furthermore, the use of a stepwise feature selection helps to boost the MLP accuracy in the regression step.

A normal universal steganalysis procedure consists of extracting relevant features and implementing a supervised algorithm for classification or regression. For that, we implement an MLP neural network (with the same architecture implemented in the second part of DeepStego) to estimate the hidden message

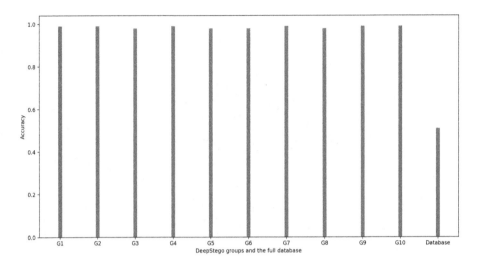

Fig. 6. Performance comparison between DeepStego (G1 to G10) and a normal universal steganalysis procedure (average score on the MNIST database).

length on the full database MNIST. In Fig. 6, we show a comparison between the results of a normal universal steganalysis procedure and DeepStego. The average score on the MNIST database for the universal steganalysis procedure is inferior to each score obtained by DeepStego in all the 10 groups. Thus, DeepStego gives better results (in all the 10 groups) than a universal steganalysis procedure on the full database.

However, the highest accuracies of the universal steganalysis approaches proposed in the literature, are often in the range [0.95, 0.97]. It rarely reaches 0.9. With DeepStego, we get accuracy which turns around 0.9 in all groups showing the interest of the proposed deep learning approach for the estimation of the hidden message length in steganography.

6 Conclusion

In this paper, we addressed the cover-source mismatch problem that prevents the utilization of regression for universal image steganalysis. For this, we need to group similar images into clusters before applying it. When the extracted feature vector from the image is very large, the k-means algorithm cannot help to perform the clustering process. To address this issue, we have proposed an original method that used in its first step CNNs to group similar images and in its second step implementation of a multilayer perceptron neural network to estimate the hidden message length. Experimental results on the MNIST database provided good approximation models in all the 10 clusters. Thus, deep learning with CNN is a suitable alternative to k-means to reduce the cover-source mismatch problem in the case of universal steganalysis.

References

1. Ashu, A., Chhikara, R.: Performance evaluation of first and second order features for steganalysis. Int. J. Comput. Appl. **92**, 17–22 (2014). https://doi.org/10.5120/16093-5372
2. Bonhomme, C., al.: Stegano: a pure python steganography module (2010). https://pypi.org/project/Stegano/. Accessed 30 Mar 2020
3. Chaumont, M.: Deep learning in steganography and steganalysis since 2015, October 2018. https://doi.org/10.13140/RG.2.2.25683.22567
4. Chen, C., Shi, Y.Q.: JPEG image steganalysis utilizing both intrablock and interblock correlations. In: 2008 IEEE International Symposium on Circuits and Systems, pp. 3029–3032, May 2008
5. Chen, M., Sedighi, V., Boroumand, M., Fridrich, J.: JPEG-phase-aware convolutional neural network for steganalysis of JPEG Images. In: Proceedings of the 5th ACM Workshop on Information Hiding and Multimedia Security, pp. 75–84. IH&MMSec 2017. ACM, New York (2017)
6. Cortez, P.: Data mining with multilayer perceptrons and support vector machines. In: Holmes, D.E., Jain, L.C. (eds.) Data Mining: Foundations and Intelligent Paradigms. Intelligent Systems Reference Library, vol. 24, pp. 9–25. Springer, Heidelberg (2012). https://doi.org/10.1007/978-3-642-23241-1_2
7. Deng, L.: The MNIST database of handwritten digit images for machine learning research [best of the web]. IEEE Signal Process. Mag. **29**(6), 141–142 (2012)
8. Dwivedi, Y.P., Bera, M.S., Sharma, M.M.: Universal steganalysis techniques based on the feature extraction in transform domain (2017)
9. Gomis, F.K., Camara, M.S., Diop, I., Farssi, S.M., Tall, K., Diouf, B.: Multiple linear regression for universal steganalysis of images. In: International Conference on Intelligent Systems and Computer Vision (ISCV), pp. 1–4, April 2018
10. O'Shea, K., Nash, R.: An introduction to convolutional neural networks. ArXiv e-prints, November 2015
11. Quach, T.T.: Extracting hidden messages in steganographic images. Digital Invest. **11**, S40–S45 (2014). https://doi.org/10.1016/j.diin.2014.05.003

An Adaptive Deep Learning Algorithm Based Autoencoder for Interference Channels

Dehao Wu[1]([⊠]), Maziar Nekovee[1,2], and Yue Wang[3]

[1] Centre for Advanced Communications, Mobile Technology and IoT,
University of Sussex, Brighton, UK
{dehao.wu,m.nekovee}@sussex.ac.uk
[2] Quantrom Technologies LTD., London, UK
[3] Samsung Electronics R&D Institute, Communications House,
South Street, Staines, Middlesex TW18 4QE, UK
yue2.wang@samsung.com

Abstract. Deep learning (DL) based autoencoder (AE) has been proposed recently as a promising, and potentially disruptive Physical Layer (PHY) design for beyond-5G communication systems. Compared to a traditional communication system with a multiple-block structure, the DL based AE provides a new PHY paradigm with a pure data-driven and end-to-end learning based solution. However, significant challenges are to be overcome before this approach becomes a serious contender for practical beyond-5G systems. One of such challenges is the robustness of AE under interference channels. In this paper, we first evaluate the performance and robustness of an AE in the presence of an interference channel. Our results show that AE performs well under weak and moderate interference condition, while its performance degrades substantially under strong and very strong interference condition. We further propose a novel online adaptive deep learning (ADL) algorithm to tackle the performance issue of AE under strong and very strong interference, where level of interference can be predicted in real time for the decoding process. The performance of the proposed algorithm for different interference scenarios is studied and compared to the existing system using a conventional DL-assist AE through an offline learning method. Our results demonstrate the robustness of the proposed ADL-assist AE over the entire range of interference levels, while existing AE fail to perform in the presence of strong and very strong interference. The work proposed in this paper is an important step towards enabling AE for practical 5G and beyond communication systems with dynamic and heterogeneous interference.

Keywords: Deep learning · Physical layer · Autoencoder · Interference channel

© IFIP International Federation for Information Processing 2020
Published by Springer Nature Switzerland AG 2020
S. Boumerdassi et al. (Eds.): MLN 2019, LNCS 12081, pp. 342–354, 2020.
https://doi.org/10.1007/978-3-030-45778-5_23

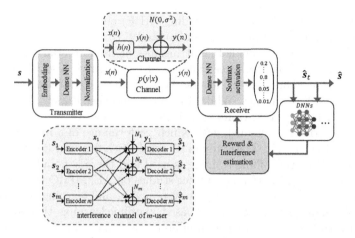

Fig. 1. System block diagram of an ADL algorithm based AE for a wireless communication interference channel with m-user

1 Introduction

Communication networks and services are becoming more intelligent with the novel advancements and unprecedented levels of computational capacity that is available for processing locally or in the cloud. AI, including machine learning (ML) and deep learning (DL), has been widely used for the design and management of communication systems, and has been shown to significantly enhance the system performance and reduce the operational cost, hence has raised great interest in standard [1], as well as in research. There has been a number of examples of using AI in communication systems in the literature, for example, for channel estimation [2], complex multiple-input and multiple-output (MIMO) detection [3], channel decoding [4], joint channel estimation and detection [5], joint channel encoding and source encoding [6].

In a conventional communication system, the channel propagation is often modeled mathematically, which may not correctly reflect the channel in practical scenarios and the dynamic nature of the changing. DL based approaches demonstrate a useful and insightful way of fundamentally rethinking the communication system design problem and hold the promise for performance enhancement in complex scenarios that are difficult to characterize with tractable mathematical models. Compared to a traditional communication system with a structure consisting multiple functional blocks, autoencoder provides a new paradigm with a pure data-driven and end-to-end learning based solution. For example, a DL based AE is proposed in [7], where the deep neural networks (DNNs) based reconstruction transceiver block jointly optimizes all the functions in a single process. The work in [8] presents end-to-end learning of a communications system without a channel model. In [9], authors propose a deep reinforcement learning approach for training a link with noisy feedback, for both additive white Gaussian noise (AWGN) and Rayleigh block-fading (RBF) channels. Also, two practical DL-based systems are implemented in [10] and [11].

All the work above provided great insights of the potential performance of applying AE for interference-free channels. However, it is also revealed in [12] that AE can be vulnerable to adversarial and jamming attacks, compared to conventional coding schemes. While [13] shows that such drawbacks can be mitigated through adversarial training, it is not clear how AE will behave under a multi-user interference channel, with which performance of a multi-user system is often impaired [14,15]. The study in [7] considers a two-user link with interference for AE. However, offline training is used and there is no adaptive training for different levels of interference. Other studies on AE, MIMO channel learning [16], channel estimation in an OFDM system [17], and learning to optimize for interference management [18], are all based on offline learning, therefore does not cope well with the situation when interference is dynamic, can be from different sources, and can vary in real time.

In this work, we characterize the tolerance of a conventional AE under a Gaussian interference channel, with respective to different interference levels. Our results demonstrate that although the offline trained AE approach has reasonable robustness for noisy to moderate interference channel, performance of AE suffers substantially under a strong or very strong interference channel. To date, there has been little work on DL-based AE in the presence of an interference channel with a variety of interference strengths, even less so to address the issue for allowing AE in practical dynamic and heterogamous interference scenarios. In this paper, we proposed an adaptive deep learning (ADL) algorithm based AE. The interference strength is predicted through an adaptive deep learning process, where real time online learning is performed to obtain the knowledge of the real time interference level for the subsequent decoding process, through an updated DNNs layer. We demonstrate that the proposed AE works robustly for all interference levels. In particular, the performance improvement compared to conventional AE [7] is more notable for the strong and very strong interference scenarios.

2   System Model

2.1   System Description

The proposed ADL algorithm based AE system for a wireless communication interference channel with m-user is shown in Fig. 1. It has three main blocks: transmitter, channel, and receiver. Compared to a conventional communication system with a number of blocks, this proposed diagram recast the block diagram as an end-to-end optimization task and represent the system as a simplified AE system by using a DL based neural network (NN) layer. For basics of DNN, an introduction is given in [24]. The NN layer stacks one on top of another. In general, the NN layers considered in this work transform an input data l_{in} into an output l_{out} as follows:

$$l_{\text{out}} = f(\mathbf{w}l_{\text{in}} + \mathbf{b}) \tag{1}$$

where $\mathbf{w}$ and $\mathbf{b}$ are weights and trainable parameters and $f(.)$ is a non-linear function [24]. The weights of the whole layers are optimized jointly. Let s as the input, and the training set contains all the possible values of s. In an AE during the training, the targets values are equal to the inputs ie $\hat{s}_i = s_i$, where s_i is a realization of s. The network is trained to optimize the reconstruction error, which is given:

$$L(s, \hat{s}) = -\log p(s|\hat{s}) \tag{2}$$

The reconstruction error here is the cross entropy loss, which is given [24]:

$$L(s, \hat{s}) = -\sum_k (s(k)\log \hat{s}(k) + (1 - s(k))\log(1 - \hat{s}(k))) \tag{3}$$

where $\hat{s}(k) = P(s(k) = 1|\hat{s})$. $s(k)$ stands for bit k of s and $\hat{s}(k)$ stands for bit k of $\hat{s}$. The training of the network is performed by solving the following optimization problem:

$$\arg \min_P \mathbb{E}_{s,N,\theta}[L(s, \hat{s})] \tag{4}$$

where P is denote the set of trainable parameters. N and θ are generated noise and phase by the channel layer each time it is used.

For the transmitter side, the transmitted messages s is reconstructed, and $s_i \in M = \{1, 2, \ldots, M\}$, where $M = 2^k$ is the dimension of M with k being the number of bits per message. The message is passed to the transmitter. The transmitter applies a transformation by a DNN layer $f : M \rightarrow \mathbb{R}^{2n}$ to the message s_i to generate the transmitted signal $x = f(s_i) \in \mathbb{R}^{2n}$. Note that the output of the transmitter is an n-dimensional complex vector which is transformed to a $2n$ real vector. We use 'one-hot vector' with size of M to reconstruct s_i for

Table 1. The structure of the MLP AE

Block name	Layer name	Output dim
	input:	M
Block name	Dense+eLu	M
	Dense+Linear	$2n$
	nomalization	$2n$
Channel	Noise	$2n$
Decoder	Dense+ReLU	M
	Dense+Softmax	M

Name	$[\sigma(u)]_i$	range
ReLU	$\max(0, U_i)$	$[0, \infty)$
Tanh	$\tanh(U_i)$	$(-1, 1)$
Softmax	$\frac{e^{u_i}}{\sum_j e(u_j)}$	$(0, 1)$

DNN layer. Following the similar definition in [7], the transmitter is constrained by an average power: $\mathbb{E}\left[|x_i^2|\right] \leq 0.5 \forall i$. In this work, we use a m-user interference channel with AWGN.

2.2   Model of Interference Channel with m-user

In Fig. 1, a m-user Gaussian AWGN interference channel is illustrated within the dashed-line rectangle block. The interference channel has m transmitter-receiver pairs that simultaneously communicate in blocks of size m. Each transmitter communicates to its own receiver a message $s \in m = \{1, 2, \ldots, m\}$. Let x^n and y^n denote the input and output signal of the nth user, respectively. $N^n \sim CN(0, 1)$ is independent and identically distributed Gaussian noise that impairs receiver n. Each x_n has an associated average power constraint P^n so that $\frac{1}{m}\sum_{n=1}^{m}|x_m^n|^2 \leq P^n$. Receiver n observes $\hat{y}^n$ and estimates the transmitted message $\hat{x}^n$. The average probability of error for user n is $\epsilon_m^n = \mathbb{E}[P(\hat{s}^n \neq s^n)]$, where expectation is over the random choice of message. The channel output at each receiver is a noisy linear combination of its desired signal and the sum of the interfering terms, of the form [19]:

$$y^n = x^n + \sqrt{\frac{\text{INR}}{\text{SNR}}} \sum_{j=1, j\neq n}^{m} x^j + N^n, \forall j, n = 1, 2, \ldots, m \tag{5}$$

where y^n and N^n are the channel output and AWGN respectively, at the nth receiver and the x^n is the channel input symbol at the nth transmitter. All symbols are real and the channel coefficients are fixed. The AWGN is normalized to have zero mean and unit variance and the input power constraint is given by [19]:

$$E[(x^n)^2] \leq \text{SNR}, \ \forall n \in m. \tag{6}$$

The INR is defined through the parameter α [19]:

$$\frac{\log(\text{INR})}{\log(\text{SNR})} = \alpha \ \rightarrow \ \text{INR} = \text{SNR}^\alpha \tag{7}$$

Note that the definition of INR ignores the fact that there are m-1 interferers observed at each receiver. This is for two reasons. First, this definition parallels that of the two-user case [20], which will make it easier to compare the two rate regions. Second, the receivers will often be able to treat the interference as stemming from a single effective transmitter, via interference alignment. This is not the case when the receiver treats the interference as noise. In this work, the introduced parameter $\alpha > 0$ defined by INR = SNR$^\alpha$; this coupling parameter α is used to specify the corresponding linear deterministic model in [21].

In this work, we address the interference scenarios including noisy, weak, moderate, strong, and very strong interferences. The definition of the classification for the interference is proposed in [19]. The degrees-of-freedom (GDoF) of

---

Algorithm 1: ADL algorithm to predict the interference

Input: • AE model and specifications: n, k, batch size, epochs number, optimizer, learning rate, etc
      • the training data set l_{in}
      • the variance of channel noise σ^2
Output: • • the estimated interference parameter α

1 Initialize:
2 Set AE model parameters (e.g., $n \leftarrow 4$, $k \leftarrow 4$, $M \leftarrow 4$)
3 **for** i in range (training data samples) **do**
4   Set $x = f(s_i) \in \mathbb{R}^{2n}$, $s_i \in \{1, 2 \ldots M\}$, encoding
5   Create and Set $\hat{y}(n)$ for receiver layer
6   **for** i in range (numble of guessing α) **do**
7   DNN layer to training data set (settings in Table I)
8   Recovery pilot signal $\hat{s}_i$ according to a guessing α
9   Calculate *reward* $\hat{R}_i$ according to Eqs. (5) and (6)
10 Set confidence interval of $\hat{R}_i$ and predict α
11 Update DNN layer with α according to Eqs. (7) to (10)

---

the symmetric m-user interference channel is identical to that of the multiple-user channel, except for a singularity at $\alpha = 1$, as follows:

$$
d(\alpha) = \begin{cases}
1 - \alpha, & 0 \leq \alpha < \frac{1}{2} \text{ (noisy)} \\
\alpha, & \frac{1}{2} \leq \alpha < \frac{2}{3} \text{ (weak)} \\
1 - \frac{\alpha}{2}, & \frac{2}{3} \leq \alpha < 1 \text{ (moderate)} \\
\frac{1}{K}, & \alpha = 1 \\
\frac{\alpha}{2}, & 1 < \alpha < 2 \text{ (strong)} \\
1, & \alpha \geq 2 \quad \text{(very strong)}
\end{cases}
\tag{8}
$$

2.3   ADL Algorithm at Receiver Blocks

As shown in Fig. 1, at the receiver side, $y(n)$ is the received signal after propagating through an AWGN channel, which includes the original transmitted signal, the channel response, AWGN noise as well as the interference from other sources. Here, the received n-dimensional signal $y(n)$ noised by a channel represented as a conditional probability density function $p(y|x)$, and the DNNs receiver subsequently learns it with multiple dense layers. The last layer of the receiver is a Softmax activation layer that outputs an M-dimensional probability vector p, in which the sum of its elements is equal to 1. The receiver first applies the transformation $f : \mathbb{R}^{2n} \to M$ to decode the signal, creating a signal $\hat{s}_i$ to recover the original transmitted signal s_i.

To enable the comparability of the results implemented in different scenarios, we set $n = 4$ and $k = 4$ throughout this work. For other setups of the AE, to allow a benchmark for comparison, we use the similar AE structure and settings as in

[7], which are based on a multi-layer perceptron (MLP) AE. The specifications are listed in Table 1. We train the AE in an end-to-end manner using the Adam optimizer, on the set of all possible messages $s_i \in M$, using the cross-entropy loss function. ReLU and Softmax are used in DNNs layer.

As shown in Fig. 1, we design and propose an adaptive learning processing, integrating with the DNNs based receiver block, named *ADL algorithm*, to estimate the interference coupling parameter α. With the Predicted α, we obtain an updated channel function, according to Eqs. (5) to (7). Then the DNN layer is updated with this knowledge by substituting α into Eq. (5). This process includes two stages. Firstly, we utilize multiple group of pilot signals for online DNN training to predict the real-time α. Then with the knowledge of the channel, we update the interference channel function, decode signals with DNN layers.

It assumes that the signals consist of two parts. The first part is pilot signal, as the training data set. The second part is the transmitted signal, which has the same structure as it's in a DL based OFDM system [17]. However, we utilize the pilot signals here for both estimating interference and the DNN training. We introduce and explain our proposed ADL algorithm in Algorithm 1. At the initialization stage, we set the specifications of an ($n = 4$, $k = 4$) AE and load the input training data set. Then, the DNN layer encodes the data for propagating through an AWGN channel. The DNNs based receiver block first captures a group of signals, and then the reinforcement block starts to train the pilots simultaneously. By process of reward computation, the block normalizes the reward regarding different guessing values of α. Then we determines the optimum α range with regarding the a predefined confidence interval. Based on the plot of the reward according to the guessing values of α. We compute the mean, as the predicted α. Next, the estimated α is substituted back into the DNNs block for the decoding process with an updated DNNs layer. For this prediction process based on the reward performance, we will give more details in the Section of Numerical Evaluation. In this work, the normalized reward is defined as follows:

$$\hat{R}_i = \frac{R_i}{||R_i||} \tag{9}$$

where

$$R_i = \frac{1}{\overline{BER}|_{Pilot \leftarrow (1,\dots,i)}} \tag{10}$$

R_i is defined as the reciprocal of the mean bit error rate (BER) value for i pilots signals.

3   Numerical Results and Discussion

In this section, numerical simulation is carried out under the environment of Python 3.0, with the libraries of PyTorch, TorchNet and TQDM. Training was done at a fixed value of $E_b/N_0 = 7$ dB using Adam [22] with a learning rate of 0.001. Activation functions rectified linear units (ReLU) [23] and Softmax are

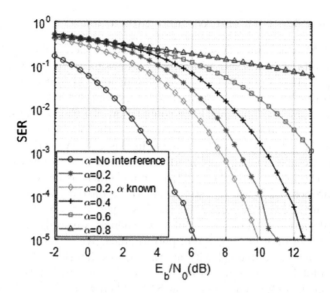

Fig. 2. SER versus SNR performance of an AE (4, 4): no interference, interference $\alpha = (0.2\ 0.8)$ with blind training, $\alpha = 0.2$ with knowledge of α.

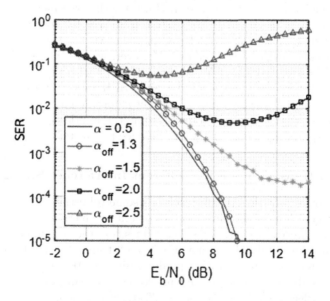

Fig. 3. SER versus SNR performance of an AE (4, 4): weak interference $\alpha = 0.5$ (at training) with offset up to $\alpha_{\text{off}} = 2.5$ (received).

used in our DNNs layer. The details are listed in Table 1. Detailed explanation of these can be found in [24]. The pilot symbol ratio we used in our simulation is 0.01. The group number of the bit streams is 30, which is used for jointly training and estimating the interference α.

3.1 Comparsion with and Without Interference

A DL based AE with different settings of (n, k) (n is the number of channel use, and k is the bits of the signal) are studied and evaluated in [7]. It compares the performance between the M-QAM modulation and AE with similar settings. It demonstrates that the AE (4, 4) and (4, 8) outperforms the 4-QAM and 16-QAM. To enable a benchmark for comparison, we choose the setting (4, 4) throughout all scenarios. However, we evaluate the performance according to our proposed interference model, as shown in Eqs. 5–8. We verify our algorithm through an example of a two-user interference channel case. For other multi-user case, the methodology is similar, and the enhancement is more significant.

In the proposed DL based system, the AE reconstruct and compressed the data with 'one hot vector' format for the NN layer. For a fair comparison to a conventional system with other modulation schemes, we study the symbol error rate (SER) for evaluating the system performance. We first simulate an AE (4, 4) system in an ideal channel without taking any interference, as a reference point. The plot is illustrated in Fig. 2. It shows that without the interference, the system works well, even under a low SNR. Then we evaluate a blind training with interference. The blind training is defined as that the system does not have any knowledge that it is an interference channel. Therefore the system trains a model without interference α. However, the true received signal has a certain value of α. We evaluate the system from $\alpha = 0.2$ to $\alpha = 0.8$, in Fig. 2. The results show that with a blind training, the AE has some robustness even without any knowledge of the channel. However, when α increases beyond 0.6, then the AE doesn't work well. We also plot the case with the knowledge of α for comparison. When $\alpha = 0.2$, we could achieve SER $\sim 10^{-3}$ at $E_b/N_0 = \sim 7$ dB, and this is assuming that we know the exact α for training. The comparison in Fig. 2 indicates that it is possible to overcome the interference effect if we have an efficient approach to predict the interference parameter α.

3.2 Robustness of an AE for Different Interference Strengths

We demonstrate that the AE approach has some robustness when it applies in an interference channel. However, we also want to characterize the robustness for difference interference strengths. It assumes that the system knows the interference channel generalized formula (Eq. 5) and it applies a DL training for the decoder. We train the model with a predetermined α. However, we assume that α may change dynamically in a real time scenario and we want to evaluate how robust of the decoder when α has some offset, denote as α_{off}.

Following the definition in Eqs. 5 to 8 , we simulate for weak ($\alpha = 0.5$) and very strong interference ($\alpha = 2$) respectively. Results are plotted in Figs. 3 and 4. It shows that the AE approach is quite robust for a weak interference. The system works even under a very large offset: 3 times of the training α. However, the situation is slightly different for very strong interference, where $\alpha = 2$. The result in Fig. 4 indicates that the system is quite sensitive to the offset under a very strong interference channel. For this scenario, it does require a technique to deal with the interference. To address this, we apply the proposed ADL algorithm and the performance evaluation is given in the next section.

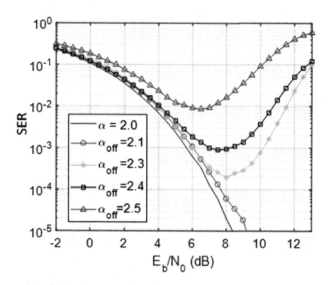

Fig. 4. SER versus SNR performance of an AE (4, 4): very strong interference $\alpha = 2$ (at training) with offset up to $\alpha = 2.5$ (received).

3.3 Evaluation with the Proposed Learning Algorithm

Recall the proposed ADL algorithm in sect. 2. We evaluate the ADL algorithm to estimate α in different interference strengths. We also carried more groups of study as in section B, and we found that for strong ($\alpha = 1.5$) and very strong ($\alpha = 2$) interference, the offset of α becomes more critical. Therefore, we address this and implement our algorithm for these cases. With the same setting in Figs. 3 and 4, we plot the normalized reward versus a predicted α (different values at training), in Fig. 5. for $\alpha = 1.5$ and $\alpha = 2$, respectively. We can see that the peak value of the normalized reward appears around 1.5 (actual value), and it reduces gradually to both sides of the actual value. By contrast, for the very strong interference, where $\alpha = 2$, we can also found out the peak value of the normalized reward appears around the real value of α. However, it decreases rapidly towards both sides of the actual value, which agree with the achievement that it's more sensitive to the offset. As the fluctuation is quite large in Fig. 5, here we define 40% offset as the confidence interval of the reward, to estimate α. We use the mean α for evaluating the performance, as we introduced in Sect. 2. Furthermore, the reward is computed according to the instant SNR condition. For this simulation, we use $E_b/N_0 = 7$ dB as an example. To evaluate the performance with and without applying the proposed ADL algorithm, we plot the SER performance for weak, strong and very strong interference channels for comparison, as shown in Fig. 6. In this simulation, we take a large interference

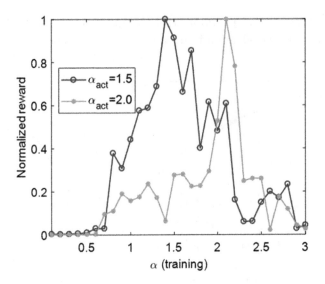

Fig. 5. Normalized reward versus predicted α: strong interference $\alpha = 1.5$ and very strong interference $\alpha = 2$.

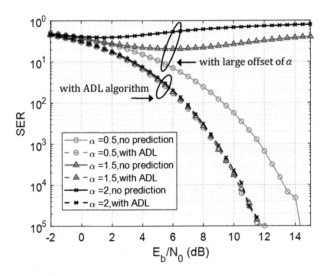

Fig. 6. SER versus SNR: comparison for strong and very strong interference channel, with and without the proposed ADL algorithm.

effect as an example, $\alpha_{\text{off}} = 2\alpha$, to demonstrate the improvement achieved by our algorithm. Two groups of data are highlighted in Fig. 6. We can see that the SER significantly degrades due to the large offset of α. In particular, for the strong and very strong interference cases, the system does not work without the knowledge of α. However, with applying the ADL algorithm, the result shows that with an efficient interference prediction, the ADL algorithm based AE is capable of robust performance over the entire range of interference levels, even for the worst case in a very strong interference channel.

4   Conclusion

An ADL algorithm based AE is proposed for interference channel with unknown interference. With the proposed online learning, interference can be estimated and predicted, which is then subsequently used for decoding of the signals using DNN. The proposed algorithm is shown to significantly enhance the robustness of the interference channel, and provides an AE system that is adaptable to real-time interference, for the entire range of interference levels. The enhancement is more notable for strong and very strong interference scenarios, compared to performance of conventional AE with offline learning.

We believe that our proposed approach is an important step towards enabling AE for 5G and beyond communication systems with dynamic and heterogeneous interference. Our future work aims at improving computational efficiency of our online learning scheme, and the implementation on real-life platforms.

References

1. Wang, Y., et al.: Network management and orchestration using artificial intelligence: overview of ETSI ENI. IEEE Commun. Stand. Mag. **2**(4), 58–65 (2018)
2. Farsad, N., Goldsmith, A.: Neural network detection of data sequences in communication systems. IEEE Trans. Signal Process. **66**(21), 5663–5678 (2018)
3. Huang, H., Yang, J., Huang, H., Song, Y., Gui, G.: Deep learning for super-resolution channel estimation and DOA estimation based massive MIMO system. IEEE Trans. Veh. Technol. **67**(9), 8549–8560 (2018)
4. Nachmani, E., Be'ery, Y., Burshtein, D.: Learning to decode linear codes using deep learning. In: 54th Annual Allerton Conference on IEEE Communication, Control, and Computing, pp. 341–34, Allerton (2016)
5. Ye, H., Li, G.Y., Juang, B.-H.: Power of deep learning for channel estimation and signal detection in OFDM systems. IEEE Wirel. Commun. Lett. **7**(1), 114–117 (2018)
6. Farsad, N., Rao, M., Goldsmith, A.: Deep learning for joint source-channel coding of text, arXiv preprint arXiv:1802.06832 (2018)
7. O'Shea, T., Hoydis, J.: An introduction to deep learning for the physical layer. IEEE Trans. Cognit. Commun. Netw. **3**(4), 563–575 (2017)
8. Aoudia, F.A., Hoydis, J.: End-to-end learning of communications systems without a channel model, arXiv preprint arXiv:1804.02276 (2018)
9. Goutay, M., Aoudia, F.A., Hoydis, J.: Deep reinforcement learning autoencoder with noisy feedback, arXiv preprint arXiv:1810.05419 (2018)
10. Dörner, S., Cammerer, S., Hoydis, J., Brink, S.: Deep learning based communication over the air. IEEE J. Sel. Top. Signal Process. **12**(1), 132–143 (2018)
11. O'Shea, T.J., Roy, T., West, N., Hilburn, B.C.: Physical layer communications system design over-the-air using adversarial networks, arXiv preprint arXiv:1803.03145 (2018)

12. Sadeghi, M., Larsson, E.G.: Physical adversarial attacks against end-to-end autoencoder communication systems. IEEE Commun. Lett. **23**(5), 847–850 (2019)
13. Goodfellow, I.J., Shlens, J., Szegedy, C.: Explaining and harnessing adversarial examples, arXiv (2018)
14. Gomadam, K., Cadambe, V.R., Jafar, S.A.: A distributed numerical approach to interference alignment and applications to wireless interference networks. IEEE Trans. Inf. Theory **57**(6), 3309–3322 (2011)
15. Geng, C., Naderializadeh, N., Avestimehr, A.S., Jafar, S.A.: On the optimality of treating interference as noise. IEEE Trans. Inf. Theory **61**(4), 1753–1767 (2015)
16. Erpek, T., O'Shea, T.J., Clancy, T.C.: Learning a physical layer scheme for the MIMO interference channel. In: 2018 IEEE International Conference on Communications (ICC), pp. 1–5 (2018)
17. Balevi, E., Andrews, J.G.: One-bit OFDM receivers via deep learning. IEEE Trans. Commun. **67**(6), 4326–4336 (2019)
18. Sun, H., Chen, X., Shi, Q., Hong, M., Fu, X., Sidiropoulos, N.D.: Learning to optimize: training deep neural networks for interference management. IEEE Trans. Signal Process. **66**(20), 5438–5453 (2018)
19. Jafar, S.A., Vishwanath, S.: Generalized degrees of freedom of the symmetric Gaussian K user interference channel. IEEE Trans. Inf. Theory **56**(7), 3297–3303 (2010)
20. Etkin, R.H., David, N., Wang, H.: Gaussian interference channel capacity to within one bit. IEEE Trans. Inf. Theory **54**(12), 5534–5562 (2008)
21. Ordentlich, O., Erez, U., Nazer, B.: The approximate sum capacity of the symmetric Gaussian K-user interference channel. IEEE Trans. Inf. Theory **60**(6), 3450–3482 (2014)
22. Kingma, D.P., Ba, J.: Adam: a method for stochastic optimization (2014), arXiv preprint arXiv:1412.6980
23. Jin, X., Xu, C., Feng, J., Wei, Y., Xiong, J., Yan, S.: Deep learning with S-shaped rectified linear activation units. In: AAAI, vol. 3, no. 2, p. 3.2 (2016)
24. Goodfellow, I., Bengio, Y., Courville, A., Bengio, Y.: Deep Learning. MIT Press, Cambridge (2016)

A Learning Approach for Road Traffic Optimization in Urban Environments

Ahmed Mejdoubi[1,2], Ouadoudi Zytoune[1,3], Hacène Fouchal[2(✉)],
and Mohamed Ouadou[1]

[1] LRIT, Associated Unit to CNRST (URAC 29), Faculty of Science,
Mohammed V University, Rabat, Morocco
ouadou@fsr.ac.ma

[2] CReSTIC, Université de Reims Champagne-Ardenne, Reims, France
{ahmed.mejdoubi,hacene.fouchal}@univ-reims.fr

[3] ENSA, Ibn-Tofail University, Kenitra, Morocco
zytoune.ouadoudi@uit.ac.ma

Abstract. In many urban areas where road drivers are suffering from the huge road traffic flow, conventional traffic management methods have become inefficient. One alternative is to let road-side units or vehicles learn how to calculate the optimal path based on the traffic situation. This work aims to provide the optimal path in terms of travel time for the vehicles seeking to reach their destination avoiding road traffic congestion and in the least possible time. In this paper we apply a reinforcement learning technique, in particular Q-learning, that is employed to learn the best action to take in different situations, where the transiting delay from a state to another is used to determinate the rewards. The simulation results confirm that the proposed Q-learning approach outperformed the greedy existing algorithm and present better performances.

Keywords: C-ITS · VANETs · Reinforcement learning · Distributed traffic management · Travel time

1 Introduction

Nowadays, emerging and developed countries suffer from the immense road traffic flow, especially in urban environments, because of the continuous increase in the number of vehicles traveling every day in parallel with the continued population growth, but much faster than transportation infrastructure. Consequently, this huge amount of vehicles will become a serious problem leading to traffic congestion, air pollution, fuel consumption [1] and excessive traffic delays. Therefore, intelligent transport systems is becoming a primary need to deal with these problems and to accommodate the growing needs of transport systems today.

Cooperative Intelligent Transport System, or C-ITS [2,3], is a new transportation system which aims to provide intelligent solutions for a variety of road

© IFIP International Federation for Information Processing 2020
Published by Springer Nature Switzerland AG 2020
S. Boumerdassi et al. (Eds.): MLN 2019, LNCS 12081, pp. 355–366, 2020.
https://doi.org/10.1007/978-3-030-45778-5_24

traffic problems, as congestion and traffic accidents, by linking vehicles, roads and people in an information and communications network through cutting-edge technologies. It applies advanced technologies of computers, communications, electronics, control and detecting and sensing in all kinds of transportation system in order to improve safety and mobility, efficiency and traffic situation via transmitting real-time traffic information using wireless technology. C-ITS focuses on the communication between vehicles (vehicle-to-vehicle), vehicle with the infrastructure (vehicle-to-infrastructure) or with other systems.

The C-ITS system have attracted both industry leaders and academic researchers. These systems are considered as a solution for many road traffic issues and as an efficient way to enhance travel security, to avoid occasional traffic jams and to provide optimal solutions for road users. In this system (C-ITS), vehicles can exchange information with each other (V2V) or with road-side units (V2I). These communications are handled through a specific WIFI called IEEE 802.11p [4].

The main contribution of this paper is to minimize the total traveling time for drivers by providing optimal paths suggestion to reach their pretended destination. The proposed solution highlights vehicular communications between vehicles and road-side units in order to collect and exchange current traffic status. The remainder of the paper is organized as follows. Section 2 presents a review of some works related to transport traffic management. Section 3 details the proposed approach. Section 4 presents the evaluation and performance of our proposed solution, and Sect. 5 concludes the paper and highlights future works.

2 Related Works

A group routing optimization approach, based on Markov Decision Process (MDP) [5], is proposed in [6]. Instead of finding the optimal path for individual vehicles, group routing suggestion will be provided using vehicle similarities and V2X communications to reduce traffic jams. The authors are studied the learning method of this approach and how it is going to work with their proposed prototype. The MDP is a type of mathematics model used for studying optimization problems solved via dynamic programming [7] and reinforcement learning [8]. MDP is characterized by a set of actions that can lead to a certain state depending on what you want to achieve. The selection of the most appropriate actions is induced by MDP rewards.

In [9], and based on the Vehicular Ad-hoc Network (VANET) architecture, the authors present a predictive road traffic management system named PRTMS. The proposed system uses a modified linear prediction (LP) algorithm to estimate the future traffic flow at different intersections based on a vehicle to infrastructure scheme. Based on the previous estimation results, the vehicles can be rerouted in order to reduce the traffic congestion and minimise their journey time. However, the proposed system relies mainly on a centralised architecture to exchange road traffic information with vehicles, which can lead to a significant overhead costs and power resources.

In order to find the shortest path, Dijkstra [10] proposed a static algorithm based only on the distance from the source node to all other nodes without considering external parameters such as density, congestion, or average vehicle speed. However, this algorithm is not practical enough in the case of continuous changes over time in road traffic network. Thus, vehicle routing optimization should always consider a continuous adaptation of routes for each vehicle to reach their destinations in the least possible time.

Nahar and Hashim [11] introduced an ant-based congestion avoidance system. This later use the average travel speed prediction of roads traffic combined with the map segmentation to reduce congestion using the least congested shortest paths to the destination. Real-time traffic information is collected from vehicles and road side units (RSU) in order predict the average travel speed. Their studies have been conducted in fixing the ACO (Ant colony optimization) variables [12] to reduce vehicle congestion on the roads. Their results show that the number of ants is directly correlated with the algorithm performance. However, the proposed method does not perform well when there is only a small number of ant-agents (under 100).

Kammoun et al. [13] proposed an adaptive vehicle guidance system. It aims to find the best route by using real-time data from a vehicular network. In order to improve driver request management and ensure dynamic traffic control, the proposed method used three different ant-agents city agent, road supervisor agent and intelligent vehicle-ant agent are three different ants, namely, city agent, road supervisor agent and intelligent vehicle-ant agent. However, the proposed method is faced with a limitation at managing a large and complex urban transportation network.

The authors in [14] come up with two algorithms named GREEDY and Probabilistic Data Collection (PDC) for vehicular multimedia sensor networks. The proposed algorithms can provide data redundancy mitigation under network capacity constraints by using submodular optimization techniques. They assume that vehicles are equipped with cameras and they continuously capture images from urban streets. The proposed algorithm is evaluated by using NS-2 simulator and VanetMobiSim to generate the mobility traces. One major drawback is that when many vehicles attempt to upload their data at the same time, quality of service can highly decrease.

Based on the literature reviews and previous studies, both traditional and centralized road traffic management solutions have become inefficient depending on road traffic demands in urban areas and the high overhead costs they consume. Also, predicting and calculating the shortest path is not always reliable due to the continuous changes of road traffic flow over time. Our proposed approach aims to enable an efficient traffic flow management by providing optimal paths suggestion and reducing the total travel time of vehicles using reinforcement learning and based on a vehicular ad-hoc network architecture (VANET).

3   The Proposed Approach

3.1   System Architecture

The system architecture is presented in Fig. 1. It is composed of two main components: Vehicles and RSUs. RSUs are placed at the intersections to collect information from vehicles. Each vehicle exchange its current traffic information with the closest RSU.

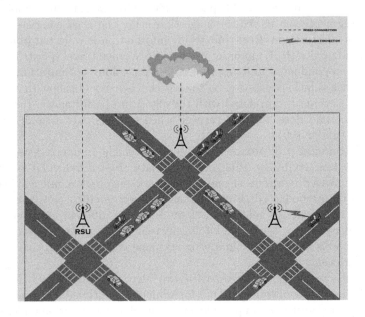

Fig. 1. System architecture

We used two types of communications in our system: wireless communication using ITS G5 (IEEE 802.11p) that handles exchanges between vehicles and the RSUs, and wired communications to handle exchanges between RSUs. As shown in Fig. 3, the transport network consists of Manhattan street topology of overall 40 segments and a grid map of 5 × 5 junctions. There are 12 RSUs placed at different intersections, the distance between two adjacent intersections is set to 0..1 km, and the maximum speed of vehicles is 60 km/h. The travel time on each segment varies according to the road traffic status and ranges from 5 s to 1 h.

3.2   Machine Learning

Machine Learning (ML) is a science that get computer systems to learn through data, observations and interacting with the world, and improve their learning over time to act without being explicitly programmed. It gives the computer to learn as well as humans do or better.

Machine learning can generally be classified into 4 main categories according to the learning style:

- **Supervised learning:** Learning is supervised when the model is getting trained on a labeled data-set (i.e. which have both input and output parameters) and the algorithms must use it to predict the future result. For example, you can give the system a list of customer profiles containing purchasing habits, and explain to it which are regular customers and which ones are occasional. Once the learning is finished, the algorithm will have to be able to determine by itself from a customer profile to which category this one belongs. The margin of error is thus reduced over the training, with the aim of being able to generalize its learning to new cases.
- **Unsupervised learning:** the learning process is completely autonomous. Data is communicated to the system without providing the examples of the expected output results. It is much more complex since the system will have to detect the similarities in the data-set and organize them without pre-existing labels, leaving to the algorithm to determine the data patterns on its own. It mainly deals with the unlabeled data. Although, unsupervised learning algorithms can perform more complex processing tasks compared to supervised learning.
- **Semi-supervised Learning:** This type is a combination of the supervised and the unsupervised categories, in which both labeled and unlabeled data are used, typically a large amount of unlabeled data with a small amount of labeled data.
- **Reinforcement Learning:** in this type of learning, the algorithms try to predict the output of a problem according to a set of parameters. Then, the calculated output becomes an input parameter and a new output is calculated until the optimal output is found. Artificial Neural Networks (ANN) and Deep Learning, which will be discussed later, use this learning style. Reinforcement learning [15] is mainly used for applications such as resource management, robotics, helicopter flight, skills acquisition and real-time decisions.

Figure 2 shows a typical reinforcement learning scenario in which an agent performs an action on the environment, this action is interpreted as a reward and a representation of the new state, and this new representation is forwarded to the agent.

3.3   Q-Learning Algorithm

Traffic routing management can be considered as a MDP while junctions states represent the system states and the process of selecting directions across the junctions represent the actions. When passing across a junction, the vehicle observes a delay that can represents the reverse of a reward. Then, the objective is to select at each junction the optimal direction in order to reduce the total traveling time. Two methods can be used to address this problem as stated in the previous section. However, instead of using dynamic programming, the

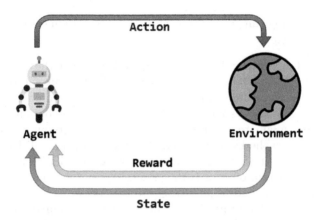

Fig. 2. A typical reinforcement learning scenario

reinforcement learning can operate in case of unknown environment. In this work, we consider that the vehicle driver is traveling in an unknown environment i.e. he has no information about junctions delay. The driver will try to minimize the cumulative long term transit delay (i.e. maximizing a reward given by the reverse of the transition delay) by experimenting actions according to the observation of current states and rewards.

Q-learning method is considered as an off policy reinforcement learning algorithm, which tries to find the best action to take in the current state. No policy is imposed, but the Q-learning algorithm learns from actions that seek to maximize the total reward. In this sub-section, we consider the driver reorientation in the case of a model-free system environment. We propose the use of a reinforcement learning approach to solve our optimization problem. Then, each junction i in the road network is represented by a state in our system representation, denoted as s_i. Let S be the set of possible states. We assume that in each state s_i the vehicle driver can take one action of the set $A = \{$turn left, turn right, go forward, go backward$\}$. When a vehicle goes across a junction, a delay time is observed. In our proposition we look for minimizing the total travel time from a source to a destination, so that our reward, that we try to maximize, will be considered as the inverse value of the delay time.

We can summarize the reinforcement learning steps as follows:

– Observes the state at the iteration n: $S_n = s_j \in S$,
– Selects and applies an action $a_n = a_i \in A$,
– Go to the next state $S_{n+1} = s_k \in S$ and observes the immediate reward $R_{a_i}(s_j, s_k)$,
– Updates the Q function using the following Equation as in [14]:

$$Q_n(S_j, a_i) \leftarrow Q_{n-1}(S_j, a_i) + \alpha_n[R_{a_i}(s_j, s_k) + \gamma\max_{a_j \in A}(Q(S_{n+1}, a)) - Q_{n-1}(S_k, a_i)]$$

Where α_n is a learning rate factor and γ is the discount factor with $\gamma \in [0, 1]$. The Q-learning algorithm is given in Algorithm 1.

Algorithm 1. Q-learning algorithm

Initialize $Q(s, a), \forall s \in S, a \in A(s)$, arbitrarily, and $Q(terminal - state, .) = 0$

Repeat (for each episode):

    Initialize S

    Repeat (for each step of episode):

        Choose A from S using policy derived from $Q(e.g., \epsilon - greedy)$

        Take action A, observe R, S'

        $Q(S, A) \leftarrow Q(S, A) + \alpha[R + \gamma max_a Q(S', a) - Q(S, A)]$

        $S \leftarrow S'$

    Until S is terminal

The learning rate α defines how much newly acquired information replaces old information. When $\alpha = 0$ that makes the agent exploiting prior knowledge, and $\alpha = 1$ makes the agent ignore prior knowledge and consider only the most recent information to explore other possibilities. However, the discount factor γ determines how the future rewards are important. When $\gamma = 0$ that will make the agent considering only the current rewards, and while γ approaching 1 will make it strive to get a long-term high reward, but if the discount factor exceeds 1 the action values may diverge [16]. The ϵ-greedy method is used for exploration during the training process. This means that when an action is selected in training, it is either chosen as the action with the highest q-value (exploitation), or a random action (exploration).

4   Evaluation and Performance Analysis

The proposed approach has been tested on a network that contains 25 intersections and 40 two-way links using the Matlab platform [17]. It is considered as a programming platform designed specifically for scientists and engineers, in which we can analyze data, create models or develop algorithms, etc. It can used for a range of applications including deep learning and machine learning, control systems, test and measurement, computational finance and biology [18], and so on. We have performed many simulations in order to compare the proposed approach with the greedy algorithm that seeks to find the path with the largest sum of the crossed nodes value. The simulations aim to determine the total travel time of a vehicle in the network for different traffic scenarios and to check how the traffic will be improved by suggesting optimal paths to the vehicles based on the Q-learning approach.

The transport network topology consists of a grid map of 5 × 5 intersections as shown in Fig. 3, in which the vehicles are supposed to move according to the Manhattan mobility model [19]. The network has 12 RSUs placed at different intersections, the distance between two adjacent intersections is set to 0..1 km, and the maximum speed of vehicles is 60 km/h. We assume that the time required for a vehicle to cross a link between two intersections is between 5 s and 1 h depending on traffic situation.

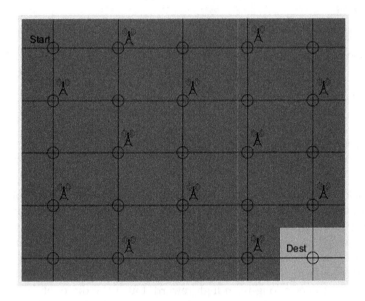

Fig. 3. Simulation system network

Table 1. Network configuration parameters

Parameters	Values
Number of intersections	25
Number of links	40
Number of actions	4
α: learning rate	[0, 1]
γ: discount rate	[0, 1]
Maximum number of iterations	3000
Mobility model	Manhattan
Number of RSUs	12
Wireless transmission range	500 m
Wireless links	ITS G5 (802.11p)

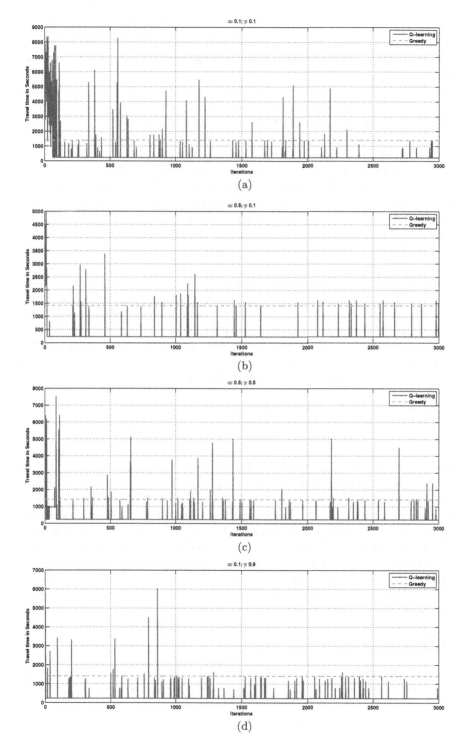

Fig. 4. Traveling time comparison between our approach and the greedy algorithm.

The vehicles can communicate with road-side units through periodic messages in order to collect the traffic status information around junctions by using ITS G5 protocol (802.11p). The system configuration parameters are shown in Table 1.

In this simulation, the total travel time is used as performance indicator for the evaluation. This parameter represents the cumulative time spent to travel from the starting node to reach the destination one. Different values of the parameters α and γ are experienced to fined the optimal combination that gives the best results. For this assessment, we take the parameter $\epsilon = 0.01$. Figure 4 shows the obtained results by varying the parameters α and γ.

It is important to remainder that α represents the learning rate i.e. how much newly acquired information replaces old information ($\alpha = 0$ implies exploiting prior knowledge and $\alpha = 1$ means ignoring prior knowledge and considering the last recent information in order to explore other possibilities). The parameter γ represents the discount factor that determines how the future rewards are important. When γ becomes close to 0 this implies that is important to find a best path to use immediately, but when γ is near to 1, the driver prefers to find the best path even if this path will take more traveling episodes. The results presented in this figure show that the learning approach gives better results than using the shortest path for searching to travel. The results are specially important when $\gamma = 0.9$ in which the proposed approach gives a traveling time always better than the greedy approach. For other values of γ, we can see that our proposition is almost better than greedy solution.

5   Conclusion

In this work, we have proposed a learning approach for traffic optimization in urban environments. The vehicles seeking to reach their destination can have the ability to learn mainly in the purpose to provide the optimal path in terms of travel time, which leads to reduce the total travel time and minimize congestion in transport network. The proposed method is based on a reinforcement learning technique, in particular Q-learning, that is used to learn the best action to take into account according to various traffic situations. The simulation results showed that the proposed Q-learning approach outperformed the greedy algorithm with better performances in terms of transit delay. As further works, we intend to improve the proposed algorithm by considering other use-cases, for example using a dynamic transition delay at the junctions or either exchanging learning data between vehicles to accelerate the process of finding the optimal path.

Acknowledgments. This work was supported by the National Center for Scientific and Technical Research (CNRST) of Morocco, and by Campus France - AAP 2017 (Appel à Projet Recherche au Profit des CEDocs).

References

1. Choudhary, A., Gokhale, S.: Urban real-world driving traffic emissions during interruption and congestion. Transp. Res. Part D Transp. Environ. **43**, 59–70 (2016)
2. Festag, A.: Cooperative intelligent transport systems standards in Europe. IEEE Commun. Mag. **52**(12), 166–172 (2014)
3. Sjoberg, K., Andres, P., Buburuzan, T., Brakemeier, A.: Cooperative intelligent transport systems in Europe: current deployment status and outlook. IEEE Veh. Technol. Mag. **12**(2), 89–97 (2017)
4. IEEE 802.11p - IEEE Standard for Information Technology - Local and Metropolitan Area Networks - Specific Requirements - Part 11: Wireless LAN Medium Access Control (MAC) and Physical Layer (PHY) Specifications Amendment 6: Wireless Access in Vehicular Environments
5. White, C.C.: Markov Decision Processes, pp. 484–486. Springer, US (2001)
6. Sang, K. S., Zhou, B., Yang, P., Yang, Z.: Study of group route optimization for IoT enabled urban transportation network. In: 2017 IEEE International Conference on Internet of Things (iThings) and IEEE Green Computing and Communications (GreenCom) and IEEE Cyber, Physical and Social Computing (CPSCom) and IEEE Smart Data (SmartData), pp. 888–893. IEEE (June 2017)
7. Puterman, M.L.: Markov Decision Processes: Discrete Stochastic Dynamic Programming. Wiley, Hoboken (2014)
8. Sutton, R.S., Barto, A.G.: Reinforcement Learning: An Introduction. MIT Press, Cambridge (1998)
9. Nafi, N.S., Khan, R.H., Khan, J.Y., Gregory, M.: A predictive road traffic management system based on vehicular ad-hoc network. In: 2014 Australasian Telecommunication Networks and Applications Conference (ATNAC), pp. 135–140 (November 2014)
10. Dijkstra, E.W.: A note on two problems in connexion with graphs. Numer. Math. **1**(1), 269–271 (1959)
11. Nahar, S.A.A., Hashim, F.H.: Modelling and analysis of an efficient traffic network using ant colony optimization algorithm. In: 2011 Third International Conference on Computational Intelligence, Communication Systems and Networks, pp. 32–36. IEEE (July 2011)
12. Dorigo, M., Maniezzo, V., Colorni, A.: Ant system: optimization by a colony of cooperating agents. IEEE Trans. Syst. Man Cybern. Part B Cybern. **26**(1), 29–41 (1996)
13. Kammoun, H.M., Kallel, I., Alimi, A.M., Casillas, J.: An adaptive vehicle guidance system instigated from ant colony behavior. In: 2010 IEEE International Conference on Systems, Man and Cybernetics, pp. 2948–2955. IEEE (October 2010)
14. Sutton, R.S., Barto, A.G.: Reinforcement Learning: An Introduction. MIT Press, Cambridge (2018)
15. Kaelbling, L.P., Littman, M.L., Moore, A.W.: Reinforcement learning: a survey. J. Artif. Intell. Res. **4**, 237–285 (1996)
16. François-Lavet, V., Fonteneau, R., Ernst, D.: How to discount deep reinforcement learning: towards new dynamic strategies. arXiv preprint arXiv:1512.02011 (2015)
17. MatlabOTB: MATLAB Optimization Toolbox. 2016 Version 9.0.0.341360. The MathWorks, Natick (2016)

18. Mathworks: What is MATLAB? (2019). https://in.mathworks.com/discovery/what-is-matlab.html. Accessed 01 Oct 2019
19. Bai, F., Sadagopan, N., Helmy, A.: IMPORTANT: a framework to systematically analyze the Impact of Mobility on Performance of RouTing protocols for Adhoc NeTworks. In: INFOCOM 2003. Twenty-Second Annual Joint Conference of the IEEE Computer and Communications. IEEE Societies, vol. 2, pp. 825–835. IEEE (March 2003)

CSI Based Indoor Localization Using Ensemble Neural Networks

Abdallah Sobehy[1], Éric Renault[2(✉)], and Paul Mühlethaler[3]

[1] Samovar, CNRS, Télécom SudParis, University Paris-Saclay,
9 Rue Charles Fourier, 91000 Évry, France
`sobehy@telecom-sudparis.eu`
[2] LIGM, Univ. Gustave Eiffel, CNRS, ESIEE Paris,
77454 Marne-la-Vallée, France
`eric.renault@esiee.fr`
[3] Inria Roquenourt, BP 105, 78153 Le Chesnay Cedex, France
`Paul.Muhlethaler@inria.fr`

Abstract. Indoor localization has attracted much attention due to its many possible applications e.g. autonomous driving, Internet-Of-Things (IOT), and routing, etc. Received Signal Strength Indicator (RSSI) has been used extensively to achieve localization. However, due to its temporal instability, the focus has shifted towards the use of Channel State Information (CSI) aka channel response. In this paper, we propose a deep learning solution for the indoor localization problem using the CSI of an 8×2 Multiple Input Multiple Output (MIMO) antenna. The variation of the magnitude component of the CSI is chosen as the input for a Multi-Layer Perceptron (MLP) neural network. Data augmentation is used to improve the learning process. Finally, various MLP neural networks are constructed using different portions of the training set and different hyperparameters. An ensemble neural network technique is then used to process the predictions of the MLPs in order to enhance the position estimation. Our method is compared with two other deep learning solutions: one that uses the Convolutional Neural Network (CNN) technique, and the other that uses MLP. The proposed method yields higher accuracy than its counterparts, achieving a Mean Square Error of 3.1 cm.

Keywords: Indoor localization · Channel State Information · MIMO · Deep learning · Neural networks

1 Introduction

Localization is the process of determining the position of an entity in a given coordinate system. Knowing the position of devices is essential for many applications: autonomous driving, routing, environmental surveillance, etc. The localization system depends on multiple factors. The environment, whether indoors or outdoors, is one of the most dominant factors. In outdoor environments, the Global Positioning System (GPS) is widely used to localize nodes. In [1], the

© IFIP International Federation for Information Processing 2020
Published by Springer Nature Switzerland AG 2020
S. Boumerdassi et al. (Eds.): MLN 2019, LNCS 12081, pp. 367–378, 2020.
https://doi.org/10.1007/978-3-030-45778-5_25

authors present a location service which targets nodes in a Mobile Ad-hoc Network (MANET). The aim is to use the GPS position information obtained from each device and disseminate this information to other nodes in the network while avoiding network congestion. Each node broadcasts its position with higher frequency to nearby nodes and lower frequency to more distant nodes. The notion of closeness is determined by the number of hops. In this way, nodes have more updated position information of nearby nodes that is sufficient for routing applications, for instance. While the GPS satisfies the requirements of many outdoor applications, it is not functional in indoor environments. Consequently, in such scenarios, other measurements have to be exploited to overcome the absence of GPS.

One family of localization methods is known as range-based localization. In this method, a physical phenomenon is used to estimate the distance between nodes. Then, the relative positions of nodes within a network can be computed geometrically [2]. One of the most used phenomena is the Received Signal Strength Indicator (RSSI). RSSI is an indication of the received signal power. It is mainly used to compute the distance between a transmitter and a receiver since the signal strength decreases as the distance increases. In [3], the distances between nodes along with the position information of a subset of nodes, known as the anchor nodes, are used to locate other nodes in a MANET. This is achieved using a variant of the geometric triangulation method. The upside of RSSI is that it does not need extra hardware to be computed and is readily available. Another physical measure to compute the distance between devices is the Time-Of-Arrival (TOA) or Time-Difference-Of-Arrival (TDOA). Here, the time taken by the signal to reach the receiver is used to estimate the distance between devices. Using TOA in localization proves to be more accurate than RSSI, but requires external hardware to synchronize nodes [4]. In the case where only the distance information is available, a minimum of three anchor nodes with previously known positions are needed to localize other nodes with unknown positions. In order to relax this constraint, the Angle-Of-Arrival (AOA) information can be used in addition to the distance. Knowing the angle makes it possible to localize nodes with only one anchor node [5]. However, the infrastructure needed to compute AOA is more expensive than TOA both in terms of energy and cost.

RSSI has been extensively used in indoor localization [6]. However, it exhibits weak temporal stability due to its sensitivity to multi-path fading and environmental changes [6]. This leads to relatively high errors in distance estimation which, in turn, deteriorates the accuracy of position estimation. With the data rate requirements of the 5G reaching up to 10 Gbps, the communication trend is switching to the use of MIMO antennas where signals are sent from multiple antennas simultaneously [7]. Furthermore, with orthogonal frequency-division multiplexing (OFDM), each antenna receives multiple signals on adjacent subcarriers. This introduces the possibility of computing a finer-grained physical phenomenon at the receiver, which is known as Channel State Information (CSI). In other words, as opposed to getting one value per transmission with RSSI, with CSI, it is possible to estimate CSI values which are equal to the number of antennas multiplied by the number of subcarriers. CSI represents the change that occurs to the signal as it passes through the channel between the transmitter and the receiver e.g. fading, scattering and power loss [8]. Equation (1) specifies the relation between the transmitted signal $T_{i,j}$ and the received signal $R_{i,j}$ at the ith antenna and the jth subcarrier. The transmitted signal is affected by both the channel through the $CSI_{i,j}$, which is a complex number, and the noise N.

$$R_{i,j} = T_{i,j} \cdot CSI_{i,j} + N \tag{1}$$

In Sect. 2, we present state-of-the-art solutions that use CSI to achieve Indoor Localization. In Sect. 3, the building blocks of the proposed solution are introduced. First, the choice of magnitude component and the data preprocessing steps are briefly explained. Second, the data augmentation step is presented, followed by the ensemble neural network technique. The localization accuracy of our solution is compared with two state-of-the-art solutions in Sect. 4. Finally, the conclusion and future work are discussed in Sect. 5.

2   Related Work

One of the very first attempts to use CSI for indoor localization is FILA [9]. With 30 subcarriers, the authors compute an effective CSI which represents the 30 CSI values at each of the subcarriers. Then, they present a parametric equation that relates the distance to the effective CSI. The parameters of the equation are deduced using supervised learning. Finally, using a simple triangulation technique, the position is estimated. In [10], the authors carry out an experiment where a robot carrying a transmitter traverses a 4×2 meter table and communicates with an 8×2 MIMO antenna. The transmission frequency is 1.25 GHz and the bandwidth is 20 MHz. Signals are received at each of the 16 subantennas over 1024 subcarriers from which 10% are used as guard bands. Using a convolutional neural network (CNN), the authors use the real and imaginary components of the CSI as an input to the learning model to estimate the position of the robot. The authors publish the CSI and the corresponding positions ($\approx$17,000 samples) readings which are used as a test bed for our algorithm. Therefore, the comparison with their results is fair since both algorithms process the same data. Figure 1 shows the experimental setup as well as a sketch of the MIMO antenna and the position of its center. The lower part of the figure shows the table which is traversed through the experiment and the MIMO antenna. The upper part shows a sketch of the MIMO antenna showing its center at (3.5, −3.15, 1.8) m in the local coordinate system. The distance between adjacent antennas is $lambda/2$ which is computed from the carrier frequency.

Another solution that is tested on the same data set is NDR [11] which is based on the magnitude component of the CSI. First, the magnitude values are preprocessed by fitting a line through the points. Then a reduced number of magnitude points are chosen on the fitted line to represent the whole spectrum of the CSI. By achieving this, both the dimensionality of the input and the noise are reduced. Since the proposed solution is based on a similar approach, we will provide a brief explanation of the preprocessing step.

Fig. 1. Experimental setup (bottom) and a sketch of the MIMO antenna (top) [10]

3   Methodology

3.1   Input of the Learning Model

Since the CSI is a complex number, it can be represented in both polar and Cartesian forms. Thus, there are a total of four components to represent the CSI; real, imaginary, magnitude, and phase. Equations (2) and (3) show the conversion from one form to another.

$$CSI_{i,j} = Re + iIm \tag{2}$$

$$Mag = \sqrt{Re^2 + Im^2}$$
$$\phi = \arctan(Re, Im) \tag{3}$$

A good input feature is one that is stable for the same output. In other words, if the transmission occurs multiple times from the same position, the feature values are expected to be similar. In order to examine the behaviour of the components, the four components are plotted for four different transmissions from the same position. Figure 2 shows the CSI components of four different transmissions from the same position. The example shown is for antenna 0 at position (3.9, −0.44, −0.53) m. Each sub-figure shows the four CSI components of one transmission. It should be noted that the phase values are all scaled to be in the same order as the other components. With careful inspection, it can be noticed that the magnitude component shows the highest stability. This conclusion is further supported by the analysis performed in [11,12]. Consequently, we chose the magnitude to be the input component to the deep learning model.

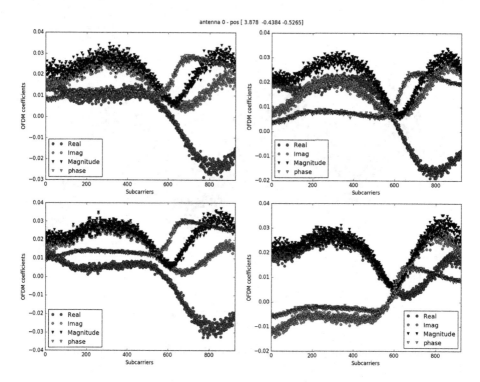

Fig. 2. Real, Imaginary, Magnitude and Phase components estimated from 4 transmissions from the same position.

3.2 Data Preprocessing

After choosing the magnitude as the input feature, the number of input values to estimate one position is 924×16. As seen in Fig. 2, the magnitude points appear to follow a continuous line with noise scattering points around this line. The first step is to retrieve a line that passes through the magnitude values. Since this process has to be performed for each of the 15k training samples multiplied by the 16 antennas, it has to be relatively fast. This process is achieved by polynomial fitting on four sections over the subcarrier spectrum [11]. Figure 3 shows the four batches, each in a different color, and the degree of the polynomial used to fit the line. The polynomial degree is chosen by attempting several values and choosing the degree that yields the highest accuracy. Dividing the spectrum into four sections increases the accuracy of the overall fitting. More details of the fitting process can be found in [11].

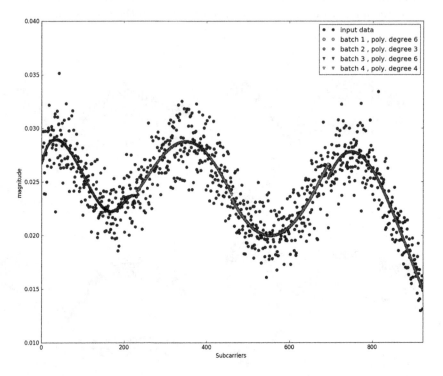

Fig. 3. Fitting a line through the magnitude values. (Color figure online)

The following step is to use a reduced number of points along the fitted lines instead of the whole set of 924 points. This mitigates the noise that leads to the scattering around the line. In addition, the reduction allows us to build a more complex MLP that can be trained in less time. In [11], 66 points are used to represent the whole spectrum. We chose the input feature to be the difference (slope) between two consecutive points. For the 66 points, there are 65 slope values. Using the same MLP structure as in [11], the mean square error is reduced from 4.5 cm to 4.2 cm over a 10-fold cross validation experiment. Even though this is a slight improvement of around 7%, it shows that the absolute value of the magnitude is not the decisive factor to determine the position, but rather the variation of the magnitude along the spectrum.

3.3 Data Augmentation

Data augmentation is a method to increase the training set which is the main driver of the learning process. The artificially created training samples are constructed from the available samples with some mutation. In an image recognition context; blurring, rotating, zooming in or out are all ways to generate a new training sample from an existing one.

In our case, a training sample is composed of a set of 924×16 magnitude points and the corresponding position. In our solution, we propose the mutation to the input and the output sides of the training sample. As for the output position, we know the measuring error of the tachymeter used to calculate the given position, which is around 1 cm [10]. We model this error by a Gaussian distribution with zero mean, which is the given position, and a standard deviation which is 1/3 cm. Thus, the position of the augmented sample is computed using this distribution. As for the magnitude points, first, a line is fitted through the points using the previously mentioned method. Next, the standard deviation of the absolute error between the line and the values is computed. The augmented sample is then calculated by scattering the points around the fitted line with a Gaussian distribution with zero mean and twice the computed standard deviation. This can be seen as an equivalent to the blurring in the image recognition context. Figure 4 shows an example of an augmented magnitude sample in red from an original training sample in blue using a fitted line in black.

In order to test the effect of data augmentation on localization accuracy, an MLP neural network is constructed with the hyperparameters listed in Table 1. Training the MLP is then executed with different percentages of augmented data. Figure 5 shows the effect of the number of augmented data samples on the mean square error of the position estimation.

It is worth mentioning that the larger the augmented data set, the better the localization accuracy. The mean square error is reduced from 8 cm to 6.7 cm using an augmented sample from each training sample.

3.4 Ensemble Neural Networks

The last part of our algorithm is to construct several neural networks with different characteristics. The difference between MLPs can be in the hyperparameters or the samples used to train the model. For instance, the neural networks used to plot Fig. 5 are different since they are trained on different training sample sizes. Also, in k-fold cross validation, neural networks are trained on the same data size but on different samples. Moreover, changing any of the hyperparameters shown in Table 1 leads to different results.

Mixing the prediction of each neural network with different characteristics can lead to a significant increase in accuracy. We examine different ways to mix the prediction results of the MLPs:

1. Mean: The simplest way to mix the results is to compute the arithmetic mean position of all the predictions.

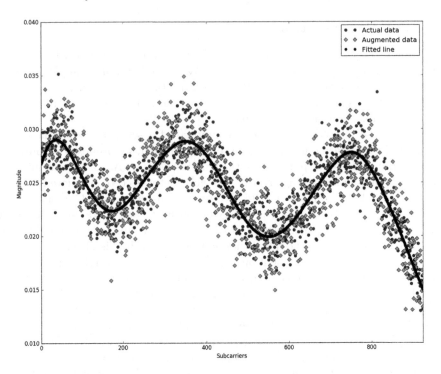

Fig. 4. Data augmentaion sample from an original training sample. (Color figure online)

2. Weighted mean: Each of the MLPs is given a weight that is proportional to its individual localization accuracy. Thus, the higher the accuracy, the higher the weight. Then the final prediction is a weighted average of the individual prediction.
3. Weighted power mean: The effect of weights is further magnified by raising them to a certain power before computing the weighted average.
4. Median: The idea is to pick one of the predictions that is closest to all other predictions. This makes sense when the ensemble has three or more MLPs. This mitigates the effect of the large errors of some predictions.
5. Random: The final prediction is a randomly selected individual prediction.
6. Best pick: This is used as an indication of the best possible result one can attain with the given ensemble. The final prediction is the closest individual prediction to the actual position. This is not feasible since in normal cases the actual position is not given.

Figure 6 shows the effect of adding one or more neural networks to the ensemble. The x-axis represents the number of neural networks in the ensemble. Beside the number of the NNs, there is a number between brackets representing the mean square error of the added neural network. This means that the first MLP has a Mean Square Error (MSE) of 3.9 cm. This is the best individual MLP that

Table 1. Hyperparameters selection.

Hyperparameter	Value
Number of layers	5
Units per layer	512
Epochs	100
Activation function	relu
Learning rate	0.0005
Optimizer	Adam
L2 regularization	Without L2
Dropout percentage	0%
Batch sizes	[128, 256, 512, 1024, 2096]

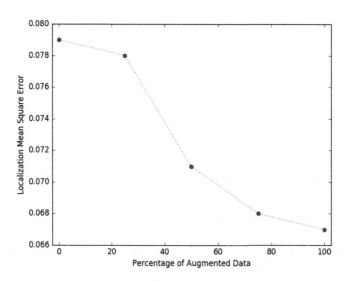

Fig. 5. Effect of data augmentation on localization accuracy.

was constructed using the data augmentation technique. The y-axis shows the MSE for each type of prediction mixing. One of the types in Fig. 6 is labeled "median + wght" meaning the average prediction of both mixing methods. It can be seen that with different mixing techniques, except for the random pick, the estimation accuracy can be improved even if the added MLP has a higher individual error. The best accuracy achieved is 3.1 cm MSE using 11 MLPs. This result outperforms [10] which uses a CNN learning from real and imaginary components and achieves an error of 32 cm. It also outperforms [11] which uses an MLP learning from the magnitude values and achieves an error of 4.5 cm. It is to be noted that the accuracy can be further increased if there is a way to select the best individual estimation of the MLPs ensemble.

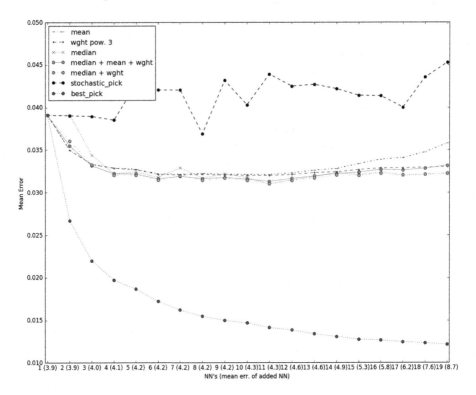

Fig. 6. Mixing the predictions of a neural network ensemble.

4   Experimental Results

In this section, we compare the localization accuracy of the proposed Ensemble NN method based on the variation of magnitude and data augmentation with the NDR [11] and CNN [10] methods. NDR is an MLP where the input is the magnitude values, and the hyperparameters are chosen emperically to get the lowest mean square error estimation. CNN is a convolutional network where the input is the real and imaginary components of the CSI. The estimation results in NDR and CNN are presented while varying the number of antennas used (Fig. 7).

As expected, a lower number of antennas is used the estimation error is high. In all solutions, the estimation improves with more data provided from the added antennas. The proposed Ensemble NN technique outperforms NDR and CNN. The error of CNN is much higher than NDR and Ensemble NN, probably due to the high temporal instability of the real and imaginary CSI components. While the error difference between Ensemble NN and NDR methods seems small, the improvement is relatively significant. When using 16 antennas, NDR achieves an MSE of 4.5 cm while the proposed Ensemble technique achieves 3.1 cm which is an improvement of ≈30%.

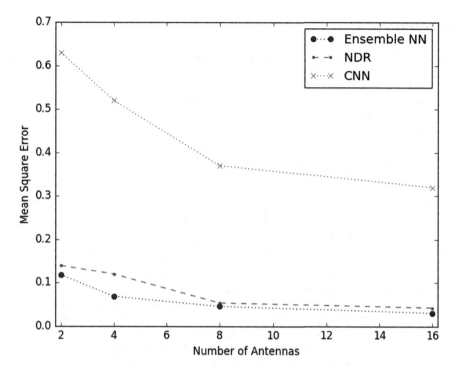

Fig. 7. Comparing Ensemble NN with NDR [11] and CNN [10]

5   Conclusion

In this work, we propose a deep learning solution for the indoor localization problem based on the CSI of a 2×8 MIMO antenna. The variation of the magnitude component is chosen to be the input feature for the learning model. Using the magnitude variation instead of the absolute values improves slightly the estimation. This shows that the focus should be on better describing the change in magnitude along the subcarrier spectrum rather than the absolute values. Data augmentation is then used to further increase the estimation accuracy. Finally, an ensemble neural network technique is presented to mix results of different MLPs and achieves an accuracy of 3.1 cm, outperforming two state-of-the-art solutions [10,11]. This work can be improved through the detection and correction of outliers as some of the errors are much larger than the mean error. The possibility of using another learning layer to detect outliers or select the best individual MLP estimation from the ensemble might enhance the estimation accuracy.

References

1. Renault, E., Amar, E., Costantini, H., Boumerdassi, S.: Semi-flooding location service. In: 2010 IEEE 72nd Vehicular Technology Conference Fall (VTC 2010-Fall), pp. 1–5. IEEE (2010)
2. Čapkun, S., Hamdi, M., Hubaux, J.-P.: GPS-free positioning in mobile ad hoc networks. Cluster Comput. 5(2), 157–167 (2002)
3. Sobehy, A., Renault, E., Muhlethaler, P.: Position certainty propagation: a localization service for ad-hoc networks. Computers 8(1), 6 (2019)
4. Nandakumar, R., Chintalapudi, K.K., Padmanabhan, V.N.: Centaur: locating devices in an office environment. In: Proceedings of the 18th Annual International Conference on Mobile Computing and Networking, pp. 281–292. ACM (2012)
5. Cidronali, A., Maddio, S., Giorgetti, G., Manes, G.: Analysis and performance of a smart antenna for 2.45-GHz single-anchor indoor positioning. IEEE Trans. Microw. Theory Tech. 58(1), 21–31 (2009)
6. Yang, Z., Zhou, Z., Liu, Y.: From RSSI to CSI: indoor localization via channel response. ACM Comput. Surv. (CSUR) 46(2), 25 (2013)
7. Jungnickel, V., et al.: The role of small cells, coordinated multipoint, and massive MIMO in 5G. IEEE Commun. Mag. 52(5), 44–51 (2014)
8. He, S., Gary Chan, S.-H.: Wi-Fi fingerprint-based indoor positioning: recent advances and comparisons. IEEE Commun. Surv. Tutor. 18(1), 466–490 (2015)
9. Wu, K., Xiao, J., Yi, Y., Gao, M., Ni, L.M.: FILA: fine-grained indoor localization. In: 2012 Proceedings IEEE INFOCOM, pp. 2210–2218. IEEE (2012)
10. Arnold, M., Hoydis, J., ten Brink, S.: Novel massive MIMO channel sounding data applied to deep learning-based indoor positioning. In: 12th International ITG Conference on Systems, Communications and Coding (SCC 2019), pp. 1–6. VDE (2019)
11. Sobehy, A., Renault, E., Muhlethaler, P.: NDR: noise and dimensionality reduction of CSI for indoor positioning using deep learning. In: GlobeCom, Hawaii, United States, Dec 2019. (hal-023149)
12. Wang, X., Gao, L., Mao, S., Pandey, S.: DeepFi: deep learning for indoor fingerprinting using channel state information. In: 2015 IEEE Wireless Communications and Networking Conference (WCNC), pp. 1666–1671. IEEE (2015)

Bayesian Classifiers in Intrusion Detection Systems

Mardini-Bovea Johan[1,2], De-La-Hoz-Franco Emiro[3(✉)], Molina-Estren Diego[3],
Paola Ariza-Colpas[3], Ortíz Andrés[4], Ortega Julio[5], César A. R. Cárdenas[6],
and Carlos Collazos-Morales[6]

[1] Universidad de la Costa, Barranquilla, Colombia
[2] Universidad del Atlántico, Barranquilla, Colombia
[3] Computer Science and Electronics Department, Research Group Software
Engineering and Networks, Universidad de la Costa, Barranquilla, Colombia
edelahoz@cuc.edu.co
[4] Communications Engineering Department, University of Málaga, Málaga, Spain
[5] Computer Architecture and Technology Department, CITIC,
University of Granada, Granada, Spain
[6] Vicerrectoría de Investigaciones, Universidad Manuela Beltrán, Bogotá, Colombia

Abstract. To be able to identify computer attacks, detection systems
that are based on faults are not dependent on data base upgrades unlike
the ones based on misuse. The first type of systems mentioned generate a knowledge pattern from which the usual and unusual traffic is
distinguished. Within computer networks, different classification traffic
techniques have been implemented in intruder detection systems based
on abnormalities. These try to improve the measurement that assess the
performance quality of classifiers and reduce computational cost. In this
research work, a comparative analysis of the obtained results is carried
out after implementing different selection techniques such as Info.Gain,
Gain ratio and Relief as well as Bayesian (Naïve Bayes and Bayesians
Networks). Hence, 97.6% of right answers were got with 13 features.
Likewise, through the implementation of both load balanced methods
and attributes normalization and choice, it was also possible to diminish the number of features used in the ID classification process. Also, a
reduced computational expense was achieved.

Keywords: Naïve Bayes · Bayesian networks · Feature selection · IDS

1 Introduction

Cyber-attacks keep being a big problem in the current productive context. They
can lead to the loss of sensitive information employed to make decisions within
organizations. Thus, the necessity to develop tools to mitigate vulnerabilities in
computer environments comes up. Several systems protect from malware data
have emerged. However, when the database is not updated frequently, these

© IFIP International Federation for Information Processing 2020
Published by Springer Nature Switzerland AG 2020
S. Boumerdassi et al. (Eds.): MLN 2019, LNCS 12081, pp. 379–391, 2020.
https://doi.org/10.1007/978-3-030-45778-5_26

systems may not be fully effective. As new attacks are created, inadequate management of vulnerabilities may generate catastrophic situations. Therefore, the detection of fraudulent actions has become one of the research priorities of information security. For this reason, algorithms have been integrated to many Intrusion Detection Systems - IDS based on data mining techniques for identification of anomalous traffic. [11–14,16,17]. The CERT (Information Security Incident Response Center) [12], has analyzed and classified in its database over 10,000 viruses. The viruses identified show that some remain over time and others evolve and adapt to new operating systems expanding their actions. It is feasible to provide users with more optimal tools for protection of different threats that may arise. INTECO has documented an important number of viruses that affect the operating systems of mobile devices, PCs and servers.

Some simulation environments have been developed which allow the design and implementation of new IDS Intrusion Detection Systems based on intelligent techniques [14,17]. The proposed models are evaluated through different experimental works in which quality measures are analyzed to be then implemented in productive environments, [3,8]. Some of the techniques considered are Naïve Bayes, J48 and PART classifiers and Chi Square selection techniques and Consistency [15], the IBK classifier and the combination of Symmetric and Gain ratio selection techniques [19], assembled vector support classifiers and non-linear projection techniques [7], Bayesian authorizing maps [3], Hybridization of statistical techniques and SOM performing feature selection with PCA + FDR [19], a wrapper-based method, applied using a multi-objective approach and using the GHSOM classifier [19]. This work focuses on the bi-class classification processes because of their relevance in real application situations where possible attacks are sought. In addition, it would be required to take corrective actions against the anomalous behavior that has been identified. Selection techniques have been applied based on information filtering: Info.Gain [3], Gain ratio [1] and Relief [10]. The main purpose is to identify the attributes that contribute the most to the classification process. Then, an appropriate selection technique is identified and applied [6,9]. A comparative analysis of the quality metrics generated is made.

2   Pre-processing

The simulation process used to validate the detection rates of a classifier implies the execution of a series of phases: pre-processing, selection, training/classification and evaluation of the performance of the classifier. The pre-processing phase involves the use of a data set from which the data to be analyzed comes from. In this type of research, the DARPA NSL-KDD data set has been used as it is widely supported by the scientific community that evaluates related studies. According to [13], there are some improvements that NSL-KDD has over its predecessors. The fact that it does not include redundant records in the data collection for training. There are no duplicated records in the data collections proposed for the tests. The number of records selected from each group of difficulty level is inversely proportional to the percentage of records in the original set of KDD data. Also, the number of records in the data collections for training and testing is reasonable. In the different simulation contexts described later, the NSL-KDD data set created from the KDD'99 [11] was used. The size of the NSL-KDD is smaller than the one of KDD'99 because the records of redundant connections have been eliminated. The NSL-KDD is made up of the KDDTrain+, KDDTrain+20Percent, KDDTest+ and KDDTest-21 files which are in both TXT and ARFF formats.

To be able to adjust the NSL-KDD set, techniques such as pre-processing, load balancing and normalization were applied. The load balancing is intended to level the number of normal connections and the number of attacks to avoid bias. Classifiers that are trained with unbalanced data sets, tend to classify data instances as part of the main class and ignoring the low representation of the minority class. Table 1 shows the amount of both the normal connections and the connections that represent attacks. They are contained in the NSL-KDD. 53.46% of the connections are normal exceeding by 6.92% the connections corresponding to the attacks. Thus, a load balancing technique called Synthetic Minority Oversampling Technique - SMOTE [12] was implemented. According to [18], this technique is responsible for adding random information to the training process of the data set generating new data instances. In this research work, SMOTE gives new instances of the "attack" data class by 14.86% of the current ones in the training data set NSL-KDD. Each new instance is computed from the average of the five closest neighbors and with a seed set to one.

Table 1. Connection distribution for NSL-KDD train

Training		
Connections	Qty	%
Normal	67.343	53.46
Attacks	58.630	46.54
Total	**125.973**	**100.00**

Regarding standardization, 41 features of the NSL-KDD data set are used in the different classification techniques. Therefore, the variables scale is very important to determine the topological organization of the structures used by these techniques. If the range of values of a variable is greater than the others, it will probably dominate the organization of the classifier structure. Normalization prevents one characteristic from contributing more than another to the measurement of distances. In [5] six standardization methods are presented and have been evaluated in this proposal. According to the results acquired, it has been demonstrated that the technique with the best performance is the one called normalization at zero mean and unit variance. The continuous variables are normalized with zero mean and unit variance by using Eq. 1. On the other hand, all the variables are scaled at the interval [0...1]. The symbolic features and the binary ones are not normalized. The normalization technique employed is a simple linear transformation as shown in the following equation:

$$\hat{x} = \frac{x - \underline{x}}{\sigma} \tag{1}$$

Where x and $\underline{x}$ are the mean and the standard deviation of the variable x. This is equivalent to express the x variable as the distance between the number of deviations and its mean. After refining the data set through the pre-processing techniques mentioned previously, the different features selection methods are evaluated. The purpose is to reduce the complexity of the appraising process which will be then executed by the classifier.

3   Feature Selection Techniques

The feature selection phase is essential for an efficient analysis of the data contained in the data set since this usually contains information that adds noise to the generation process of the model. Because of this issue, there is some degradation of the quality of the patterns to be detected. The redundant variables and the irrelevant ones make it difficult to get significant patterns from the data. In [15], it is stated that the ability to use a feature selection is essential to perform an effective analysis because the data contains information that is not necessary for the generation of the model. It is affirmed in [19] that the features selection allows to reduce the entries of the data to an appropriate size for processing and analysis. Therefore, attributes or features must be selected or discarded depending on their usefulness for the analysis. Every selection process of attributes has a starting point, which can be the complete set of attributes, the empty set or any intermediate state. After the first subset is evaluated, other subsets will be examined based on a search direction that can be forward, backward, random or any variation of the above. The process will finish when the entire space is covered or when a stop condition is fulfilled depending on the search strategy followed. There are other methods of attribute selection which are based on the transformation of input values providing information related to: how relevant is each variable as a whole?

It is possible to discard the ones that are irrelevant or those that are below a certain threshold of relevance. According to [8], filtering-based selection techniques are used to find the best subset of features of the original set. The filtering methods seem to be optimal for the selection of a subset of data. These do not depend on the classification algorithm and the computational cost is lower for large data sets. The wrapper-based choice features techniques (wrappers) also defined in [3], use the prediction capability of the classification algorithm to select the optimal subset of features. In this study the filtering-based selection techniques known as Info.Gain [3], Gain ratio [7] and Relief [4] have been used. In the references review, it was noticed that promising results can be got when applied to detect faults. During the experimental works carried out, Bayesian classifiers and networks were utilized to analyze the performance measurements obtained from the proposed models [1,3]. The following is a detailed description of both the features selection techniques Info.Gain, Gain ratio, Relief and the classification methods Naïve Bayes, Bayesian networks which are based upon the suggested model.

3.1 Info.Gain

As presented in [2], Info.Gain is a filter-based features choice method. It is also known as information gain and is used to identify the importance of the features of a data set collection. The attribute with the largest gain is chosen as the division feature for the umpteenth node. This trait minimizes the required information to classify the couples in the resulting allocation. It reflects very small defects among these partitions.

3.2 Gain Ratio

As studied in [7], Gain ratio belongs to the category of filtering-based traits selection techniques which is applied to analyze the features of big size data sets. When there are many different values, the gain information relationship is used to consider these features. This approach is widely applied due to the good results that can be obtained. Additionally, these results can be employed during the classification phase. Its main distinctive feature is the modification of the information gain that reduces the error. Gain ratio considers the number and size of branches to choose from a characteristic.

3.3 Relief

This is an algorithm that determines significant features and allow to easily distinguish between instances of diverse classes [4]. Based on this approach, it defines the weight for each feature. However, the Relief genuine version limits its application field to two-class problems. Hence, for weight allocation purposes just the closest neighbor of a different class is utilized.

4 Classification and Training Techniques

At this stage, firstly, the classifier is trained. This process is done from the learning algorithm chosen and by using the normalized data set which is reduced to the most important features. Hence, an efficient learning is created. Once the training is performed, the classifier determines normal traffic and attacks through a subsequent classification of every connection within the data set. Then, the quality measures are computed to assess the classification technique performance. Bayesian classifiers were used. These are based on the Bayes theorem [3].

$$p(A|B) = \frac{p(A, B)}{p(B)} = \frac{p(A)\, p(B|A)}{p(B)} = \frac{p(A)\, p(B|A)}{\Sigma A' p(A')\, p(B|A')} \qquad (2)$$

Being A and B two random events whose possibilities are denoted as $p(A)$ and $p(B)$ respectively and taking $p(B) > 0$. The A and B event possibilities previously known are supposed to be true. Likewise, the probability is subjected to event B to be true assuming that A is too $p(B|A)$. Finally, $p(A|B)$ is the possibility of A to be true considering that B is also true.

4.1 Naïve Bayes

As stated in [3], Naïve Bayes is a descriptive and predictive classification technique based upon the probability theory which comes from the Byes theorem analysis. This theory suggests both an infinite sample and an independent statistics among variables. In this case referring to the characteristics not to the class. Under these conditions, the probability distributions of each class can be calculated to establish the relationship between the traits and the class. If $x = (x_1, ..., x_n)$ is given, where x_i is the observed value for the umpteenth feature. Hence, the possibility for a class ym with k possible values $(y1, ..., yk)$ to occur, results from the Bayes rules as shown in Eq. 3.

$$P(y_m|x_1, x_2, \ldots, x_n) = \frac{p(y_m) \prod_{i=1}^{n} p(x_i|y_i)}{p(x_1, x_2, \ldots, x_n)} \tag{3}$$

In the above equation, $p(ym)$ is the class proportion of ym in the data set. Also, $p(x_i|y_i)$ is computed from the examples amount with a x_i value whose class is ym. Therefore, it can be inferred that to compute $p(x_i|y_i)$ makes the x_i values be discrete. So if there is some continuous feature, it should be discretized in advanced. The assorting of a new class ""x"" is done by calculating the conditioned possibilities of each class and choosing the best option. If $Y = (y1, y2, ..., yk)$ is the current class data sets, it will be sorted with the class that satisfies Eq. 4.

$$\forall i \neq j, \quad P(y_i|x_1, x_2, \ldots, x_n) > P(y_j|x_1, x_2, \ldots, x_n) \tag{4}$$

Although the Bayesian classifier is a fast and simple method, it is required to go all over the training set to compute $P(y_m|x_1, x_2, ..., x_n)$. This calculation is unfeasible for a large number of examples. So, a simplification is needed. Therefore, the conditional independence hypothesis is considered for decomposing purposes of the probability.

4.2 Bayesians Networks

As stated in [1], a Bayesian network is a defined, directed and labeled acyclic graph, which describes the joint probability distribution that governs a set of random variables. Let $X = x1, x2, ..., xn$ be a set of random variables, a Bayesian Network for X is a pair $B =< G, T >$ in which:

- G is a directed acyclic graph in which every node represents a variable $x_1, x_2, ..., x_n$ and every arc symbolizes direct dependence relationships among the variables. The arcs direction shows that the variable pointed by the arc depends on the variable placed at its origin.

- T is a parameters set to quantify the network. It contains the probabilities $PB(xi|Xi)$ for each possible x_i value of each variable X_i and each possible value of n which denotes the parents set of X_i in G. Hence, a Bayesian network B defines a joint probability distribution over X as given in [1] and as indicated in Eq. 5.

$$P_B\left(x_1, x_2, \ldots, x_n\right) = \prod_{i=1}^{n} P\left(X_i \Big| \prod x_i\right) \tag{5}$$

5   Methodology

The proposal in this work was initially to take the NSL-KDD (raw data) data set and to apply the pre-processing techniques: load balancing by data instances (through the implementation of Synthetic Minority Oversampling Technique - SMOTE) and normalization (applying standardization to zero average and unit variance). When the purified data is obtained a series of features selection techniques are applied to identify the attributes that affect the performance of the classifier. After filtering data, two Bayesian classification techniques are employed. The test process was performed through cross-validation using 10-folds. The results got from it were represented in the respective confusion matrices allowing the calculation of the quality metrics of each of the experimental scenarios. From this, the techniques (deselection and classification) which provide the best results were identified (Fig. 1).

6   Simulations and Results

Two sets of experimental tests are involved in the development of this research. For the first set of tests, the Naïve Bayes classifier is used to change the features selection techniques (Info.Gain, Gain ratio and Relief). Once the corresponding feature selection technique is applied, the priority order of the attributes can be identified. Based on this, a series of experimentation scenarios are simulated in which the number of attributes for each of the selection techniques implemented are varied. See Table 2 and Fig. 2(a), (c) and (e). For the second set of tests the Bayesian networks classifier is applied, and the features selection techniques are also varied. After identifying the priority order of the attributes, the experimental scenarios are carried out. In these simulations the traits number are modified for each of the choice techniques implemented. (See Table 3 and Fig. 2(b), (d) and (f)). For the set of tests carried out with the Naïve Bayes classifier, the best results have been obtained by using the selection technique of Relief features with 20 attributes. An accuracy of 91.27% was reached as shown in Table 2. For the tests carried out with Bayesian Networks, the best results are obtained with Gain ratio using only 13 attributes with a success rate of 97.56% (See Table 3). The most significant simulation scenarios are provided in Table 3. The compilation described does not intend to indicate that Bayesian Network + Gain

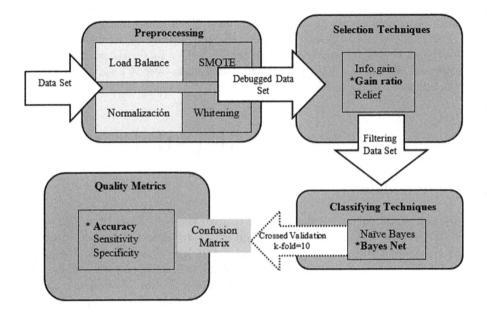

Fig. 1. Suggested methodology

ratio is the best solution. The goal is to give a performance perspective of the proposed procedure compared to the results provided by previous studies.

The methods shown can be classified into: (1) Methods that do not use features selection, (2) filtering-based methods and (3) wrapping-based methods. All the experimental work is performed by using a MacBook Air mid 2015 with an Intel processor, 1.6 Ghz and an 8 gb RAM DDR3 at 1600 Mhz. Each experiment is completed 10 times. Thanks to this, metrics values are obtained, and it allows to evaluate the quality of the processes. See Tables 2 and 3 in which each quality metric by technique of selection of features and by classification method is shown with their respective standard deviation. In the classifying process of both sets of tests, the crossed to 10 folds validation technique is used. It is applied to the NSL KDD data set of training. Simulations that allow to evaluate the network traffic with a behavior like real computer attacks are generated. When each experimental scenario is solved, the evaluation of the proposed functional models is performed. Hence, the metrics of accuracy, sensitivity and specificity can be computed.

7   Related Works

In [15] an analysis of the features selection techniques for a data set of network traffic like the one proposed here was done. The Naïve Bayes, J48 and PART classifiers were utilized. The performance of each of these classifiers was assessed with the entire data set NSL-KDD and with subsets of data identified from

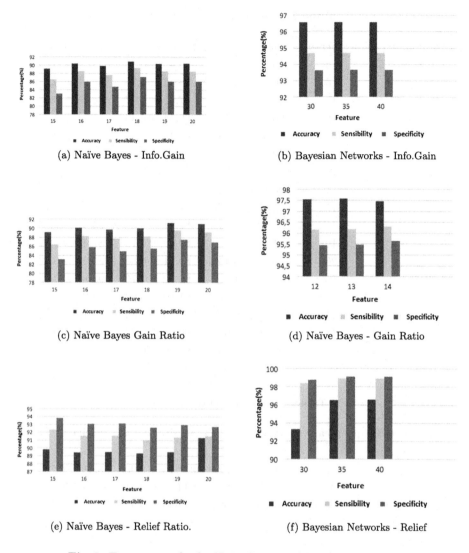

(a) Naïve Bayes - Info.Gain

(b) Bayesian Networks - Info.Gain

(c) Naïve Bayes Gain Ratio

(d) Naïve Bayes - Gain Ratio

(e) Naïve Bayes - Relief Ratio.

(f) Bayesian Networks - Relief

Fig. 2. Tests set results for Naïve Bayes and Bayesian networks

the application of different features selection techniques. The best results were obtained from the PART classifier (97.57% of accuracy). The techniques Chi Square (30 features) and Consistency (14 features) were individually applied. The results acquired were very similar to the ones got with Bayesian network + Gain ratio (97.56%) and just 13 attributes were used. See Table 3. Moreover, in this study, tests with Naïve Bayes and Gain ratio feature selection techniques are also performed (89.03%) and Info.Gain (93.49%). Both tests with 30 features. In the scenarios set up for this research with Naïve Bayes experimentation + Gain ratio + 19 features, a correct rate is obtained 91.22%. Naïve Bayes + Info. Gain

Table 2. Naïve Bayes classifier tests

Selection technique	Features	Percentage (%)	Sensitivity (%)	Specificity (%)
Info.Gain	15	89.14 ± 0.35	86.47 ± 0.35	83.070 ± 0.43
	16	90.41 ± 0.35	88.46 ± 0.35	85.91 ± 0.42
	17	89.80 ± 0.37	87.61 ± 0.37	84.74 ± 0.43
	18	90.86 ± 0.41	89.34 ± 0.43	87.13 ± 0.32
	19	90.30 ± 0.35	88.48 ± 0.37	85.97 ± 0.37
	20	90.35 ± 0.43	88.54 ± 0.43	86.05 ± 0.47
Gain ratio	15	89.07 ± 0.36	86.460 ± 0.43	83.090 ± 0.82
	16	90.09 ± 0.52	88.35 ± 0.65	85.84 ± 0.4
	17	89.67 ± 0.47	87.72 ± 0.48	84.96 ± 0.77
	18	90.03 ± 0.39	88.17 ± 0.47	85.57 ± 0.78
	19	91.22 ± 0.41	89.63 ± 0.73	87.48 ± 0.42
	20	90.95 ± 0.43	89.21 ± 0.81	86.92 ± 0.49
Relief	15	89.8 ± 1.1	92.32 ± 0.76	93.81 ± 0.78
	16	89.45 ± 0.97	91.50 ± 0.93	93.08 ± 0.87
	17	89.49 ± 0.81	91.56 ± 0.84	93.14 ± 0.76
	18	89.30 ± 0.93	90.99 ± 0.74	92.61 ± 0.84
	19	89.47 ± 0.75	91.36 ± 0.59	92.94 ± 0.67
	20	91.27 ± 0.82	91.50 ± 0.53	92.74 ± 0.71

Table 3. Bayesian networks tests

Selection technique	Features	Percentage (%)	Sensitivity (%)	Specificity (%)
Info.Gain	–	–	–	–
	30	96.55 ± 1.02	94.670 ± 0.93	93.61 ± 0.73
	35	96.56 ± 1.21	94.690 ± 0.51	93.64 ± 0.49
	40	96.56 ± 1.02	94.690 ± 0.47	93.64 ± 0.51
	–	–	–	–
Gain ratio	–	–	–	–
	12	97.55 ± 0.41	96.14 ± 0.36	95.44 ± 0.47
	13	97.56 ± 0.53	96.17 ± 0.51	95.47 ± 0.49
	14	97.46 ± 0.56	96.29 ± 0.70	95.63 ± 0.53
	–	–	–	–
Relief	–	–	–	–
	30	93.30 ± 0.92	98.40 ± 0.98	98.77 ± 0.83
	35	96.54 ± 1.03	98.92 ± 0.75	99.11 ± 0.97
	40	96.56 ± 1.28	98.93 ± 0.83	99.12 ± 1.03
	–	–	–	–

+ 16 features with a success rate of 90.41%. In [19], a combination approach to feature selection techniques is proposed for intrusion detection systems. In this work the number of attributes is reduced by using different classification techniques based on feature selection and evaluation is done through ten classification algorithms that generate the most representative results. The best results are achieved with the IBK classifier and the combination of selection techniques such as Symmetric and Gain ratio with 15 features reaching a success rate of 98.5%. In [3], a method based on wrapper applied with a multi-objective approach using the GHSOM classifier is studied. It is employed with a probabilistic adaptation for the re-labeling process allowing to differentiate between normal and anomalous traffic as well as different typologies of anomalies. This proposal provided a rate of 99.12 ± 0.61 in which 25 features are analyzed.

8   Conclusions

The success rate is the most appropriate metric to evaluate the performance of a classifier (regarding the level of traffic detection in computer networks). It can be verified that better values are obtained with the Bayesian network classifier with the Gain ratio feature selection. Some of the features that best contribute to the classification process are: logged_in, srv_serror_rate, flag, serror_rate, dst_host_srv_serror_rate, diff_srv_rate, dst_host_serror_rate, dst_host_srv_diff_host_rate and wrong_fragment. The quality metrics obtained is: a success rate of 97.56%, 96.17% of sensitivity and the specificity of 95.47%. Using the thirteen most relevant characteristics of the 41 possible attributes of the NSL-KDD dataset helps to create a lighter IDS.

An important improvement in the detection rate of attacks and normal traffic in computer networks has been identified. A lower proportion of features and less computational resources are applied. It may be useful for a later solution on equipment with lower performance and if necessary, for a real time analysis. an exhaustive comparison would be required which is currently not possible because the only available performance results refer to the success rate. Most of the similar research works did not present statistical significance test from which the standard deviation of the success rates could be extracted. Further, there are not specific implementations to be able to execute and compare the results obtained in a more detailed way.

9   Future Work

As future research work, an exhaustive review combining features selection techniques with different classifying techniques is proposed. So, this will allow to determine the optimal number of characteristics to acquire the best results and the appropriate classifier. In addition, a choice technique based upon wrapper will be developed in which an optimal classification technique is integrated and identified from the experimental processes suggested.

References

1. Hota, H.S., Shrivas, A.K.: Data mining approach for developing various models based on types of attack and feature selection as intrusion detection systems (IDS). In: Mohapatra, D.P., Patnaik, S. (eds.) Intelligent Computing, Networking, and Informatics. AISC, vol. 243, pp. 845–851. Springer, New Delhi (2014). https://doi.org/10.1007/978-81-322-1665-0_85

2. de la Hoz Franco, E., Ortiz García, A., Ortega Lopera, J., de la Hoz Correa, E., Prieto Espinosa, A.: Network anomaly detection with bayesian self-organizing maps. In: Rojas, I., Joya, G., Gabestany, J. (eds.) IWANN 2013. LNCS, vol. 7902, pp. 530–537. Springer, Heidelberg (2013). https://doi.org/10.1007/978-3-642-38679-4_53

3. De La Hoz-Franco, E., De La Hoz-Correa, E., Ortiz, A., Ortega, J., Martínez-Alvarez, A.: Feature selection by multi-objective optimisation: application to network anomaly detection by hierarchical self-organising maps. Knowl. Based Syst. **71**, 322–338 (2014)

4. De La Hoz-Correa, E., De La Hoz-Franco, E., Ortiz, A., Ortega, J., Prieto, B.: PCA filtering and probabilistic SOM for network intrusion detection. Neurocomputing **164**, 71–81 (2015)

5. Barry, B., Chan, C.H.: Intrusion detection systems. In: Stavroulakis, P., Stamp, M. (eds.) Handbook of Information and Communication Security, pp. 193–215. Springer, New York (2010). https://doi.org/10.1007/978-3-642-04117-4_10. Chapter 10

6. Sun, Y., Wu, D.: A RELIEF based feature extraction algorithm. In: Proceedings of the 2008 SIAM International Conference on Data Mining, pp. 188–195. Society for Industrial and Applied Mathematics, Atlanta-Georgia (2008)

7. De La Hoz-Franco, E., Ortiz Garcia, A., Lopera, J.O., De La Hoz-Correa, E., Mendoza-Palechor, F.: Implementation of an intrusion detection system based on self organizing map. J. Theor. Appl. Inf. Technol. JATIT **71**, 324–334 (2015)

8. de la Hoz, E., Ortiz, A., Ortega, J., de la Hoz, E.: Network anomaly classification by support vector classifiers ensemble and non-linear projection techniques. In: Pan, J.-S., Polycarpou, M.M., Woźniak, M., de Carvalho, A.C.P.L.F., Quintián, H., Corchado, E. (eds.) HAIS 2013. LNCS (LNAI), vol. 8073, pp. 103–111. Springer, Heidelberg (2013). https://doi.org/10.1007/978-3-642-40846-5_11

9. Bayes, T., Price, R., Canton, J.: An essay towards solving a problem in the doctrine of chances. Royal Society of London (1763)

10. Karegowda, A.G., Manjunath, A.S., Jayaram, M.A.: Comparative study of attribute selection using gain ratio and correlation based feature selection. Int. J. Inf. Technol. Knowl. Manag. **2**, 271–277 (2010)

11. Li, K., Deolalikar, V., Pradhan, N.: Big data gathering and mining pipelines for CRM using open-source. In: 2015 IEEE International Conference on Big Data. Ed. by IEEE Xplore, pp. 2936–2938. IEEE (2015)

12. None: Instituto Nacional de Ciberseguridad de España, June 2017. https://www.certsi.es/respuesta-incidentes

13. Karna, N., Supriana, I., Maulidevi, N.: Social CRM using web mining for Indonesian academic institution. In: 2015 International Conference on Information Technology Systems and Innovation (ICITSI). Ed. by IEEEXplore, vol. 15, pp. 1–6. IEEE, Bandung (2015)

14. Atkinson, M., Baxter, R., Brezany, P., Corcho, O.: Analytical platform for customer relationship management. In: The DATA Bonanza: Improving Knowledge Discovery in Science, Engineering, and Business. Ed. by Sons John Wiley, vol. 1, pp. 287–300. Wiley-IEEE Computer Society Press (2013)
15. Singh, R., Kumar, H., Singla, R.K.: Analysis of feature selection techniques for network traffic dataset. In: International Conference on Machine Intelligence and Research Advancement. Ed. by Society IEEE Computer, pp. 42–46 (2013)
16. Ummugulthum, S., Baulkani, S.: Customer relationship management classification using data mining techniques. In: International Conference on Science, Engineering and Management Research. Ed. by IEEEXplore, pp. 27–29. IEEE (2014)
17. Rotovei, D., Negru, V.: A methodology for improving complex sales success in CRM systems. In: 2017 IEEE International Conference on Innovations in Intelligent Systems and Applications (INISTA). Ed. by IEEEXplore, vol. 17, pp. 1–6. IEEE (2017)
18. Ippoliti, D., Zhou, X.: A-GHSOM: an adaptive growing hierarchical self organizing map for network anomaly detection. J. Parallel Distrib. Comput. **72**, 1576–1590 (2012)
19. Garg, T., Kumar, Y.: Combinational feature selection approach for network intrusion detection system. In: International Conference on Parallel, Distributed and Grid Computing, pp. 82–87 (2014)

A Novel Approach Towards Analysis of Attacker Behavior in DDoS Attacks

Himanshu Gupta[(✉)], Tanmay Girish Kulkarni[(✉)], Lov Kumar[(✉)],
and Neti Lalita Bhanu Murthy[(✉)]

BITS Pilani, Hyderabad Campus, Hyderabad, India
{f20150339,f20150647,lovkumar,bhanu}@hyderabad.bits-pilani.ac.in

Abstract. Traditionally, research in Network Security has largely focused on Intrusion Detection and the use of Machine Learning techniques towards identifying malicious agents as well as work on methods towards protecting ourselves from such attacks. In this paper, we wish to make use of the same techniques to analyze the profile of the attacker in the case of a DDoS attack on a distributed honeypot.

Keywords: Distributed denial of service attacks · Honey pot · Machine learning · Clustering algorithms · Attacker profiling

1 Introduction

The username password combination is one of the primary methods of authentication in most of the organizations portals. Many methods such as the man in the middle attack [3], DNS spoofing [6], and phishing attacks [16] are used to obtain username password combinations. All of these activities are examples of penetration attacks as they allow an attacker to intercept the connection and make them believe that they are on the right website [1]. In the aforementioned approaches, the user is fooled into giving their access credentials. Here, we analyze another type of attack, known as a brute force attack. In this approach, the attacker attempts to guess the username and password with the help of tools that make use of dictionaries of a username and password combinations. This approach leads to an increase in load on the server, which in turn block the actual user from logging in, this is an example of a denial of service attack. In the scenario in which, such an attack is distributed, it is an example of a distributed denial of service attack [5,7,13].

In this paper, we make use of Kippo honeypot [4,10], which helps us log brute force attacks and help us understand the behavior patterns of the hacker. The hacker attempts to gain access with the help of a Secure Shell session. Here, we have made use of the data obtained from a honey pot deployed within the Information Security Lab of BITS Pilani, Hyderabad Campus [14,17].

The primary reason for targeting SSH sessions is due to the fact that a significant number of servers are not well maintained and often make use of weak

© IFIP International Federation for Information Processing 2020
Published by Springer Nature Switzerland AG 2020
S. Boumerdassi et al. (Eds.): MLN 2019, LNCS 12081, pp. 392–402, 2020.
https://doi.org/10.1007/978-3-030-45778-5_27

credentials which make a perfect target for malicious agents [12]. A preliminary analysis of credentials and passwords on SSH remote login servers from secure-honey.net gave the following results (Table 1):

Table 1. Most common SSH usernames and passwords.

Username	Frequency	Password	Frequency
Root	89%	123456	41%
Test	6%	Admin	19%
User	2%	Password	11%
Admin	2%	Root	15%

The primary motive of our research is to find out how data with respect to login credentials propagates [15], once a hacker has been successful in obtaining access to an SSH server. Figure 1 shows how successful attacks on the honey tend to be clustered around certain locations [9].

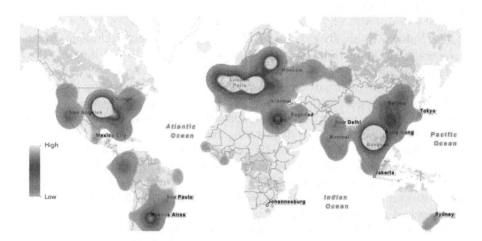

Fig. 1. From the above heatmap, we can see that most of the successful attacks seem to be stemming from North America, Europe, and Southeast Asia.

We also have an image that shows us a zoomed-in perspective in China, from which the majority of the attacks had originally originated. As we can see from the image it appears as if all the attacks appear in pockets, which lends some preliminary support to the hypothesis that data of the credentials appears to spread in the vicinity of the original successful attempt. In the remainder of the paper, we make use of a variety of clustering methods to catch patterns that may escape the human eye (Fig. 2).

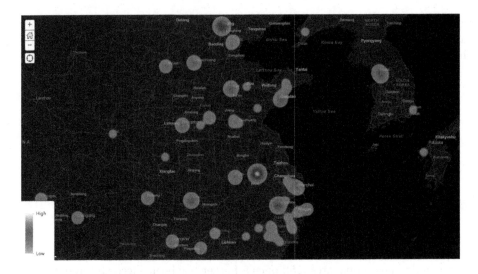

Fig. 2. Distribution of login attempts from China.

2   Related Work

Babak Nabiyev in his work on the application of Clustering Techniques for the detection of DDoS attacks had made use of the KDD CUP 99 dataset which had been developed by DARPA. He attempted to differentiate between Normal Traffic and DDoS traffic with the help of K-Means and EM Clustering techniques. He had clubbed together six cases of DoS attacks as a single type and he defined normal traffic flow to be the other type of behavior. Consquently, he made use of these two classes for the final clustering analysis [8].

Shi Zhong also had made use of different clustering techniques for intrusion detection. In addition, he had also made use of the DARPA intrusion detection project for his dataset. Furthermore, he had done a comparative study on different clustering algorithms for intrusion detection, in which he concluded that unsupervised clustering algorithms performed better than supervised learning methods. Out of all the clustering algorithms, his proposed self-labeling heuristic performed the best with an overall accuracy of 93.6% [19].

Nikolskaia Kseniia analyzed IP traffic with the help of clustering on IP packet headers. He considered multiple parameters such as the classification parameters based on packet and transmission properties, choice of clustering methods and the number of clusters. He concluded that real-time data is too complex to dynamically change features or clustering algorithms. A hybrid neural network approach showed the best results with about 95% correctness [11].

Jie Wang argues that clustering algorithms may not work very properly for intrusion detection because the similarity level of data points cannot be controlled. He proposes a two seed expanding algorithm that splits the attacks into different phases. The preprocessing includes creating a network flow and changing continuous-valued features to binary features. Based on these features, the algorithm selects seeds until all flows are divided into clusters. Their experiments show that two seed expanding algorithm performs better than the k-means and other clustering methods [18].

Geoff Boeing used k-means clustering and dbscan techniques to cluster 1759 points of latitude and longitude data and they were reduced to 138 points and obtained 92% compression, without losing out on the key features of the information that had been spatially represented within the dataset [2].

3   Research Framework

Experimental Setup. We have deployed honey pots with the distributed architecture as shown in Fig. 3.

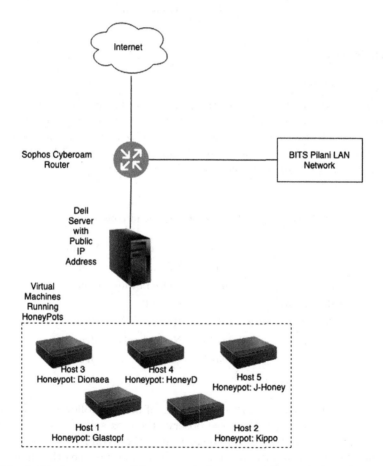

Fig. 3. The honey pot architecture which was used for the D-DoS Attack.

The hypervisor runs five virtual machines, each of which runs a mini-Ubuntu 16.04. Each instance, in turn, runs a different honeypot. The traffic to the virtual machines is controlled with the help of a firewall and Network Address Translation(NAT) to assist us to communicate with the outside world. The server runs within the Information Security Laboratory of BITS, Pilani-Hyderabad campus network. The server continuously monitors the activity that occurs on the public IP addresses (Table 2).

Table 2. Spec table of the honeypot used — Kippo.

Components	Specs
Processor	Intel Xeon
RAM	8GB
Hard Disk	400 GB
Operating System	Ubuntu 16.04

4   Analysis

4.1   Attackers Origin

The origin of the attacker refers to the country or the city location from which the attack is being initiated. The source of their IP address help determines the location of the attacker. We made use of the urllib2 library to find the location of the attackers. However, IP addresses do not prove to be useful if the attacker makes use of a VPN or Tor Network. The results of the analysis have been mentioned in Table 3:

Table 3. Successful attempts city and country wise.

City	Attempts	Country	Attempts
Ho Chi Minh	3225	Vietnam	3586
Kansas City	1237	United States	1368
Radomsko	521	Poland	522
Saint Petersburg	306	Russia	354
Prague	251	Netherlands	326
Hanoi	193	China	323

We observe that there seem to be clusters of activity followed by patches of inactivity as seen in Fig. 4. Here, we observe there as spikes of activity in the second week and the last week of June as well as the second week of July as well as the end of October and the beginning of November. On the other hand, there seem to be very less attacks initiated in the months of August and September and hence they were not accommodated in the graph.

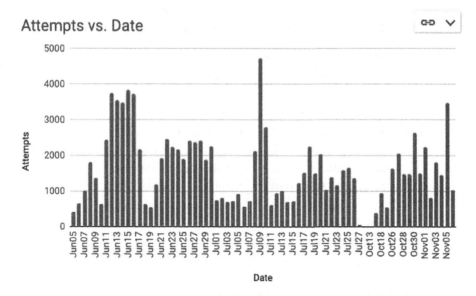

Fig. 4. Attempts distribution over 6 months

Table 4. Most popular passwords and number of attempts

Username	Counts	Username	Counts
Root	190791	1234	1033
Admin	27161	Guest	838
Ubnt	4056	Test	816
Support	3597	Usuario	740
User	2533	pi	730

Table 5. Most popular passwords and number of attempts

Password	Attempts	Password	Attempts
Admin	13802	12345	2285
Ubnt	4653	123456	2146
1234	4508	User	1962
Support	3179	Default	1690
Password	2707	Admin123	1341

4.2 Traffic Analysis

We had segmented the data into files of 1MB size and had a total of 250MB data. The configuration had allowed at most 21 attempts from a particular IP before the IP was banned. Total 870 usernames and 9027 unique passwords were attempted.

The most attempted username was "root" and the most attempted password was "admin". In addition to the popular combination of 'root' and 'admin' we also get to see that the attackers tried other popular default passwords such as ubnt (as we made use of the Ubuntu operating system) as well as 1234, support and password. Furthermore, the hackers had also made use of popular usernames such as admin, user and guest. This analysis shows something as simple as setting a strong username password combination can reduce the number of successful breaches in security. Finally, we observe that an overwhelming

Table 6. Two Day of Interactions for the Ho Chi Minh City, Vietnam on 26th June and 27th June, 2018—Obtained by 2 g Clustering Approach

IP	Attempts	City	Country
116.31.116.20	20350	Guangzhou	China
58.218.198.147	18817	Nanjing	China
58.218.198.153	7478	Nanjing	China
103.207.36.213	4545	Ho Chi Minh	Vietnam
58.218.198.167	2590	Jiangsu	China
91.211.1.100	2161	Vabalninkas	Lithuania
58.218.198.170	1969	Nanjing	China
31.207.47.50	1653	Amsterdam	Netherlands
116.31.116.21	1640	Guangzhou	China
58.218.198.172	1542	Nanjing	China

majority of attacks on the distributed honeypot system appear to be coming from China (Tables 4, 5 and 6).

4.3   Machine Learning Analysis

On this data, we have made use of three clustering methods which has helped us gain insight on the attacker's profile after obtaining access to the system. Here, we have pooled the data in a manner that is similar to that used within n-gram models of Natural language processing. Thus, the data comes in three forms-

– Single day data
– Two days at a time
– Three days at a time

We have made use of 3 different clustering algorithms to gain a better insight on the information presented through the data. From the Figs. 5, 6 and 7 we observe that most of the attacks seem to be concentrated only in certain parts of the world. This means that the information gained by the attacker seems to be spreading only to the vicinity to the earliest attack, rather than spreading randomly over the world.

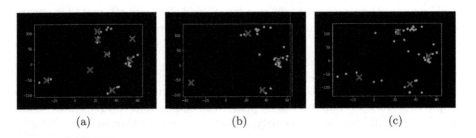

(a)   (b)   (c)

Fig. 5. Mean shift clustering (a) 1 g (b) 2 g (c) 3 g

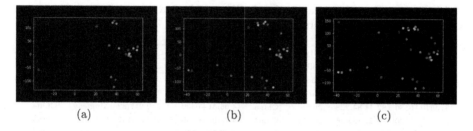

Fig. 6. GMM clustering (a) 1 g (b) 2 g (c) 3 g

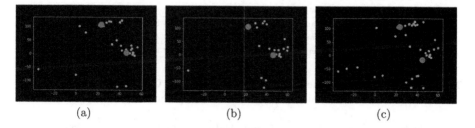

Fig. 7. Kmeans clustering (a) 1 g (b) 2 g (c) 3 g

All three techniques seem to give us the **similar results-**

- All techniques give cluster centers which are very close to one another.
- The cluster centers obtained are similar across 1 g, 2 g and 3 g

On the other hand there seem to be some **key differences-**

- The mean shift algorithm appears to be more susceptible to outliers, which causes it to detect a greater number of clusters.
- On the other hand, the algorithm behaves better when we increase the number of data points as in the case of 2 g and 3 g.

To better understand why the clustering algorithms have singled out these locations, we have probed the data from 1 g, 2 g and 3 g on specific geographic locations so as to search for patterns that could help us better understand how the attack seems to propogate.

In the 1 g analysis for Table 7, we observe that all the successful attacks have appeared to have taken place one after another after short intervals of time. In addition, we can see that once an attacker gains access, it seems like the others in the vicinity gain access after a short interval of time.

In Table 8, we observe the following observation. The set of IP addresses that make a successful attempt on the first day are the same as those which are obtained on the following day. However, we notice that now there is a new IP from the same location that is now able to successfully gain access to the honeypot. This means either the attacker has gained access to a new IP or another attacker has received information about the same from another attacker in the same geolocation.

Table 7. One Day of interaction for the Date 27th October, 2018 from China on - Obtained from the 1 g clustering approach

Time	IP	City	Country	Time	IP	CIty	Country
13:45:41+0530	60.182.212.131	Jinhua	China	13:58:45+0530	112.236.177.74	Qingdao	China
13:46:03+0530	60.255.146.181	Chengdu	China	14:04:46+0530	125.92.182.50	Shiqi	China
13:46:11+0530	110.184.170.247	Zhongba	China	14:05:30+0530	121.14.7.244	Guangzhou	China
13:49:10+0530	182.44.84.228	Jinan	China	14:06:51+0530	110.249.217.82	Xingfeng	China
13:52:27+0530	210.51.191.26	Beijing	China	14:07:03+0530	58.218.198.147	Nanjing	China
13:52:53+0530	58.48.178.200	Shuiguo	China	14:07:52+0530	218.60.136.106	Chaoyang	China
13:54:21+0530	60.185.214.42	Zhoushan	China	14:08:52+0530	111.121.192.6	Guiyang	China
13:54:49+0530	222.47.26.139	Hangzhou	China	14:10:41+0530	153.34.109.60	Chaowai	China
13:55:29+0530	124.243.216.102	Beijing	China	14:14:27+0530	122.190.252.82	Xiangfan	China
13:58:08+0530	218.108.124.26	Hangzhou	China	14:15:32+0530	113.122.34.247	Jinan	China
13:58:11+0530	113.122.5.6	Jinan	China	14:19:16+0530	113.206.115.125	Beiwenquan	China
13:58:11+0530	210.51.191.26	Beijing	China	14:23:03+0530	123.149.128.181	Henan	China

Table 8. Two day of interactions for the Ho Chi Minh City, Vietnam on 26th June and 27th June, 2018 — Obtained by 2 g clustering approach

Date	IP	Count	Date	IP	Count
2018-06-26	142.54.189.114	90	2018-06-27	142.54.189.114	104
2018-06-26	173.208.187.66	2	2018-06-27	192.187.103.2	223
2018-06-26	192.187.103.2	204	2018-06-27	192.69.95.132	4
2018-06-26	69.197.135.10	23	2018-06-27	69.197.135.10	15

Table 9. 3 Days of interactions for the country of Vietnam from 6th June to 8th June 2018 — Obtained by 3 g clustering approach

Date	IP	Count	Date	IP	Count	Date	IP	Count
2018-06-06	103.207.36.117	1	2018-06-06	116.98.0.212	1	2018-06-07	116.103.77.175	2
2018-06-06	103.207.36.9	2	2018-06-06	117.3.47.59	2	2018-06-07	116.105.225.86	2
2018-06-06	103.207.37.239	4	2018-06-06	117.5.195.121	1	2018-06-07	117.3.47.59	1
2018-06-06	103.207.39.43	3	2018-06-06	123.16.32.196	3	2018-06-07	125.212.226.227	1
2018-06-06	103.207.39.54	1	2018-06-06	123.19.170.93	1	2018-06-07	14.176.232.175	5
2018-06-06	103.79.141.153	2	2018-06-06	14.167.67.203	1	2018-06-07	27.70.150.55	1
2018-06-06	103.79.141.39	1	2018-06-06	14.176.232.175	3	2018-06-07	58.186.98.43	1
2018-06-06	103.79.143.136	6	2018-06-06	27.78.21.103	2	2018-06-08	103.207.39.159	2
2018-06-06	103.89.88.11	9	2018-06-06	42.118.152.107	1	2018-06-08	103.79.141.153	1
2018-06-06	113.170.210.40	1	2018-06-07	103.207.37.239	4	2018-06-08	103.89.88.11	9
2018-06-06	113.22.152.202	2	2018-06-07	103.207.39.228	1	2018-06-08	116.103.147.230	6
2018-06-06	116.104.79.5	1	2018-06-07	103.207.39.43	2	2018-06-08	116.98.44.241	1
2018-06-06	116.105.225.86	4	2018-06-07	103.79.141.153	2	2018-06-08	14.176.232.175	4
2018-06-06	116.110.160.11	2	2018-06-07	103.79.143.136	4	2018-06-08	27.70.151.209	2
2018-06-06	116.97.24.95	1	2018-06-07	103.89.88.11	9	2018-06-08	27.78.21.103	2

In Table 9, the pattern in the data obtained from the 3 g analysis further strengthens the observations that we had made in the case of 2 g. Here, we can clearly observe that the same set of IP addresses make attack in regular intervals of time. In addition, to those we see additional IP addresses which originate from the same or nearby locations which gives weight to the argument that the information about the credentials is spreading to the geographical vicinity.

5   Conclusion

We would like to draw the conclusion that attacks appear to be concentrated in certain regions. Furthermore, it appears as if the data with respect to the access credentials does not seem to spread randomly rather, it appears as if the success with respect to successful attacks seems to spread in the near vicinity of the first attack.

References

1. Bacudio, A., Yuan, X., Chu, B., Jones, M.: An overview of penetration testing. Int. J. Netw. Secur. Appl. **3**, 19–38 (2011). https://doi.org/10.5121/ijnsa.2011.3602
2. Boeing, G.: Clustering to reduce spatial data set size. arXiv preprint arXiv:1803.08101 (2018)
3. Callegati, F., Cerroni, W., Ramilli, M.: Man-in-the-middle attack to the https protocol. IEEE Secur. Priv. **7**(1), 78–81 (2009)
4. Doubleday, H., Maglaras, L., Janicke, H.: SSH honeypot: building, deploying and analysis. Int. J. Adv. Comput. Sci. Appl. **7** (2016). https://doi.org/10.14569/IJACSA.2016.070518
5. Feinstein, L., Schnackenberg, D., Balupari, R., Kindred, D.: Statistical approaches to DDOS attack detection and response. In: Proceedings DARPA information survivability conference and exposition, vol. 1, pp. 303–314. IEEE (2003)
6. Klein, A., Golan, Z.: System and method for detecting and mitigating DNS spoofing trojans, 11 September 2012, US Patent 8,266,295
7. Mirkovic, J., Reiher, P.: A taxonomy of DDOS attack and DDOS defense mechanisms. ACM SIGCOMM Comput. Commun. Rev. **34**(2), 39–53 (2004)
8. Nabiyev, B.: Application of clustering methods network traffic for detecting DDOS attacks. Probl. Inf. Technol. **09**, 98–107 (2018). https://doi.org/10.25045/jpit.v09.i1.11
9. Najafabadi, M.M., Khoshgoftaar, T.M., Kemp, C., Seliya, N., Zuech, R.: Machine learning for detecting brute force attacks at the network level. In: 2014 IEEE International Conference on Bioinformatics and Bioengineering, pp. 379–385. IEEE (2014)
10. Nawrocki, M., Wählisch, M., Schmidt, T.C., Keil, C., Schönfelder, J.: A survey on honeypot software and data analysis. arXiv preprint arXiv:1608.06249 (2016)
11. Nikolskaia, K.: Network attacks detection based on cluster analysis, September 2017
12. Owens, J., Matthews, J.: A study of passwords and methods used in brute-force SSH attacks (2008)
13. Schuba, C.L., Krsul, I.V., Kuhn, M.G., Spafford, E.H., Sundaram, A., Zamboni, D.: Analysis of a denial of service attack on TCP. In: Proceedings. 1997 IEEE Symposium on Security and Privacy (Cat. No. 97CB36097), pp. 208–223. IEEE (1997)

14. Sochor, T., Zuzcak, M.: Study of internet threats and attack methods using honeypots and honeynets. In: Kwiecień, A., Gaj, P., Stera, P. (eds.) CN 2014. CCIS, vol. 431, pp. 118–127. Springer, Cham (2014). https://doi.org/10.1007/978-3-319-07941-7_12

15. Tatlı, E.: Cracking more password hashes with patterns. IEEE Trans. Inf. Forensics Secur. **10**, 1 (2015). https://doi.org/10.1109/TIFS.2015.2422259

16. Thomas, K., et al. (eds.): Data Breaches, Phishing, or Malware? Understanding the Risks of Stolen Credentials (2017)

17. Valli, C., Rabadia, P., Woodward, A.: Patterns and patter-an investigation into SSH activity using kippo honeypots (2013)

18. Wang, J., Yang, L., Wu, J., Abawajy, J.H.: Clustering analysis for malicious network traffic. In: 2017 IEEE International Conference on Communications (ICC), pp. 1–6. IEEE (2017)

19. Zhong, S., Khoshgoftaar, T.M., Seliya, N.: Clustering-based network intrusion detection. Int. J. Reliab. Qual. Saf. Eng. **14**(02), 169–187 (2007)

Jason-RS, A Collaboration Between Agents and an IoT Platform

Hantanirina Felixie Rafalimanana[1(✉)], Jean Luc Razafindramintsa[1],
Sylvain Cherrier[2], Thomas Mahatody[1], Laurent George[2],
and Victor Manantsoa[1]

[1] Laboratory for Mathematical and Computer Applied to the Development
Systems, University of Fianarantsoa, Fianarantsoa, Madagascar
dhaliahfeli@gmail.com,
razafindramintsa.jeanluc@yahoo.fr,
tsmahatody@gmail.com
[2] Université Paris-Est, Institut Gaspard Monge (LIGM),
77454 Marne-la-Vallée Cedex 2, France
{sylvain.cherrier,laurent.george}@u-pem.fr

Abstract. In this article we start from the observation that REST services are the most used as tools of interoperability and orchestration in the Internet of Things (IoT). But REST does not make it possible to inject artificial intelligence into connected objects, i.e. it cannot allow autonomy and decision-making by the objects themselves. To define an intelligence to a connected object, one can use a Believe Desire Intention agent (BDI an intelligent agent that adopts human behavior) such as Jason Agentspeak. But Jason AgentSpeak does not guarantee orchestration or choreography between connected objects. There are platforms for service orchestration and choreography in IoT, still the interconnection with artificial intelligence needs to be built. In this article, we propose a new approach called Jason-RS. It is a result of pairing Jason BDI agent with the web service technologies to exploit the agent capacity as a service, Jason-RS turn in Java SE and it does not need any middleware. The architecture that we propose allows to create the link between Artificial Intelligence and Services choreography to reduce human intervention in the service choreography. In order to validate the proposed approach, we have interconnected the Iot BeC^3 platform and the REST agent (Jason-RS). The decision-making faculty offered by Jason-RS is derived from the information sent by the objects according to the different methods of REST (GET, POST, PUT, and DELETE) that Jason-RS offers. As a result, the objects feed the inter-agent collaborations and decision-making inside the agent. Finally, we show that Jason-RS allows the Web of Objects to power complex systems such as an artificial intelligence responsible for processing data. This performance is promising.

Keywords: BDI agent · BeC^3 · Internet of Thing · Jason

© IFIP International Federation for Information Processing 2020
Published by Springer Nature Switzerland AG 2020
S. Boumerdassi et al. (Eds.): MLN 2019, LNCS 12081, pp. 403–413, 2020.
https://doi.org/10.1007/978-3-030-45778-5_28

1 Introduction

The tendency towards the Web of Object (WoO) does not cease to increase from time to time. Given the number of objects to facilitate the daily life, the number of possible interactions and the applications that one can imagine become more powerful and rich. But what are the ways to make these application in a generic way? The object web orientation is considered as a great resource not only for managing objects [1] but also for questions of flexibility of exchanges. In the Web of Object (WoO) one is rather directed towards the use of Web Service [1].

We note that the most used Web Service type is REST[1] [1–3]. REST (Representational State Transfer) or RESTful is an architecture style for building applications (Web, Intranet, Web Service). It is a set of conventions, rules and good practices to respect and not a technology in its own right. Not only REST is stateless [4] (meaning that the service consumer does not have to store information about the way he is using the service); it is design to reduce the coupling between software pieces. In the Iot, REST may allow reconfiguration of the object, it reuse in a different application for a multiplatform application, REST service structure is easy to describe. REST allows machine to machine communication and is simplier to do than SOAP. The communication is done through the HTTP protocol. Nevertheless the possibility of using the resource via the web given by the REST, it still lacks intelligent data processing. Our contribution aim to add a complex data processing of the object web. We then opt for the reuse of our search fruit in [5] which we call Jason-RS or Jason-Rest for more precision.

Jason-RS gives the REST Web Service ability to the BDI agent Service [5]. Jason is a multi-agent platform based on the Agentspeak-L language [6, 7]. It is a platform for BDI agent, that is to say goal-oriented software agent, it is a smart agent. We have to look at the choice of the BDI agent because it is close to human behavior [8, 9]. So the connexion of the latter with the REST Web Service allows him to publish his capacity as a Web Service. The processing of a complex action can be done inside Jason (prediction, composition of services, constraint logic programming) and could be published as a Web Service. Jason-RS can serve as an intelligent treatment center.

We focus in the collaboration between agent and Iot platform cause connected object had no intelligence, designed differently, so with this collaboration we can open in the intelligence of Iot with a sensor less, little human intervention. In our approach, we collect the information from the connected object and by Jason-RS we delegate the intelligent agent works. All change in the object behavior is sent to the agent BDI (as a perception) in order to provide a decision or to change the agent comportment. Through this perception that the agent can decide and delegate a task to agent or service choreography to object.

The rest of this article is organized as follows: Sect. 2 will discuss the related work of our work. Section 3.1 describes in general the architecture of our contribution. In Sect. 3.2 we will evaluate our work through an architecture. In Sect. 4 we discuss a discussion and open challenge. We will conclude this article with a conclusion (Sect. 5).

---

[1] https://www.supinfo.com/articles/single/5642-qu-est-ce-qu-une-api-rest-restful.

2 Related Work

In this article we propose a new form of Web Service that the WoO can use. The use of REST in WoO is not an innovative work as researchers have already done research for the implementation of the latter to allow connected objects to trade [10, 11]. The Internet of Things is a system of physical objects that can be discovered, monitored, controlled or interact with electronic devices that communicate on various network interfaces and can potentially be connected to the wider Internet.

The webs of object is a concept say recent in computing that offers a future with which object of the daily life can integrate into the web [12]. The Web of Object is considered as a subset of the Internet of Things. WoO applies to standards and software infrastructures such as REST, HTTP and URIs to create applications and services that combine and interact with a variety of network devices. The key point of the WoO is that its implementation does not involve the reinvention of the means of communication given that they use the existing standards. It also gives an easy access to a wide population of programmer that used to handle such an architecture. It facilitates the integration of the IoT in the internet. It is also a solution that can open the connection of the Industry 4.0 [13].

View the exponential increase of the connected object, to reduce the cost, the required performance and to facilitate communication in the object internet, the W3C recently launched the working group on the Web of Objects. They aim to fight the fragmentation of IoT. They developed the initial standards for the Web of Objects [14]. To facilitate service choreography for Iot Cherrier describes in his thesis a new approach based on a platform to standardize the interoperability of heterogeneous objects [15].

Castellani and Al focused on the implementation of both CoAP and EXI technology for communication between objects by adopting the REST approach [16]. They implemented the CoAP protocol to enable communication between distributed connected objects. Then they also implemented the EXI library for compressing XML data into binary.

In the paper [2], they discuss how to create an event SOA. They found a new approach for coordinating IoT's services with the resource service. They named EDSOA the result of their research. It is an asynchronous SOA and responsive it can be considered as an event based service. We have inspired by this approach for the technical aspect of Jason-RS.

The Web of Object is used in almost any IoT field even in the field of health. On what [17] proposed a new instance of cloud server and set up data collection services on the heterogeneous devices they use. This aims to improve online health service by properly implementing the architectural web object principle for remote control.

Based on WoO, [4] describes the object-based web architecture that adopts the principle of Resource Oriented Architecture (ROA), hence the basis of RESTful. They implemented a smart gateway for smart meters. This allows the connected object management servers to keep track of the status and movements of the objects. [2] the HTTP protocol is used to facilitate access to the service. The answer emitted is in JSON format. The architecture proposed by [4] allows developers to use web programming languages such as (HTML, Python, …).

All of the research cited above are interesting from the point of view of exposure and communication capacity in the Internet of Things. Each with its strong point in communication and interconnection in the world of IoT. They all almost used the REST Web Service. But for us, a simple service is not enough not only to bathe the connected objects in the artificial intelligence but also to treat in intelligent way the data captured at the level of the objects. This treatment helps automatic decision-making thanks to the intelligent BDI agent that has the same reasoning mode as humans. We focus our research on the intelligent Web Service, that is, a Web Service associated with artificial intelligence in order to reduce human intervention to the orchestration of services in connected objects and decision-making. We have exploited here our previous research work, Jason-RS. An intelligent Web Service with BDI agents that perform the resolution and automation of certain tasks thanks to their ability to cooperate and exchange using the AgenSpeak (L) agent language. These agents may then exhibit functionality as a Web Rest Service. Jason-RS has the methods used in REST (GET, PUT, POST, and DELETE) to ensure the dialog with heterogeneous clients.

3   Proposed Architecture

3.1   General Architecture

In this section we discuss the general architecture of our approach. As we are in the world of the IoT there is a mean of communication: the network. There are many types of networks today for the connectivity of objects such as wifi, Bluetooth for common devices, LoraWAN for IoT devices using long distance communication, and ZigBee, 6LowPan, EnOcean, Zwave for short range communication devices, etc. As connected objects do not have enough memory space and very limited energy to feed themselves permanently so it is important to outsource most of its function. Although the design of its objects is different (i.e. the objects are heterogeneous), a standardization platform is important to allow inter-object choreography. Taking into account all these parameters we have our control center server constituted by the Jason-RS server and the IoT server illustrated by Fig. 1. Generally the IoT server is a centralized platform of (Cloud, big data) whether it is an IoT server that administers the objects.

Jason plays a complex problem solver role as well as the complex system such as optimization … The autonomous and goal-oriented capability through the BDI agent that it implements allows it to perform a task reasonably. This capability is now exploitable through a distributed component as a service [5] through Jason-RS. Then the objects can use it not only to send the sensor data but also to initiate a decision-making process at the agent level.

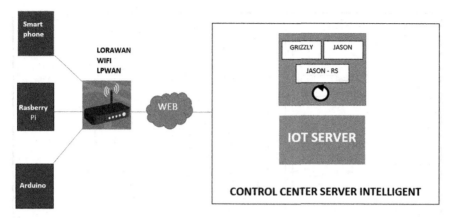

Fig. 1. General architecture of the JASON-RS implementation in the Web of Object

Communication between the IoT and the web is done through the HTTP protocol. With Jason-RS we can use all the existing methods of the REST and most importantly there is an interaction with the agent. As a result we can send data to the central control servers via the POST method. We can also retrieve the data or decision made by the GET method. JASON-RS recognizes directly the task that will send back to the agent.

At the agent level as soon as the service is called by the objects or other platform, the agent reacts and changes state in relation to the perception it receives from it environment.

We have already studied in our article [5] the publication of Jason's BDI agent capability as a Web Service. We have proposed an approach allows deploying and running Web Services (REST and SOAP) pairs with a Jason BDI agent in java SE environment. We didn't use a modern Web-App server or application server, and we didn't developed a different middleware cause it requires a lot of time. Deploying an agent inside a server is a tedious task. Then in our strategy, we reused the existing Java frameworks called Non-Blocking Input Output. A small illustration for brightening the architecture is shown in Fig. 2.

Jason-RS offers an opportunity to exploit the world-class logic, which is embedded in Jason. It can associate them with other elements or heterogeneous application. When we can manipulate first-order logic, we can imagine a kind of possibility of complex problem solving as well as the common problem in artificial intelligence. What is interesting here is that the reasoning mode of Jason agent is near the human reasoning mode to achieve a goal. Jason-RS aims to outsource the capacity present in Jason as a service.

This initial Jason-RS that we have exploited with the collaboration with the Iot platform to intelligently process the data received from the connected object. By exploiting the horn clause and the first-order logic included in Jason we will use this ability to decide the action of objects. Then we take into account the behavior of the objects as perception for the agents.

To validate our approach we tested the interoperability of Jason-RS and heterogeneous objects choreographed with BeC^3.

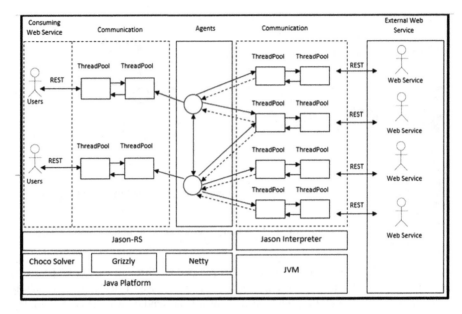

Fig. 2. General architecture of the JASON-RS micro-service [5]

3.2 Application of the Architecture with a Python Object and BeC^3

Scenario. To illustrate and motivate our approach, we introduce a concrete application using the following scenario. We take the example of the disposal of household waste in our paper [5]. Waste disposal agents have waste disposal costs that vary according to time constraints, people, distances traveled and evacuation loads. A decision-maker who gives these opinions to optimize the evacuation of waste thanks to the information that the evacuator agents gives him. Information of each of its agents could be changed thanks to the perception emitted by different connected objects. The Fig. 3 illustrates the application of this scenario. We used BeC^3 for the service Mashup. BeC^3 is a composer who allows a user to choreograph heterogeneous objects. In our experimentation we have created sensor objects choreographed in BeC^3 with other objects. This sensor sends the data captured via Jason-RS in POST for the agents to change the perception of the Jason agent (receiver agent). A decision-maker analyzes the present data and the cost of performing the tasks for each of his agents in order to decide which one is durable, and which one can perform a task at lower cost.

The sensor is controlled by the Python version of service compatible with BeC^3. When the service is launched, it can send a POST request to BeC^3 to identify itself and then another POST request to set up its feature handled by the device. Any mashup designed by a user can then be assured by a communication the service on each object. The decision made by the agent could be consumed via a GET request to the Jason-RS.

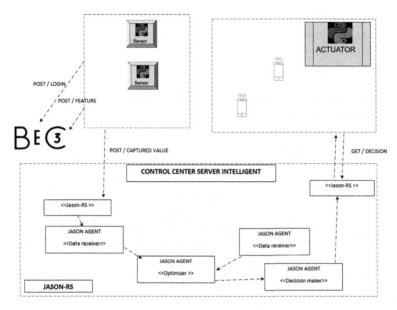

Fig. 3. Application interaction between object choreography and Jason-RS

4 Result and Discussion

We tested this approach in two different objects: a python object, a mobile phone. We used Jason-RS to communicate the BeC^3 platform with choreographed objects and BDI agents who take the place of a decision maker instead of a human being (such as it is originally designed in BeC^3.

There is a demo script (in Python) APIDemo.py in Pylite project (resources dir). We have used xmpp[2] server for login, its specificity is the presence. So as we describe in the Sect. 3 we must log in BeC^3 in order to create a virtual programmable object:

```
POST/login with content
{
    "username": {xmpp login},
    "password": {xmpp password},
    "service": {xmpp service (for example im.bec3.com)}
}
```

After login, create a BeC^3 object with the following JSON format:

```
POST /feature with content
{
    "name": "tot2",
    "path" : "truc/bidul21",
```

---

[2] https://xmpp.org/.

```
    "type": "switch",
    "details": "ooo",
    "widget": "none",
    "mqtt": false
}
```

Available types are accelerometer, button, buzzer, gauge, gps.
These object can be delete, update.

```
DELETE /feature/{id}
PUT /feature/{id}
{
    "data": {data (dependent of object type}
}
```

One object is associate with one BDI agent. To connect BeC^3 an URL is available from Jason-RS. To change the agent BDI perception a POST request is send by this url format:

```
POST jason_rs_based_url/object_agent/
{
    "data":value_from_python_object
}
```

To receive a decision from BDI agent, Jason-RS provide an URL that we can consume:

```
GET jason_rs_based_url/agent_decider/decision
```

The following sequence diagram presented in Fig. 4 shows the exchange between these two platforms.

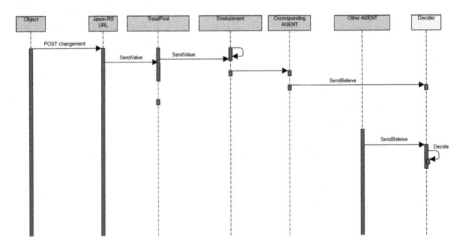

Fig. 4. Sequence diagram for communication between object a Jason agent by Jason-RS

We analyzed the performance of our approach via a browser (Table 1).

Table 1. The calculation time of the exchanges between the objects and Jason-RS

Device	Method REST	Task	Duration (millisecond)
Python Object	GET	Consumation WS	200–450
Smart Phone	POST, GET	Sending data	428–1000

We tested our approach on both devices. During our experiment the return is almost ≤ 1000 ms. Our test area are done through the WIFI network so the discussion field is open for other types of networks available for the Internet of Things.

Agent-based internet of things is an interested field of research now [18], in the face of these works, we focus in the technical side of things, and we compare in Table 2 what differentiates our approach compared to the state of the art.

Table 2. Comparison with a similar research domain work

Work in field Jason agent and IoT	Agent flexibility in internet	Horn logic	Taking in count Heterogeneity of IoT
[19]	?	yes	?
[20]	?	?	yes
[21]	?	yes	?
Jason-RS and IoT Platform	yes	yes	yes

So in our approach we taking in count the heterogeneity of IoT by using a platform BeC^3. To permit an automatic decision making we exploit the horn logic provided in Jason. This collaboration permit a flexibility in interacting between agent and heterogeneous connected object.

5 Conclusion

This article describes our contribution on service improvement in the world of Web of Object. We have integrated our previous Jason-RS research work into the sensor device connected to the web. Object of the IoT do not support the execution of artificial intelligence within themselves because there resources are very limited. Our approach is towards the decentralization of this AI treatment with Jason-RS. Not only is Jason-RS a Web Service but it has an artificial intelligence BDI engine running. So the complex data processing in the connected web object is provided by the Jason. The communication is done through HTTP protocol as it is more flexible and commonly adopted in the world of web. We tested our approach with a Mashup service platform for heterogeneous connected objects called BeC^3. Our approach therefore allows the object connected to the web to open up to data processing with artificial intelligence as well as complex systems itself. As perspective we are considering the automatic injection of service inside the connected object to reduce the human task of defining the service choreography.

References

1. Laine, M.: Restful Web Services for the Internet of Things. Saatavilla. http://media.tkk.fi/webservices/personnel/markku_laine/restful_web_services_for_the_internet_of_things.pdf
2. Gupta, P., Mokal, T.P., Shah, D.D., Satyanarayana, K.V.V.: Event-driven SOA-based IoT architecture. In: Dash, S.S., Das, S., Panigrahi, B.K. (eds.) International Conference on Intelligent Computing and Applications. AISC, vol. 632, pp. 247–258. Springer, Singapore (2018). https://doi.org/10.1007/978-981-10-5520-1_24
3. Wilde, E.: Putting things to REST (2007)
4. Guinard, D., Trifa, V., Wilde, E.: A resource oriented architecture for the Web of Objects. In: IoT, pp. 1–8, November 2010
5. Rafalimanana, H.F., Razafindramintsa, J.L., Ratovondrahona, A.J., Mahatody, T., Manantsoa, V.: Publish a Jason agent BDI capacity as web service REST and SOAP. In: Bouhlel, M.S., Rovetta, S. (eds.) SETIT 2018. SIST, vol. 146, pp. 163–171. Springer, Cham (2020). https://doi.org/10.1007/978-3-030-21005-2_16
6. Rao, A.S.: AgentSpeak(L): BDI agents speak out in a logical computable language. In: Van de Velde, W., Perram, J.W. (eds.) MAAMAW 1996. LNCS, vol. 1038, pp. 42–55. Springer, Heidelberg (1996). https://doi.org/10.1007/BFb0031845
7. Ricci, A., Bordini, R.H., Hübner, J.F.: AgentSpeak(ER): An Extension of AgentSpeak(L) Improving Encapsulation and Reasoning About Goals, p. 3 (2018)
8. Norling, E.: Capturing the quake player: using a BDI agent to model human behaviour. In: Proceedings of the Second International Joint Conference on Autonomous Agents and Multiagent Systems, pp. 1080–1081. ACM, July 2003
9. Kamphorst, B., van Wissen, A., Dignum, V.: Incorporating BDI agents into human-agent decision making research. In: Aldewereld, H., Dignum, V., Picard, G. (eds.) ESAW 2009. LNCS (LNAI), vol. 5881, pp. 84–97. Springer, Heidelberg (2009). https://doi.org/10.1007/978-3-642-10203-5_8
10. Ciortea, A., Boissier, O., Ricci, A.: Beyond physical mashups: autonomous systems for the Web of Objects. In: Proceedings of the Eighth International Workshop on the Web of Objects, pp. 16–20. ACM, October 2017
11. Kovatsch, M., Hassan, Y.N., Mayer, S.: Practical semantics for the Internet of Things: physical states, device mashups, and open questions. In: 2015 5th International Conference on the Internet of Things (IOT), pp. 54–61. IEEE, October 2015
12. Guinard, D.: A Web of Objects application architecture: integrating the real-world into the web. Ph.D. Thesis, ETH Zurich (2011)
13. Kim, J.H.: A review of cyber-physical system research relevant to the emerging IT trends: industry 4.0, IoT, big data, and cloud computing. J. Ind. Integr. Manag. **2**(03), 1750011 (2017)
14. W3C Web of Objects at W3C. https://www.w3.org/WoO/
15. Cherrier, S., Salhi, I., Ghamri-Doudane, Y.M., Lohier, S., Valembois, P.: BeC^3: behaviour crowd centric composition for IoT applications. Mob. Netw. Appl. **19**(1), 18–32 (2013). https://doi.org/10.1007/s11036-013-0481-8

16. Castellani, A.P., Gheda, M., Bui, N., Rossi, M., Zorzi, M.: Web services for the Internet of Things through CoAP and EXI. In: 2011 IEEE International Conference on Communications Workshops (ICC), pp. 1–6. IEEE, June 2011
17. Pescosolido, L., et al.: An IoT-inspired cloud-based web service architecture for e-Health applications. In: 2016 IEEE International Smart Cities Conference (ISC2), Trento, Italy, p. 1–4 (2016)
18. Savaglio, C., Ganzha, M., Paprzycki, M., Bădică, C., Ivanović, M., Fortino, G.: Agent-based Internet of Things: state-of-the-art and research challenges. Future Gener. Comput. Syst. **102**, 1038–1053 (2020)
19. Ciortea, A., Boissier, O., Zimmermann, A., Florea, A.M.: Responsive decentralized composition of service mashups for the internet of things. In: Proceedings of the 6th International Conference on the Internet of Things, pp. 53–61. ACM, November 2016
20. Singh, M.P., Chopra, A.K.: The Internet of Things and multiagent systems: decentralized intelligence in distributed computing. In: 2017 IEEE 37th International Conference on Distributed Computing Systems (ICDCS), pp. 1738–1747. IEEE, June 2017
21. Novais, P., Konomi, S.: From raw data to agent perceptions for simulation, verification, and monitoring. In: Intelligent Environments 2016: Workshop Proceedings of the 12th International Conference on Intelligent Environments, vol. 21, p. 66. IOS Press, October 2016

Scream to Survive(S2S): Intelligent System to Life-Saving in Disasters Relief

Nardjes Bouchemal[1,2(✉)], Aissa Serrar[1], Yehya Bouzeraa[1], and Naila Bouchmemal[3]

[1] University Center Abdelhafid Boussouf, Mila, Algeria
n.bouchemal.dz@ieee.org,
aissa.serrar95@gmail.com, bouzeraayahiasc@gmail.com
[2] University of Constantine 2, Constantine, Algeria
[3] ECE Ecole d'Ingénieur de Paris, Paris, France
naila.bouchemal@ece.fr

Abstract. Disasters are becoming more and more common around the world, making technology important to guarantee people's lives as much as possible.

One of the most modern advances of recent years is how AI is used in disaster relief. Researchers propose works based on new technologies (IoT, Cloud Computing, Blockchain, etc.) and AI concepts (Machine Learning, Natural Language Processing, etc.). But these concepts are difficult to exploit in low and middle socio-demographic index (SDI) countries, especially as most disasters happen in.

In this paper we propose S2S intelligent system, based on voice recognition to life-saving in disaster relief. Generally, a disaster victim is enable to access to his Smartphone and ask help, with this system, saying "help" will be enough to send automatically alerts to the nearest Emergency Operation Services (EOS).

S2S is composed of two parts: Intelligent application embedded on citizens and victims Smartphones, and S2S System for the Emergency Operation Services.

Keywords: Intelligent system · Voice recognition · Life-saving · Disaster relief

1 Introduction

In countries around the world, natural disasters have been much in the news. Indonesia tsunami in 2004, Wenchuan (China) earthquake in 2008, freezing rain disaster in southern China in 2008, devastating 2011 earthquake in Japan, flood disaster in India in 2013. China severe flood in 2016, 2018 Earthquake and Tsunami in Indonesia and the 2019 tropical cyclone in Mozambique, Zimbabwe and Malawi.

Natural disasters caused by climate change, extreme weather, and aging and poorly designed infrastructure, among other risks, represent a significant risk to human life and communities.

© IFIP International Federation for Information Processing 2020
Published by Springer Nature Switzerland AG 2020
S. Boumerdassi et al. (Eds.): MLN 2019, LNCS 12081, pp. 414–430, 2020.
https://doi.org/10.1007/978-3-030-45778-5_29

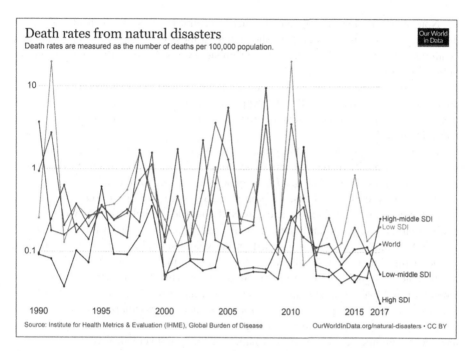

Fig. 1. Link between poverty and deaths from natural disasters (Death rates from natural disasters)

A natural disasters is the encounter between a hazard of natural origin and human, economic or environmental issues. We talk about major risk when the damage and the number of victims are important.

According to the World Health Organization, 160 million people are affected by natural disasters and around 90,000 people are killed every year, [1].

However, with advances in technology, more and more deaths are becoming preventable, encouraging researchers to develop new methods of responding to natural disasters.

Behind this improvement has been the enhancement in living standard and effective response systems. These factors have been driven by an increase in incomes across the world.

What remains true today is that populations in low-income countries, those where a large percentage of the population still live in extreme poverty, or score low on the Human Development Index are more vulnerable to the effects of natural disasters.

We see this effect in the visualization shown in Fig. 1. This chart shows the death rates from natural disasters the number of deaths per 100,000 population of countries grouped by their socio-demographic index (SDI). SDI is a metric of development, where low-SDI denotes countries with low standards of living, [33].

What we see is that the large spikes in death rates occur almost exclusively for countries with a low or low-middle SDI. Highly developed countries are much more resilient to disaster events and therefore have a consistently low death rate from natural disasters.

Note that this does not mean low-income countries have high death tolls from disasters year-to-year: the data here shows that in most years they also have very low death rates. But when low-frequency, high-impact events do occur they are particularly vulnerable to its effects, [33].

Furthermore, the first three days after a natural disaster are the most critical when it comes to saving human life. People trapped on rooftops, under rubble, or in isolated areas need to be found and rescued before they succumb to the effects of the disaster that imperiled them.

For human responders on the ground, this is an almost impossible challenge and what has traditionally made disasters so deadly throughout human history; *"you cannot save people you do not know need help or whom you cannot reach"*, [22].

To deal with these problems, we propose S2S system: Scream to Survive. We used technologies available in middle and low SDI countries to help disaster victims, generally weak and helpless and unable to make call or ask for help in traditional way.

The S2S intelligent system is based on the voice recognition algorithm embedded on the victim Smartphone. That is justified by the fact that an individual in crisis could not be able to signal urgency to intervention team. This is a most helpless situation where affected people need help, but they do not have the ability to look for it.

Furthermore, Smartphone attracts the users and its popularity is increasing in worldwide due to its powerful processing and wireless network capabilities. It enables users to communicate and share information in easily convenient way, [2, 3].

S2S is composed of two parts: Smartphone Intelligent Application installed on citizen Smartphone and intelligent system for Emergency Operation Services (EOS).

Finally, the disaster management consists of four fundamental steps such as mitigation, preparedness, response, and recovery. Among these steps, the emphasis of our work are response and recovery because most of disasters deaths happen in these two steps.

The rest of the paper is organized as follow: Sect. 2 summarizes the related work, based on new technologies. Section 3 presents the proposed system. In Sect. 4 we present implementation concepts, in Sect. 5 we evaluate the S2S performances and Sect. 6 concludes the paper.

2 Related Work

In this section, we will present and discuss disaster management systems based on new technologies and those based on the use of smart phones. The goal is to discuss advantages and inconveniences of each system to motivate our proposition based on the combination of intelligent systems embedded on Smartphone.

(1) **New Technologies for Disasters Management**

Emerging technologies present greater opportunities to make emergency management systems intelligent, protected, and efficient. Today, artificial intelligence (AI), the Internet of Things (IoT), cloud computing and blockchain offer the potential to generate, transmit and read emergency-related data for better decision-making in disasters management, [23].

a. **Blockchain**

Blockchain is in the earliest stages of development, but is a tool that some claim will be transformational for how we transact data. Blockchain is a distributed and immutable digital ledger, secured by cryptography, which can be programmed to record a series of transactions. Its most scalable application today is bitcoin, a cryptocurrency and payment system still growing in its use around the world. A blockchain solution enables the key players/organizations during a disaster management situation to communicate effectively and act on time, [24].

b. **Natural Language Processing**

Natural Language Processing is the technology used to aid computers to understand the human's natural language. Natural Language Processing, usually shortened as NLP, is a branch of artificial intelligence that deals with the interaction between computers and humans using the natural language. The use of NLP to understand social, political, and economic processes aspects in disaster management has become popular with the increase in the volume of data about human communication, including text, audio, and video [4].

Example applications include automatic extraction of international events from political context [5], public opinion measurement from social media posts [6], sense of place [7], and community happiness [8]. There are a growing number of uses of NLP methods to understand topics of disasters, [9, 10].

c. **Machine Learning in Disaster Management**

AI and machine learning can help public safety officials refine strategies over time, getting smarter about planning and response. AI can be used to analyze event data for patterns, identify current at-risk areas and populations, and model future needs, based on population growth, development, and climate change, among other variables. Government leaders can use these insights to craft policies that reduce the impact of disasters on communities, like planning new buildings in less vulnerable areas, [25].

(2) **Intelligent Systems Based IoT and Cloud Computing in Disaster Management**

IoT refers to a network of physical objects embedded with sensors and software that collect data and communicate with one another.

As it relates to emergency management, IoT can be used to enhance data collection from the physical environment and quickly communicate this data to different city departments, [26].

During a crisis, IoT technology can help by continually updating which evacuation routes are no longer available and what transit options are up and running, for safer, faster mass people movement, [21]. Say there's a fire in a building or a stadium: IoT-powered systems can help direct individuals to all approved exits, while providing updates on which to avoid, [25].

Authors in [11] proposed an approach based on the ant system algorithm and Internet of Things. IoT was used to consider smoke concentration, temperature carbon monoxide concentration. Then, they apply ant colony algorithm for intelligent evacuation. The purpose of intelligent evacuation is achieved.

Furthermore, based on the research foundation of building data model construction, intelligent evacuation application, indoor location, shortest path solution and other issues, an intelligent evacuation system for large public buildings based on mobile terminal is constructed.

The proposed project in [12] is based on the powerful spatial analysis function of GIS, and uses the IoT, sensor network and artificial intelligence algorithm to analyze events in the intelligent space processing system, to support the development of intelligent evacuation systems for large public buildings. Large public building intelligent evacuation system takes mobile terminal as carrier, and install sensors, RFID tags, etc. in the interior space of the building, aiming to provide technical services such as emergency evacuation guidance and escape rescue for the personnel in the disaster.

In [13], the sensor network, which will be installed around 47 volcanoes that the Japanese government has selected for around-the-clock observation, will measure several different variables. In addition to the seismic activity that almost always occurs before an eruption, the sensors will monitor gas emissions, topography changes, and vibrations in the air caused by rocks and ash spewing from the volcano.

The information gathered by the sensors will be transmitted via LoRa [27], gateways to manned monitoring stations located 5–10 km away from the volcanoes. LoRa, also known as LoRaWAN, operates using a chirp spread spectrum radio scheme, sending data through a series of gateways that serve as a bridge between the sensors and network servers.

BRINCO system is the first IoT-enabled beacon that is designed to notify its user about possible earthquake or tsunami in personal-aware mode. The sensor system comprises of accelerometer, signal processing unit and audio alarm units. It works as follows. If it perceives a vibration of the ground, it sends this information to the Brinco Data Center (BDC), a private cloud service. This DC assimilates this information with other seismic networks information to obtain its perception. Finally, if the judgment is good enough, it makes alarming sound and sends push notifications to it users smart phone (Android or iOS) instantly. Further, this information can be shared among the local as well as global community utilizing social network sites, [28].

BRCK It is versatile IoT-enabled device meant to be used in poor infrastructures. This gives it power to connect with low connectivity areas where 2G communication still exists. It is also em-powered with its private cloud service where environment data could easily be transmitted and fetched on. It is capable to work with solar energy, hence very much suitable for disastrous sites where flawless power is a main constraint. The rugged design makes Brck the most suitable product to be deployed in disaster management scenario. Users having smart phone can easily connect with it and share the information to other WiFi-enabled local devices, [29].

a. **Discussion**

Intelligent systems have some characteristics making them difficult to apply in middle and low SDI countries, where infrastructures are also poor. We present bellow some of these characteristics.

Data Cannot be Effectively Collected

How to realize the integration of disaster data becomes an urgent and necessary key problem.AI related data include meteorological data, urban waterlogging data, socio-economic data, and other sources, and the amount of data is huge. Furthermore, as the data come from different departments such as water conservancy, meteorology, urban management, operators and Internet, the spatial and temporal scales are not compatible with each other, and the format standards are not unified, which poses a great obstacle to the AI for natural disasters, [14, 15].

Incomplete Information

Decision is a question of timing, and this is particularly prominent in intelligent systems because of the sudden, rapid evolution of disasters. Short time emergency decision face the restriction of personnel, resources, information and other factors, therefore, decision information is discredited and incomplete, [16, 17]. How to deal with the incomplete information constraint is a difficult problem faced by intelligent systems.

Data Unavailability

Intelligent information processing techniques based on AI and machine learning such as big data mining, remote sensing and GIS are promising methods, especially when applied with a combination of conventional forecasting approaches working to update dynamic demand information. However, its application is constrained due to the lack of data availability from governments concerning risk and safety issues during the urgent and limited time after unconventional emergency events have occurred. From this perspective, the access to open source data from governments should be properly unimpeded, [18].

Prediction Problem

In emergency situations there is an inherent demand uncertainty, requiring a large scale of data sources to explore the characteristics of the target prediction case. A great deal of crucial information required for demand predictions is difficult to obtain in the hours immediately after an emergency event. Additionally, in order to save as many lives as possible, analysis of large-scale data requires information processing techniques and methods to be rapid and efficient, making the demand prediction problem based on information processing techniques unique and challenging, [34].

(3) **Smartphone Applications for Disasters Management**

When a disaster happens, the Smartphone is generally used to send information report. If a natural disaster happens, disaster information, including time, location, classification, degree, trend of disaster, etc., need to be collected and sent to the emergency management center through Smartphone. Among information, the geometry location of the disaster is provided by the Smartphone's location based services, [18, 19].

a. **Smart Rescue**

The basic notion of Smart Rescue is to use Smartphone technology to assist in delay phase in the initial crisis times. Smart Rescue technology maps threats and help people in the case of emergency. If many Smartphones are sensing the environment surrounding the people then those phones are used as input sources for getting threat pictures and allow and inform people to take necessary actions to avoid any hazards in the affected area [1].

b. **FEMA**

The FEMA Application (Smartphone app for mobile devices) contains disaster safety tips, interactive lists for storing your emergency kit and emergency meeting location information, and a map with open shelters and open FEMA Disaster Recovery Centers (DRCs). Bellow some FEMA functionalities [31].

- Receive real-time alerts from the National Weather Service for up to five locations nationwide.
- Share real-time notifications with contacts via text, email and social media.
- Learn emergency safety tips for over 20 types of disasters, including fires, flooding, hurricanes, snowstorms, tornadoes, volcanoes and more.
- Locate open emergency shelters and disaster recovery centers in the area where user can talk to a FEMA representative in person.
- Prepare for disasters with a customizable emergency kit checklist, emergency family plan, and reminders.
- Connect with FEMA to register for disaster assistance online.
- Upload and share disaster photos through Disaster Reporter.
- Toggle between English and Spanish.
- Follow the **FEMA blog** to learn about the work FEMA does across the United States.

c. **First Aid Application (FA)**
FA is developed to give some preliminary instructions for taking care of users in Android Smartphones; basically navigation system uses Google API (maps) for searching an appropriate or suitable way or path to the nearest hospital. In the case of any emergency this function is activated on user's Smartphone to navigate victims through the shortest path to the hospital [2]. This application gives some useful instructions or precautionary measures about taking initial care of the patients before sending them to the doctors or hospitals.

d. **Fire Ready (FR)**
FSC (Fire Service Commissioner) has launched the fire ready application is the official Victoria government app for Country Fire Authority (CFA), Metropolitan Fire Bridge (MFB) and Department of environment, Land, Water, Planning (DELWP). This fire warnings and information system, notifies users of fire dangers in affected area and sends photographs of bushfire activity. Application is managed by Victoria emergency management on behalf of the fire agencies, supported by the department of Justice and Regulations, [32].

e. **Automatic Crash Notification (ACN)**
Christopher Thompson presented another innovative application that is Automatic Crash Notification system. The aim of this application is to save lives by reducing time required for emergency teams to arrive to victims. ACN sensor network in automobiles is used to detect car accidents, it also communicates with a checking or monitoring station through radio cellular network. Sensor devices provide useful information to detect auto car destruction. Wreck watch server utilize ACN system to detect car accidents that are displayed on Smartphone devices and the user is instantly allowed to access accident information through webpage [20].

f. MYSHAKE

It is an APP-based service for the detection of seismic activities. This APP has initially to be installed on users smart phone which whenever perceives a ground vibration through the phones accelerometer, performs a match operation with the vibrational profile of the quake. If matched, the information along with the present GPS coordinate (received from the Smartphone) is sent for analysis to the Berkeley Seismological Laboratory (BSL) for final check. This has opened a way to develop a cost-effective, distributed and crowd sourced quake monitoring system that is obviously a demand of time, [30].

3   Scream to Survive System (S2S)

S2S system combines the use of Smartphones, widely spread even in developing countries, and voice recognition to help victims. Generally, a disaster victim is enable to access to his Smartphone and ask help, with this system, saying "help" will be enough to send automatically alerts to the nearest Emergency Operation Services (EOS).

The essential feature of this system is the ability to analyze victims voices, words, sentences, etc., in order to detect the emergency situation and immediately transmit victims GPS location, identification, etc.

In EOS side, after the first alert, the system save victim GPS coordinates and collect instantly all near users locations (potential victims). It proceeds to alert EOS staff, family contacts and eventually nearest volunteers registered in the system.

Some of the basic objectives of S2S are:

- It automatically detects any disaster by using voice recognition algorithms.
- It responds in the time of critical situations by using real time system, right records are sent at right time. Also, the other stakeholders are informed automatically by S2S and the alerts are communicated through notifications, SMS, email etc.
- It responds to emergency situation with minimal human/manual intervention and interaction.
- It takes into account potential victims, by collecting all near users having S2S system on their smart phones.

S2S system is composed of two parts:

(1)   S2S for victims and citizens (Android or IOS Smartphone)

This application collects voices, words or sentences, said by the victim, through Smartphone microphone and converts them to text in order to deduce the emergency situation and its type (fire, earthquake, flood, etc.).

After that, the application sends an alert message containing victim information, his last location and risk type to the EOS.

(2)   S2S for EOS

For EOS, we propose S2S part that allows officers and employees to receive alerts sending by victims and select the appropriate action plan according to emergency type.

They can consult all medical and personnel information from database and send help to the exact location of the victim.

The system collects all near users positions. This information will be very useful for victims relief, in case of earthquake for example.

Furthermore, the system sends victims location to nearest volunteers and family contacts to increase surviving chance. The process is detailed in Fig. 2.

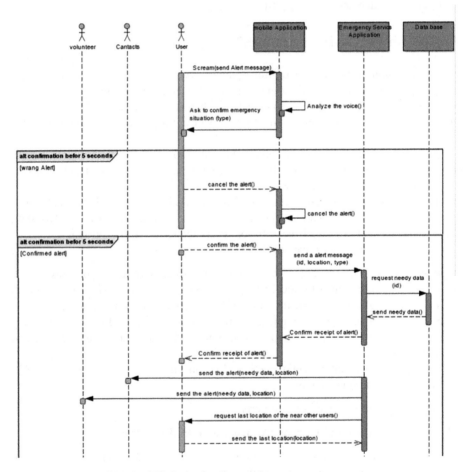

Fig. 2. S2S Sequence diagram between system actors

4   Implementation

In this section, we will describe implementation principals and tools. Then we present some screens of S2S system.

(1) Implementation Tools

a. Flutter

Flutter is Google's mobile app SDK that gives developers an easy way to build and deploy visually attractive, fast mobile apps on both Android and iOS platforms.

Flutter is an app SDK for building high-performance, high-fidelity apps for iOS, Android, and web from a single codebase.[1] Flutter combines a Dart framework with a high-performance engine.

The Flutter Engine is a portable runtime for high-quality mobile applications. It implements Flutter's core libraries, including animation and graphics, file and network I/O, accessibility support, plug-in architecture, and a Dart runtime and tool chain for developing, compiling, and running Flutter applications.

Dart[2] is a client-optimized language for fast apps on any platform, made by Google, it is:

- Optimized for UI: Develop with a programming language specialized around the needs of user interface creation.
- Productive development: Make changes iteratively, use hot reload to see the result instantly in your running app.
- Fast on all platforms: Compile to ARM & x64 machine code for mobile, desktop, and backend. Or compile to JavaScript for the web.

To develop the UI of S2S system, we choose Flutter for many reasons, here some of them:

- Faster code writing: With flutter we can make changes in the code and see them straight away in the app, this is called Hot reload, which usually takes milli-seconds and help teams add features, fix bugs and experiment faster.
- Faster apps: Flutter apps work in a smooth and fast way, without hanging and cutting while scrolling.
- Same app UI on older devices: Your new app will look the same, even on old versions of Android and iOS systems. Flutter runs on Android Jelly Bean and newer, as well as iOS 8 and newer.
- And the most important reason why we choose flutter is **one code for two platforms:** this means that we can code once and get same app for two platforms (Android and iOS), that helps us to spread S2S system in both platforms and target more users.

b. TensorFlow Lite

TensorFlow Lite is a set of tools to help developers run TensorFlow models on mobile, embedded, and IoT devices. It enables on-device machine learning inference with low latency and a small binary size.[3]

S2S system will be used in emergency situation and with that a lot of pressure is put on the device to act accordingly. The biggest problem we are dealing with is weak

---

[1] Official flutter documentation: https://flutter.dev/docs/resources/technical-overview.

[2] Official dart website: https://dart.dev/.

[3] Official TensorFlow website: https://www.tensorflow.org/lite/guide.

internet connection (maybe loss of internet connection). Which will limit the process of being able to send data back and forth from the server. On the other hand, we offer a lot of benefits that comes with an on-device machine learning such as:

- Privacy: Data will not leave the device.
- Connectivity: Internet connection isn't required
- Power consumption: Network connections are power hungry

(2) **Principles of voice recognition implementation**

When a catastrophe occurs, the victim is in a state of fear and panic, so he begins to shout and ask for help in order to survive. The first step we do is to build datasets words said during disasters as well as surrounding sounds.

- The first dataset is a sample of a human words during a disaster.
- The second dataset is a sample of surrounding voices when a disaster strike.
- The third is a combination of the first and the second.

The second step is building files with the extension **.tflite** models (created with TensorFlow Lite) from our datasets and integrate them in S2S mobile application where the captured voice will be compared with models.

(3) **Prototyping**
We realize S2S prototypes: for the victim and the EOS.

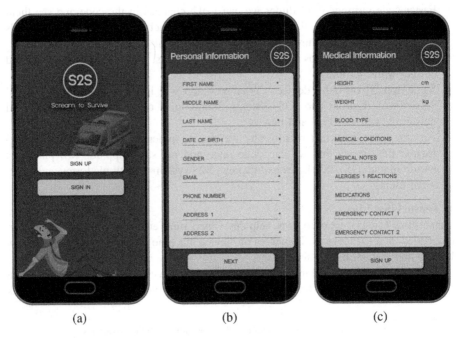

Fig. 3. a. Signing screen b. Personal information c. Medical information

- **S2S system for the victim**

Some screens of S2S application for the victim are shown in Fig. 3. User login to the application to update his information. In emergency situation, user access directly. The system collects medical information which will be used to help him in emergency situation. Furthermore, the user enter emergency contacts.

When the application detects an emergency (after voice recognition process and translating to words and/or texts), it will ask user to confirm the sending or cancel it during 5 s then it sends the alert message automatically to the EOS. The message contains user id, type of emergency and last location. The goal of the confirmation is to avoid errors and conflict situation (one can discuss with friends about earthquakes) (Fig. 4).

- **S2S system for Emergency Operation Services**

In the side of EOS, when officers receive the alert as notification and a red label will appear in the map, as shown in Fig. 5. Using the received information about victim, S2S EOS gives the possibility to consult all victims personal and medical information. EOS employees will send helps to the exact locations.

Fig. 4. Voice recognition and sending process

Furthermore, we propose messages to emergency contacts and to nearest volunteers, Fig. 6. This is called: Crisis Mode.

5 Performances Evaluation

All systems, presented in this paper, have been grouped together in Table 1 in order to compare and analyze them. We have Intelligence and new technologies based systems and Smartphone based systems.

We can observe that most of presented intelligent systems are successful in countries where infrastructures are adapted, because they are based on new technologies (IoT, permanent Internet connection, cloud, WSN, etc.).

As for communication, it is done through advanced technologies such as WSN, internet, satellite telemetering technology, high-precision air-to-land observation technology.

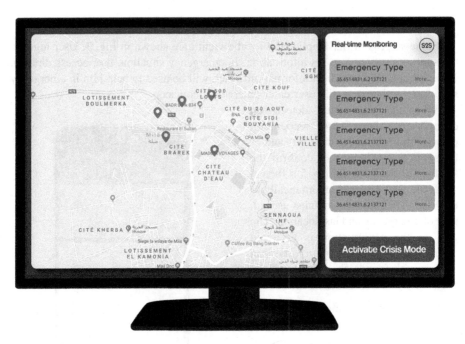

Fig. 5. S2S EOS: Victims list with the last positions (Red Sparrows) (Color figure online)

Fig. 6. S2S EOS: Crisis Mode Activated: Collecting all surrounding victims coordinates

Table 1. Comparison between Intelligent based Systems, Smartphone based Systems and S2S.

Works based on New Technologies

Disaster management solution	Used technologies	Automatic activation	Feasibility low and middle SDI countries	Taking other victims in consideration	Communication	Disaster management levels
Mobile fire evacuation system [11]	Iot, Indoor location (Ant Algorithm)	Yes	Medium	No	WiFi, WSN Internet	Recovery
GIS and IoT for evacuation [12]	Iot, Powerful spatial analysis function Of GIS, AI algorithms	Yes	Low	No	WiFi, WSN, Satellite telemetering Internet	Recovery
WSN for active Volcanoes [13]	Sensors, Analysis function of GIS, AI algorithms, Cloud	Yes	Low	Yes	Wifi, Servers Network, WSN, Satellite telemetering, Internet, Lora	Preparedness
BRINCO [28]	Sensors, Big data	Yes	Low	Yes	Accelerometer, Signal Processing, WiFi, WSN, Internet, Smartphones	Preparedness, Response
BRCK [29]	Iot, Smart devices, Cloud, Wifi	No	Good	No	WiFi, WSN, Internet, Smartphones	Response, Recovery

Smartphone applications for disasters management

Smart rescue [1]	Sensors network, WiFi	No	Medium	Yes	WiFi, WSN, Internet, Smartphones	Preparedness, Response
FEMA [31]	GIS, Cloud	No	Medium	Yes	WiFi, WSN, Internet, Smartphones	Response, Recovery
First Aid Application [2]	GIS,	No	Good	No	WiFi, WSN, Internet, Smartphones	Response,
Fire Ready [32]	GIS, WiFi, Sensor network	Yes	Low	No	WSN, Internet, Smartphones	Mitigation preparedness, Response
Automatic crash notification [20]	Sensor network,	Yes	Medium	No	Internet, Smartphones, Wifi, WSN	Response

S2S: Scream to Survive System

| S2S | GIS, Voice recognition algo | Yes | Good | Yes | Internet, Smartphones | Response, Recovery |

In disaster situation they are automatically triggered or activated. and take into consideration several victims at a time. But most of these systems are not feasible in SDI-reduced countries.

For their part, Smartphone-based systems are mostly feasible in these countries, because Smartphone today is widely used. But, most of these systems are not activated automatically and do not take the other users in consideration (only those who have the application installed on their Smartphone).

In addition they do not take into account the rescue level which is very important to save lives after any disaster.

Our system is a combination between smart systems and mobile apps on Smartphone. On the one hand we used simple artificial intelligence algorithms for voice recognition.

On the other hand, communication will be done either by telephonic network or Internet (depending on the situation of these networks) to inform the EOS. Furthermore, our system takes into consideration a maximum of victims by collecting their coordinates as soon as the disaster happens.

Finally, the system takes into consideration the two levels where most lives could be saved: response and recovery.

6 Conclusion

In recent years the disaster impact on human and material losses are considerable, especially in middle and low SDI countries.

There are many systems designed for EOS in case of disasters to help the victims. But these systems are often expensive for middle and low SDI countries. That is why we have proposed in this paper Scream to Survive system (S2S) based on Smartphone, as they are increasingly used, and artificial intelligence technology for voice recognition. With S2S system, it is sufficient for the victim to say words about disasters or for asking help. The system is automatically activated to send notifications (sms, notifications, etc.) to the EOS, to the victims' contacts and to the volunteers pre-registered in S2S system. Furthermore, in the case of disasters like earthquakes, the system proceeds to the collection of close people positions. It will be very useful in the recovery relief.

References

1. Mokryn, O., Karmi, D., Elkayam, A., Teller, T.: Opportunistic smart rescue application and system. In: The 11th Annual Mediterranean Ad Hoc Networking Workshop (Med-Hoc-Net) (2012)
2. Surachat, K., Kasikri, S., Tiprat, W., Wacharanimit, A.: First aid application on mobile device. Int. Sch. Sci. Res. Innov. 7(5), 361–366 (2013)
3. Mythili, S., Shalini, E.: A comparative study of smart phone emergency applications for disaster management. Int. Res. J. Eng. Technol. (IRJET) 03(12), 392–395 (2016). e-ISSN 2395-0056

4. Kongthon, A., Sangkeettrakarn, C., Kongyoung, G., Haruechaiyasak, C.: Implementing an online help desk system based on conversational agent. In: MEDES 2009: The International Conference on Management of Emergent Digital EcoSystems. ACM, France (2009). https://doi.org/10.1145/1643823.1643908
5. O'Connor, B., Stewart, B.M., Smith, N.A.: Learning to extract international relations from political context. In: Proceedings of the 51st Annual Meeting of the Association for Computational Linguistics, vol. 1, Long Papers (2013)
6. O'Connor, B., Balasubramanyan, R., Routledge, B.R., Smith, N.A.: From tweets to polls: linking text sentiment to public opinion time series. In: Proceedings of the 4th International AAAI Conference on Weblogs and Social Media (2010)
7. Adams, B., Raubal, M.: Identifying salient topics for personalized place similarity. Research@Locate (2014)
8. Ramirez-Esparza, N., Chung, C.K., Sierra-Otero, G., Pennebaker, J.W.: Cross-cultural constructions of self-schemas: Americans and Mexicans. J. Cross Cult. Psychol. 43(2), 233–250 (2012)
9. Lin, Y.-R., Margolin, D.: The ripple of fear, sympathy and solidarity during the Boston bombings. EPJ Data Sci. 3(1), 31 (2014)
10. Cohn, M.A., Mehl, M.R., Pennebaker, J.W.: Linguistic markers of psychological change surrounding September 11, 2001. Psychol. Sci. 15(10), 687–693 (2004)
11. Jiang, H.: Mobile fire evacuation system for large public buildings based on artificial intelligence and IoT. IEEE Access (2019). Special section on data mining for internet of things. https://doi.org/10.1109/ACCESS.2019.2915241
12. Liu, S.J., Zhu, G.Q.: The application of GIS and IOT technology on building evacuation. Procedia Eng. 71, 577–582 (2014)
13. Morra, J.: Wireless sensor networks monitor active volcanoes in Japan, Electronic Design (2016). Accessed 27 Aug 2017. http://electronicdesign.com/iot/wireless-sensor-networks-monitor-activevolcanoes-japan
14. Zhou, L.: Emergency decision making for natural disasters: an overview. Int. J. Disaster Risk Reduction (2017). http://dx.doi.org/10.1016/j.ijdrr.2017.09.037
15. Wu, X.H., Xiao, Y., Li, L.S., Wang, G.J.: Review and prospect of the emergency management of urban rainstorm waterlogging based on big data fusion. Chin. Sci. Bull. 62 (2017), 920–927 (2017)
16. Xu, Z.S., Zhang, X.: Hesitant fuzzy multi-attribute decision making based on TOPSIS with incomplete weight information. Knowl. Based Syst. 52(6), 53–64 (2013)
17. Ergu, D., Kou, G., Peng, Y., et al.: Estimating the missing values for the incomplete decision matrix and consistency optimization in emergency management. Appl. Math. Model 40(1), 1–14 (2015)
18. Zhu, X., Zhang, G., Sun, B.: A comprehensive literature review of the demand forecasting methods of emergency resources from the perspective of artificial intelligence. Natural Hazards 97, 65–82 (2019). https://doi.org/10.1007/s11069-019-03626-z
19. Baoquan, Y., Ruizhi, S., Hongjun, Y.: A study of smartphone based disaster information reporting system under disaster environment. Int. J. Smart Home 9(1), 45–52 (2015)
20. Thompson, C., White, J., Dougherty, B., Albright, A., Schmidt, D.C.: Using smartphones to detect car accidents and provide situational awareness to emergency responders. In: International Conference on Mobile Wireless Middleware, Operating Systems, and Applications, pp. 29–42 (2012)
21. Ray, P.P., Mukherjee, M., Shu, L.: Internet of things for disaster management: state-of-the-art and prospects. IEEE Access 5, 18818–18835 (2017)
22. https://interestingengineering.com/technology-used-in-disaster-relief-saving-more-lives-every-year

23. https://datasmart.ash.harvard.edu/news/article/three-emerging-technologies-improve-emergency-management
24. https://developer.ibm.com/articles/disaster-management-using-blockchain-iot/
25. https://azure.microsoft.com/es-es/blog/using-ai-and-iot-for-disaster-management/
26. https://www.ibm.com/blogs/internet-of-things/what-is-the-iot/
27. https://www.electronicdesign.com/industrial-automation/lorawan-made-iot
28. https://atmelcorporation.wordpress.com/2015/08/03/brinco-is-a-personal-early-warning-system-for-earthquakes-and-tsunamis/
29. https://www.brck.com/
30. MyShake. http://myshake.berkeley.edu
31. https://www.fema.gov/
32. https://www.emv.vic.gov.au/our-work/victorias-warning-system/fireready-app/
33. https://ourworldindata.org/natural-disasters
34. https://www.nap.edu/read/11793/chapter/5#50

Association Rules Algorithms for Data Mining Process Based on Multi Agent System

Imane Belabed[1]([⊠]), Mohammed Talibi Alaoui[2], Jaara El Miloud[1], and Abdelmajid Belabed[1]

[1] University Mohammed The First, Oujda, Morocco
Belabedimane@gmail.com
[2] Faculty of Science and Technology, University Mohammed Ben Abdellah, Fez, Morocco

Abstract. In this paper, we present a collaborative multi-agent based system for data mining. We have used two data mining model functions, clustering of variables in order to build homogeneous groups of attributes, association rules inside each of these groups and a multi-agent approach to integrate the both data mining techniques. For the association rules extraction, we use both apriori algorithm and genetic algorithm.

The main goal of this paper is the evaluation of the association rules obtained by running apriori and genetic algorithm using quantitative datasets in multi agent environment.

Keywords: Association rules · Apriori · Clustering · Multi agent system · Genetic algorithm

1 Introduction

In recent years, more researchers have been involved in research on both agent technology and data mining. A clear disciplinary effort has been activated toward removing the boundary between them, which form the interaction and integration between agent technology and data mining.

DM (Data Mining) has evolved to become a well-established technology field with subfields such as classification, clustering, and rule mining.

In fact, the clustering encompasses a number of different algorithms and methods for grouping objects of similar kind into respective categories. Such algorithms or methods are concerned with organizing observed data into meaningful structures.

Otherwise, the association rules aims at finding strong relations between attributes.

In order to integrate the two data mining techniques, we used a multi agent system which is the combination of multiple agents. Indeed, an agent is a computer system that is capable of autonomous action on behalf of its user or owner. It is capable to figure out what it is required to be done, rather than just been told what to do.

In this work, we developed a multi agent data mining framework to extract useful rules from real data sets, relying on clustering of variables to build homogeneous groups of attributes and mine supervised association rules inside each cluster.

© IFIP International Federation for Information Processing 2020
Published by Springer Nature Switzerland AG 2020
S. Boumerdassi et al. (Eds.): MLN 2019, LNCS 12081, pp. 431–443, 2020.
https://doi.org/10.1007/978-3-030-45778-5_30

For the clustering step, we used the K-Means algorithm for variables and for the association rules step we compare the results obtained by using two association rules algorithm, the apriori algorithm and the genetic algorithm knowing that we deal only with quantitative datasets.

2 Related Work

In this section, we will briefly review of the previously proposed studies in autonomous intelligent agent systems or multi-agent systems for data mining and knowledge discovery in database. The techniques used in these studies include association rule mining, associative classification mining, computational intelligence and rule generation algorithms. The proposed approach is mainly related to two areas of research, knowledge extraction from large dataset and knowledge modeling using multi-intelligent agent system. Warkentin, Sugumaran, and Sainsbury produce a study in which they discuss the role of intelligent agents and data mining in electronic partnership management. The procedures of data mining used in this process can be enhanced by using intelligent agents [13]. Nahar, Imam, Tickle, and Chen discussed a paper in which they used association rule mining and a computational intelligence to identify the factors which contributes to heart diseases for males and females. This research presents rule extraction experiments on heart disease data using three rule generation algorithms apriori, predictive apriori and tertius [14]. Ait-Mlouk, Agouti and Gharnati propose an approach to discover a category of relevant association rules based on multi-criteria analysis to avoid redundant rules, they use multi agent system to manage and model the quality measurement according to six agents working in cooperation [15].

3 Data Mining Process

Currently, enormous volumes of data are being produced and stored in computer systems around the world. So, data mining techniques are adequate to address the problem of analyzing and understanding the massive datasets [11].

In this work, we use firstly, a combination of K-Means clustering for variables and supervised association rules i.e. the right part of the rule are always known (the variables to predict) Table 1. Secondly we automate the process by relying on a multi agent system. Through the research we were faced with several limitations such:

- Using K-Means for clustering variables.
- Using association rules algorithm for quantitative datasets.

To deal with the first limitation, we choose to deal only with quantitative datasets and transpose the data in order to cluster the variables instead of individuals.

For the second limitation, we choose to compare two approach of rule mining. The first one concerns the use of apriori algorithm, this after the discretization of all datasets. The second approach is using genetic algorithm for quantitative datasets.

We will apply the proposed system on a real dataset to illustrate how the proposed system can extract a set of rules from real dataset to construct knowledge base.

- Heart datasets: Heart disease.
- Pima datasets: Pima Indians Diabetes Database.
- Vehicle datasets: Use vehicle silhouette to predict the model of a vehicle.

The data used in this work comes from UCI archives, internet.

Table 1. Datasets description

Datasets	Number of variables	Number of lines	Variable to predict
Heart datasets	13	244	Presence of heart disease
Pima datasets	8	692	Presence of diabetes
Vehicle datasets	15	677	The model of vehicle

4 Overview of Basic Techniques Used in the Data Mining Process

4.1 K-Means Clustering

The K-Means algorithm takes two input parameters: the dataset of n objects, and k, the number of clusters to be created. The algorithm partitions the dataset of n objects into k clusters. Cluster similarity is measured by taking the Euclidean distance between objects. In this way, K-Means finds spherical or ball shaped clusters. The mean value of the objects in a cluster can be viewed as the cluster's center of gravity.

Formally, the K-Means clustering algorithm follows the following steps:

Step 1: Choose a number of desired clusters, k.

Step 2: Choose k starting points to be used as initial estimates of the cluster centroids. These are the initial starting values.

Step 3: Examine each point in the dataset and assign it to the cluster whose centroid is nearest to it.

Step 4: When each point is assigned to a cluster, recalculate the new k centroids.

Step 5: Repeat steps 3 and 4 until no point changes its cluster assignment, or until a maximum number of passes through the dataset is performed.

4.2 Discretization Pre-processing

Discretization is a data preprocessing technique which transforms continuous attributes into discrete ones by dividing the continuous values into intervals, or bins. In this work, we based our discretization process on Class-Attribute Interdependence Maximization (CAIM) which is a discretization algorithm of data where the classes are known. In fact the CAIM algorithm works in a greedy top down manner. It starts with a single interval and divides it iteratively, using for the division the boundary that gave the highest values of the CAIM criterion. The algorithm assumes that every discretized attribute needs at least number of intervals equal to the number of classes.

Let us assume that we have a training data set consisting of M examples, and that each example belongs to only one of the S classes. F will indicates any of the continuous attributes. Then there exists a discretization scheme D on F, which discretizes the continuous domain of attribute F into n discrete intervals bounded by the pairs of numbers: [1].

$$D : \{[d_0, d_1], (d_1, d_2], \ldots (d_{n-1}, d_n\}$$

where d_0 is the minimal value d_n is the maximal value of attribute F, and the values are arranged in the ascending order. These values constitute the boundary set $\{d_1, d_2, d_3, \ldots d_{n-1}, d_n\}$ for discretization D.

• Caim criterion

Given the quanta matrix defined in Fig. 1, the Class-Attribute Interdependency Maximization (CAIM) criterion that measures the dependency between the class variable C and the discretization variable D for attribute F is defined as:

$$\text{CAIM (C, D|F)} = \sum_{i=1}^{n} \frac{\frac{max_i^2}{M_{ir}}}{n} \tag{1}$$

Where:
n is the number of interval
i iterates through all intervals, i.e. i = 1,2,..n.

max_i is the maximum value among all q_{ir} values (maximum value within the ith column of the quanta matrix), r = 1,2,...S (see Fig. 1).

M_{ir} is the total number of continuous values of attribute F that are within the interval $[d_r, d_{r-1}]$.

Class	Interval					Class Total
	$[d_0, d_1]$	...	$(d_{r-1}, d_r]$	...	$(d_{n-1}, d_n]$	
C_1	q_{11}	...	q_{1r}	...	q_{1n}	M_{1+}
:	:	...	:	...	:	:
C_i	q_{i1}	...	q_{ir}	...	q_{in}	M_{i+}
:	:	...	:	...	:	:
C_S	q_{S1}	...	q_{Sr}	...	q_{Sn}	M_{S+}
Interval Total	M_{+1}	...	M_{+r}	...	M_{+n}	M

Fig. 1. Quanta matrix

4.3   Association Rules

The most popular task of DM is to find trends in data that show associations between domain elements. This is generally focused on transactional data such as a database of purchases at a store. This task is known as Association Rule Mining (ARM). It was first introduced in Agrawal et al. [2]. Association rules identify collections of data attributes that are statistically related in the underlying data. An association rule is of the form $X \rightarrow Y$ where X and Y are disjoint conjunctions of attribute value pairs. The confidence of the rule is the conditional probability of Y given X, $Pr(Y|X)$, and the support of the rule is the prior probability of X and Y, $Pr(X \cap Y)$. Here probability is taken to be the observed frequency in the dataset.

The traditional ARM problem can be described as follows. Given a database of transactions, a minimal confidence threshold and a minimal support threshold, find all association rules whose confidence and support are above the corresponding thresholds.

Apriori Algorithm. The apriori algorithm iteratively identifies frequent itemsets FIs, in data by employing the "closure property" of itemsets in the generation of candidate itemsets, where a candidate (possibly frequent) itemset is confirmed as frequent only when all its subsets are identified as frequent in the previous pass. The closure property of itemsets can be described as follows: if an itemset is frequent then all its subsets will also be frequent; conversely if an itemset is infrequent then all its supersets will also be infrequent.

- Apriori

```
Input: (a) A transactional database Dt;
(b) A support threshold s;
Output: A set of frequent itemsets S;
1: begin:
2: k ←1;
3: S← an empty set for holding the identified frequent itemsets;
4: generate all candidate 1-itemsets from Dt;
5: while (candidate k-itemsets exist) do
6: determine support for candidate k-itemsets from Dt;
7: add frequent k-itemsets into S;
8: remove all candidate k-itemsets that are not sufficiently supported to
give frequent k-itemsets;
9: generate candidate (k + 1)-itemsets from frequent k- itemsets using
closure property;
10: k←k + 1;
11: end while
12: return (S);
13: end Algorithm
Note: A k-itemset represents a set of k items.
```

Genetic Algorithm. Genetic algorithms are stochastic search methods that mimic the metaphor of natural biological evolution [4]. At each generation, a new set of approximations is created by the process of selecting individuals according to their level of fitness in the problem domain and breading them together using operators borrowed from natural genetics. Genetic algorithm take as an input the following elements: population, selection according to fitness, crossover to produce new off-spring, and random mutation of new offspring.

- Initial population: The initial population of individuals is generated as follows: in the first individuals, the intervals $[l_i, u_i]$ represent the whole domain of the ith numeric attribute, and the following individuals encode intervals with decreasing amplitudes (length of intervals) until they reach a minimum support in the dataset. Once the amplitudes are fixed for an individual, the bounds l_i and u_i are chosen at random.
- Mutation and crossover: The crossover operator consists in taking two individuals, called parents, at random and generating new individuals: Each attribute the interval is either inherited from one of the parents or formed by mixing the bounds of the two parents [5]. Mutation works on a single individual and increases or decreases the lower or upper bound of its intervals respectively. Moving interval bounds is done so as to discard/involve no more than 10% of tuples already covered by the interval [6].
- Fitness function: The fitness function used is based on the gain measure [7]. If the gain is positive (the confidence of the rule exceeds the minimum confidence threshold), we take into account the proportions of the intervals (defined as the ratios between the amplitudes and the domains). Moreover, rules with low supports are penalized by decreasing drastically their fitness values by the number of tuples in the database [8].

Algorithm. The algorithm starts with a set of rule templates and then looks dynamically for the "best" intervals for the numeric attributes present in these templates. An optimization criterion based on both support and confidence is used to keep only high quality and interesting rules [4]. The algorithm follows a prototypical genetic algorithm scheme. The inputs are the minimum support (MinSupp), the minimum confidence (MinConf), the population size (PopSize), the number of generations (GenNum), the fraction of population to be replaced by crossover (Cross) and the mutation rate (MutR).

```
Input: A dataset composed of NbTuples, PopSize, GenNb, CR, MR, MinSupp,
MinConf
Output: Quantitative association rules R
Select a set of attributes
Let Rt a set of rule templates defined on these attributes
foreach r ∈ Rt do
Generate a random population P of PopSize
While ≤ GenNum do
Form the next generation of population by mutation and crossover w.r.t.
MutR and Cross.
```

```
Extract the itemsets that satisfy the best fitness to constitute the asso-
ciation rule values
i++
Return R= max (fitness (r)); r belongs to P
```

4.4   Agent and Multi Agent System

Agents and multi-agent systems are an emergent technology that is expected to have a significant impact in realizing the vision of a global and informational rich services network to support dynamic decision making [3].

Agents. Agents are defined by Wooldridge [9] as computer systems that are situated in some environment and are capable of autonomous action in this environment in order to meet their design objectives.

Multi Agent Systems. By combining multiple agents in one system to solve a problem, the resultant system is a multi-agent system (MAS). These systems are comprised of agents that individually solve problems that are simpler than the overall system problem. They can communicate with each other and assist each other in achieving larger and more complex goals [12].

5   Multi Agent Framework for Data Mining Process

The proposed mining framework comprises four categories of agent Fig. 3:

- User agent: User agent is charged by the communication with the user interface.
- Coordinator agent: Coordinator Agent is focused on the correct message transmission among the agents. It takes the requirements (data, number of clusters...) and sends them to the corresponding agent.
- Data agent: Data Agent is in charge of a data source; it interacts and allows data access. There is one data agent per data source.
- Clustering agent: Clustering agent is concerned with a clustering K-Means algorithm.
- Association rules Agent: Association rules agent is in charge of extracting supervised rules though the genetic algorithm or apriori algorithm inside each cluster.

The sequence of operation between different agents constituted the system given in Fig. 3. This diagram shows the sequence of operations during the execution of the proposed multi-agent system. However in this work, we will focus on the execution of association rules agent. Indeed there are two scenarios for the quantitative datasets. The first is to execute the apriori algorithm and this includes the use of the discretization preprocessing. The second is the use of the genetic algorithm Fig. 2.

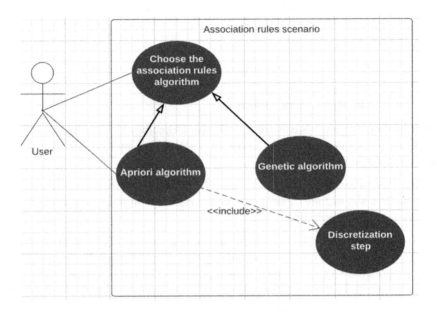

Fig. 2. Association rules use case.

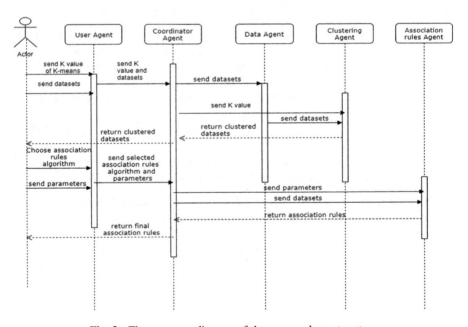

Fig. 3. The sequence diagram of the proposed agent system.

5.1   Overview of the Implementation of the Data Mining Process

The implementation of the Multi-Agent System for centralized Data Mining framework was done by using java platform through Agent-Oriented Programming paradigm (AOP). In order to allow inter-agents communication, agents must share the same language, vocabulary and protocols so; we have followed the recommendations of the standard Foundation for Intelligent, Physical Agents (FIPA).

We have developed our proposed framework with Java Agent Development (JADE) [10] which is FIPA-compliant middleware that enables the development of applications based on the agent paradigm and is adequate to process large amounts of data with a data mining approach.

6   Results

In this section, we will illustrate the results obtained by running our data mining process on the three datasets; heart, pima and vehicle datasets.

As perquisite, for clustering part, we choose to fix the K value of K-Means into 3. Also, for association rules part, we fix the confidence threshold at 60%, the support threshold at 10%; particularly for genetic algorithm we fix the population size at 250, the crossover rate at 50%, generation number at 100 and the mutation rate at 40%.

1. Heart datasets

Table 2. Result of heart datasets.

Association rules agent				
Heart datasets	Number of rules with apriori	Confidence	Support	Execution time
Cluster 1	1	70%	35%	3.27E-4 s
Cluster 2	101	>62%	>10%	
Cluster 3	10	>63%	>10%	
Total of rules	112	–	–	
Heart datasets	Number of rules with genetic	Confidence	Support	Execution time
Cluster 1	1	72%	27%	2.46E-4 s
Cluster 2	45	>63%	>10%	
Cluster 3	6	>65%	>10%	
Total of rules	52	–	–	–

For the heart datasets, we find that, first; the association rules agent with apriori algorithm generates more rules than with genetic algorithm. Second we notice that there is a small difference in the execution time between the two algorithms (Table 2).

However, in the case where the number of the rules generated is the same (cluster 1), we find that the confidence of the rule with genetic algorithm is higher than the one with apriori algorithm.

In this part and the next part, we specify that the execution time exclude the execution time of the dicretization in the case of apriori algorithm.

2. Pima datasets

Table 3. Result of Pima datasets.

Association rules agent				
Pima datasets	Number of rules with apriori	Confidence	Support	Execution time
Cluster 1	6	>65%	>39%	2.79E-4 s
Cluster 2	2	>71%	>16%	
Cluster 3	10	>66%	>45%	
Total of rules	18			
Pima datasets	Number of rules with genetic	Confidence	Support	Execution time
Cluster 1	10	>60%	>10%	2.62E-4 s
Cluster 2	2	>81%	>11%	
Cluster 3	14	>74%	>10%	
Total of rules	26			

The results of the pima datasets are the opposite of the results of heart datasets. In addition of the high confidence compared to apriori algorithm, the genetic algorithm extracts more rules than the apriori algorithm. Also, in the case where the rules extracted are the same (cluster 2), we notice that the confidence of rules with genetic algorithm is higher compared to apriori algorithm. Moreover, the execution time of genetic algorithm is lower than with apriori algorithm (Table 3).

3. Vehicle datasets

Table 4. Result of vehicle datasets.

Association rules agent				
Vehicle data	Number of rules with apriori	Confidence	Support	Execution time
Cluster 1	31	>61%	>10%	3.65E-4 s
Cluster 2	32	>60%	>10%	
Cluster 3	53	>60%	>11%	
Total of rules	116			
Vehicle data	Number of rules with genetic	Confidence	Support	Execution time
Cluster 1	25	>79%	>10%	0.003 s
Cluster 2	26	>60%	>10%	
Cluster 3	107	>60%	>10%	
Total of rules	158			

For the vehicle datasets, we conclude that genetic algorithm generates more rules than apriori algorithm, this with high confidence compared to apriori. However in this case the execution time of genetic algorithm is 10 times more than apriori algorithm (Table 4).

From the results presented in this section, we can conclude that genetic algorithm is more performing than the apriori algorithm. In the majority of cases this is due to the quality of the discretization phase.

In fact, some rules with apriori algorithm are redundant, does not brings new information, this is due to the dicretization intervals.

Example:

Support = 42   (17%),   confidence = 60%:   MaxHeartRate = [71.0–147.5]   and SerumCholestoral = [126.0–272.0] –> class = presence of disease.

Support = 26   (10%),   confidence = 86%:   MaxHeartRate = [71.0–147.5]   and SerumCholestoral = [272.0–417.0] –> class = presence of disease

On the other hand, in the three datasets, we find that genetic algorithm brings new rules that involve more attributes.

As a conclusion, we find that genetic algorithm is more adequate taking into consideration to use of multi agent system. Firstly, it generates more significant rules; secondly it avoids the discretization step which means that the gain of execution time in whole process including clustering and association rules will be considerable.

7 Discussion

The proposed approach illustrated in this work is more efficient compared to previous works such Ait-Mlouk (2016) [15]. In fact our approach deals with redundant rules by using genetic algorithm instead of multi criteria approach. That allows decreasing the number of agent in the association rules step. We have one association rules agent rather than six agents in that work.

Also our proposed approach surpasses that of Nahar [14], mainly in the execution time knowing that the both works processes heart datasets with the same number of variables.

The developed framework in this paper is presented with three real test cases from different domains such health and industry; however the framework is applicable to any other datasets. The use of multi agent system allows us to take advantage of four features: reactivity, autonomy, interaction and initiative. This makes our work extending to distributed and parallel paradigm.

8 Conclusion

We have proposed a new approach based on a Multi Agent framework for data mining process that includes genetic algorithm for extracting association rules, JADE frameworks and five different types of agents (user agent, clustering agent, association rule agent, data agent and coordinator agent).

In this work we focus on the association rules step, because we propose two scenarios for extracting rules. The experimental results proved that the extraction based on genetic algorithm is more adequate. In addition to the quality of rules extracted, the execution time of genetic algorithm is interesting because it avoids the discretization step.

Also, if we take into consideration the integration of the algorithm in data mining process using multi agent system and other algorithm that increase the execution time, such clustering, we conclude that the use of genetic algorithm is an optimization of the whole process.

As a perspective, we want to extend our approach to deal with a real time data sets from agriculture field in order to extract rules based on real time weather, ground composition. This, in order to improve the agricultural yields.

References

1. Ching, J.Y., Wong, A.K.C., Chan, K.C.C.: Class-dependent discretization for inductive learning from continuous and mixed mode data. IEEE Trans. Pattern Anal. Mach. **17**, 641–651 (1995)
2. Agrawal, R., Imielinski, T., Swami, A.N.: Mining association rules between sets of items in large databases. In: Proceedings of the 1993 ACM SIGMOD International Conference on Management of Data (1993)
3. Marwala, T., Hurwitz, E.: Multi-agent modeling using intelligent agents in a game of Lerpa. eprint arXiv:0706.0280 (2007)
4. Pei, M., Goodman, E.D., Punch, F.: Feature Extraction using genetic algorithm, Case Center for Computer-Aided Engineering and Manufacturing W. Department of Computer Science (2000)
5. Guo, H., Zhou, Y.: An algorithm for mining association rules based on improved genetic algorithm and its application. In: 3rd International Conference on Genetic and Evolutionary Computing, WGEC 2009, pp. 117–120 (2009)
6. Gonzales, E., Mabu, S., Taboada, K., Shimada, K., Hirasawa, K.: Mining multi-class datasets using genetic relation algorithm for rule reduction. In: IEEE Congress on Evolutionary Computation, CEC 2009, pp. 3249–3255 (2009)
7. Tang, H., Lu, J.: Hybrid algorithm combined genetic algorithm with information entropy for data mining. In: 2nd IEEE Conference on Industrial Electronics and Applications, pp. 753–757 (2007)
8. Dou, W., Hu, J., Hirasawa, K., Wu, G.: Quick response data mining model using genetic algorithm. In: SICE Annual Conference, pp. 1214–1219 (2008)
9. Wooldridge, M.: An Introduction to Multi-Agent Systems. Wiley, Hoboken (2003). (Chichester, England)
10. Bellifemine, F., Poggi, A., Rimassi, G.: JADE: a FIPA-compliant agent framework. In: Proceedings Practical Applications of Intelligent Agents and Multi-Agents (1999). http://www.jade.tilab.com
11. Han, J., Kamber, M.: Data Mining: Concepts and Techniques, 2nd edn. Morgan Kaufman Publihers, San Francisco (2006)
12. Popa, H., Pop, D., Negru, V., Zaharie, D.: AgentDiscover: a multi-agent system for knowledge discovery from databases. In: Ninth International Symposium on Symbolic and Numeric Algorithms for Scientific Computing, Timisoara, pp. 275–281. IEEE (2008)

13. Warkentin, M., Sugumaran, V., Sainsbury, R.: The role of intelligent agents and data mining in electronic partnership management. Expert Syst. Appl. **39**(18), 13277–13288 (2012)
14. Nahar, J., Imam, T., Tickle, K., Chen, Y.: Association rule mining to detect factors which contribute to heart disease in males and females. Expert Syst. Appl. **40**(4), 1086–1093 (2013)
15. Ait-Mlouk, A., Agouti, T., Gharnati, F.: Multi-agent-based modeling for extracting relevant association rules using a multi-criteria analysis approach. Vietnam J. Comput. Sci. **2016**(3), 235–245 (2016). https://doi.org/10.1007/s40595-016-0070-4

Internet of Things: Security Between Challenges and Attacks

Benali Cherif[1], Zaidi Sahnoun[1], Maamri Ramdane[1],
and Bouchemal Nardjes[1,2(✉)]

[1] Lire Laboratory- A. Mehri, Constantine, Algeria
{cherif.benali, zaidi.sahnoun,
ramdane.maamri}@univ-constantine2.dz,
n.bouchemal.dz@ieee.org
[2] University Center Abdelhafid Boussouf of Mila, Mila, Algeria

Abstract. In recent years, the fast developments in hardware, software, networking and communication technologies have facilitated the big emergence of many technologies such as Internet of things. Measurement and collecting data from physical world, and then sending it to digital world is base of this technology. The transmitted data are stocked, processed and then possibly used to act upon the physical world. IoT adds intelligence and autonomy to many domains (e.g. health care, smart transportation and industrial monitoring). As a result, it makes human life more comfortable and simple. However, as all emerging technologies, IoT is suffering from several security challenges and issues, especially that most of IoT devices and sensors are resources- constrained devices. As security issues and attacks could put systems in dangerous and could threat human life too, this paper treats these problems. We will provide an overview about IoT technology, and we will present various security issues that target the perception and the network levels. Moreover, we will discuss how each layer is damaged by harmful and malicious purposes. Most of recent papers use the three layers architecture (which is an old architecture) to present security problems; but this paper uses one of the new reference architectures to study security threats and attacks.

Keywords: IoT architectures · Security · Challenges

1 Introduction

The internet of things (IoT) could be seen as the second version of Internet, where large number of physical objects (e.g. intelligent devices, sensors, actuators etc.) have the ability to sense, collect data, and communicate with each other without any human assistance .This technology gives many services in several application domains such as health care, smart industry, and smart homes [12]. Nevertheless, with the great benefits of IoT, there are many problems, challenges and issues of security which require deep and serious thinking. Nodaway, security problems are increasing seriously [3], where IoT has not only the same security issues of its construction technologies, but it has more [1].

© IFIP International Federation for Information Processing 2020
Published by Springer Nature Switzerland AG 2020
S. Boumerdassi et al. (Eds.): MLN 2019, LNCS 12081, pp. 444–460, 2020.
https://doi.org/10.1007/978-3-030-45778-5_31

Today, IoT architecture is very important, because a good architecture is the main key to create a secure IoT system. But, there is no universal architecture used by all the constructors to shape an IoT system [8].

For that, this paper provides an overview about IoT technology and presents the key problems of security. It reposes on three main phases: In the first one we give an overview about IoT technology and we present two main IoT architectures. The second phase presents IoT security challenges that face the implementation of security policies. It presents also the security feathers in IoT (CIA security triad).Finally, the third part is reserved to present the most important security attacks and issues of perception (sensing) and network levels. In order to analyze the IoT security issues and attacks in more details, this part presents and classifies them using IBorgia et al [15]. ++ this paper is organized

2 IoT Overview

In 1999, Kevin Ashton was the first person that used the term Internet of things (IoT). IoT uses a set of sensor nodes and intelligent devices to collect data from physical world (environment), and then send it to the digital world. RFID and WSN are the two main technologies used to collect and send data in network level. After that, the data get processed and delivered to final application and end-users [5].

IoT may be defined as a dynamic worldwide network infrastructure of intelligent devices and sensor nodes, which are able to configure themselves automatically and they can make their own decisions without human intervention. Each IoT device has a unique identifier that allows this device to communicate with others (IoT devices use many types of communication protocols) [13].

There are many application domains of IoT such as the following: [7]

- Smart energy, smart homes, Smart Buildings, smart cities.
- Internet connected cars and buses (smart transportation),health care and fitness monitoring(smart watch and bracelets)
- Earth supervision and environment monitoring (water quality, fire detection, air pollution monitoring etc.), industrial monitoring.
- Smart devices like tablets and smart phones.

2.1 The Three Layers Architecture

It presents the first IoT architecture which is composed of three layers: Perception layer, Network Layer and Application layer [6], Fig. 1.

1. *Perception Layer.* Known also as physical layer, is the responsible layer of interconnecting and identifying the different IoT devices [7]. It uses a very large number of smart devices and

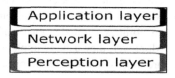

Fig. 1. Three layers architecture of IoT.

sensor nodes to collect data from physical word (environment) [5, 6]. To connect with other devices, each device must be identified with unique identifier [7].

2. *Network Layer.* The main objective of this layer is gathering information that is obtained from physical layer, and then transfers it to application layer. WSN and RFID are the main two technologies used to collect and send data.

   This layer is the responsible of the communication between different devices, using many communication protocols (e.g. MQTT, CoAP...) and technologies (e.g. ZigBee, Bluetooth, WI-Fi...) [5, 6].

3. *Application Layer.* It presents the top layer of this architecture, which takes two main responsibilities: data storage and processing, and provide a set of services to different applications (final users) [6]. This layer is service-oriented that offers data to different kind of final users and applications, to satisfy their needs. There are many applications domains such as smart transportation and healthcare [4, 7].

2.2 IoT Layered Architecture of IBorgia and Al.

Borgia and al. propose an IoT architecture that is very helpful to solve the interoperability and security issues. It has six different layers, presented in Fig. 2. [8, 15].

From the bottom to up we have:

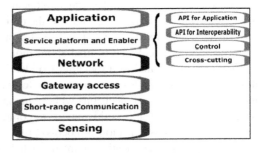

Fig. 2. IBorgia and al IoT architecture.

- Sensing layer is responsible layer to percept and collect data from physical world using a large number of sensor and device nodes.
- The three layers *Short-range Communication, Gateway access* and *network*, serve as Communication Bridge between *Sensing* and *Service platform and enabler* layers. They use many standards and technologies to exchange data [11]. The idea of splitting the network level into three layers comes from the fact that the existing internet protocols (such as HTTP) require a memory size and power capabilities, which is an issue for small devices, We have to point out that most of IoT devices are small and weak devices [10].
- Most of IoT devices and sensor nodes are characterized with low processing capabilities, limited storage and constrained memory.

   Moreover, they usually implement only two or three bottom layers of OSI, and they are mostly not directly compatible with TCP/IP. Gateways can solve this problem, because they use HTTP protocol to communicate with each other and they support specialized protocols and technologies to interact with physical sensor and devices [9].

They enables their connection with high bandwidth networks (the network layer) [8], and they support aggregation, processing, and bridging [10]. In Short-range Communication, IoT devices are usually interconnected through a short range wireless network (WSN), where several technologies are used (e.g. Bluetooth, Z-wave, ZigBee).

- IoT Service platform and Enabler: The fifth layer includes software and services to control the IoT system (storage, processing etc.). It guarantees many non-functional requirements such as security, safety, and availability.
- Application layer is the top layer of this architecture. It is a service-oriented layer that offers services to final applications and users, such as services and software devoted to smart transport, Health care and energy monitoring.

3   Security Challenges in IoT

This section will present the main challenges that face implementation of security policies. Security trends in IoT will be also presented.

3.1   IoT Security Challenges

Security in IoT domain has many challenges that complicate the construction of security solutions and policies, such as the following:

- The limitation of resources: IoT devices have usually limited resources such as low processing power, limitations of energy and memory. These limitations complicate the implementation of powerful encryption algorithm in IoT systems [9, 13]. Moreover, most of devices are resource constrained and they have not enough hardware and software to support TCP/IP protocol and security protocols [12].
- Heterogeneity of devices and network technologies: IoT use many types of sensors, devices and network technologies and this can result many security problems. It complicates also the creation of powerful security policies [12].
- Lack of standardization: there are not unique standards that all the constructors of IoT devices use. Each vendor uses his own standards, protocols and technologies [12].
- The integration of the physical and cyber domains exposes the system to attacks. Cyber attacks may target the cyber domain and paralyzes the physical domain (IoT devices) [14].

- IoT devices are placed everywhere, so they can easily be damaged, stolen, and get unauthorized access [11].
- The proposed techniques and security methods are essentially based on traditional network security methods. However, IoT system is more challenging than traditional networks, due to the heterogeneity of devices and protocols [14].
- Millions of devices could be used in an IoT system (e.g. a system to measure the temperature all around the country), which result unmanageable amount of data [2].

3.2 Security Trends and Feathers in IoT

Security includes many trends or feathers, but in this section we present the three main trends and the security triad CIA (confidentiality, integrity and availability) [4, 11].

1. *Confidentiality*
   It is a security characteristic and it means that just the sender and the receiver can read the exchanged information. So, data must be protected in all communication process: in sender and receiver sides, and during data transportation in network [11].
2. *Integrity*
   It refers to the absence of unauthorized data changing (modification) .So, in all process of communication; the data must not get modified in the sender side, the receiver side and between them. The unauthorized data modification compromises this security trend [11].
3. *Availability*
   It means that the system or the service (or a device) is available and accessible to his clients, and everything is offered correctly. The availability is stolen if the target system or service is inaccessible, or the client couldn't even make a communication with it [11].

4 Related Work

This section will present three propositions to solve the security problem in IoT.

4.1 Ioannis Andrea and Al Classification of IoT Security Attacks

According to the authors [7], this contribution is a new classification of different types of attacks. Compared to other classifications, this one is unique, because it uses four distinct classes to divide the current different attacks. The four classes are: Physical, Network, Software and Encryption attacks. We have to note that this classification is based on the target point of attacks to classify them. So the attack can target the system physically (IoT devices), or its network, or from applications (that are running on devices) on the system, and finally from encryption schemes.

1. *Physical Attacks*

   In this type of attacks, the physical components (devices or things) are the target of attacker. The goal of this type of attack is to compromise security feathers as availability. It can be just to harm the target component(s) (the functional roles) or as an enter point to harm all the system. To make the attack works, the attacker has to be in the IoT system (as a foreign element) or physically close. Many attacked could be mentioned such as: Malicious Node Injection, Physical Damage, *and Node jamming (in WSNs).*

2. *Network Attacks*

   Contrary to the previous type, the attacker doesn't have to be close or near the IoT system, he can make the attack works remotely. This class contains a set of attacks which threat the level network of the IoT system. The communication between the different physical devices is guaranteed by the network level (layer), so network attacks are very dangerous for information confidentiality and privacy. There are many attacks but the most important are: Traffic Analysis Attacks, Routing Information Attacks, RFID Unauthorised Acces.

3. *Software Attacks*

   In this type of attack, the software part of IoT system is the source of vulnerabilities. The attack is basing of the use of deferent types of malicious programs to steal information, change and tamper the system data, deny of service and even harm the IoT system devices. The main tools (malicious programs) that are used in this class are: worms, Trojan horses, spywares, viruses and malicious scripts. The main attacks in this class are: *Phishing Attacks Malicious, Script Attack,* and *Denial of Service.*

4. *Encryption Attacks*

   The IoT system uses encryption scheme to protect the exchanged data between devices. This class gathers a set of attacks that try to break the encryption scheme of IoT system (generally, the goal of attack is to obtain the encryption key that has being used for encrypting and decrypting data. *Side channel Attacks* and *Cryptanalysis Attacks are the main* encryption scheme *attacks.*

   A summarized representation of this classification is shown in table below:

4.2 Abdul W.A and Al Classification of IoT Security

In this classification, the four layers architecture of IoT has been used to classify the different attacks (the aim of the paper is to discuss security of four layered architecture of IoT). So, in each layer, this classification presents the possible attacks that could be, as shown in the next figure [5]:

The four types of attacks are [5]:

1. *Physical Layer Attacks*

   The *Physical Layer* is the responsible layer of collecting information from the physical word by using a set of sensor nodes and intelligent devices, and ensures the communication between these physical equipments. Those devices (hardware parts of an IoT system) are the target of the physical layer attacks to: cause damages on the physical node, steal the data confidentiality and integrity, and deny the access to services. To achieve his attack, the adversary has to be close to IoT system. There are many physical attacks such as *Node Tempering*, Unauthorized Access to the Tag and Tag cloning.

2. *Network Layer Attacks*

   In this type, the attacker concentrates on the network level of the IoT system, which presents the communication

CLASSIFICATION OF IOT ATTACKS

Physical Attacks	Network Attacks	Software Attacks	Encryption Attacks
Node Tampering	Traffic Analysis Attacks	Virus and Worms	Side Chanel Attacks
RF Interference	RFID Spoofing		Cryptanalysis Attacks:
Node Jamming	RFID Cloning	Spyware and Adware	a) Ciphertext Only Attack
Malicious Node Injection	RFID Unauthorised Access		b) Known Plaintext Attack
Physical Damage	Sinkhole Attack		c) Chosen Plaintext
Social Engineering	Man In the Middle Attack	Trojan Horse	or Ciphertext Attack
Sleep Deprivation Attack	Denial of Service	Malicious scripts	
	Routing Information Attacks		Man In the Middle Attack
Malicious Code Injection on the Node		Denial of Service	
	Sybil Attack		

Fig. 3. A summarized representation of AIoannis A and Al's

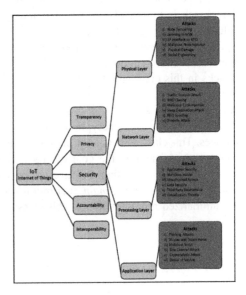

Fig. 4. Abdul W.A and Al classification [5]

bridge between different physical devices and sensor nodes. The network layer gathers information which is obtained from physical layer (collected by devices), and then transfers this data to processing layer, so attackers find it as a good part or level to steal information. There are many network attacks such as: *RFID Spoofing, RFID Unauthorized Access* (Fig. 4).

3. *Processing Layer Attacks*

The processing layer is the responsible of the storage, the processing, and data analysis, as a result, this qualifies it to be a good level to practice several malicious activities by attacker. Most of attacks are inherited from the used technologies (such as cloud computing attacks).This type of attacks gathers many attacks such as: *Malicious Insider, Virtualization threats.*

4. *Application Layer Attacks*

The application layer is service oriented layer which provides the processed information to the final users (applications such as healthcare, smart homes etc.) as services. In this layer, the attackers use malicious programs to harm the systems, such as viruses, spywares, Trojan horse and worms. The application layer attacks present a serious type of attacks, they are used to: steal private and confidential data, altering data, damage the IoT devices, and get unauthorized access. There are many attacks like: Virus, Worms, Trojan Horse and Spyware attacks, Malicious Scripts attacks, and Denial of Service.

4.3 A Systemic Approach for IoT Security

In the paper [16], the aim of authors is the exploration of a new approach to design security mechanisms and deployment in IoT context. They propose a systemic (and cognitive) approach to ensure the IoT security, and to explore each actor's role and its interactions with the other principal actors of the proposed scheme. The paper [16] sees the IoT system as a complex system in which people interact with intelligent devices.

In this proposed approach, the set of connections between different nodes have a specific character depending on complex nature of IoT environment. Moreover the paper [16] takes into consideration the dynamic and complex nature of this proposed model. It presents its perspective in respect of the main elements illustrated in the approach which are "nodes" and "tensions".

The interactions between nodes are represented by tensions. The nodes are the origination and destination actors of a tension. This approach takes into consideration the environment complexity. The approach is presented in the Fig. 5.

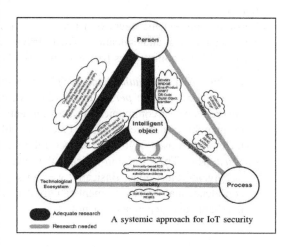

Fig. 5. A systemic approach for IoT security [16]

In order to explain this model, we will describe each node and its functions briefly . The tensions between different nodes need a special study and discussion; we will try just to explain them shortly:

1. *Nodes*

   There are four nodes: Person, Process, Intelligent Object, Technological ecosystem

   - *Person.* The human resources play a principal role in the IoT security, because they are responsible for security rules management that includes: the definition of security rules and practices, ensure efficiency of rules, auditing and verification practices. This vital node plays an essential role in the management and enhancement of security. So, the person node should be able to analyze all the context of IoT.
   - *Process.* This node refers to a resources or a means that are used to accomplish tasks, and to guarantee security requirements. In order to ensure the security of the environment at different levels, the process has to be conformable and compliant with the security policies. Furthermore, there is a big difficulty to implement security processes, because the model is complex and the existence of several interactions originating from the process node. According to practices, security process has to face many requirements such as requirements of standards, requirements of strategies, requirements of policies etc.
   - *Intelligent Object.* This node presents the heart of this approach; it refers to an "object" enhanced with electronic capabilities to communicate with other objects in his environment (intelligent devices). An object can exchange information, cooperate and connect with other objects.
   - *Technological Ecosystem.* The technological choices (technologies) that have been made to ensure the security of IoT is represented by this node. There are many categories of information security technology (or technologies) such as Identification and Authorization, and Security Design and Configuration.

2. *Tensions*

   Tensions represent the interaction between nodes. The paper presents 7 tensions: Identification and authentication, Trust, Privacy, Responsibility, Autoimmunity, Safety, and Reliability. This part wills discuss them:

   - *Identification and Authentication.* This tension attaches the two nodes: *intelligent object* with the *person.*In IoT context, each entity must be identified, to ensure a correct communication between entities, and to guarantee the absence of unauthorized access. Radio Frequency Identification (RFID) is the main technologies used in IoT to connect different devices.
   - *Trust.* The "Trust" tension attaches the technological ecosystem node with the intelligent object node. Basically, we can say that Trust represents the level of confidence that the environment can grantee to the intelligent object (if the level is reliable and dependable or not).

- *Privacy.* The tension that attaches the person with the technological ecosystem is "privacy". The ubiquitous characteristic of the IoT environment make the privacy an important tension in the systemic model of IoT security.
- *Responsibility.* The "Responsibility" tension attaches the *process* node with the intelligent object. It means the set of access rights and privileges, which have to be clearly specified and defined evidently, depending on privacy constraints. Moreover, in order to avoid dangers when the object regulates a process; the set of rules of liabilities and responsibilities for each entity must be taken in consideration.
- *Autoimmunity*
  The tension that attaches the *intelligent object* in self loop (with its self) is "Autoimmunity". Proposing an artificial immune system solution for IoT is the aim of this tension.
- *Safety*
  The "*safety*" tension attaches the two nodes: person with process. Ensuring safety when an unexpected problem (egg: failure, attack …) appears, is one of the main security challenges that the IoT system has to face (and overcome it). So, the reduce damage possibility is considered by safety
- *Reliability*
  The tension that attaches the process node with the technological ecosystem node is "Reliability". The goal of this tension is to guarantee the availability of data and information, using efficient ways of managing data repositories. It deals with communications management and data.

5   Security Attacks and Threats in IoT

IBorgia and al. architecture offers an interesting functional view for IoT system, and it satisfies the recent requirements of IoT system. It catches the main features of an IoT system that are: the interaction between the local and personal networks of sensors nodes on one side and the interaction between high-bandwidth networks with computation power systems in the other side.

Basing on these considerations, we adopt this architecture as a mould (model), to analyze security issues and attacks in IoT system. The main security attacks are presented in the Fig. 3.

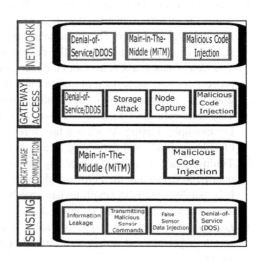

Fig. 6. Some security attacks in IoT (using IBorgia architecture)

Sensor-based threats present a serious family of IoT security threats [2], which could be classified into four categories, basing on intentions and nature of these threats. These categories are: (1) Information Leakage (2) Transmitting malicious sensor commands (3) false sensor data injection (4) denial of-service (DOS) [1] Fig. 6.

1. *Information Leakage*

IoT sensors could stock sensitive data like login, passwords, and credit card information; and the steal of this data puts the user privacy and IoT system security in danger. IoT attacker can use a sensor information to achieve his attack (or information from multiple sensors to achieve a more complex attack).

Fig. 7. Information leakage method sensing layer

In this category, four methods could be used: keystroke inference, task inference, location inference, or eavesdropping [1], Fig. 7.

- *Keystroke Inference.* In this method, the attacker try to deduce the keystrokes entered in the IoT device.

  When a user types (or gives) inputs to his device, tilts it, or turns it, a set of deviations are resulting. These deviations are used later by the attacker to infer the entered data. Keystroke Inference can be performed on the device itself or by using nearby sensor.

  This attack can be performed using magnetic Sensors, light Sensors, audio Sensors, and video Sensors [1].

- *Task Inference.* This type of attack is based on the deduction (the reasoning), in which the attacker tries to extract information about the ongoing task or application inside the target device. This information presents the state of the device and used to start an attack, without alerting the device security policies.

  The idea of this attack starts from the fact that sensors show deviation in the reading process for various tasks running on the devices, and this deviation can be used to infer the running process or application inside this device.

  Task inference can be performed using magnetic Sensors, Power Analysis etc. [1]. For example, Timing Attack is a task inference attack, which enables the discovery of vulnerabilities and extracting information about security policy.

  Timing attack is done by observing the responding time for different inputs and queries to determine the cryptographic algorithms implemented in the system.

  It is usually used with small devices that have weak computing capabilities [3, 5]. This attack threats the data confidentiality.

- *Location Inference.* This type of attack is used to determine the victim location, which is private and sensitive information in itself, and use it to launch another attack.

This attack steals the location-privacy. The attacker use acoustic information embedded in an audio source (e.g. audio messaging) to identify sensitive locations of the target entity. For example, this attack could be used to compromise location privacy of participant in anonymous session. This information is used to produce a location fingerprint [1].

- *Eavesdropping.* In this type of attack, a malevolent application uses an audio sensor (e.g.: microphone) to listen to a private conversation secretly. After that, the attacker tries to extract confidential information from this conversation (e.g. social- security number and credit card information).

  The attacker can record the conversation on a storage device or listen to it in real-time [1, 3].

  For example, replay attack (or play back attack) uses the eavesdropping to steal authentication information from the sender and then use it to send a request message (Identity stealing) [3].

2. *Transmitting   Malicious   Sensor Commands*

   Today, most of IoT devices and sensors allow the creation of unexpected communication channel with other entities. This weak point could be used by attacker to create a communication channel, and then he launches his attack. This attack could change critical parameters of the target sensor (e.g. light intensity), or even transmit malicious commands (trigger messages) to

Fig. 8. Transmitting malicious sensor commands method

activate a pre- planted malware [1]. The malicious program (virus or malware) could be inserted into the device physically, or via Malicious Code Injection attack. As a result, the attacker gains a full access to that node, and then he can control all the IoT system [1]. There are many methods to transmit signals and malevolent commands such as using a audio sensors, light sensors, or a magnetic sensors [1], Fig. 8.

3. *False Sensor Data Injection*

   IoT system uses different devices and sensors to collect very important and sensitive data. We could not imagine the results if a patient data in a hospital have been altered or faked.

   False sensor data injection is an attack where the sensor data is forged (faked), or even to inject false data. It's used to perform malicious

Fig. 9. False sensor data injection method

activities. The attacker use specific commands to change the real information or to modify the device's actions. This attack needs a physical access to the target device or a remote access by using various communication medium (Wi-Fi, Bluetooth, etc.) [1]. For example, Malicious Fake Node attack belongs to this type, in which the attacker uses a fake node to inject false data [3], Fig. 9.

4. *Denial of-Service (DoS)*

In this section, we talk about Denial-of-Service (DoS) for a sensing and perception device. DoS for a device is a type of attack to deny maliciously the normal operation of this device, and to forbid the access to it.

There are two types of DoS attacks: active and passive attacks. In active attacks, the access to an application, a task or a device is refused effectively. However, if one

Table 1. The stolen security trends of each attack type

	Confidentiality	Integrity	Availability
Information leakage	YES	NO	NO
Transmitting malicious sensor commands	YES	YES	YES
False sensor data injection	NO	YES	YES
Denial of-Service (DOS)	NO	NO	YES

application has been attacked to stop another ongoing task on the device, we call this a passive attack [1]. DoS attack could have an after-effect to exhaust the system resources, such as battery and memory resources [3]. For example, DoS attack is used with gyroscopes of drones and accelerometers to shut the device down [1].This attack will be more explained in a next part.

From the explication above for each method, we conclude that each type can threat one or more security trends. This is represented in the next table Table 1.

Note that, results of a type of attack (or all the attack) could be used to launch another attack (The second attack can threat another security trends). That is called composition of attacks.

5.1   Short-Range Communication, Gateway Access and Network Layers

Short-range communication, Gateway access and network represent together the network layer of the three layers architecture [8, 11]. They have many common attacks, but with some specifications in each one.

That is why this section treats them together, and it presents the attack specification in each layer. The network level has many attacks but the main ones are:

1. *Denial of Service (DoS)*

It is an attack to deny authentic users to access a device or a network resource. The attacker accomplishes this attack by flooding the targeted component with redundant requests. He inundates the network traffic by sending a large amount of data, and this results massive consumption of system resources. The flooding process makes the system or the target device inaccessible or difficult to use by some or all authentic users [3], Fig. 10.

The DoS attack has a distributed version called distributed DoS (DDoS). DDoS attack is defined as a set of concurrent DoS attacks. The attacker could use botnet army, which is an army of IoT devices that are infected with malwares. DoS and DDoS attacks may cause energy dissipation issues and physical damage [4], Fig. 11.

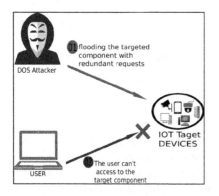

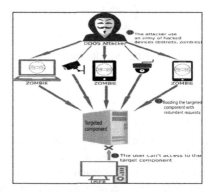

Fig. 10. Denial of-Service (DOS) attack

Fig. 11. Distributed Denial of-Service (DDoS) attack

2. *Man-in-The-Middle (MiTM)*

In this attack, the hacker plays secretly a role of a mediator between the sender and the receiver who believe they have a direct communicating with each other.

He becomes the controller of all the communication; therefore he can capture, change and manipulate the communication information in real time according to his needs.

It is a serious security threat that steals the integrity of information [3]. MITM is also known as Malicious Node Injection because the attacker injects (plants) a new malicious node between the sender and receiver, to control all the exchanged data [5], Fig. 12.

3. *Storage Attack*

In this attack, the hacker tries to get the stored data and information inside the target node. For example, the gateway node can store sensitive user information, and that make it a good target for attackers. The gateway can be attacked to change or delete his stored information [3], Fig. 13.

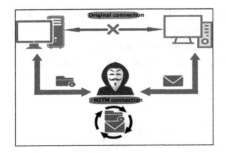

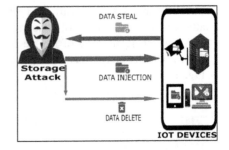

Fig. 12. Man-in-The-Middle (MiTM) attack

Fig. 13. Storage attack

4. *Node Capture*

   Is a serious attack faced the IoT system, in which an attacker gets the full control over a key node, like a gateway node. The attacker can steal many private information such as communication information between a device and the gateway, a communication security key, and many sensitive information stored in the gateway's memory [3]. Moreover, the attacker can add a duplicate node to the network to send malicious data; as a result he threats the data integrity and confidentiality [5], Fig. 14.

5. *Malicious Code Injection*

   As we presented earlier, the injected malicious code (or malware) gives the attacker the full control over the infected node. He could activate the injected malware by *transmitting malicious command attack*.

   The attacker can use the infected nodes (devices) to gain a full control over the IoT network, affect the IoT network, or even block it completely. This type of attack can really cause serious problems in the IoT system [5], Fig. 15.

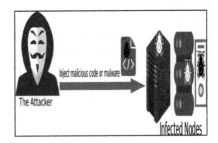

Fig. 14. NodeCaptureAttac attack

Fig. 15. Malicious code injection attack

5.2 Discussion

As we said earlier, the three layers Short-range communication, Gateway access and network have many common attacks, but with some specifications or differences in each one. The next table explains the specifications (properties) of each attack in each layer (if the Attack could be per formed), Table 2.

6 Conclusion

The increasing popularity of IoT and its applications is bringing attention towards their security issues, threats and attacks. This paper has presented the IoT technology and its main architectures and then it focused a very important aspect in IoT: the security.

As a perspective of this paper, some points will be discussed in an extension paper for this work such as:

Security issues in the last two layers of IBorgia and al architecture.

Table 2. Main attacks in network level

	Short-range communication layer	Gateway access layer	Network layer
DOS/D.DOS (attacks to compromise the availability)	——	Deny the access to the gateway (devices could not access to the gateway)	Deny of access to the gateway from "service platform and enabler layer"(or the opposite sense)
Main-in-The-Middle (MiTM)	The attacker Intercepts and alters the communication information, which is sent between a device and the gateway	——	Intercepts and alters the information between the gateway and capabilities of "service platform and enabler layer" (e.g. cloud)
Storage Attack	——	Steal, change, or delete the gateway's stored information	——
Node Capture	——	Get the full control over the gateway	——
Malicious Code Injection	The attacker affects and controls the communication. He could block it completely	The attacker could control or block the Gateway node (as a result all the IoT system)	The attacker affects and controls the entire network. He could block it completely

- Current security mechanisms to prevent security threats and attacks.
- Several security solutions and approaches.
- Some security implementation attempts, counter measures like Software Defined Networking (SDN) and Blockchain.

References

1. Sikder, A.K., Petracca, G., Aksu, H., Jaeger, T., Uluagac, A.S.: A survey on sensor-based threats to Internet-of-Things (IoT) devices and applications, pp. 01–08 (2018)
2. Jing, Q., Vasilakos, Athanasios V., Wan, J., Lu, J., Qiu, D.: Security of the Internet of Things: perspectives and challenges. Wireless Netw. **20**(8), 2481–2501 (2014). https://doi.org/10.1007/s11276-014-0761-7
3. Burhan, M., Rehman, R.A., Khan, B., Kim, B.S.: IoT elements, layered architectures and security issues: a comprehensive survey. Sensors **18**(9), 2796 (2018)
4. Mendez, D.M., Papapanagiotou, I., Yang, B.: Internet of Things: Survey on Security and Privacy. Inf. Secur. J. Global Perspect. **27**(3), 162–182 (2018)
5. Ahmed, A.W., Ahmed, M.M., Khan, O.A., Shah, M.A.: A comprehensive analysis on the security threats and their countermeasures of IoT. IJACSA J. **8**(7), 489–501 (2017)
6. Vijayalakshmi, A.V., Arockiam, L.: A study on security issues and challenges in IoT. IJESMR **3**(11), 1–9 (2016)
7. Andrea, I., Chrysostomou, C., Hadjichristofi, G.: Internet of Things: security vulnerabilities and challenges, pp. 180–187 (2015)
8. Di Martino, B., Rak, M., Ficco, M., Esposito, A., Maisto, S.A., Nacchia, S.: Internet of Things reference architectures, security and interoperability: a survey. Internet of Things **1**, 99–112 (2018)
9. Efremov, S., Pilipenko, N., Voskov, L.: An integrated approach to common problems in the Internet of Things. Procedia Eng. **100**, 1215–1223 (2015)
10. Fremantle, P: A reference architecture for the Internet of Things, WSO2 White paper, pp. 02–04 (2015)
11. Radovan, M., Golub, B.: Trends in IoT Security. In: MiPro 2017. Daimler AG, Stuttgart, pp. 1302–1308 (2017)
12. Ara, T., Gajkumar Shah, P., Prabhakar, M.: Internet of things architecture and applications: a survey. Indian J. Sci. Technol. **9**(45), 01–06 (2016)
13. Sadeeq, M.A.M., Subhi, R.M., Qashi, Z.R., Ahmed, S.H., Jacksi, K.: Internet of Things security: a survey. In: ICOASE Conference, pp. 162–166 (2018)
14. Mohamad Noor, M.B., Hassan, W.H.: Current research on Internet of Things (IoT) security: a survey. Comput. Netw. **148**, 283–294 (2019)
15. Borgia, E.: The internet of things vision: key features, applications and 560 open issues. Comput. Commun. 54, 1–31 (2014). arXiv:1207.0203. https://doi.org/10.1016/j.comcom.2014.09.008
16. Riahi, A., Natalizio, E Challal, Y., Chtourou, Z.: A systemic approach for IoT security, pp. 351–355 (2013)

Socially and Biologically Inspired Computing for Self-organizing Communications Networks

Juan P. Ospina[1]([✉]), Joaquín F. Sánchez[2], Jorge E. Ortiz[1],
Carlos Collazos-Morales[2], and Paola Ariza-Colpas[3]

[1] Universidad Nacional de Colombia, Bogotá, Colombia
{jpospinalo,jeortiz}@unal.edu.co
[2] Universidad Manuel Beltran, Bogotá, Colombia
{joaquin.sanchez,carlos.collazos}@umb.edu.co
[3] Departamento Ciencias de Computación y Electrónica,
Universidad de la Costa-CUC, Barranquilla, Colombia
pariza@cuc.edu.co

Abstract. The design and development of future communications networks call for a careful examination of biological and social systems. New technological developments like self-driving cars, wireless sensor networks, drones swarm, Internet of Things, Big Data, and Blockchain are promoting an integration process that will bring together all those technologies in a large-scale heterogeneous network. Most of the challenges related to these new developments cannot be faced using traditional approaches, and require to explore novel paradigms for building computational mechanisms that allow us to deal with the emergent complexity of these new applications. In this article, we show that it is possible to use biologically and socially inspired computing for designing and implementing self-organizing communication systems. We argue that an abstract analysis of biological and social phenomena can be made to develop computational models that provide a suitable conceptual framework for building new networking technologies: biologically inspired computing for achieving efficient and scalable networking under uncertain environments; socially inspired computing for increasing the capacity of a system for solving problems through collective actions. We aim to enhance the state-of-the-art of these approaches and encourage other researchers to use these models in their future work.

Keywords: Self-organization · Natural computing · Complex systems · Ad hoc networks

1 Introduction

During the last decades, the number of services and technologies available for networking applications has increasing significantly. These developments have shown a direct relationship with different aspects of human society like the

© IFIP International Federation for Information Processing 2020
Published by Springer Nature Switzerland AG 2020
S. Boumerdassi et al. (Eds.): MLN 2019, LNCS 12081, pp. 461–484, 2020.
https://doi.org/10.1007/978-3-030-45778-5_32

economy, education, politics, and quality of life. Computational devices seem to be ubiquitous and are present in almost all aspect of our daily life. This trend is promoting a technological integration that has already gone beyond of what traditional networking paradigms can do regarding scalability, dynamic environments, heterogeneity and collaborative operation. As a result, these conditions impose several challenges for building the envisioned future networking technology and show the need to explore new engineering approaches.

The next generation of communication networks will be composed of ubiquitous and self-operating devices that will transform our immediate environment into an intelligent computational system. New technological developments like self-driving cars, wireless sensor networks, drones swarm, Internet of Things, Big Data, and Blockchain are promoting an integration process that will bring together all these technologies in a large-scale heterogeneous network. All these applications involve a set of autonomous components (with possibly conflicting goals) interacting asynchronously, in parallel, and peer-to-peer without a centralized controller; they should be easily accessible by users and operate with minimum human intervention.

Given those conditions, it is necessary that all computational devices can operate autonomously and collaborate with others to offer services through collective actions. Besides, the future communication networks will require high levels of self-organization for both, face challenges related to scalability, heterogeneity, and dynamic environments, and minimize centralized control and human intervention during the processes of planning, deployment, and optimization of the network. Indeed, these requirements cannot be faced using traditional approaches; they are not able to deal with scale, heterogeneity, and complexity of the future networking applications, making necessary to explore novel paradigms for designing and implementing communication systems that can operate under those conditions.

Accordingly, our aim in this paper is to introduce and overview the biologically and socially inspired computing used as technological solutions in networking and artificial systems. The principal idea is to show that it is possible to create analogies between living and artificial systems that enable us to inspire mimetic solutions (biological, social, economic or political) and translate those principles into engineering artifacts. Living systems show desirable properties like adaptation, robustness, self-organization, and learning, all of them required to handle the complexity of the future networking systems. In this regard, we can analyze biological and social phenomena as a source of inspiration for new technological developments; biologically inspired computing for achieving efficient and scalable networking under uncertain environments, and socially inspired computing for increasing the capacity of a system for solving problems through collective actions. In this work, we expect to provide a better comprehension of the opportunities offered by these models and encourage other researchers to explore these approaches as part of their future work.

The rest of the article is organized as follows: in Sect. 2 we present a historical review of the scientific and technological development of communication systems. In Sect. 3 we summarize the most challenging issues of the next generation of communication networks from the perspective of biologically and socially inspired computing. Section 4 introduces a general method for developing these models; the main idea is to expose how to create a technological solution from properties and behaviors observed in living systems. Section 5 concludes the article.

2   Self-organizing Communication Networks: A Historical Review

In this Section, the need of using self-organization as control paradigm for the next generation of communication networks is discussed. First, we provided a historical review of the scientific paradigms used for studying and building communications systems. Second, we show complexity signs related to traffic, topologies and chaotic behaviors because of interactions among users, nodes, and applications. Third, a comparison of the current control and management paradigms used for designing, controlling, and developing artificial systems is presented. Finally, we depict some properties required for the future communications systems based on self-organizing properties.

2.1   Scientific Paradigms in Communications Networks Development

Traditionally, the scientific paradigm used for communications networks development has been reductionism. Engineers conceived communication systems as a hierarchical structure that allows offer services through protocols and distributed algorithms; each layer was studied individually, and a communication interface among them was used to provide functionalities during the network operation [21]. Devices, protocols, and applications were designed separately, and linear behavior in the whole system was expected. This idea arose from the first mathematical models used for planning and dimensioning communications systems, in which engineers used stochastic models and queue theory to compute the average traffic and assign resources according to the users demands [48,90]. This approach played an essential role in traditional telephone networks in which there was only one service and the performance required for all users was the same. Thus, it was easy to combine the traffic flows and take advantage of their homogeneous features for analytical purposes. However, an increasing amount of networking technologies and also more complex software applications changed the linear behavior expected inside communications networks [25].

During the last decades, integration of services and technologies available for networking applications have occurred. Nowadays is possible to find data transfer, online games, video, email, e-commerce, and browsing, working on the same network infrastructure [86]. Also, we can find different transmission technologies like wired connections, optical fiber, IEEE 802.11, WiMAX or Bluetooth, and the performance required for each application (bandwidth, delay, and errors handling) is different every case [85]. As a consequence, this increasing number of services and technologies changed the design principle on which engineers based the networking development: linearity. Communication networks do not have linear behaviors anymore, and it is necessary to see them as complex systems if we want to design algorithms and control mechanisms capable of operating in a dynamic environment with non-linear properties [39,65].

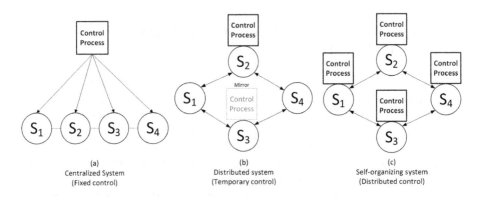

Fig. 1. Control paradigms evolution in artificial systems [25]

2.2   Complexity Signs: Self-similar Traffic, Chaos and Scale-Free Topologies

Because of technologies and services integration, communications networks started showing complexity signs like self-similar traffic, chaotic behaviors and scale-free topologies. Although none of these properties were in the initial conceptual framework used by engineers for design and building communications systems, nowadays there is enough evidence to consider them as an inherent part of the communication networks. A brief overview of these complexity signs is exposed below.

Usually, traffic is modeled as a stochastic process that shows the amount of data moving across a network and establishes a measure to represent the demand that users imposed on the network resources. Both requests per time unit and the incoming packets have been modeled as sequences of independent random variables (call duration, packet lengths, file sizes, etc.) to make easier their analytical treatment [48]. However, the correlation among these variables persists through several time scales and has a significant impact on the network performance [25,103]. It is important to mention that self-similarity is not a property of traffic sources; it arises as emergent behavior from interactions among users, applications, and networking protocols. Besides, traditional traffic models based on Poisson processes has proven not be suitable to describe traffic patterns in modern communications networks [3,90].

Similarly, chaotic behaviors take place in dynamical systems that are high-sensitivity to initial conditions; small differences in the system states can produce a significant number of different outcomes. Chaos theory studies these behaviors and tries to deal with the apparent randomness present in strange attractors, feedback loops, and self-similarity. For example, in communications networks, this behavior appears through interactions between TCP protocol and the RED algorithm used for queue management [88]. Other examples are presented in [70] in which chaos appear in the profile of daily peak hour call arrival and daily call drop of a sub-urban local mobile switching center, or in [93] in which chaotic patterns serve as a mobility model for an ad hoc network. More examples can be found in [53,106,110].

Finally, the scale-free property is another complexity sign that suggests self-similarity patterns in terms of the network topology [11]. The structure of the network has nodes with more connections than others, and follows a power law distribution. This pattern was found in the late 1990s when a part of the *World Wide Web* was mapped in a moment of internet connection [58]. This phenomenon could be explained analyzing the evolution of communications networks in terms of their physical and logical topologies according to the preferential connectivity principle [69]. If a web page is created, is reasonable to assume that links to highly connected sites like Google, social networks, services companies, etc., will be added. Also, the physical topology of the internet is also defined by economic and technological requirements of the Internet Service Providers (ISP) [3].

2.3   Control Paradigms Evolution

All artificial systems, including communications networks, use management and control processes to regulate their behaviors. The management process consists in manipulate subsystems, parameter updates, and verify the system state. On the other hand, control is about feedback and run-time control according to variations in the environment. Both processes define the routines to maintain, operate, and adapt the system during operation time. Figure 1 shows a historical review of the current control paradigms for artificial systems [25].

Initially, communications networks were composed by a single device and some remote terminals. There was a single control process and all parameters required for the network operation, e.g., addresses, access privileges and resources were pre-configured by default. Changes in topology and applications were possible but required a complete manual configuration of the system [25]. Figure 1a presents an example of these control paradigm through a hierarchical architecture; the root shows the control process and the leaves the subsystems it can handle. For instance, traditional telephone networks and client/server applications are classic examples of this approach [90]. Even though there are others control schemes, centralized systems are still the preferred solution due to its simplicity and effectiveness; if only a few well-known subsystems have to be managed, there is no need for the high computational cost of distributed algorithms or possibly less deterministic self-organizing methods [29,38].

The next paradigm is the distributed control [25,102]. Distributed systems are composed of a set of independent nodes that works as a single coherent system. In this case, a logical abstraction is deployed as middleware in each device to hide the internal structure and the communication process for the application layer. Figure 1b shows a scheme for this paradigm. Although the control process works in a centralized way, it is possible to locate it dynamically inside any node to improve the fault tolerance and achieve a better use of the resources. Cellular networks, distributed databases [1], orchestration software [98], and multi-agent systems like JADE [12] are examples of this control paradigm. Distributed systems offer several advantages to operate and combine resources from different nodes. However, some issues like impossible synchronization and overhead for resource management show the limits of this approach [25,71]. The need to maintain complete information about the system state and handle changes related to configuration and topology is an expensive computational task in highly dynamic environments [25].

Finally, we have self-organizing systems [38,39]. In this approach, the management and control process is completely distributed, i.e., each sub-system has its own control process. The functionalities and the system structure arise as emergent behaviors from interactions among elements. Similarly, the goals of the system should not be designed, programmed, or controlled by default; the components should interact with each other until they reach the expected configuration. Self-organizing control is flexible, adaptive, robust, and scalable, it does not need perfect coordination and can operate in dynamic environments [47]. Since each component is autonomous, it is necessary to develop additional mechanisms to promote cooperation, coordination, and synchronization among the system components. It is important to mention that self-organization is not a human invention; it is a natural principle that has been used for designing, building, and controlling artificial systems, and face limitations of centralized and distributed approaches [25]. Examples of this control paradigm can be found in Smart Grids [77], communication networks [64], transportation systems [18], and logistical processes [42].

Although self-organization increase scalability, also causes less deterministic behaviors. The system predictability is reduced due to self-organized control. Nevertheless, this is not a real disadvantage in a dynamic system with nonlinear properties in which an approximate solution can be very useful. Additionally, we are in a transition process from distributed to self-organizing systems due to changes in the network architectures, new computational technologies and the need to build large-scale communication systems [85,86]. To sum up, Table 1 describes the relationship between resources and control according to the different paradigms presented above.

2.4   Current Self-organizing Communications Networks

The increasing use of mobile devices, pervasive computing, wireless sensor networks (WSNs), and cloud computing establish new requirements for future communications systems (See Sect. 3). New applications like self-driving cars [28], drones swarm [116], Internet of Things [59], Big Data and Blockchain [30] are

Table 1. Control vs resources: a comparison for artificial systems.

Control paradigms for artificial systems		
Centralized systems	Resources	Centralized
	Control	Centralized
Distributed systems	Resources	Distributed
	Control	Centralized
Self-organizing systems	Resources	Distributed
	Control	Distributed

promoting a technological integration that will bring together all these applications in a large scale heterogeneous network. As a result, the future networking applications will require high levels of self-organization for both, face challenges related to scalability, heterogeneity, and dynamic environments, and minimize centralized control and human intervention during the processes of planning, deployment, and optimization of the network. These challenges may be faced through a set of networking functionalities based on self-organizing properties [85–87]:

- *Self-configuration:* in this context, configuration refers to how the network is set up. Nodes and applications should configure and reconfigure themselves automatically under any predictable or unpredictable condition with minimum human intervention. Self-configuration expects to reduce the effects of networking dynamics to users.
- *Self-deployment:* preparation, installation, authentication, and verification of every new network node. It includes all procedures to bring a new node or applications into operation. Also, self-deployment try to find strategies to improve both coverage and resource management in networking tasks.
- *Self-optimization:* it refers to the use of measurements and performance indicators to optimize the local parameters according to global objectives. It is a process in which the network settings are autonomously and continuously adapted to the network environment regarding topology, resources, and users.
- *Self-healing:* execution of routines that keep the network in the steady state and prevent problems from arising. These methods can change configuration and operational parameters of the overall system to compensate failures.

3   Networking Challenges

Indeed, the majority of the requirements for the next generation of communication networks cannot be faced using traditional approaches [27,38,86]. In this Section, we present some of those challenges and their possible relationship with biological and social phenomena. It is important to mention that this Section is not a full reference of challenges in networking but could be seen a list we can address through biologically and socially inspired computing.

3.1 Scalability

One of the most desirable properties in communication networks is the capacity to increase the network size and be able to receive new nodes and applications without affecting the quality of the services [73]. This property, known as scalability, is one of the leading challenges in protocols design, and it is a requirement for building large-scale communication systems. Scalability can be measured regarding applications, users, physical resources, and the network ability to react properly to unexpected conditions [71]. For example, wireless sensor networks usually need to collect data from several hundred sensors, and during this process the capacity of the network can be easily exceeded, causing loss of packets, low network reliability, and routing problems [113].

Furthermore, the decision process required to operate a large-scale network is too fast, too frequent and too complex for being handled by human operators. As a result, network components need to self-organize by themselves across different scales of time and space to adapt their behavior to any variation in the network size [38,82]. Fortunately, there are many biological and social systems with self-organization mechanisms we can learn from to inspire the design of scalable systems [105]. For instance, data dissemination based on epidemic spreading [27], routing protocols based on Ant Colony Optimization (ACO) [24], and trust and reputation models for controlling free-riders may help to face challenges related to large-scale networking [68].

3.2 Dynamic Nature

Unlike traditional communication networks in which infrastructure and applications were static, the future networking schemes will be highly dynamic regarding devices, users, resources, and operating conditions [64,86]. For example, the network topology may change according to different mobility patterns, and applications will need different levels of performance concerning bandwidth, delay, and errors handling [16]. Also, cognitive radio allows to configure the spectrum dynamically through overlapping spectrum bands, and users may decide what will be their role in the network due to the absence of centralized control [120]. Additionally, the increasing autonomy in the network components may cause unexpected behaviors, turning into a difficult task to predict the temporal evolution of the system. Under these conditions, self-organizing protocols are essential to improve adaptation, robustness, and face challenges related to highly dynamic environments [25,38].

3.3 Need for Infrastructure-Less and Autonomous Operation

The current levels of heterogeneity in communication systems in terms of users, devices and services become centralized control an impractical solution [25,86]. Moreover, there is another trend towards automation in which networking applications require to operate with minimum human intervention. For example, drone swarm [116], delay tolerant networks [35], sensor networks [113] and cognitive radio [64], demand networking protocols that can operate without a centralized control, recover from failures, and deal with highly dynamic environments. In order to address these needs, networking protocols could be equipped with self-organizing mechanisms observed in biological and social systems to develop autonomous applications and decrease the level of centralized control required for the network operation [27,50].

3.4 Heterogeneous Architectures

Future communications networks require integrating several technologies through internet-based platforms. Given the diverse range of networking components and the numerous interactions among them, it is reasonable to expect complex global behaviors. The next generation of networking applications will be composed of WSNs, ad hoc networks, wireless fidelity networks, VANETs, etc., all of them working on a large-scale communication system [85,86]. For instance, one of the emerging and challenging future networking architectures is the Internet of things (IoT) [112]. This paradigm includes the pervasive presence of network devices that through wireless connections can communicate among them, and transform our immediate environment into an intelligent large-scale computational system. Also, Wireless Mesh Networks and WiMAX are expected to be composed of heterogeneous devices and protocols [64].

Heterogeneity needs to be understood, modeled and managed regarding technologies, users, and applications if we want to take advantage of large-scale heterogeneous networks [27]. Therefore, we can analyze living systems with high levels of heterogeneity and use them to inspire technological solutions. For example, biological and social phenomena show stable behaviors through the cooperation of a heterogeneous set of subsystems, e.g., nervous system, immune system and normative social systems. This functionality is called homeostasis and can be used for designing computational mechanisms to face challenges related to heterogeneity [22].

3.5 Solving Problems Through Collective Actions

A standard requirement in self-organizing communication networks is to produce coordination, cooperation, and synchronization among the network components to achieve individual and collective goals. This process can be understood as a requirement to solve problems through collective actions, in which accomplishment of tasks depends on interaction and interoperation of unreliable and conflicting components [78]. Likewise, due to the absence of a centralized control, the network is instead relying on self-organization mechanisms to produce the system functionalities. These models are useful for resource provisioning in grid computing [79], cooperation in mobile clouds [34], platooning in vehicular networks [4] and coordination in drone swarms [116].

Collective actions are necessary to construct new levels of social organization; multicellular organisms, social insects, and human society use it to take advantage of skills and knowledge of others to achieve collective benefits [72]. Although this is a common phenomenon in living systems, it is important to mention that human society has more complex collective actions patterns than other species and we can use them as a source of inspiration for engineering developments. For example, computational justice models could be used for appropriation and distribution of resources in mobile clouds and ad hoc networks [79], cooperation models for controlling free-riders and promote collaborative work among network components [68]. Also, collective behaviors from biological systems like firefly synchronization and swarm intelligence could improve routing and network optimization [27,38].

3.6   Appropriation and Distribution of Resources

One advantage offered by the next generation of communication networks is the opportunity to share resources among nodes, users, and services, through the combination of wireless technologies, mobile devices and the network capacity to operate as a self-organizing system. For example, a mobile cloud allows to exploit distributed resources inside a network if they are wirelessly connected; energy, storage, communication interfaces and software applications can be exchanged, moved, augmented and combined in novel ways [34]. Also, grid and cloud computing provided an infrastructure based on common pool resources to support on-demand computing applications [79]. As a consequence, optimal mechanisms for resources appropriation and distribution are required [81,84]. This process may be in a stochastic or deterministic manner, and the network components need to self-organize themselves to achieve a distributed resources operation. In this regard, several challenges related to how to carry out a sustainable cooperation process in environments composed of potentially selfish components arise. One solution could use electronic institutions and social capital as a way to increase the capacity of the network to use collective actions. Applications of this approach can be found in Smart Grids [80], VANET's [37] and Multi-agent systems [76].

3.7   Security and Privacy

Since the networks become a flexible, attackers can get sensitive information analyzing the messages embedded in communications channels and relay nodes [85]. Also, according to mobility patterns the network topology may change in dynamical and unpredictable ways changing routing tables and increasing the risks of exposing crucial private information [86]. As a result, there are several security challenges such as a denial of service, black hole, resource consumption, location disclosure, wormhole, and interference [64]. For instance, the future internet of things will transfer a significant amount of private information through wireless channels, and security protocols need to defend malicious attacks to provide a relatively secure network environment [85]. One solution could use game theory to address situations where multiple players with contradictory goals or incentives compete among them. Many biological and social systems have inspired solutions to deal with security and privacy issues. For example, artificial immune systems for anomaly and misbehavior detection and trust and reputations models to control free-riders and selfish behaviors [27].

One purpose of this work is to introduce and overview the biologically and socially inspired models used as technological solutions in networking and artificial systems. The main idea is to show how an abstract analysis of living systems (biological, social, economic or political) can be made to develop computational models that may provide a suitable conceptual framework for technological developments. According to this purpose, this Section is organized as follows: first, we present a general method for developing computational models inspired by biological and social phenomena. Second, we try to classify them and present some selected examples to motivate their applications in the current networking developments. Finally, we depict the need for both biologically and socially inspired computing in the next generation of communications systems.

3.8   A General Methodology

The modeling approach presented below should not be seen as a general principle, but it may work as a guideline to design algorithms and protocols for artificial systems. It is important to mention that the proposed steps are not new and have been used by many researchers during the last years [8,27,31]. However, we try to take the essential parts of the approaches presented by Dressler for biologically inspired networking [27], Pitt for socially inspired computing [50], and Gershenson for designing and controlling of self-organizing systems [38]. Our aim is to show the necessary steps for developing biologically and socially inspired models, and also present how they may have a remarkable impact on technological developments. Figure 2 presents the steps included in this methodology. It starts with a required system functionality, i.e. what the system should do, and enables the designer to produce a protocol or an algorithm that fulfills those requirements. Also, it is not necessary to follow this steps in order; according to the designer needs, it is possible to return to an early step to make any necessary adjustment.

4   Biological and Social Computing Inspiring Self-organizing Networks Design

Identification of Analogies Between Living and Artificial Systems. In the first step, an analogy between living and technological systems must be made to identify similar patterns that help to understand and propose new computational solutions [27,50]. Analogies are the tools of the comprehension; people understand new concepts by relating them to what they already knew [107]. If we chose the right analogy, the model reaches a level of abstraction that allows people foreign to the problem get a better understanding through a well-known

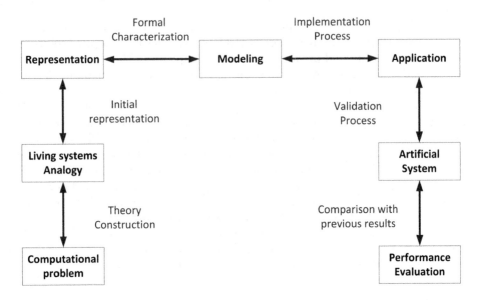

Fig. 2. A modeling approach for socially and biologically inspired computing

vocabulary. Also, create analogies among different systems will enable us to inspire mimetic solutions (biological, social, economic or political) and translate those principles into engineering artifacts. However, analogies could have a limitation regarding expressiveness; using a specific description to represent a problem, may limit its comprehension if the analogy is not good enough. Therefore, you can not use every analogy you know, it is necessary to master the selection process to get access to new interpretation tools.

Representation. In this step, a pre-formal representation that relates the observed biological or social phenomenon with a technological problem is developed. The designer should always remember the distinction between the model and the modeled; there are many representations of a system, and it is not possible to say one is better than another independently of a context [38, 50]. Similarly, the initial representation can be made in natural language or through any tool that allows us to describe variables, abstraction levels, granularity and interactions among components.

Although there is a wide diversity of systems, we can use a general method for developing an initial representation [38]. First, we need to divide the systems into components and identify their internal goals. Second, since the number of components may increase the complexity of the model, we should group them according to their dynamic, and analyze the most important based on the problem requirements. Finally, the designer should consider at least two abstraction levels to capture emergent properties and possible collectives behaviors.

Nevertheless, if the initial description has just few elements, probably the system is predictable, and we could get a better understanding using traditional approaches [39].

Modelling. In science and engineering, models should be as simple as possible and predict as much as possible; they should provide a better understanding of problems and not complicate them unnecessary [38,56,95]. Also, the quality of the model is related to the analogies we chose to describe the system; if the model becomes impractical, the selected representation should be carefully revised [107]. This stage should not be driven by implementation issues because of its primary goal is to achieve a clear understanding of the problem through a formal analysis of biological and social phenomena.

Furthermore, this stage should specify a control paradigm that ensures the expected behavior of the system. Since we are interested in self-organizing properties, the control mechanisms need to be internal and distributed. Given these conditions, several approaches like actions languages, modal logic, game theory and agent-based modeling have been extensively used to model complex systems and may help during this process. Finally, the expected result of this stage is a formal characterization that will enables us to translate biological and social principles into computational protocols [50].

Application. This step aims both to translate the current model into computational routines, and tune its parameters through different test scenarios. This process should be made from general to particular. Usually, little details take time to develop, and sometimes we will require an ideal scenario to test the central concepts involved in the model (for example through simulation techniques) [38,56]. Particular details can influence the system behavior, and they should not be included meanwhile their mechanisms and effects are not understood. According to the application results, modeling and representation stages should be improved.

Moreover, to get algorithms or protocols with acceptable computational tractability, probably we need some degree of simplification in the concepts involved in the model. However, it is a good practice to get as transparent as possible an idea of what is going to be simplified; any simplification that needs to be done should be carried out carefully with the purpose of not to dismiss essential parts of the model [50]. In an ideal scenario, application stage should not be constrained by considerations of computational tractability.

Performance Evaluation. The purpose of this step is to measure and compare the performance of resulting algorithms or protocols with the performance of previous results. This is an essential part of the method because allow integrating our results with the current scientifical and engineering developments. Also, if the system has multiple designers, they should agree on the expected functionality of the system [27,38]. According to the performance evaluation, the efforts to improve the model should continue as long as possible and even return to an early step to do any necessary adjustment.

Table 2. Categorization of biologically inspired models

Biological principle	Application fields in networking	POE	Selected references
Swarm intelligence and social insects	Distributed search and optimization; routing in computer networks especially in MANETs, WSNs, and overlay networks; task and resource allocation	E	[23, 32, 40, 43, 52, 92, 119]
Firefly synchronization	Robust and fully distributed clock synchronization	E	[13, 20, 45, 46, 51, 91, 118]
Artificial inmune system	Network security; anomaly and misbehavior detection	E	[2, 14, 49, 60, 62, 74, 104]
Epidemic spreading	Content distribution in computer networks (e.g. in DTNs); overlay networks; analysis of worm and virus spreading	PE	[19, 36, 63, 100, 111, 114, 115]
Evolutionary computing	Optimization, cooperation strategies, adaptation to dynamic conditions	PO	[66, 68, 89, 94, 121–123]

4.1 Classification and Categorizations

The majority of the proposed solutions for self-organizing networks are based on biologically inspired computing, which have successfully solved problems related to routing, synchronization, security, and coordination [27]. However, there is a new kind of socially inspired computing coming up; human society has many self-organizing mechanisms that we can learn from to enhance the capacity of artificial systems to solve problems through collective actions [50, 78, 82]. Not only these models are useful to face the tension between individual and collective rationality, but also they help to answer questions like: are the cooperation processes sustainable? Is the resources distribution efficient and fair? Can a set of rules evolve autonomously in an artificial system? Socially inspired computing tries to answer these question through a formal analysis of social phenomena. It is important to mention that neither all socially and biologically inspired models are related to self-organizing properties, nor all self-organizing behaviors arise from living systems. However, this work focus on computing models with distributed and internal control related to social and biological systems. An overview of these models is presented in the following subsections.

Biologically Inspired Computing. Biological systems exhibit a wide range of desirable characteristics, such as evolution, adaptation, fault tolerance and self-organizing behaviors. These properties are difficult to produce using traditional approaches, and make necessary to consider new methods [26]. Thus, the purpose of biologically inspired computing is design algorithms and protocols based on biological behaviors that allow artificial systems to face challenges related to optimization, collective behavior, pattern recognition and uncertain environments [15, 57]. Classical examples of these models can be found in swarm intelligence, firefly synchronization and evolutionary algorithms [27, 75]. Table 2 shows a summary of biologically inspired models successfully used in networking.

Furthermore, if we analyze living organisms, three different levels of organization are found: Phylogeny (P), Ontogeny (O), and Epigenesis (E) [97]. First, Phylogeny is related to the temporal evolution of the genetic program. This process is fundamentally non-deterministic and gives rise to the emergence of new organisms through recombination and mutation of the genetic code. Second, Ontogeny is related to the development of a single individual from its genetic material. Finally, Epigenesis is concerned about the learning process in which an organism can integrate information from the outside world through interactions with the environment. The distinction among these categories cannot be easily drawn and may be subject to discussion.

POE model can be used in the context of engineering to classify biologically inspired models and identify possibles directions for future research [15]. We can understand the POE model as follows: Phylogeny involves evolution, Ontogeny involves development and Epigenesis involves learning. In this regard, evolutionary computing can be seen as a simplified artificial counterpart of Phylogeny in nature. Multicellular automata, self-replicating and self-healing software are based on ontogeny properties. For example, when a program can produce a copy of its code or regenerate parts of itself to compensate failures. Finally, artificial neural network and artificial immune systems can be seen as examples of epigenetic processes. In Table 2 a classification of the biologically inspired models according to POE model is presented.

Socially Inspired Computing. Pitt, Jones, and Artikis introduced social inspired computing as a way to create mechanisms that allow artificial systems to solve problems through collective action [50]. Even though this is not the first attempt to use social models in computer science [8,44], from the author's knowledge is the first proposal that presents a systematic method to develop them. These models are useful in systems formed by a set of co-dependent components in which there is a tension between individual and collective rationality [54,78]. In such systems, the achievement of individual and collective goals depends on possible unreliable and conflicting components, interacting in the absence of centralized control or other orchestration forms.

Although biological processes are the foundation of social systems, they are not the core of sociability. Despite the fact that both living organisms and societies can be considered as meta-systems, the difference between them is the level of autonomy in their components; while the units of an organism have little or no independence, those of social systems have a maximum level of autonomy. As a result, new kinds of self-organizing phenomenon appear, and it is valuable to make a difference between biologically and socially inspired computing. On the other hand, human society has more complex social patterns than other species; cooperation, institutions, symbolic language and justice could be useful to inspire computational mechanisms that allows translating these principles into technological artifacts [44,82]. In Table 3 a summary of socially inspired models successfully used in ad hoc networks, smart grids, and multi-agent systems is presented.

Table 3. Categorization of socially inspired models

Social principle	Application fields in networking and artificial systems	Selected references
Trust	Cooperation mechanisms in self-organizing artificial systems: Ad hoc networks, WSNs, Smart grids, Multi-agent systems	$[41, 55, 61, 67, 101, 109]$
Justice	Resources distribution and allocation; Smart grids, VANETs, multi-agent systems; Social dilemmas	$[37, 77-79, 81, 84, 96]$
Norms and institutions	Evolution of norms. Institutions as a mechanism for collective actions. Self-organizing open systems	$[6, 7, 17, 83, 99, 117]$
Negotiation	Resources negotiation in ad hoc network and multi-agent systems	$[5, 9, 10, 33]$

4.2   The Need for Biological and Social Self-organizing Approaches

The design and development of communication networks, as well as all self-organizing artificial systems, call for a careful examination of biological and social concepts. In this section, we present the relationship between the networking challenges presented in Sect. 3 and the biologically and socially inspired models that we may use to deal with them. Although both biological and social inspired models exhibit self-organizing patterns, in each case their goals are different. Biologically inspired computing try to achieve efficient and scalable networking under uncertain environments, and socially inspired computing is useful for solving problems through collective behaviors. Therefore, the combination of these two approaches allows us to develop communication networks not only enough robust and adaptive to be able to operate in highly dynamic environments, but also with the capacity to use collective actions for solving complex problems. In Fig. 3 the relationship between the biologically and socially inspired models and the networking challenges presented in Sect. 3 is shown.

In general terms, a self-organizing network is a dynamic system of many agents (which may represent nodes, services, applications, users) working in parallel, always acting and reacting to what the other agents are doing. The control process is highly dispersed and decentralized, and any expected behavior in the network need to arise from competition, cooperation or coordination among network components [108]. Biological and social systems have dealt with similar situations for thousands of years, and we can learn from them to develop new types of computational solutions. Although biologically inspired computing has been successfully used during the last years, at this moment it is necessary to design technological artifacts able to solve problems through collective actions. Therefore, socially inspired computing turns into an opportunity for the next generation of artificial systems, giving us a route to include these properties in the future engineering developments.

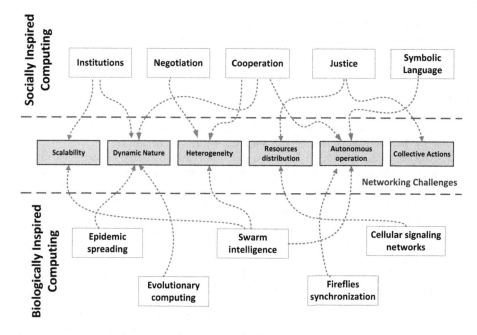

Fig. 3. Biologically and socially inspired computing for artificial systems

5   Conclusions

In this article, we have shown that is possible to use biologically and socially inspired computing for building communications systems. We argue that an abstract analysis of biological and social phenomena can be made to create a conceptual framework for developing a new kind of networking technology. Biologically inspired computing can be used for achieving efficient and scalable networking under uncertain conditions, and socially inspired computing for solving problems through collective actions. The combination of these two approaches enables us to develop communication networks not only enough robust and adaptive to operate in highly dynamic environments but also with the capacity to use collective behaviors for solving complex problems.

Furthermore, we showed the challenges of the next generation of communication networks from the perspective of biologically and socially inspired computing; we introduced a general method for developing these models and presented an overview in Tables 2 and 3. Also, we argue that the expected features of the next generation of communications networks become centralized control an impractical solution, and as a result, self-organization will take an essential role in the future networking developments.

Despite the considerable amount of ongoing advances on biologically and socially inspired computing, the research community is still quite young. There are many challenges that we need to face if we want to integrate these models with the emerging networking architectures. We expect this review will provide a better comprehension of the opportunities for biologically and socially inspired computing inside technological developments and encourage other researchers to explore these approaches as part of their future work.

References

1. Abadi, D.: Consistency tradeoffs in modern distributed database system design: CAP is only part of the story. Computer **45**(2), 37–42 (2012)
2. Alrajeh, N.A., Lloret, J.: Intrusion detection systems based on artificial intelligence techniques in wireless sensor networks. Int. J. Distrib. Sens. Netw. **9**(10), 351047 (2013)
3. Alzate, M., Monroy, A.: Introducción al tráfico autosimilar en redes de comunicaciones. Rev. Ing. **6**(2), 6–17 (2001)
4. Amoozadeh, M., Deng, H., Chuah, C.N., Zhang, H.M., Ghosal, D.: Platoon management with cooperative adaptive cruise control enabled by vanet. Veh. Commun. **2**(2), 110–123 (2015)
5. An, B., Lesser, V., Sim, K.M.: Strategic agents for multi-resource negotiation. Auton. Agent. Multi-Agent Syst. **23**(1), 114–153 (2011)
6. Artikis, A., Kamara, L., Pitt, J., Sergot, M.: A protocol for resource sharing in norm-governed ad hoc networks. In: Leite, J., Omicini, A., Torroni, P., Yolum, I. (eds.) DALT 2004. LNCS (LNAI), vol. 3476, pp. 221–238. Springer, Heidelberg (2005). https://doi.org/10.1007/11493402_13
7. Artikis, A., Pitt, J.: A formal model of open agent societies. In: Proceedings of the Fifth International Conference on Autonomous Agents, AGENTS 2001, pp. 192–193. ACM, New York (2001). https://doi.org/10.1145/375735.376108. http://doi.acm.org/10.1145/375735.376108
8. Axelrod, R., Hamilton, W.D.: The evolution of cooperation. Science **211**(4489), 1390–1396 (1981)
9. Baarslag, T., et al.: Evaluating practical negotiating agents: results and analysis of the 2011 international competition. Artif. Intell. **198**, 73–103 (2013)
10. Baarslag, T., Hindriks, K.V.: Accepting optimally in automated negotiation with incomplete information. In: Proceedings of the 2013 International Conference on Autonomous Agents and Multi-Agent Systems, pp. 715–722. International Foundation for Autonomous Agents and Multiagent Systems (2013)
11. Barabási, A.L., Bonabeau, E.: Scale-free networks. Sci. Am. **288**(5), 60–69 (2003)
12. Bellifemine, F.L., Caire, G., Greenwood, D.: Developing Multi-Agent Systems with JADE, vol. 7. Wiley, Hoboken (2007)
13. Brandner, G., Schilcher, U., Bettstetter, C.: Firefly synchronization with phase rate equalization and its experimental analysis in wireless systems. Comput. Netw. **97**, 74–87 (2016)
14. Butun, I., Morgera, S.D., Sankar, R.: A survey of intrusion detection systems in wireless sensor networks. IEEE Commun. Surv. Tutor. **16**(1), 266–282 (2014)
15. Câmara, D.: Bio-inspired Networking. Elsevier, Amsterdam (2015)
16. Camp, T., Boleng, J., Davies, V.: A survey of mobility models for ad hoc network research. Wirel. Commun. Mob. Comput. **2**(5), 483–502 (2002)
17. Carballo, D.M., Roscoe, P., Feinman, G.M.: Cooperation and collective action in the cultural evolution of complex societies. J. Archaeol. Method Theory **21**(1), 98–133 (2014)

18. Carreón, G., Gershenson, C., Pineda, L.: Improving public transportation systems with self-organization: a headway-based model and regulation of passenger alighting and boarding. PLoS ONE **12**(12), e0190100 (2017)
19. Chitra, M., Sathya, S.S.: Bidirectional data dissemination in vehicular ad hoc networks using epidemic spreading model. In: Proceedings of the International Conference on Informatics and Analytics, p. 57. ACM (2016)
20. Chovanec, M., Milanová, J., Čechovič, L., Šarafín, P., Húdik, M., Kochláň, M.: Firefly-based universal synchronization algorithm in wireless sensor network. In: Grzenda, M., Awad, A.I., Furtak, J., Legierski, J. (eds.) Advances in Network Systems. AISC, vol. 461, pp. 71–86. Springer, Cham (2017). https://doi.org/10.1007/978-3-319-44354-6_5
21. Day, J.D., Zimmermann, H.: The OSI reference model. Proc. IEEE **71**(12), 1334–1340 (1983). https://doi.org/10.1109/PROC.1983.12775
22. Dell, P.F.: Beyond homeostasis: toward a concept of coherence. Fam. Process **21**(1), 21–41 (1982)
23. Di Caro, G., Ducatelle, F., Gambardella, L.M.: AntHocNet: an adaptive nature-inspired algorithm for routing in mobile ad hoc networks. Trans. Emerg. Telecommun. Technol. **16**(5), 443–455 (2005)
24. Dorigo, M., Birattari, M.: Ant colony optimization. In: Sammut, C., Webb, G.I. (eds.) Encyclopedia of Machine Learning, pp. 36–39. Springer, Boston (2011). https://doi.org/10.1007/978-0-387-30164-8
25. Dressler, F.: Self-Organization in Sensor and Actor Networks. Wiley, Hoboken (2008)
26. Dressler, F., Akan, O.B.: Bio-inspired networking: from theory to practice. IEEE Commun. Mag. **48**(11), 176–183 (2010)
27. Dressler, F., Akan, O.B.: A survey on bio-inspired networking. Comput. Netw. **54**(6), 881–900 (2010)
28. Dressler, F., Klingler, F., Sommer, C., Cohen, R.: Not all vanet broadcasts are the same: context-aware class based broadcast. IEEE/ACM Trans. Netw. **26**(1), 17–30 (2018)
29. Dressler, F., et al.: Self-organization in ad hoc networks: overview and classification. University of Erlangen, Department of Computer Science, vol. 7, pp. 1–12 (2006)
30. Dunphy, P., Petitcolas, F.A.: A first look at identity management schemes on the blockchain. arXiv preprint arXiv:1801.03294 (2018)
31. Eigen, M., Schuster, P.: The Hypercycle: A Principle of Natural Self-Organization. Springer, Heidelberg (2012)
32. Fahad, M., et al.: Grey wolf optimization based clustering algorithm for vehicular ad-hoc networks. Comput. Electr. Eng. **70**, 853–870 (2018)
33. Faratin, P., Sierra, C., Jennings, N.R.: Negotiation decision functions for autonomous agents. Robot. Auton. Syst. **24**(3), 159–182 (1998)
34. Fitzek, F.H., Katz, M.D.: Mobile Clouds: Exploiting Distributed Resources in Wireless, Mobile and Social Networks. Wiley, Hoboken (2013)
35. Galati, A.: Delay tolerant network (2010)
36. Gan, C., Yang, X., Liu, W., Zhu, Q., Jin, J., He, L.: Propagation of computer virus both across the internet and external computers: a complex-network approach. Commun. Nonlinear Sci. Numer. Simul. **19**(8), 2785–2792 (2014)
37. Garbiso, J.P., Diaconescu, A., Coupechoux, M., Pitt, J., Leroy, B.: Distributive justice for fair auto-adaptive clusters of connected vehicles. In: 2017 IEEE 2nd International Workshops on Foundations and Applications of Self* Systems (FAS* W), pp. 79–84. IEEE (2017)

38. Gershenson, C.: Design and control of self-organizing systems. CopIt ArXives (2007)

39. Gershenson, C., Heylighen, F.: When can we call a system self-organizing? In: Banzhaf, W., Ziegler, J., Christaller, T., Dittrich, P., Kim, J.T. (eds.) ECAL 2003. LNCS (LNAI), vol. 2801, pp. 606–614. Springer, Heidelberg (2003). https://doi.org/10.1007/978-3-540-39432-7_65

40. Giagkos, A., Wilson, M.S.: BeeIP-a swarm intelligence based routing for wireless ad hoc networks. Inf. Sci. **265**, 23–35 (2014)

41. Grieco, L.A., et al.: IoT-aided robotics applications: technological implications, target domains and open issues. Comput. Commun. **54**, 32–47 (2014)

42. Grippa, P., Behrens, D.A., Bettstetter, C., Wall, F.: Job selection in a network of autonomous UAVs for delivery of goods. arXiv preprint arXiv:1604.04180 (2016)

43. Hammoudeh, M., Newman, R.: Adaptive routing in wireless sensor networks: QoS optimisation for enhanced application performance. Inf. Fusion **22**, 3–15 (2015)

44. Hatfield, U.: Socially inspired computing (2005)

45. He, J., Cheng, P., Shi, L., Chen, J., Sun, Y.: Time synchronization in WSNs: a maximum-value-based consensus approach. IEEE Trans. Autom. Control **59**(3), 660–675 (2014)

46. He, J., Li, H., Chen, J., Cheng, P.: Study of consensus-based time synchronization in wireless sensor networks. ISA Trans. **53**(2), 347–357 (2014)

47. Heylighen, F., Gershenson, C.: The meaning of self-organization in computing. IEEE Intell. Syst. **18**(4), 72–75 (2003)

48. Iversen, V.B.: Teletraffic engineering and network planning. Technical University of Denmark, p. 270 (2010)

49. Jamali, S., Fotohi, R.: Defending against wormhole attack in MANET using an artificial immune system. New Rev. Inf. Netw. **21**(2), 79–100 (2016)

50. Jones, A.J., Artikis, A., Pitt, J.: The design of intelligent socio-technical systems. Artif. Intell. Rev. **39**(1), 5–20 (2013)

51. Jung, J.Y., Choi, H.H., Lee, J.R.: Survey of bio-inspired resource allocation algorithms and MAC protocol design based on a bio-inspired algorithm for mobile ad hoc networks. IEEE Commun. Mag. **56**(1), 119–127 (2018)

52. Karaboga, D., Akay, B.: A survey: algorithms simulating bee swarm intelligence. Artif. Intell. Rev. **31**(1–4), 61 (2009)

53. Kocarev, L.: Complex Dynamics in Communication Networks. Springer, Heidelberg (2005). https://doi.org/10.1007/b94627

54. Kollock, P.: Social dilemmas: the anatomy of cooperation. Ann. Rev. Sociol. **24**(1), 183–214 (1998)

55. Kshirsagar, V.H., Kanthe, A.M., Simunic, D.: Trust based detection and elimination of packet drop attack in the mobile ad-hoc networks. Wireless Pers. Commun. **100**(2), 311–320 (2018)

56. Law, A.M.: Simulation Modeling and Analysis, vol. 3. McGraw-Hill, New York (2007)

57. Leibnitz, K., Wakamiya, N., Murata, M.: Biologically inspired networking. In: Cognitive Networks: Towards Self-Aware Networks, pp. 1–21 (2007)

58. Li, L., Alderson, D., Willinger, W., Doyle, J.: A first-principles approach to understanding the internet's router-level topology. In: ACM SIGCOMM Computer Communication Review, vol. 34, pp. 3–14. ACM (2004)

59. Li, S., Da Xu, L., Zhao, S.: 5G internet of things: a survey. J. Ind. Inf. Integr. **10**, 1–9 (2018)

60. Li, W., Yi, P., Wu, Y., Pan, L., Li, J.: A new intrusion detection system based on KNN classification algorithm in wireless sensor network. J. Electr. Comput. Eng. **2014** (2014)

61. Li, Y., Xu, H., Cao, Q., Li, Z., Shen, S.: Evolutionary game-based trust strategy adjustment among nodes in wireless sensor networks. Int. J. Distrib. Sens. Netw. **11**(2), 818903 (2015)

62. Liaqat, H.B., Xia, F., Yang, Q., Xu, Z., Ahmed, A.M., Rahim, A.: Bio-inspired packet dropping for ad-hoc social networks. Int. J. Commun. Syst. **30**(1), e2857 (2017)

63. Liu, B., Zhou, W., Gao, L., Zhou, H., Luan, T.H., Wen, S.: Malware propagations in dwireless add hoc networks. IEEE Trans. Dependable Secure Comput. **15**(6), 1016–1026 (2016)

64. Loo, J., Mauri, J.L., Ortiz, J.H.: Mobile Ad Hoc Networks: Current Status and Future Trends. CRC Press, Boca Raton (2016)

65. Maldonado, C.E.: Complejidad: ciencia, pensamiento y aplicaciones, vol. 1 (2007)

66. Nagula Meera, S.K., Kumar, D.S., Rao, S.: Ad hoc networks: route discovery channel for mobile network with low power consumption. In: Satapathy, S.C., Bhateja, V., Chowdary, P.S.R., Chakravarthy, V.V.S.S.S., Anguera, J. (eds.) Proceedings of 2nd International Conference on Micro-Electronics, Electromagnetics and Telecommunications. LNEE, vol. 434, pp. 665–671. Springer, Singapore (2018). https://doi.org/10.1007/978-981-10-4280-5_70

67. Mejia, M., Pena, N., Munoz, J.L., Esparza, O.: A review of trust modeling in ad hoc networks. Internet Res. **19**(1), 88–104 (2009)

68. Mejia, M., Peña, N., Muñoz, J.L., Esparza, O., Alzate, M.A.: A game theoretic trust model for on-line distributed evolution of cooperation inmanets. J. Netw. Comput. Appl. **34**(1), 39–51 (2011)

69. Mihail, M., Papadimitriou, C., Saberi, A.: On certain connectivity properties of the internet topology. J. Comput. Syst. Sci. **72**(2), 239–251 (2006). https://doi.org/10.1016/j.jcss.2005.06.009. jCSS FOCS 2003 Special Issue

70. Mukherjee, S., Ray, R., Samanta, R., Khondekar, M.H., Sanyal, G.: Nonlinearity and chaos in wireless network traffic. Chaos Solitons Fractals **96**, 23–29 (2017)

71. Neuman, B.C.: Scale in distributed systems. ISI/USC (1994)

72. Nowak, M.A.: Five rules for the evolution of cooperation. Science **314**(5805), 1560–1563 (2006)

73. Ospina, J.P., Ortiz, J.E.: Estimation of a growth factor to achieve scalable ad hoc networks. Ing. Univ. **21**(1), 49–70 (2017)

74. Pathan, A.S.K.: Security of Self-organizing Networks: MANET, WSN, WMN, VANET. CRC Press, Boca Raton (2016)

75. Pazhaniraja, N., Paul, P.V., Roja, G., Shanmugapriya, K., Sonali, B.: A study on recent bio-inspired optimization algorithms. In: 2017 Fourth International Conference on Signal Processing, Communication and Networking (ICSCN), pp. 1–6. IEEE (2017)

76. Petruzzi, P.E., Pitt, J., Busquets, D.: Electronic social capital for self-organising multi-agent systems. ACM Trans. Auton. Adapt. Syst. (TAAS) **12**(3), 13 (2017)

77. Pitt, J., Diaconescu, A., Bourazeri, A.: Democratisation of the smartgrid and the active participation of prosumers. In: 2017 IEEE 26th International Symposium on Industrial Electronics (ISIE), pp. 1707–1714, June 2017. https://doi.org/10.1109/ISIE.2017.8001505

78. Pitt, J.: From trust and forgiveness to social capital and justice: formal models of social processes in open distributed systems. In: Reif, W., et al. (eds.) Trustworthy Open Self-Organising Systems. AS, pp. 185–208. Springer, Cham (2016). https://doi.org/10.1007/978-3-319-29201-4_7

79. Pitt, J., Busquets, D., Riveret, R.: The pursuit of computational justice in open systems. AI Soc. **30**(3), 359–378 (2015)

80. Pitt, J., Diaconescu, A., Bourazeri, A.: Democratisation of the smartgrid and the active participation of prosumers. In: 2017 IEEE 26th International Symposium on Industrial Electronics (ISIE), pp. 1707–1714. IEEE (2017)

81. Pitt, J., Schaumeier, J.: Provision and appropriation of common-pool resources without full disclosure. In: Rahwan, I., Wobcke, W., Sen, S., Sugawara, T. (eds.) PRIMA 2012. LNCS (LNAI), vol. 7455, pp. 199–213. Springer, Heidelberg (2012). https://doi.org/10.1007/978-3-642-32729-2_14

82. Pitt, J., Schaumeier, J., Artikis, A.: The axiomatisation of socio-economic principles for self-organising systems. In: 2011 Fifth IEEE International Conference on Self-Adaptive and Self-Organizing Systems (SASO), pp. 138–147. IEEE (2011)

83. Pitt, J., Schaumeier, J., Artikis, A.: Axiomatization of socio-economic principles for self-organizing institutions: concepts, experiments and challenges. ACM Trans. Auton. Adapt. Syst. **7**(4), 39:1–39:39 (2012). https://doi.org/10.1145/2382570.2382575. http://doi.acm.org/10.1145/2382570.2382575

84. Pitt, J., Schaumeier, J., Busquets, D., Macbeth, S.: Self-organising common-pool resource allocation and canons of distributive justice. In: 2012 IEEE Sixth International Conference on Self-Adaptive and Self-Organizing Systems (SASO), pp. 119–128. IEEE (2012)

85. Pureswaran, V., Brody, P.: Device democracy: saving the future of the internet of things. IBM Corporation (2015)

86. Qiu, T., Chen, N., Li, K., Qiao, D., Fu, Z.: Heterogeneous ad hoc networks: architectures, advances and challenges. Ad Hoc Netw. **55**, 143–152 (2017)

87. Ramiro, J., Hamied, K.: Self-Organizing Networks (SON): Self-Planning, Self-Optimization and Self-Healing for GSM. UMTS and LTE. Wiley, Hoboken (2011)

88. Ranjan, P., Abed, E.H., La, R.J.: Nonlinear instabilities in TCP-RED. IEEE/ACM Trans. Netw. **12**(6), 1079–1092 (2004)

89. Reina, D.G., Ruiz, P., Ciobanu, R., Toral, S., Dorronsoro, B., Dobre, C.: A survey on the application of evolutionary algorithms for mobile multihop ad hoc network optimization problems. Int. J. Distrib. Sens. Netw. **12**(2), 2082496 (2016)

90. Robertazzi, T.G.: Computer Networks and Systems: Queueing Theory and Performance Evaluation. Springer, New York (2012)

91. Roh, B., Han, M.H., Hoh, M., Park, H.S., Kim, K., Roh, B.H.: Distributed call admission control for DESYNC-TDMA in mobile ad hoc networks. In: Proceedings of the 9th EAI International Conference on Bio-inspired Information and Communications Technologies (formerly BIONETICS), pp. 411–413. ICST (Institute for Computer Sciences, Social-Informatics and Telecommunications Engineering) (2016)

92. Saleh, S., Ahmed, M., Ali, B.M., Rasid, M.F.A., Ismail, A.: A survey on energy awareness mechanisms in routing protocols for wireless sensor networks using optimization methods. Trans. Emerg. Telecommun. Technol. **25**(12), 1184–1207 (2014)

93. San-Um, W., Ketthong, P., Noymanee, J.: A deterministic node mobility model for mobile ad hoc wireless network using signum-based discrete-time chaotic map. In: 2015 International Telecommunication Networks and Applications Conference (ITNAC), pp. 114–119. IEEE (2015)

94. Sayed, A.H., Tu, S.Y., Chen, J., Zhao, X., Towfic, Z.J.: Diffusion strategies for adaptation and learning over networks: an examination of distributed strategies and network behavior. IEEE Signal Process. Mag. **30**(3), 155–171 (2013)

95. Shalizi, C.R., et al.: Causal architecture, complexity and self-organization in the time series and cellular automata. Ph.D. thesis, University of Wisconsin-Madison (2001)

96. Shi, H., Prasad, R.V., Onur, E., Niemegeers, I.: Fairness in wireless networks: issues, measures and challenges. IEEE Commun. Surv. Tutor. **16**(1), 5–24 (2014)

97. Sipper, M., Sanchez, E., Mange, D., Tomassini, M., Pérez-Uribe, A., Stauffer, A.: A phylogenetic, ontogenetic, and epigenetic view of bio-inspired hardware systems. IEEE Trans. Evol. Comput. **1**(1), 83–97 (1997)

98. Smith, R.: Docker Orchestration. Packt Publishing Ltd., Birmingham (2017)

99. Sørensen, E., Torfing, J.: Theories of Democratic Network Governance. Springer, Heidelberg (2016)

100. Sugihara, K., Hayashibara, N.: Collecting data in sensor networks using homesick Lévy Walk. In: Barolli, L., Enokido, T., Takizawa, M. (eds.) NBiS 2017. LNDECT, vol. 7, pp. 779–786. Springer, Cham (2018). https://doi.org/10.1007/978-3-319-65521-5_70

101. Sugumar, R., Rengarajan, A., Jayakumar, C.: Trust based authentication technique for cluster based vehicular ad hoc networks (VANET). Wireless Netw. **24**(2), 373–382 (2018). https://doi.org/10.1007/s11276-016-1336-6

102. Tanenbaum, A.S., Van Steen, M.: Distributed Systems: Principles and Paradigms. Prentice-Hall, Upper Saddle River (2007)

103. Taqqu, M.S., Willinger, W., Sherman, R.: Proof of a fundamental result in self-similar traffic modeling. ACM SIGCOMM Comput. Commun. Rev. **27**(2), 5–23 (1997)

104. Tiwari, S., Mishra, K.K., Saxena, N., Singh, N., Misra, A.K.: Artificial immune system based MAC layer misbehavior detection in MANET. In: Sulaiman, H.A., Othman, M.A., Othman, M.F.I., Rahim, Y.A., Pee, N.C. (eds.) Advanced Computer and Communication Engineering Technology. LNEE, vol. 362, pp. 707–722. Springer, Cham (2016). https://doi.org/10.1007/978-3-319-24584-3_60

105. Tsuchiya, T., Kikuno, T.: An adaptive mechanism for epidemic communication. In: Ijspeert, A.J., Murata, M., Wakamiya, N. (eds.) BioADIT 2004. LNCS, vol. 3141, pp. 306–316. Springer, Heidelberg (2004). https://doi.org/10.1007/978-3-540-27835-1_23

106. Vaseghi, B., Pourmina, M.A., Mobayen, S.: Secure communication in wireless sensor networks based on chaos synchronization using adaptive sliding mode control. Nonlinear Dyn. **89**(3), 1689–1704 (2017)

107. Videla, A.: Metaphors we compute by. Commun. ACM **60**(10), 42–45 (2017)

108. Waldrop, M.M.: Complexity: The Emerging Science at the Edge of Order and Chaos. Simon and Schuster, New York (1993)

109. Wang, B., Chen, X., Chang, W.: A light-weight trust-based QoS routing algorithm for ad hoc networks. Pervasive Mob. Comput. **13**, 164–180 (2014)

110. Wang, S., Chen, Y., Tian, H.: An intrusion detection algorithm based on chaos theory for selecting the detection window size. In: 2016 8th IEEE International Conference on Communication Software and Networks (ICCSN), pp. 556–560. IEEE (2016)

111. Wang, Y., Wen, S., Xiang, Y., Zhou, W.: Modeling the propagation of worms in networks: a survey. IEEE Commun. Surv. Tutor. **16**(2), 942–960 (2014)

112. Wortmann, F., Flüchter, K.: Internet of things. Bus. Inf. Syst. Eng. **57**(3), 221–224 (2015)

113. Yang, K.: Wireless Sensor Networks. Principles, Design and Applications (2014)
114. Yang, L.X., Yang, X., Liu, J., Zhu, Q., Gan, C.: Epidemics of computer viruses: a complex-network approach. Appl. Math. Comput. **219**(16), 8705–8717 (2013)
115. Yang, X., Mishra, B.K., Liu, Y.: Computer virus: theory, model, and methods. Discrete Dyn. Nat. Soc. **2012** (2012)
116. Yanmaz, E., Yahyanejad, S., Rinner, B., Hellwagner, H., Bettstetter, C.: Drone networks: communications, coordination, and sensing. Ad Hoc Netw. **68**, 1–15 (2018)
117. Yu, C., Zhang, M., Ren, F., Luo, X.: Emergence of social norms through collective learning in networked agent societies. In: Proceedings of the 2013 International Conference on Autonomous Agents and Multi-Agent Systems, pp. 475–482. International Foundation for Autonomous Agents and Multiagent Systems (2013)
118. Zhang, F., Trentelman, H.L., Scherpen, J.M.: Fully distributed robust synchronization of networked lur'e systems with incremental nonlinearities. Automatica **50**(10), 2515–2526 (2014)
119. Zhang, X., Zhang, X., Gu, C.: A micro-artificial bee colony based multicast routing in vehicular ad hoc networks. Ad Hoc Netw. **58**, 213–221 (2017)
120. Zhang, Y., Zheng, J., Chen, H.H.: Cognitive Radio Networks: Architectures, Protocols, and Standards. CRC Press, Boca Raton (2016)
121. Zhang, Z., Huangfu, W., Long, K., Zhang, X., Liu, X., Zhong, B.: On the designing principles and optimization approaches of bio-inspired self-organized network: a survey. Sci. China Inf. Sci. **56**(7), 1–28 (2013)
122. Zhang, Z., Long, K., Wang, J.: Self-organization paradigms and optimization approaches for cognitive radio technologies: a survey. IEEE Wirel. Commun. **20**(2), 36–42 (2013)
123. Zhang, Z., Long, K., Wang, J., Dressler, F.: On swarm intelligence inspired self-organized networking: its bionic mechanisms, designing principles and optimization approaches. IEEE Commun. Surv. Tutor. **16**(1), 513–537 (2014)

Author Index

Printed in the United States
By Bookmasters